Fundamentals of Operational Amplifiers and Linear Integrated Circuits

Howard M. Berlin and Frank C. Getz, Jr.

Merrill Publishing Company
Columbus Toronto London Melbourne

Cover Photo: Doug Pulsipher/Stock Solution

Published by Merrill Publishing Company
Columbus, Ohio 43216

This book was set in Times Roman.

Administrative Editor: David Garza
Production Editor: Carol Sykes
Art Coordinator: Mark Garrett
Cover Designer: Russ Maselli

Library of Congress Catalog Card Number: 89–63843
International Standard Book Number: 0–675–21002–X
Printed in the United States of America
1 2 3 4 5 6 7 8 9—94 93 92 91 90

Preface

The development of the integrated circuit has dramatically changed the design and performance of electronic circuits. Networks that were virtually impossible to implement even a decade ago now seem simple to accomplish by comparison. Of the integrated circuits that are available, perhaps the most widely used, if not the most versatile, is the operational amplifier. It is inexpensive (some cost less than 50¢), and easy to use, and it requires a minimum of algebra.

This text is primarily aimed at the technician, or 2-year associate degree level, although this book could also be used successfully at the 4-year technology level. By the time the student encounters a course on operational amplifiers and integrated circuits, prerequisite training in basic AC and DC electricity as well as devices (diodes and transistors) is assumed.

The thirteen chapters cover virtually every basic aspect of operational amplifiers and the major classes of linear integrated-circuit devices. The technical content of this book goes beyond the basic material found in most texts to include an entire chapter devoted to single-supply biasing of operational amplifiers and Norton amplifiers (Chapter 5) as well as one on audio applications (Chapter 7). In addition, Chapter 6 presents unique operational amplifier devices, such as programmable operational amplifiers, CMOS and BIFETs, and the operational transconductance amplifier, along with suitable circuits. Finally, the last chapter is devoted solely to the phase-locked loop, and with the exception of the 566 voltage controlled oscillator discussed in Chapter 12, covers the entire 560 series of phase-locked loop devices as well as the 4046 CMOS phase-locked loop.

The intent of *Fundamentals of Operational Amplifiers and Linear Integrated Circuits* is to show (1) that the operational amplifier is a versatile building block and (2) that there are, in some cases, better ways to build the proverbial mousetrap using specially designed linear integrated circuits. The chapters are logically divided by overall function; linear amplifiers, differentiators and integrators, single-supply operation, unique devices, audio applications, nonlinear circuits, voltage and current regulators, oscillators and function generators, active filters, converters, and the phase-locked loop. Linear integrated circuits that can perform the same function as operational amplifier circuits (such as comparators, voltage regulators, waveform generators, and sample/hold amplifiers), are discussed and compared with their corresponding operational amplifier circuits. In situations where there are no functions that can conveniently be built with operational amplifiers, such as phase-locked loops, analog multipliers, and compandors, the functions are discussed simply as additional devices in the linear integrated-circuit family.

Each chapter includes a list of instructional objectives, a summary, review questions, and numerical exercises. A glossary in each chapter defines the important terms that are highlighted in the text by bold type. As an additional aid to the reader, each major equation is numbered and enclosed in a box.

The wealth of material included in this text should be sufficient to fulfill the needs of either a single- or two-term course covering operational amplifiers and linear integrated-circuit devices. In addition, a laboratory workbook is available, which contains a wide range of laboratory experiments on virtually every aspect of operational amplifiers and many of the integrated circuits discussed in this text.

Acknowledgments

We would like to acknowledge the following reviewers who have provided us with helpful comments and suggestions: Andrew Anderson, Brookdale Community College, Lincroft, New Jersey; Phillip Chiarelli, Electronics Institute, Kansas City, Missouri; Ken Engh, North Dakota School of Science, Wahpeton, North Dakota; Bob Hinders, Western Iowa Technical College, Sioux City, Iowa; Malcolm Skipper, Midlands Technical College, Columbia, South Carolina; John Slough, DeVry Institute of Technology, Irving, Texas; Ronald Way, Muskegon Community College, Muskegon, Mississippi; and Ron Worley, San Diego Mesa College, San Diego, California. Furthermore, the assistance of the following manufacturers in providing us with technical material and permitting us to reproduce data sheets of some of their products is gratefully appreciated: Exar, National Semiconductor, Raytheon, RCA, and Signetics.

Howard M. Berlin
Frank C. Getz, Jr.

Contents

Chapter 8 Nonlinear Signal Processing Circuits *184*

Chapter 9 Voltage and Current Regulators *231*

Chapter 10 Oscillators and Function Generators *264*

1 Operational Amplifiers and Integrated Circuits: An Introduction

Instructional Objectives

Upon completing this chapter, you will be able to

- Define the terms *monolithic integrated circuit, linear integrated circuit, operational amplifier,* and *DIP*.
- Describe how the leads of both DIP and TO-type integrated-circuit devices are numbered.
- Briefly describe how to decode important information about a given device from its part number.
- Wire a bipolar, or split, power supply.
- List several precautions that should be observed when working with linear integrated circuits as well as most electronic circuits.

Introduction

In 1947, at New Jersey's Bell Laboratories, the transistor was born. The three men who were awarded the Nobel Prize for Physics in 1956 for that achievement started a technological revolution that still shows no sign of ending. The simple transistor that William Shockley, John Bardeen, and William Brattain created has paved the way for the creation of advanced solid-state devices that were once never thought possible. These include the operational amplifier as well as the larger class of linear integrated circuits.

The term **operational amplifier,** or *op-amp,* was originally applied to a high-performance DC differential amplifier using vacuum tubes that formed the basis of the analog computer capable of solving differential equations. As early as 1948, the vacuum tube operational amplifier was capable of performing mathematical functions, or *operations,* such as addition, subtraction, arithmetic sign inversion, averaging, scaling (i.e., multiplication or division by a constant), integration, differentiation, comparisons, logarithms, and inverse logarithms. Because of its size and power requirements, the vacuum tube soon gave way to the transistor and eventually the integrated circuit, as improvements in technology were introduced.

Figure 1–1
The 741 integrated circuit. (a) Photograph of monolithic silicon chip. (b) Schematic of internal equivalent circuit.

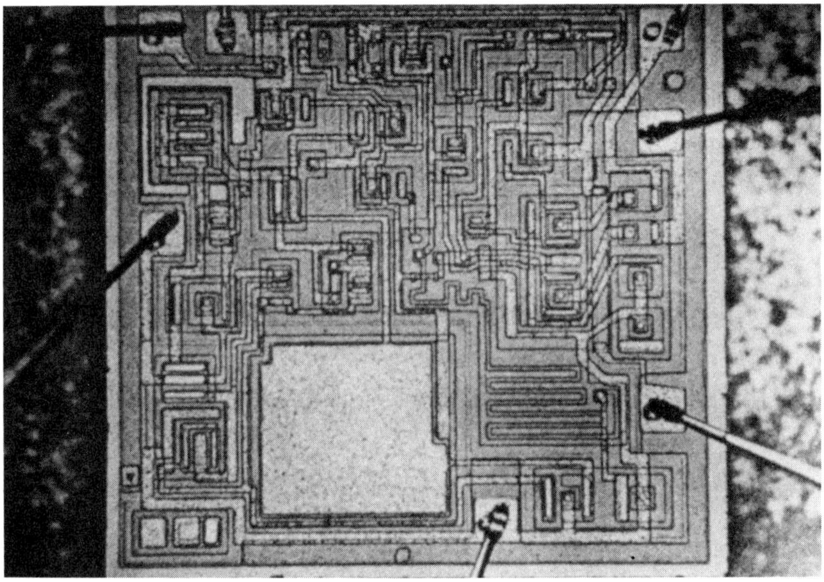

(a)

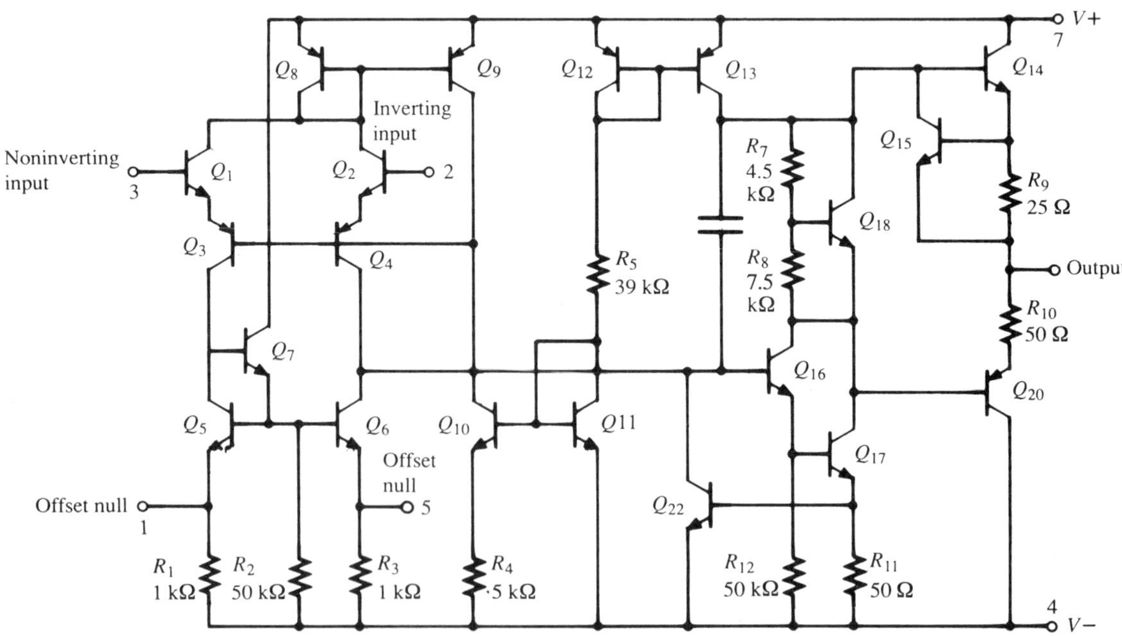

(b)

Table 1–1
Evolution of the integrated-circuit operational amplifier

Date	Manufacturer	Device	Remarks
1963	Fairchild Semiconductor	μA702	First integrated-circuit operational amplifier
1965	Fairchild Semiconductor	μA709	Second-sourced by more manufacturers than any other type
1967	National Semiconductor	LM101	Second-generation design
1968	National Semiconductor	LH101	Hybrid design with 30-pF compensation capacitor
1968	Fairchild Semiconductor	μA741	Short-circuit protection
1968	National Semiconductor	LM101A	Improved 101 version
		LM107	LM101A with internal compensation capacitor
1969	Fairchild Semiconductor	μA748	741 design with external compensation

Except for a marked reduction in size and cost, today's operational amplifier is little changed from its first-generation version.

Based on the early vacuum tube operational amplifiers and coupled with improvements in integrated-circuit technology, the first series of commercially available operational amplifiers was made in the mid-1960s. In 1963, Fairchild Semiconductor introduced the first commercially available operational amplifier, the μA702; this was followed two years later by the μ709. In 1965, National Semiconductor introduced the LM101, while Fairchild in 1967 unveiled the ever-popular μA741, whose integrated circuit chip photograph is shown in Figure 1–1a. Table 1–1 summarizes the evolution of the integrated-circuit operational amplifier.

Instead of being made up of individual transistors wired with solid-state resistors, capacitors, and diodes, the integrated-circuit operational amplifier, such as the 741, is an example of a **monolithic integrated circuit.** Here, all the individual transistors and resistors of the equivalent circuit of Figure 1–1b are fabricated from a single piece of solid-state material into an integrated circuit. From the operational amplifier, itself in the broad class of **linear integrated circuits,** improvements in technology allowed for the development of integrated-circuit voltage regulators, comparators, timers, function generators, phase-locked loops, sample-hold amplifiers, A/D and D/A converters, as well as a variety of very specialized functions.

1–1 Device Identification

Like other semiconductor devices, such as transistors and diodes, the particular part number of an integrated-circuit device generally carries no indication of why it is different from another device. For example, there is no technical information that can be gained from the number 741 (an operational amplifier) that would tell us why it is different from either a 318 (another operational amplifier) or a 309 (a voltage regulator). In fact, this situation is similar to most models of automobiles. By its name alone, there is nothing to indicate why a Camaro (made by Chevrolet) is different from a Mustang (made by Ford). On the other hand, BMW and Mercedes car models are logically designated to provide information about body style and engine size. Unless you are familiar with each device (or car), there is no way to identify what each device does or looks like. The numbers for integrated-circuit devices were assigned by their original manufacturers, either in some logical sequence known only to them or in a haphazard, random fashion that makes no sense to the user.

Despite the confusion that may occur to the first-time user, there nevertheless are several aspects about the device-identification system that are somewhat straightforward. From a given identification code marked on the device, it is usually possible to determine the manufacturer, product grade, package material, and date of manufacture, all of which try to answer the questions who, what, how good, and when.

Manufacturer Prefixes

A given device, such as the popular type-741 operational amplifier, can be made by more than one manufacturer. To identify the manufacturer, a prefix is added to the device number marked on the package. As examples, a type-741 operational amplifier may be marked in one of the following ways: μA741, SN741, NE741, CA741, MC1741, or RC741. The most commonly encountered manufacturer prefixes are listed in Table 1–2, which notes that several prefixes (e.g., μA, NE, SE) are used by more than one manufacturer. To distinguish themselves, some manu-

Table 1–2
Manufacturer identification prefix codes

Manufacturer	Prefixes
Analog Devices	AD
Burr-Brown	BB
Exar	XR
Fairchild	NE, μA
Harris Semiconductor	HA
Intersil	ICL, ICM, NE, SE
Lithic Systems	LC
Motorola	MC
National Semiconductor	LF, LM, LP
Precision Monolithics	OP
Raytheon	RC, RM
RCA	CA, CD
Signetics	NE, SE, μA
Silicon General	SG
Texas Instruments	SN

facturers also include their distinctive *logo,* or corporate symbol on the device. For example, the logo for Texas Instruments is the outline of the state of Texas with the letters *TI*. Some of the more common logos that appear on operational amplifiers and linear integrated circuits are shown in Figure 1–2.

Some prefixes are obvious abbreviations for the manufacturer's name (e.g., *AD* for Analog Devices), whereas some give more insight as to function. For example, the *μA* prefix used by Fairchild and Signetics probably stands for microamplifier. National Semiconductor uses the *LM* prefix to designate that the device is a *linear-monolithic* device (e.g., an LM309 voltage regulator), whereas the *LF* prefix is used by National Semiconductor to refer to *linear-FET,* those operational amplifiers having FET-input stages, such as the LF356. Furthermore, the *LP* prefix refers to *linear-low-power*. On the other hand, some companies manufacture both analog- and digital-type integrated circuits. For example, RCA uses the *CA* prefix to designate its analog devices (e.g., a CA3080 operational transconductance amplifier) and the *CD* prefix for digital devices, such as the CD4046 phase-locked loop.

Grades

In most cases, a given device is available in different grades. The top grade is for military or high-reliability applications. This is sometimes called a "mil-spec" device.* On the other hand, the lowest grade is the "commercial" or "hobby-grade" version. In appearance, both grades for a given device have the same package style and the same pin configuration. The only difference is the specifications. For operational amplifiers, the offset parameters for the mil-spec version are usually better. For all integrated-circuit devices, the allowable operating and storage temperature ranges are much wider. Of course, this does not come without its price. Mil-spec-graded devices are more expensive than their commercial or hobby-grade counterparts. Typically, this premium may be approximately five times the cost of the commercial device.

To distinguish the various performance grades, some manufacturers use distinctive prefixes, others use appropriate suffixes, and still others use a number code. For example, Raytheon uses the *RC* and *RM* prefixes to identify its commercial (e.g., RC4151) and military (RM4151) versions, respectively. The *SE* and *NE* prefixes used by some manufacturers respectively refer to the military (e.g., SE555) and commercial or hobby-grade (NE555) versions. On the other hand, some manufacturers such as Signetics use the letter *C* as a suffix to indicate commercial-grade devices. A μA741C is the commerical-grade type-741 operational amplifier, whereas the μA741 is its designation for the military-grade 741 device. National Semiconductor further designates its devices in terms of a 100, 200, or 300 series. These series refer to the military, commercial, and hobby-grade specifications. For example, the same Norton amplifier manufactured by National

*Applicable military standards for the manufacture and reliability performance of semiconductor and integrated-circuit devices include MIL-M-38510 and MIL-STD-883.

Figure 1–2
Commonly encountered logos of semiconductor manufacturers.

ANA – Analog Devices Inc.

EXR – Exar Integrated Systems

FSC – Fairchild Camera & Instrument Corp.

HAS – Harris Semiconductor

HITJ – Hitachi Ltd.

ITL – Intel Corp.

INL – Intersil Inc.

ITTG – ITT Semiconductors (Intermetall)

MATJ – Matsushita Electronics Corp.

MITJ – Mitsubishi Electric Corp.

MOS – Mostek Corp.

MOTA – Motorola Semiconductor Products Inc.

NSC – National Semiconductor

NECJ – Nippon Electric Co.

PHIN – N.V. Philips Gloeilampenfabrieken

RTN – Raytheon Co.

RCA – RCA Corp.

SIC – Signetics Corp.

TII – Texas Instruments Inc.
TIIB – Texas Instruments Ltd.

Semiconductor may be designated as either an LM1900, LM2900, or LM3900. Another example is a voltage regulator, such as the LM109, LM209, or LM309. As a general rule, the 300-hobby-grade series number is generally used as a generic number to describe a particular device. Most users would then say "use a 309 regulator" to mean that either an LM109, LM209, or LM309 could be used as appropriate. As if National Semiconductor didn't already have enough variations, the type 111/211/311 comparator, for example, is available in the following designations:

LM111, LM211, LM311:	standard versions
LF111, LF211, LF311:	FET-input versions
LP311:	low-power versions

Package Material Suffixes

To accurately complete the description of a particular device, a suffix code letter is added to signify the type of material from which the package is made. The most common letters are *C* (for ceramic) and *P* (for plastic). Other codes are usually found in the manufacturer's data sheet. In most circumstances, however, these codes are of minor importance.

Date of Manufacture

Analog devices and, to a greater extent, digital devices are marked to indicate their date of manufacture. Very simply, the date is represented by a four-digit code representing the year and number of the week. For exampale, the code *8521* indicates manufacture during the 21st week in 1985. Since the week number cannot be greater than 52 (or possibly 53 in certain calendar numbering sequences), it is impossible to get the year and week codes confused. When we reach the year 2000, perhaps a new method will be devised to eliminate confusion. Other than for curiosity's sake, the date of manufacture is not that important.

1–2 Package Styles

Integrated-circuit devices are currently available in a wide variety of package styles (Figure 1–3). Of these, the most popular is referred to as the **DIP,** which is the acronym for *dual-in-line package*. The DIP configuration always has an equal number of pins spaced 0.1 in. apart in two parallel rows, which makes them perfect for use with solderless breadboarding sockets as well as printed-circuit-board layouts. Most integrated circuits have either 8, 14, 16, 24, 28, or 40 pins. Since the 8-pin DIP is the smallest, it is often referred to as a *mini-DIP*.

Using the 14-pin DIP of Figure 1–4 as a guide, the pins are numbered in a counterclockwise fashion starting with the number 1 pin located in the upper left-hand corner. It is very important that the number 1 pin be correctly identified because the device can be easily destroyed if the power supply is connected to the

Figure 1–3
Various package styles of integrated circuits.

wrong terminals. One end of the package is usually different. It has either a semicircular notch, or channel, or a dot in one corner. If this end is then designated as the *top*, then the number 1 pin is the top pin of the left-hand row of pins. This holds true no matter how many pins there are in a DIP package.

Besides the DIP configuration, many integrated-circuit devices are available in various **TO** package styles. The abbreviation *TO* stands for *t*ransistor *o*utline and forms the basis of a numbering system created by the Joint Electron Device Engineering Council (JEDEC). These body styles are designated by numbers such

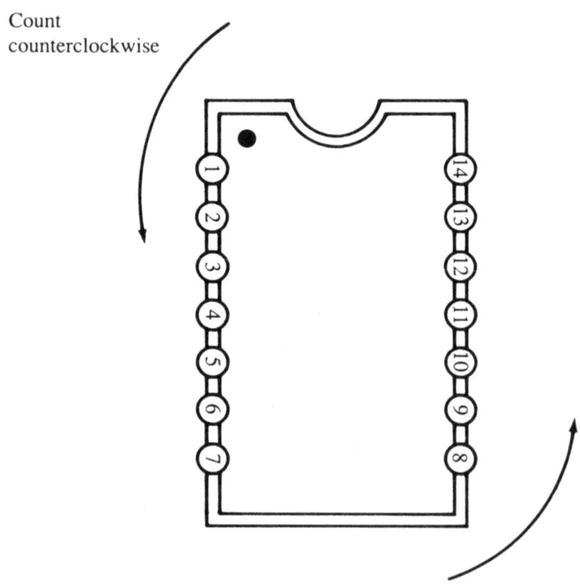

Figure 1–4
Determining the pin numbers for a 14-pin DIP device.

Figure 1–5
Determining the pin numbers for an 8-pin, TO-99 package.

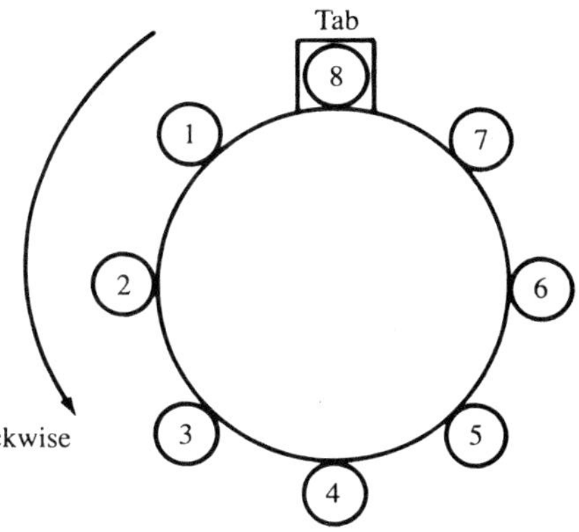

as TO-3, -5, -39, -78, -97, -99, -100, -101, -202, and -220. Of these, the TO-3 (3 leads), TO-99 (8 leads), TO-100 (10 leads), and TO-220 (3 leads) are the most commonly encountered. The standard 14-pin DIP is also referred to as a TO-116-style package.

These devices generally look like transistors, except they have more leads, which are arranged in a cylindrical formation. For the 8-pin, TO-99-style metal can shown in Figure 1–5, the highest pin number is the one located underneath the tab, as viewed from the top of the case. Like the DIP configuration, the pins are then numbered in a counterclockwise fashion, starting with the number 1 pin located immediately to the left of the tab.

Besides the TO package designations, many package styles are sometimes designated by a single-letter suffix. For example, the LM309 5-V voltage regulator, as manufactured by National Semiconductor, comes in two package styles: the *H* and *K* packages, which are equal, respectively, to JEDEC's TO-39 and TO-3 package styles. As another example, the LM317 adjustable voltage regulator (National Semiconductor) is available in three packages: *H* (TO-39), *K* (TO-3), and *T* (TO-220). For these two examples, the different package styles allow for different power and output ratings, as follows:

Device	Package	Maximum Power Dissipation	Output Load Current
LM317H	TO-39	20 W	1.5 A
LM317K	TO-3	2 W	0.5 A
LM317T	TO-220	15 W	1.5 A

Unfortunately, not all manufacturers play by the same rules. Table 1–3 summarizes some of the more popular package styles among manufacturers.

Table 1–3
Package style cross-references

Package	Manufacturers						
	Fairchild	Motorola	NS	Raytheon	RCA	Signetics	TI
8-, 14-, 16-pin ceramic DIP	D, R	U	J	DC, DD	—	F	J
8-, 14-, 16-pin plastic DIP	P, T	P	N	N	E	V	P, N
TO-3 aluminum	K	K	KC	LK, TK	—	DA	K
TO-3 steel	—	KS	K	K	—	—	—
TO-39/TO-99/TO-100	H	G	H	H, T	V	K, L, T	L

1–3 Power Supplies

Integrated circuits are powered by reasonably well regulated power supplies. Depending on circumstances, however, these power supplies may be either driven by the AC power line or battery-powered. Many integrated-circuit devices, such as operational amplifiers and comparators, are normally powered from a *split,* or *bipolar supply* voltage, which can range from approximately ±5 V to ±15 V. As shown in Figure 1–6a, a split supply is formed by connecting two DC voltage sources in series. The junction of the two voltage sources, which are usually equal in magnitude, is the *common,* or *ground,* lead and is the point from which all voltages are referenced. On the other hand, many devices, as well as certain operational amplifier circuits, are capable of being powered by a single supply voltage (Figure 1–6b). Depending on the complexity of a given circuit, a power supply may be required to deliver from several milliamperes of current up to several amperes.

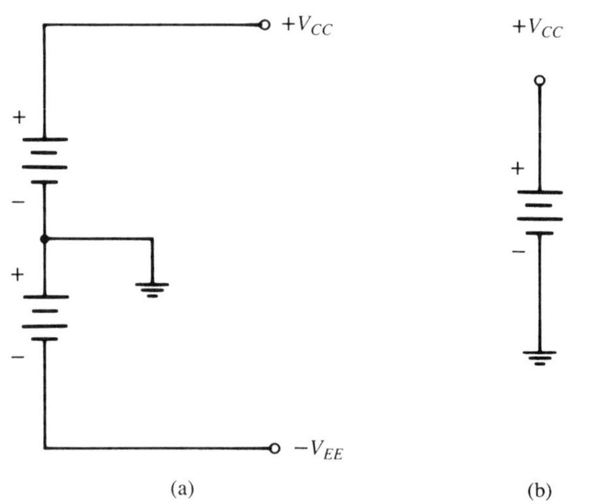

Figure 1–6
Schematic representations for power supplies used with integrated circuit devices. (a) Dual-supply voltage. (b) Single-supply voltage.

1–4 Some Precautions

As with any type circuit, certain rules should be followed so that a circuit will operate as it was intended and so that possible destruction of the circuit or components will be prevented.

- Always wire circuits or make component or wiring changes with the power off.
- Connect signal sources only after DC power has been applied to the circuit. Conversely, disconnect signal sources before removing or turning off DC power from the circuit.
- Make sure that the power-supply polarities to the integrated-circuit device are correct. In general, these connections should be made first when breadboarding so they will not be forgotten later.
- Many times either an operational amplifier or other type circuit will oscillate or the noise originating from the power line or adjacent signals will appear at its output. Generally, a 0.1-μF bypass capacitor should be connected between the power supply pins of the device and ground. For devices using split supplies, bypass capacitors are connected between both the positive supply pin and ground and the negative supply pin and ground (Figure 1–7).

For those devices powered from a single supply voltage, the bypass capacitor is connected only between the positive supply pin and ground. However, it should be strongly emphasized that these connections should be made as close as possible to the device's supply pins and its ground pin. Furthermore, all connections should be made as short as possible for optimum effectiveness in decoupling extraneous signals from the DC power supply lines.

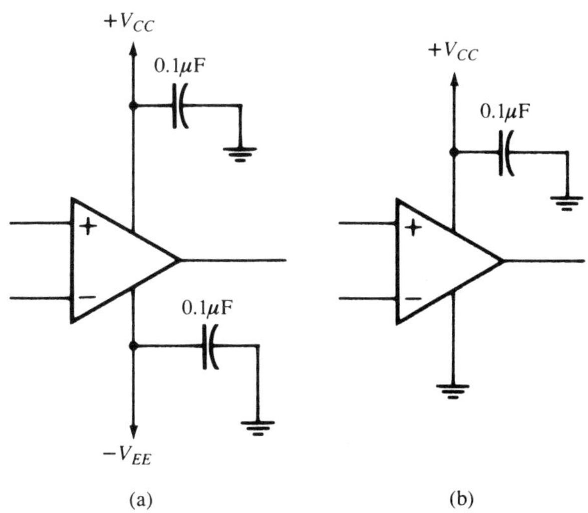

Figure 1–7
Using 0.1-μF capacitors to bypass power supply pins to reduce oscillations and self-coupled noise from the power supply leads. (a) Dual-supply voltage connections. (b) Single-supply voltage connections.

Summary

This first chapter briefly introduced the origin and evolution of the integrated circuit, using the operational amplifier as its most notable example. As well as its historical significance, information concerning package styles, numbering conventions, power supply requirements, and precautions in using integrated circuits were also discussed.

Glossary

DIP An acronym for *d*ual-*i*n-line *p*ackage, a package configuration having an equal number of pins spaced 0.1 in. apart in two parallel rows.

linear integrated circuit Any integrated circuit device that responds to analog signals, as opposed to digital signals.

monolithic integrated circuit An integrated circuit whose individual transistors and resistors are fabricated from a single piece of solid-state material.

operational amplifiers Also called an *op-amp,* a high-gain DC differential amplifier capable of performing a wide range of functions by using external feedback.

TO An abbreviation for *t*ransistor *o*utline, a metal configuration package.

Review Questions

1. Explain the origin of the term *operational amplifier.*
2. Explain how most manufacturers differentiate their integrated circuits from those of other manufacturers, especially if the devices are otherwise the same.
3. Explain the differences between commercial and military integrated circuits and how they are designated in the devices' serial numbers.
4. What information is represented by the suffix code of a device number?
5. For what does *DIP* stand? How are the pins of DIP devices numbered?
6. For what does *TO* stand? How are the pins of TO devices numbered?

2 The Operational Amplifier

Instructional Objectives

Upon completing this chapter, you will be able to

- Identify by name the *input, output,* and *power supply terminals* of a typical operational amplifier from its schematic symbol.
- List the characteristics of an ideal operational amplifier.
- Describe the SP and PP forms of negative feedback.
- Explain the differences among *open-loop gain, loop gain,* and *closed-loop gain.*
- Draw the circuit and determine the output voltage of the following circuits:
  - Noninverting and inverting amplifiers
  - Voltage follower
  - Inverting and noninverting summing amplifiers (adders)
  - Difference amplifier

Introduction

The basic purpose of an electronic amplifier is to increase the size of a signal. Besides voltage, the input signal parameter that is to be increased may also be current or power. A **linear amplifier** not only increases the signal's level but also produces an output that is a faithful reproduction of the input. A small signal applied to the input terminals of a perfect linear amplifier will produce a larger signal of *exactly* the same waveform at its output terminals.

The operational amplifier, often called an *op-amp,* is such a device; it lends itself to the construction of very good linear amplifiers as well as many nonlinear circuits. To keep things simple initially, this chapter first discusses the *ideal* operational amplifier and how it is used to form voltage followers as well as noninverting, inverting, summing, and difference amplifiers. In Chapter 3, we take into account many of the parameters that place some limits on the operational amplifier's performance.

2–1 Fundamentals of the Ideal Operational Amplifier

Originally, the term *operational amplifier* was used to describe a series of high-performance DC differential amplifiers using vacuum tubes that formed the basis of the analog computer. As a result of improvements in technology, today's operational amplifier is a solid-state, integrated-circuit device that uses external feedback networks to control its response.

As shown by the two common schematic symbols of Figure 2–1, the operational amplifier has two input terminals. One input is called the *noninverting input* and is indicated on the schematic symbol by a plus sign (+). The other input terminal is called the *inverting input* and is indicated by a minus (−) sign. A single output is shown to the right. The operational amplifier is usually powered by a dual-voltage, or split, power supply. The positive supply voltage with respect to ground is often designated as $+V_{CC}$, whereas the negative supply voltage with respect to ground is $-V_{EE}$, although it is occasionally written as $-V_{CC}$.

There is no standard convention in regard to which input is drawn above the other. For clarity, some operational amplifier circuits have the noninverting input on top; for others, the inverting input is above the noninverting input. The left-hand triangular symbol is the preferred symbol, although the older symbol (with a curved side) is still used by some engineers as well as by some books and magazines.

As a difference, or differential, amplifier, the ideal operational amplifier amplifies any difference in voltage that appears between these two inputs. Without any type of path that feeds back part of or all the output signal to one of its two

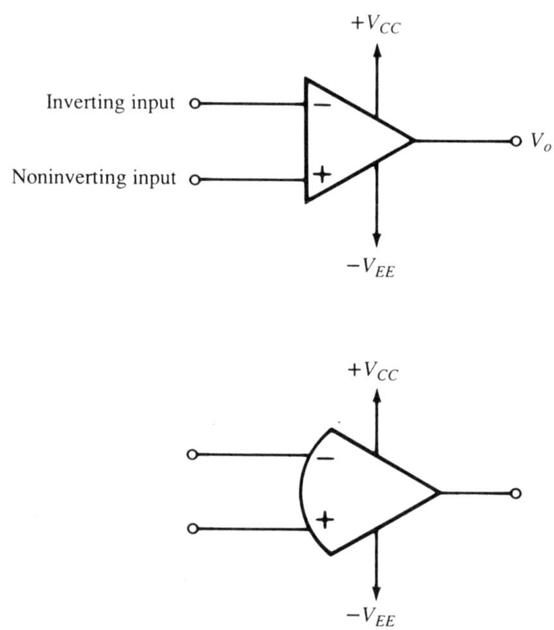

Figure 2–1
Bipolar operational amplifier schematic symbols showing inputs, output, and power supply connections.

inputs, the operational amplifier is said to be operating in the **open-loop mode.** The internal gain* of the operational amplifier in this case is called the *open-loop gain* (A_{OL}). In the open-loop mode, the operational amplifier's output voltage equals the input voltage difference times the open-loop gain:

$$V_o = A_{OL}(V_1 - V_2) \qquad (2\text{–}1)$$

where

V_1 = noninverting input voltage
V_2 = inverting input voltage

Equation 2–1 is a generalized algebraic expression in which the polarities of V_1 and V_2 may be positive or negative with respect to the circuit ground. If both inputs are at the same potential, this is referred to as the *common-mode input* (i.e., the same voltage is *common* to both inputs); $V_1 = V_2$ and the output is zero.

The ideal operational amplifier is expected to have the following major characteristics:

1. Infinite open-loop voltage gain
2. Infinite input impedance
3. Zero output impedance
4. Infinite bandwidth
5. Zero offset voltage (i.e., zero input produces zero output)
6. Infinite common-mode rejection

Unfortunately no real operational amplifier meets these ideal characteristics, but—except for infinite bandwidth—many are close enough for practical purposes.

We are often able to assume that we are working with ideal operational amplifiers, as we do in this chapter. The results obtained are usually more than adequate for the particular application.

2–2 Power Supplies for Operational Amplifiers

Most operational amplifier circuits require a dual, or split, power source for proper operation. However in Chapter 5, we will see how to operate the operational amplifier for some applications with a single supply voltage.

A split power supply (Figure 2–2) has three terminals. One terminal is positive ($+V_{CC}$) with respect to the ground, whereas the other terminal is negative ($-V_{EE}$) with respect to ground. The third, or *common,* terminal is always connected to the circuit ground. The supply voltage required by most modern operational am-

*When the term *gain* is used, the implication is that it refers to the *voltage gain* of an amplifier.

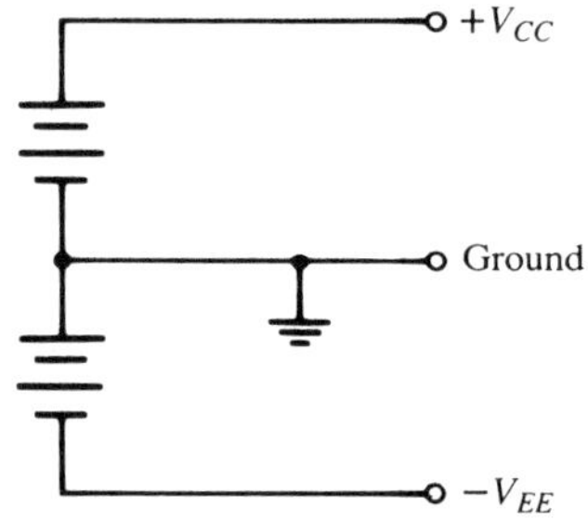

Figure 2–2
Schematic diagram of a dual, or split voltage power supply used with operational amplifiers.

plifiers falls in the range of ±3 to ±15 V. This means that the differential voltage ($V_{CC} - V_{EE}$) between the two power-supply terminals may be as high as 30 V. Although the power-supply connections are shown on these schematic symbols, it is the normal practice to omit them for clarity, as they are implied to have been made. Even though they are not shown, they must be included in an actual circuit.

It is considered good practice to capacitively decouple, or bypass, the positive and negative power-supply terminals to ground as close to the operational amplifier as possible. Usually this is done with 0.01-μF to 0.1-μF ceramic capacitors. Capacitive bypassing provides an AC path to ground for the transfer of electrical noise between the power supply and the operational amplifier. Although bypassing is not always absolutely necessary and is sometimes done partially or not at all, it nevertheless is excellent insurance against noisy and unstable operational amplifier operation.

2–3 Operational Amplifiers with Negative Feedback

As briefly mentioned earlier, the gain that the operational amplifier provides without any external circuitry connected between its output and one of its inputs is called the open-loop gain, A_{OL}. For the commonly used type-741 operational amplifier, the open-loop gain is typically 200,000 (+106 dB) and is much too large to be used in a single stage of amplification without loss of amplifier stability. Fortunately, we can decrease and stabilize the circuit gain of an operational amplifier stage and greatly enhance several other parameters that are very desirable in a linear amplifier by adding a few external components. This process is known as adding *negative feedback.* Like most things in life, you do not get something for nothing. Here, the trade-off is a loss in gain for an increase in stability, plus a few other things thrown in as a bonus. Positive feedback, on the other hand, increases the circuit gain and makes the circuit unstable, so that it will oscillate.

To apply negative feedback, a fraction of the output signal that is 180° out of phase with the incoming signal is added to the incoming signal, as shown in the block diagram of Figure 2–3. The input signal actually applied to the operational amplifier (V_d) is now the difference between the incoming signal (V_i) and the portion of the output signal being fed back from the output (V_f). By adding negative feedback, we do not actually change the gain of the operational amplifier

Figure 2–3
Block diagram of an amplifier stage with negative feedback.

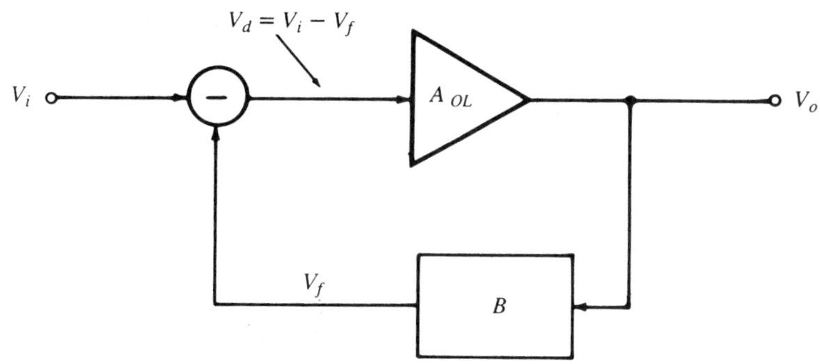

itself (A_{OL}) but only the gain of the overall stage. In effect, we have made the input signal to the operational amplifier (i.e., $V_i - V_f$) smaller than it would have been without feedback (V_i). Consequently, the output signal will also be smaller.

The feedback network usually consists of passive components such as resistors and capacitors. The *feedback factor* B is the ratio V_f/V_o; it represents the fraction of the output voltage that is fed back to the amplifier's input. The circle with the minus sign in the center represents that portion of the operational amplifier's internal circuit that subtracts the output voltage of the feedback network from the incoming signal. The voltage difference (V_d) is sometimes referred to as the *error voltage* and is equal to $V_i - V_f$. In terms of V_i, V_f, A_{OL}, and B, the output voltage of the amplifier is

$$V_o = A_{OL} V_d \tag{2–2a}$$

Since $V_d = V_i - V_f$,

$$V_o = A_{OL}(V_i - V_f) \tag{2–2b}$$

The feedback voltage (V_f) is then written as BV_o, so that Equation 2–2b becomes

$$V_o = A_{OL}(V_i - BV_o) \tag{2–2c}$$

Finally, solving for the ratio V_o/V_i, the voltage gain of the amplifier circuit with feedback is

$$\frac{V_o}{V_i} = \frac{A_{OL}}{1 + BA_{OL}} \tag{2–3}$$

Since Equation 2–3 represents the voltage gain of the circuit with a closed feedback path between the output and the amplifier's input, this voltage gain is referred to as the **closed-loop voltage gain** (A_{CL}). If no feedback were used, the circuit gain would equal the amplifier's open-loop gain. The quantity BA_{OL} is referred to as the *loop gain* (A_L) and is usually a large number much greater than 1. In terms of A_{OL} and A_{CL}, the loop gain can be shown to be

$$A_L = \frac{A_{OL}}{A_{CL}} \tag{2–4}$$

Since the loop gain is much greater than 1, the denominator of Equation 2–3 is approximately equal to BA_{OL} and the closed-loop gain is approximately equal to

$$A_{CL} \simeq \frac{1}{B} \tag{2–5}$$

As long as the loop gain is very large, the closed-loop gain of the amplifier equals the reciprocal of the feedback factor, $1/B$. The voltage gain of the stage (i.e., the closed-loop gain) is no longer dependent on the open-loop gain of the amplifying device; instead, it depends only on the elements of the feedback network. Since B is usually determined by stable components, such as resistors and capacitors, we are now able to build amplifiers with very stable and predictable values for the closed-loop gain. While the closed-loop gain of Equation 2–3 is more accurate than that of Equation 2–5, the latter is generally used, since the loop gain is always high.

Example 2–1

Negative feedback is applied to an amplifier with an open-loop gain of 100,000. The feedback factor (B) is 0.1. (a) Find the exact closed-loop gain using Equation 2–3 and (b) compare it with the approximate closed-loop gain of Equation 2–5.

Solution:

(a) Using Equation 2–3, the exact closed-loop gain for the amplifier with feedback is

$$A_{CL} = \frac{100{,}000}{1 + (0.1)(100{,}000)} = \mathbf{9.999}$$

(b) Since the loop gain (10,000) is significantly larger than 1, Equation 2–5 yields

$$A_{CL} = \frac{1}{0.1} = \mathbf{10}$$

Example 2–2

If the RMS input voltage for the feedback amplifier of Example 2–1 is 1 V, find (a) the output voltage and (b) the difference voltage, V_d.

Solution:

(a) From Equation 2–3, the closed-loop gain was found to be 9.999. The output voltage is then

$$\begin{aligned} V_o &= (9.999) \times (1 \text{ V RMS}) \\ &= \mathbf{9.999\ V\ RMS} \end{aligned}$$

(b) Since the output voltage of the feedback network is 0.1 times the output voltage, or 0.9999 V RMS, the difference voltage is then $V_i - V_f$:

$$\begin{aligned} V_d &= 1 \text{ V} - 0.9999 \text{ V} \\ &= \mathbf{0.1\ mV\ RMS} \end{aligned}$$

In Example 2–2, the differential, or error, voltage was very small (0.1 mV). As long as the loop gain is large, this will be true. As a check, the output voltage, using Equation 2–2a, is

$$\begin{aligned} V_o &= (100{,}000) \times (0.1 \text{ mV RMS}) \\ &= 10 \text{ V RMS} \end{aligned}$$

which is 0.1 mV RMS from the true value of 9.999 V, or an error of 0.001%. In many applications, this error voltage can be assumed to be zero without introducing any appreciable error in the results.

When we have an amplifier with negative feedback, any change in the level of the input signal produces a change in the error signal. In turn, the error signal at the operational amplifier's inputs produces a change in the output signal, which opposes the effects of the input signal. The net result is always to minimize the error signal. When an operational amplifier is used with negative feedback, the error voltage is the difference (or differential) voltage between the inverting and noninverting inputs. When the loop gain is very high, as is usually the case in operational amplifiers, the two inputs have essentially no voltage across them.

The negligible error voltage is consistent with the ideal operational amplifier's infinite input impedance characteristic. If the input impedance is infinite, no current flows into the operational amplifier and, consequently, there is no voltage drop. The following sections discuss two commonly encountered linear operational amplifier circuits using negative feedback: the noninverting and inverting amplifiers.

2–4 The Noninverting Operational Amplifier

One way of implementing negative feedback with an operational amplifier is shown in Figure 2–4. The input to the feedback network is connected in parallel with the output of the operational amplifier, whereas the output of the feedback

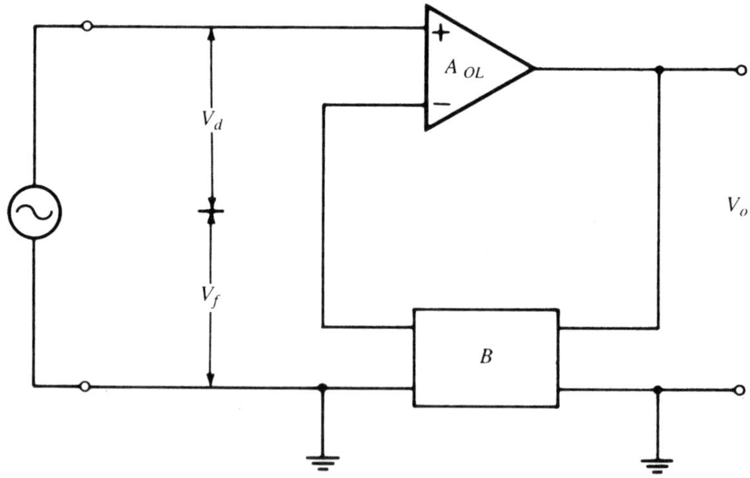

Figure 2–4
Block diagram of a series-parallel (SP) feedback network used to create a noninverting amplifier.

network is connected in series with the inverting input of the operational amplifier. This is sometimes referred to as an *SP connection* (*s*eries input, *p*arallel output). A fraction of the operational amplifier output voltage is fed back in series with the applied signal, so the differential input voltage to the operational amplifier is very small.

As described earlier, V_d is usually very small, so V_i is essentially equal to V_f. Furthermore, since $V_f = BV_o$, then V_i is approximately equal to BV_o. Using these approximations, the voltage gain for the SP feedback connection is

$$A_V = \frac{V_o}{BV_o} \tag{2–6a}$$

$$= \frac{1}{B} \tag{2–6b}$$

which is the same result as Equation 2–5.

As shown in Figure 2–5, the passive circuit elements forming the feedback section make up a simple voltage divider string consisting of two resistors (R_1 and R_2) connected in series. The fraction of the output signal (B) fed back to the operational amplifier's inverting terminal is equal to the voltage across resistor R_1. Using the voltage divider equation, V_f is

$$V_f = \frac{R_1}{R_1 + R_2} V_o \tag{2–7}$$

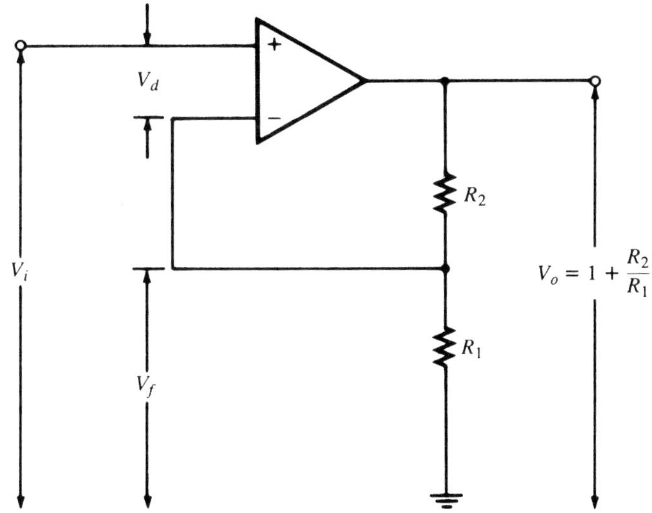

Figure 2–5
Noninverting amplifier having closed-loop voltage gain of $1 + R_2/R_1$.

Since the feedback factor is $B = V_f/V_o$, then in terms of R_1 and R_2,

$$B = \frac{R_1}{R_1 + R_2} \qquad (2\text{–}8)$$

and the closed-loop gain of this stage is equal to $1/B$, or

$$A_{CL} = \frac{V_o}{V_i} \qquad (2\text{–}9a)$$

$$= 1 + \frac{R_2}{R_1} \qquad (2\text{–}9b)$$

This SP feedback–operational amplifier circuit is known as a **noninverting amplifier.** For a given AC or DC input signal, the corresponding output signal will at all times have the *same polarity* as the input. When the input goes positive, the output goes positive; when the input goes negative, the output goes negative. For AC signals, the output signal of a noninverting amplifier is usually thought of as being *in phase* with the input signal.

Closer inspection of Equation 2–9b for the noninverting amplifier circuit of Figure 2–5 reveals that the closed-loop gain can never be less than 1. In fact, it can never be *exactly* equal to 1 either. No matter whether the value of R_1 is greater than R_2 or less than R_2, the magnitude of the output voltage will *always be greater* than the magnitude of the input voltage. However, if R_1, for example, is made 1000 times larger than R_2 (for example, 1 MΩ versus 1 kΩ), then the closed-loop voltage gain will be 1 + (1 kΩ/1 MΩ), or 1.001, which is so close to 1 that it would be extremely difficult to measure any difference after eliminating the tolerance of the resistor values.

In theory, there are no restrictions on what values R_1 and R_2 may have because the closed-loop gain is based simply on the ratio of their values. The closed-loop gain is 11 both for $R_1 = 47\ \Omega$ and $R_2 = 470\ \Omega$ and for $R_1 = 1.5\ \text{k}\Omega$ and $R_2 = 15\ \text{k}\Omega$. In practice, however, most resistors associated with operational amplifier circuits are generally made larger than 1 kΩ. To find R_2, choose a value for R_1 and then calculate R_2 for a given closed-loop gain.

In the next chapter, more guidance is given about selecting these two resistors. In this chapter, however, the emphasis is on acquainting you with operational amplifier circuits without getting involved with those features that may limit their performance.

2–5 The Voltage Follower

For the noninverting amplifier circuit of Figure 2–5, if we now short-circuit R_2 and open-circuit R_1, we get a very simple but useful special-case variation of the SP connection called the *voltage follower,* which is shown in Figure 2–6. From Equation 2–9b, with $R_1 = \infty$ and $R_2 = 0$, the closed-loop gain is 1. The feedback factor also equals 1 because all the output voltage is now fed back to the operational amplifier's inverting input.

From the concept of the ideal amplifier, the voltage follower has essentially an infinite input impedance and zero output impedance. Like the transistor emitter follower and the FET source follower, the operational amplifier serves a similar function as a *buffer amplifier*. The output equals the input in both signal amplitude and polarity (or 0° phase shift). Except for matching high-impedance signal sources to low-impedance loads, the circuit is virtually useless.

It is important to remember that operational amplifier input terminals need a DC reference to establish a bias point. For example, as shown in Figure 2–7, if we use a coupling capacitor between the signal source and the operational amplifier's noninverting input to remove any DC level on the input signal, a shunt resistor must then be placed from the noninverting input to ground in order to provide a path for a small input current. This is sometimes referred to as providing a DC return path. Even though this resistor can usually be made relatively large, it tends to lower the input impedance of the stage. Unfortunately, we have no control over this.

Figure 2–6
Operational amplifier voltage follower having unity closed-loop gain.

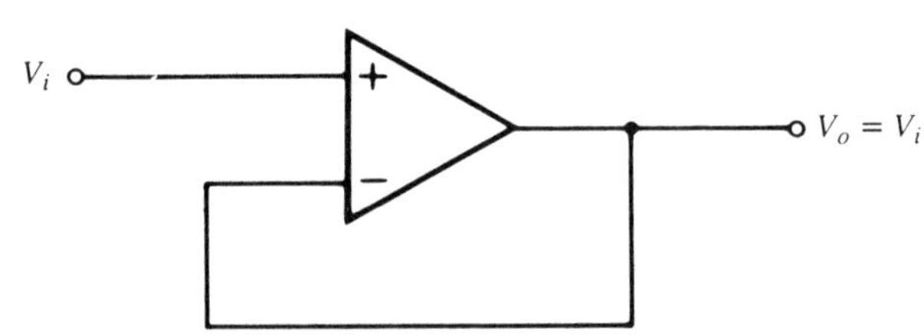

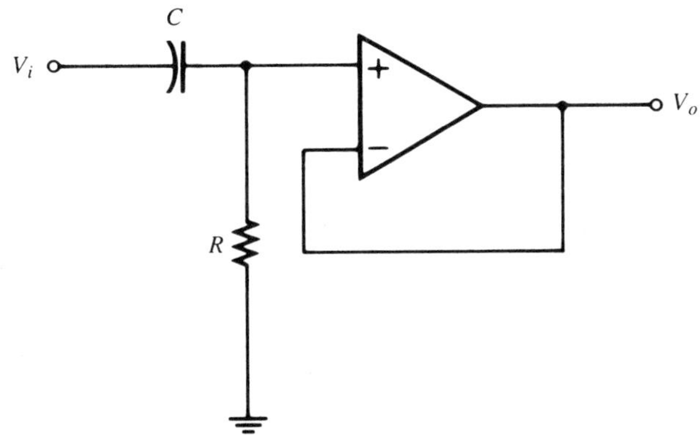

Figure 2–7
AC coupled amplifier using a shunt resistor to establish a DC return path for the input bias current of the operational amplifier.

2–6 The Inverting Operational Amplifier

In addition to the SP connection, the other way to implement negative feedback with an operational amplifier is to have the feedback network connected in parallel with both the input and output of the amplifier. As shown in Figure 2–8, this is sometimes referred to as the *PP connection* (for *p*arallel in, *p*arallel out).

In the PP-feedback connected circuit, the feedback network usually consists of a single resistor. Since the open-loop gain of the amplifier is very large, the error voltage between the two inputs is very small, and we can normally assume that both inputs are at the same potential: ground. The circuit behaves as though the inverting input is at ground potential even though it is not connected directly to ground. In this configuration, the inverting input is often described as a **virtual ground.**

Figure 2–8
Block diagram of a parallel-parallel (PP) feedback connection used to create an inverting amplifier.

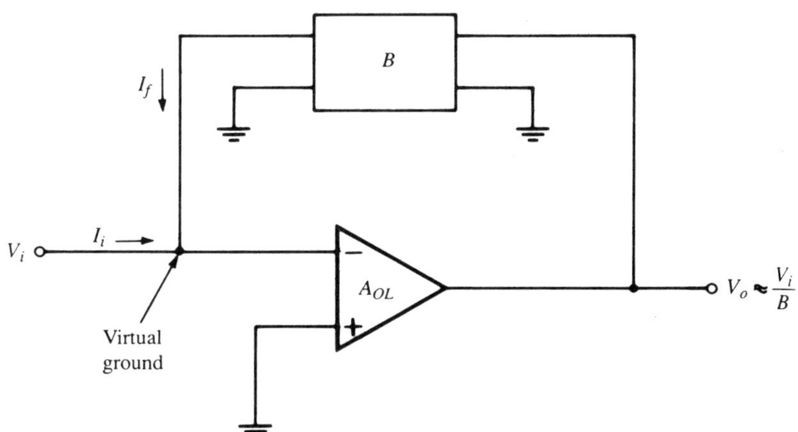

When a signal source is connected to the circuit's input, because of the virtual ground the input current I_i appears to flow directly to ground. Actually, the input current then flows through the feedback network ($I_f = -I_i$) to the output terminal of the amplifier. From Ohm's law, a voltage equal to the feedback current times the impedance of the feedback element is developed across the feedback element. For this to happen (assuming "conventional" current flow), the amplifier's output voltage must be below ground by an amount equal to this voltage drop. If the feedback network is a single resistor (R_f), then

$$V_o = -I_i R_f \tag{2–10}$$

This means that the simple PP-feedback connection of Figure 2–9 functions as a *current-to-voltage converter*. However the resultant input impedance (the short circuit from the input signal to the virtual ground of the inverting input) is such that no significant voltage is developed across it. This circuit must then be current-driven.

Figure 2–10 shows the addition of resistor R_i to the current-to-voltage circuit. To a signal source, R_i appears to go to ground, since it is connected to the virtual ground of the operational amplifier's inverting input. When an input voltage (V_i) is now applied to the circuit, a corresponding current (I_i) flows through R_i:

$$I_i = \frac{V_i}{R_i} \tag{2–11}$$

Resistor R_i, which is frequently called the *input resistor,* functions as a *voltage-to-current converter,* and when combined with the current-to-voltage action of the feedback resistor R_f, the PP-connected amplifier then functions as an *inverting* voltage amplifier. Resistors R_i and R_f now form a voltage divider between V_i and V_o. The operational amplifier produces an output voltage that keeps the operational amplifier's inverting input at the junction of R_i and R_f at nearly the same voltage as the noninverting input. As with the SP connection, this error, or difference, voltage can be considered to be zero for all practical purposes.

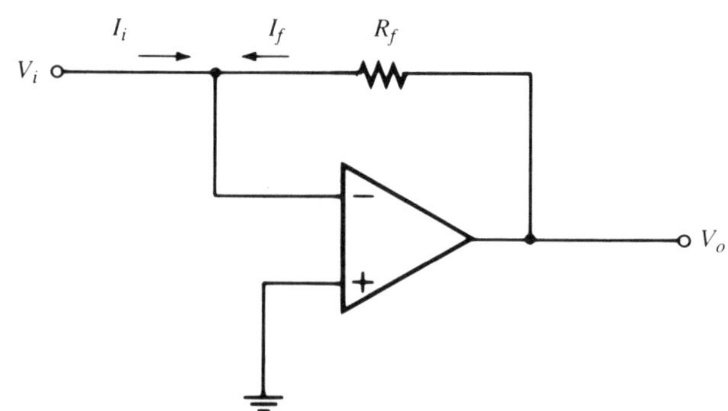

Figure 2–9
The PP feedback connection actually functions as a current-to-voltage converter.

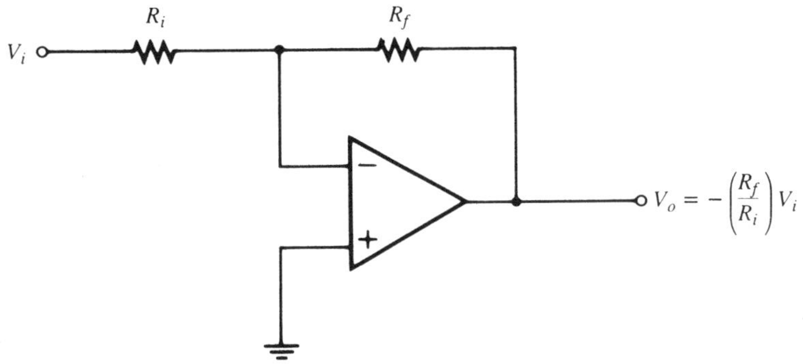

Figure 2–10
Inverting amplifier with a closed-loop gain of $-R_f/R_i$.

To illustrate these points, Figure 2–11a shows the schematic of an inverting amplifier having a closed-loop gain of 10 and a DC input voltage of +1 V. As shown by the equivalent circuit of Figure 2–11b, the 2-kΩ input resistor (R_i) converts the 1-V input to a (1 V)/(2 kΩ), or 0.5-mA, current flowing through R_i. Since the operational amplifier's inverting input is a virtual ground, the current flowing from the amplifier's output through R_f to the virtual ground has to be the same as the 0.5 mA flowing through R_i but of opposite polarity (Kirchhoff's current law). Using Ohm's law, the output voltage must then equal 20 kΩ × −0.5 mA, or −10 V.

Figure 2–11
Inverting amplifier with a closed-loop gain of 10. (a) Schematic diagram. (b) Equivalent circuit.

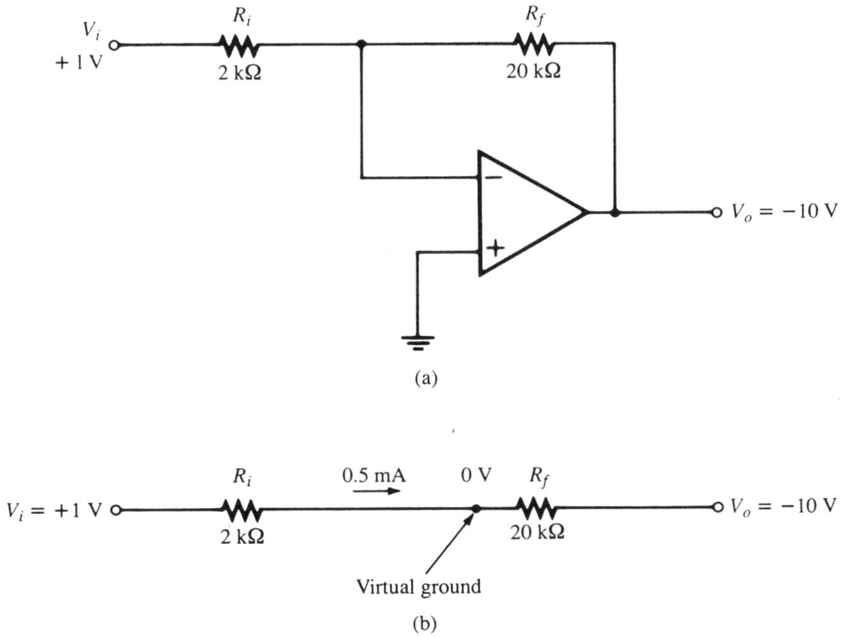

Equations 2–10 and 2–11 are then combined to give the inverting amplifier's closed-loop gain as

$$A_{\mathrm{CL}} = \frac{V_o}{V_i} \qquad (2\text{–}12a)$$

$$= -\frac{R_f}{R_i} \qquad (2\text{–}12b)$$

The minus sign indicates that the polarity of the output voltage is opposite that of the input voltage. For AC signals, this represents a 180° phase shift. Closer inspection of Equation 2–12b for the inverting amplifier circuit of Figure 2–11a reveals that the closed-loop gain can be less than 1, equal to 1, or greater than 1. Furthermore, it can be easily shown that the feedback factor for the inverting amplifier of Figure 2–11a is R_i/R_f.

As with the ideal noninverting amplifier circuit, there are, in theory, no restrictions on the values R_i and R_f may have because the closed-loop gain is based only on the ratio of their values. However, there are several practical considerations that should be kept in mind, which are discussed in Chapter 3.

2–7 Summing Amplifiers

If more inputs are added to the inverting amplifier of Figure 2–10, the result is what is known as a **summing amplifier,** or *adder* (Figure 2–12). The output of the summing amplifier is proportional to the *algebraic sum* of its inputs. Each input signal produces a component of the output signal that is completely independent of the other input signals.

Since the inverting input of the operational amplifier is still a *virtual ground,* each input signal will produce its own component of input current independently of all other inputs. For example, V_1 produces a component of the total input current equal to V_1/R_1; the remaining inputs behave similarly. Using Kirchhoff's current law, all the input current components added together are equal in magnitude (but opposite in direction) to the current through the feedback resistor R_f that produces the corresponding output voltage. Alternatively, the total output voltage could be determined by superposition, taking each input one at a time:

$$V_o = -\frac{R_f}{R_1}V_1 - \frac{R_f}{R_2}V_2 - \frac{R_f}{R_3}V_3 - \frac{R_f}{R_4}V_4 \qquad (2\text{–}13a)$$

or

$$V_o = -R_f\left[\frac{V_1}{R_1} + \frac{V_2}{R_2} + \frac{V_3}{R_3} + \frac{V_4}{R_4}\right] \qquad (2\text{–}13b)$$

Figure 2–12
Four-input inverting summing amplifier that amplifies the algebraic sum of its inputs.

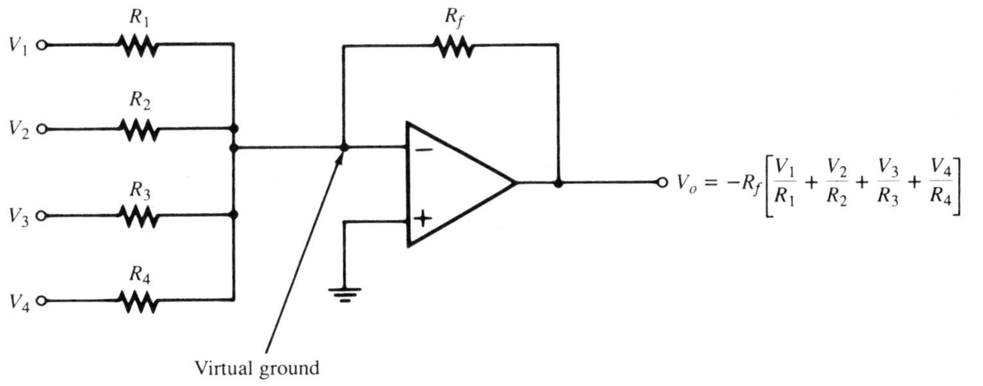

For the special case where all four input resistors are equal to the feedback resistor ($R_1 = R_2 = R_3 = R_4 = R_f$), Equation 2–13 reduces to

$$V_o = -(V_1 + V_2 + V_3 + V_4) \tag{2–14}$$

which is the inverted algebraic sum of all the inputs. The input signals can be all DC voltages, all AC voltages of the same or different waveforms and frequencies, or a mixture of AC and DC signals. The output signal will, nevertheless, be proportional to the instantaneous algebraic sum of all the input signals. When all input signals are of the audio type (20 Hz to 20 kHz), the summing amplifier is frequently called a *signal mixer*. A mixer is used to combine audio signals from several microphones, guitars, or tape recorders to provide a single output.

Figure 2–13
Summing amplifier circuit for Example 2–3.

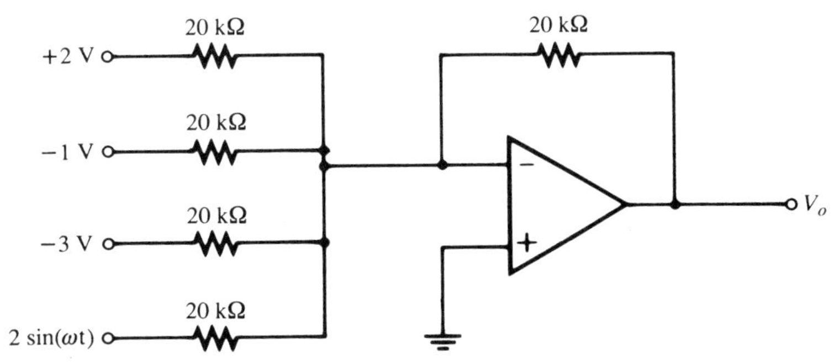

Example 2–3
Determine the output voltage for the summing amplifier of Figure 2–13 if the four input signals are given as

$$V_1 = +2\text{ V}$$
$$V_2 = -1\text{ V}$$
$$V_3 = -3\text{ V}$$
$$V_4 = 2\sin(\omega t)$$

Solution: Since all circuit resistors are equal, the output is simply the inverted algebraic sum of the three DC voltages and the one AC voltage:

$$V_o = -[2\text{ V} - 1\text{ V} - 3\text{ V} + 2\sin(\omega t)]$$
$$= 2 - 2\sin(\omega t)$$

This is nothing more than a 2-V peak sine wave that is 180° out of phase with the input sine wave and rests on an average (DC) voltage level of +2 V, as shown in Figure 2–14.

The basic inverting summing amplifier circuit can be used to take the instantaneous average of its inputs. For the four-input summing amplifier of Figure 2–13, it can be easily shown that the output voltage will represent the instantaneous average of its four inputs when all four input resistors (R_1) are equal and $R_F = R_1/4$. For an inverting summing amplifier having N inputs,

$$R_F = \frac{R_1}{N} \qquad (2\text{–}15)$$

Example 2–4
Determine the component values for an inverting summing amplifier that will take the average of three input signals.

Solution: Assuming that all three input resistors are made equal to a given standard value of 30 kΩ, then feedback resistor must then be equal to 30 kΩ/3, or **10 kΩ,** giving the final circuit of Figure 2–15.

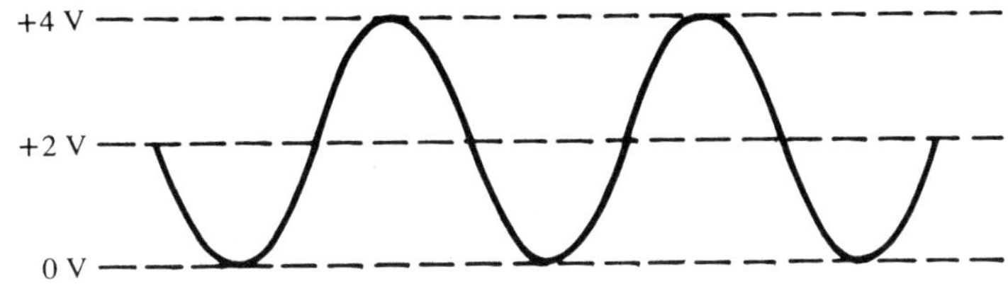

Figure 2–14
Output waveform for the circuit of Figure 2–13.

Figure 2–15
Completed inverting summing amplifier for Example 2-4.

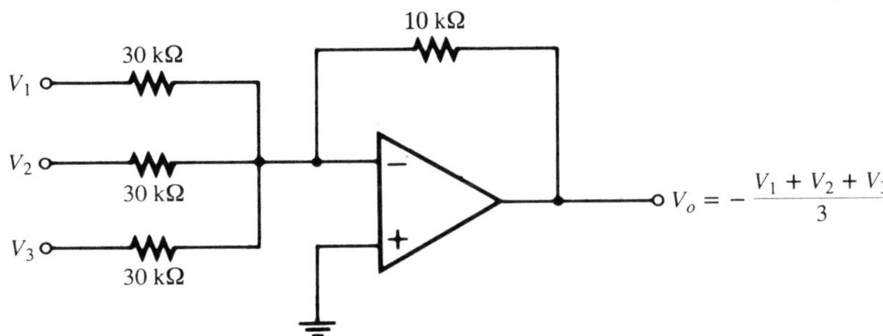

Figure 2–16
Noninverting summing amplifier.

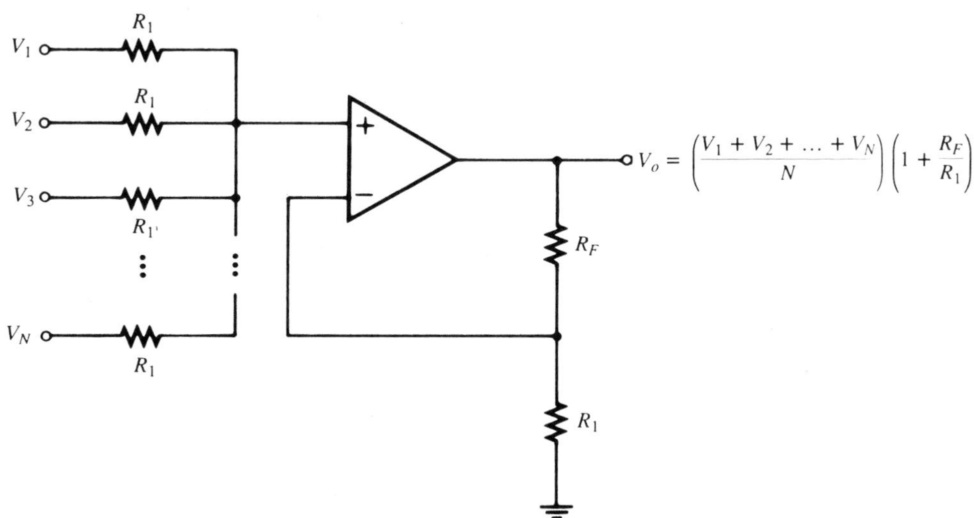

In addition to the inverting summing amplifier, it is possible to have a noninverting summing amplifier, as shown by the circuit of Figure 2–16. In actuality, such an amplifier involves nothing more than the process of passively summing the voltages at a single node followed by a conventional noninverting amplifier. If all the resistors that are in series with each of the N inputs are made equal, for convenience, to R_1, the voltage present at the operational amplifier's noninverting input is then

$$V_{total} = \frac{V_1 + V_2 + V_3 + \cdots + V_N}{N} \tag{2–16}$$

Since the output voltage of the noninverting amplifier circuit must be

$$V_o = A_{CL} V_{total} \tag{2–17a}$$

$$= \left[1 + \frac{R_F}{R_1}\right](V_{total}) \tag{2–17b}$$

then, by equating Equations 2–16 and 2–17b, the following must hold:

$$N = 1 + \frac{R_F}{R_1} \tag{2–18}$$

or

$$R_F = (N - 1)R_1 \tag{2–19}$$

Example 2–5

Determine the component values for a noninverting summing amplifier that will take the sum of three input signals.

Solution: Assuming that all three input resistors are made equal to a given standard value of 10 kΩ, then the feedback resistor must be equal to

$$R_F = (3 - 1)(10\ \text{k}\Omega)$$
$$= \mathbf{20\ k\Omega}$$

giving the final circuit of Figure 2–17.

Figure 2–17
Completed summing amplifier circuit for Example 2–5.

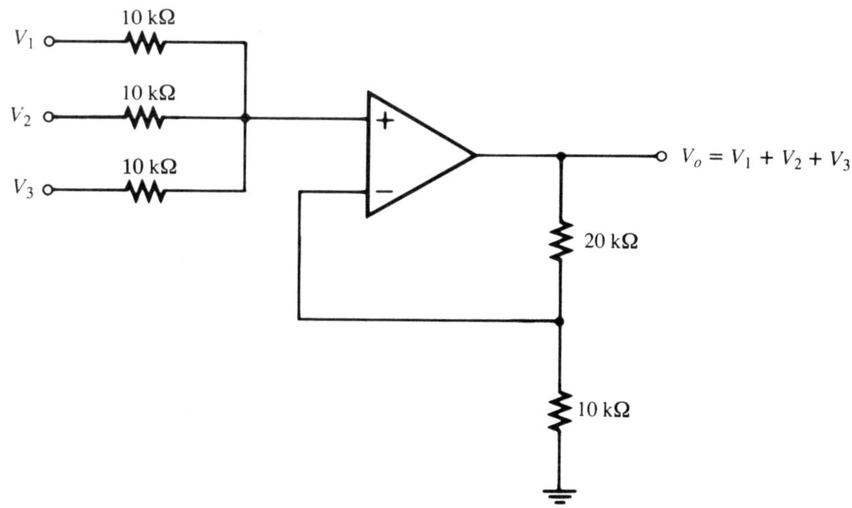

2–8 The Difference Amplifier

By using both the inverting input and the noninverting input at the same time, as shown in Figure 2–18, we get what is called a **difference,** or *differential,* **amplifier,** whose output voltage is proportional to the voltage difference $V_2 - V_1$.

Each of the two input signals will produce its own component of the total output signal, which is independent from the other input signal. We may easily determine the resulting output signal by using the principle of superposition. If V_2 is first set to zero (Figure 2–19b), we have an inverting amplifier with a closed-loop gain equal to $-R_2/R_1$. The output voltage is then

Figure 2–18
Difference amplifier that amplifies the input voltage difference $V_2 - V_1$.

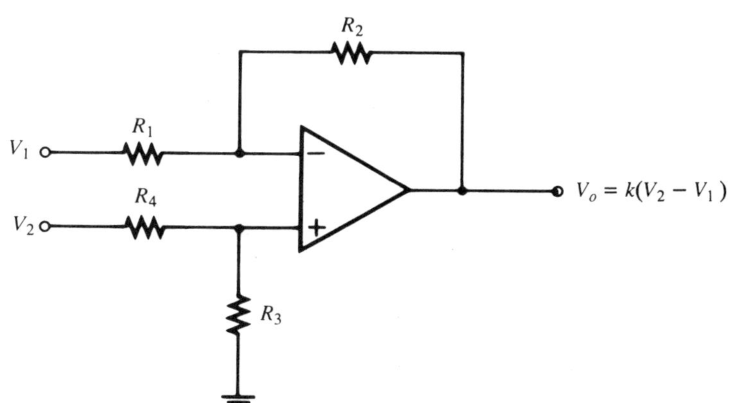

Figure 2–19
Analysis of the differential amplifier using superposition. (a) Basic circuit. (b) Equivalent circuit showing an inverting amplifier by reducing V_2 to zero. (c) Equivalent circuit showing an noninverting amplifier with voltage divider ($R_3 - R_4$) by reducing V_1 to zero.

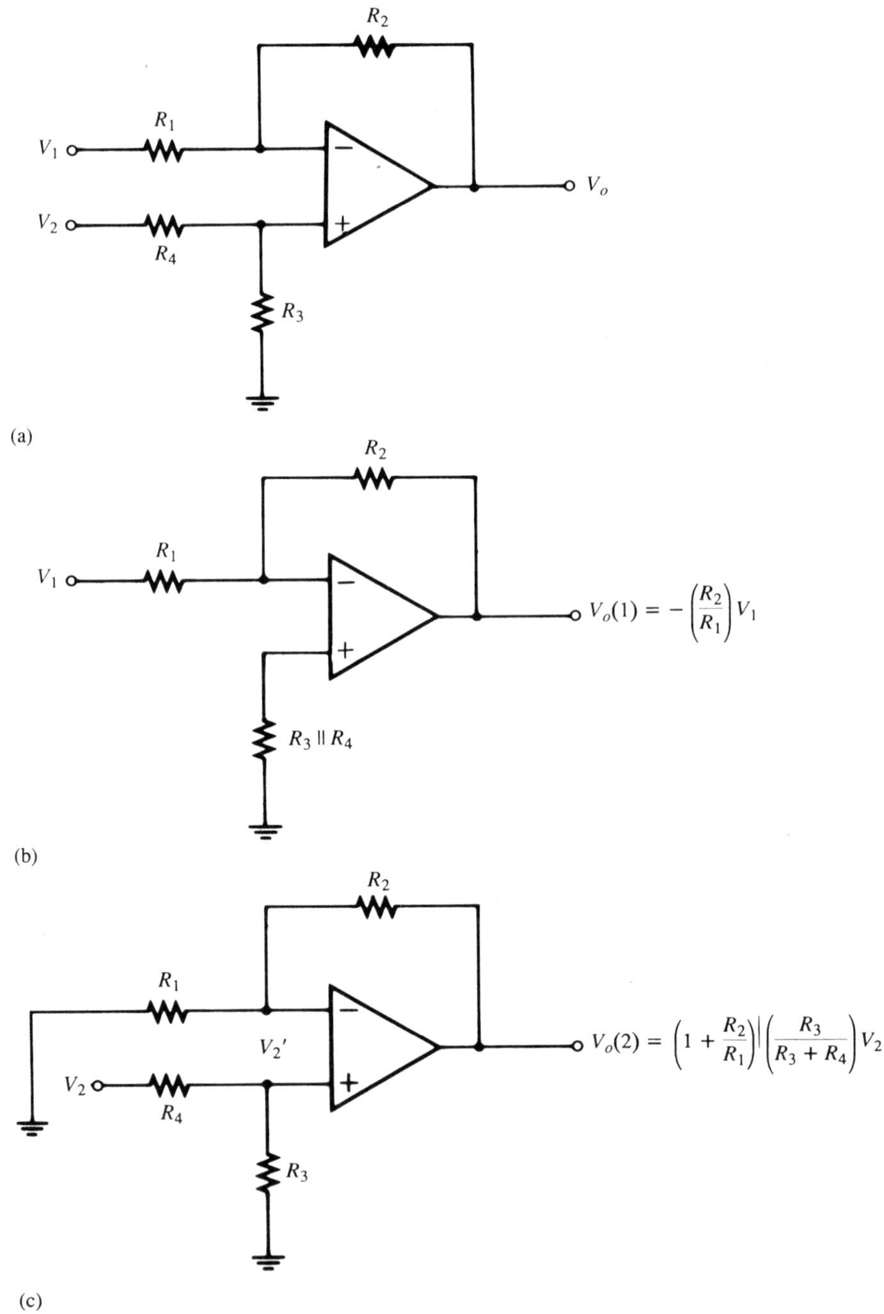

$$V_o(1) = -\frac{R_2}{R_1} V_1 \qquad \text{when } V_2 = 0 \qquad (2\text{–}20)$$

By setting only V_1 equal to zero, the circuit now resembles a noninverting amplifier (Figure 2–19c) preceded by a voltage divider composed of R_1 and R_2. The actual voltage at the noninverting input (V'_2) is equal to

$$V'_2 = \frac{R_3}{R_3 + R_4} V_2 \qquad \text{when } V_1 = 0 \qquad (2\text{–}21)$$

The output voltage of the noninverting amplifier connection is

$$V_o(2) = \left[1 + \frac{R_2}{R_1}\right] V'_2 \qquad (2\text{–}22a)$$

or

$$V_o(2) = \left[1 + \frac{R_2}{R_1}\right]\left[\frac{R_3}{R_3 + R_4}\right] V_2 \qquad (2\text{–}22b)$$

The resulting output signal from both signal sources simultaneously is the sum of Equations 2–20 and 2–22b:

$$V_o = \left[1 + \frac{R_2}{R_1}\right]\left[\frac{R_4}{R_3 + R_4}\right] V_2 - \left(\frac{R_2}{R_1}\right) V_1 \qquad (2\text{–}23)$$

From Equation 2–23, the output voltage depends on the values of four resistors. To simplify matters, two pairs are chosen so that $R_1 = R_4$ and $R_2 = R_3$. With these simplifications, Equation 2–23 reduces to

$$V_o = \frac{R_2}{R_1}(V_2 - V_1) \qquad (2\text{–}24)$$

The resistance ratio R_2/R_1 is referred to as the *differential gain* so that the output voltage is the product of the differential gain and the differential input voltage, $V_2 - V_1$. For the special case where all four resistors are equal, Equation 2–24 reduces further to

$$V_o = V_2 - V_1 \tag{2–25}$$

in which case the difference amplifier is now referred to as an *analog subtractor.*

2–9 The Instrumentation Amplifier

It is often desired to be able to vary the gain of a difference amplifier using a single potentiometer. For the difference amplifier circuit of Figure 2–18, the ratios R_2/R_1 and R_3/R_4 must always hold. If R_2 is increased by 5%, then R_3 also must be increased by 5%. As a common practice, it is desired to change the performance of a given circuit by varying the value of only one component rather than simultaneously changing two or more values.

An improvement of the basic difference amplifier is the **instrumentation amplifier,** or *IA* (Figure 2–20). It adds two operational amplifiers (A_1 and A_2) and three resistors ahead of the basic difference amplifier connection. Amplifiers A_1 and A_2

Figure 2–20
Instrumentation amplifier with a single resistance gain control.

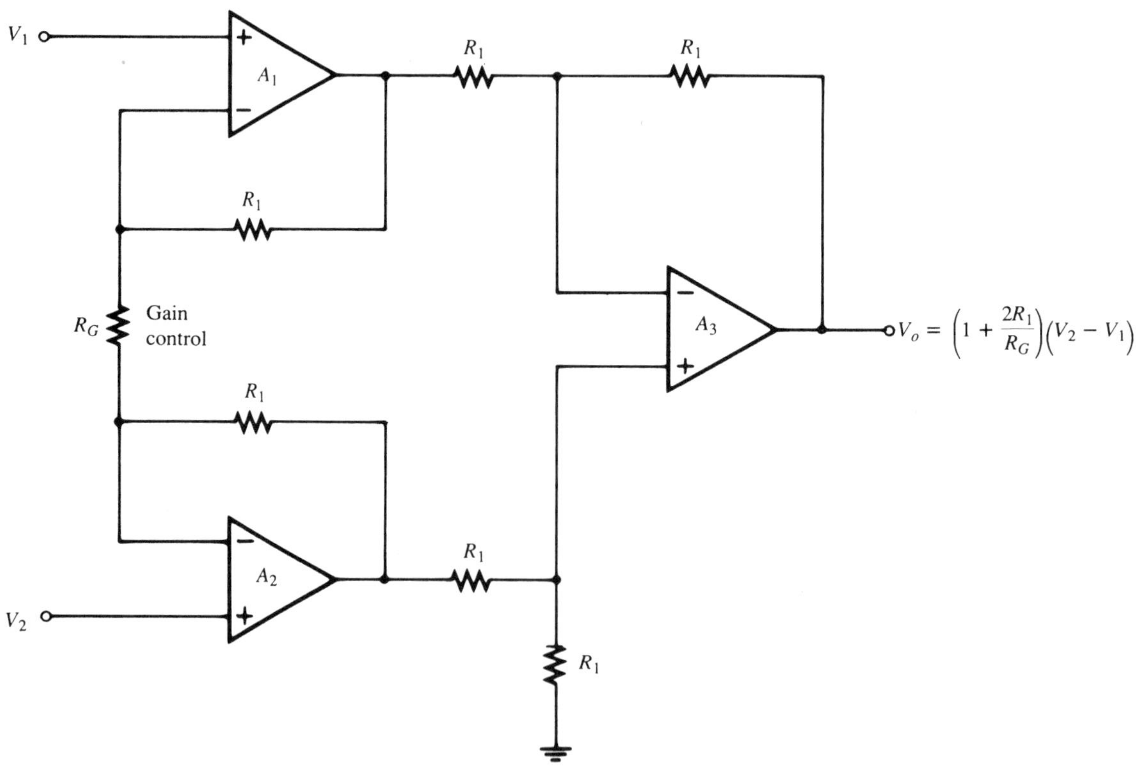

basically function as buffer amplifiers for the two input voltages V_1 and V_2. Except for R_G, all the remaining six resistors are made the same value. For this arrangement, the differential gain is

$$A_D = \frac{V_o}{V_2 - V_1} \qquad (2\text{–}26a)$$

$$= 1 + \frac{2R_1}{R_G} \qquad (2\text{–}26b)$$

Like the other six resistors, the gain-setting resistor R_G could be a fixed value so that the differential gain would be fixed. However, by varying R_G alone, we are then able to vary the differential gain smoothly. As R_G is decreased, A_D becomes larger. As R_G is increased toward infinity (an open circuit), A_D approaches 1. Instead of making an instrumentation amplifier from individual operational amplifiers and resistors like the circuit of Figure 2–20, some manufacturers have produced an integrated-circuit IA in a single package, except for the gain-setting resistor R_G. In this case, the characteristics of the amplifier sections are matched, as are the six internal resistors (R_1).

Figure 2–21
Instrumentation amplifier circuit for Example 2–6.

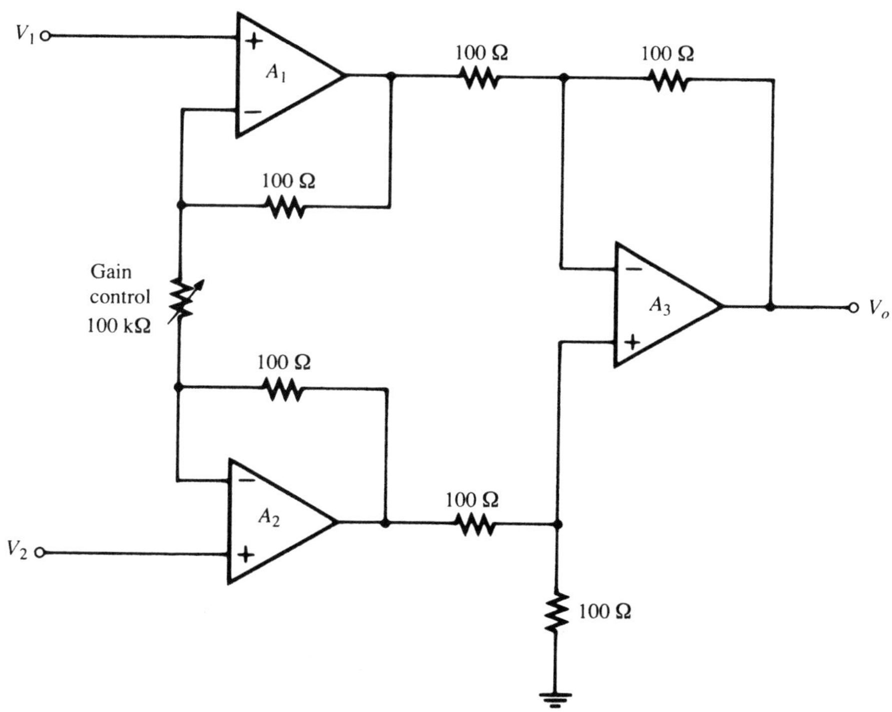

Example 2–6
For the IA circuit of Figure 2–21, determine the differential gain as the 100-kΩ potentiometer is varied over its entire range.

Solution: Resistance R_G is actually the series combination of the 100-Ω resistor and the 100-kΩ potentiometer. When the potentiometer resistance is zero, then $R_G = 100\ \Omega$, and the differential gain is

$$A_D = 1 + \frac{(2)(100\ \text{k}\Omega)}{100\ \Omega}$$
$$= \mathbf{2001}$$

When the potentiometer resistance is at its maximum value, $R_G = 100.1$ kΩ, and the differential gain is

$$A_D = 1 + \frac{(2)(100\ \text{k}\Omega)}{100.1\ \text{k}\Omega}$$
$$= \mathbf{2.998}$$

For all practical purposes, the differential gain for the circuit of Figure 2–21 would then vary from 3 to 2000.

Summary

This chapter is an introduction to the operational amplifier as an ideal linear amplifier that uses a controlled amount of negative feedback for proper operation. In this closed-loop mode, the operational amplifier can be wired to function as voltage followers (buffers) and noninverting, inverting, summing, and difference amplifiers. In addition, the instrumentation amplifier is shown to be an extension of a difference amplifier whose gain can be controlled by a single potentiometer.

In this chapter, the emphasis is on getting acquainted with *basic* concepts without concern for those operational amplifier parameters that can affect circuit performance. In Chapter 3, however, we will take into account many of the parameters that do place limits on the operational amplifier's performance in these circuits.

Glossary

closed-loop voltage gain The voltage gain of an amplifier circuit with a closed feedback path between the output and the amplifier's input.

difference amplifier An operational amplifier circuit whose output voltage is proportional to the voltage difference of its two inputs. Also called a *differential amplifier*.

instrumentation amplifier A precision difference amplifier whose gain can be varied by a single resistor.

linear amplifier Any type of amplifier producing an output that is a faithful reproduction of the input with virtually no distortion.

noninverting amplifier An operational amplifier circuit having series-parallel feedback connection whose output is in phase with the amplifier's input.

open-loop mode The situation in which an operational amplifier is operated without external feedback.

summing amplifier An operational amplifier circuit whose output is proportional to the algebraic sum of its inputs.

virtual ground The situation in which the inverting input of an operational amplifier is at ground potential even though it is not connected directly to ground.

voltage follower An operational amplifier circuit whose output is exactly the same as its input. It is frequently called a *buffer* because it isolates a load from a given signal source.

Review Questions

1. Draw the schematic symbol of an operational amplifier and identify its terminals.
2. List six major characteristics of an ideal operational amplifier.
3. Draw a schematic diagram of how a split-, or dual-, polarity power supply is connected to an operational amplifier.
4. Why are decoupling capacitors frequently used, and how are they connected?
5. What is *negative feedback*?
6. Define *open-loop gain, closed-loop gain,* and *loop gain.*
7. Why is it unsuitable for an operational amplifier to be connected in an open-loop mode for use as an amplifier?
8. Draw the schematic diagram for and list the characteristics of a noninverting amplifier.
9. Give the equation for the closed-loop gain of a noninverting amplifier.
10. Draw the schematic diagram for and list the characteristics of a voltage follower.
11. How is the voltage follower a special case of the noninverting amplifier?
12. Draw the schematic diagram for and list the characteristics of an inverting amplifier.
13. What is a *virtual ground*?
14. Give the equation for the closed-loop gain of an inverting amplifier.
15. Draw the schematic diagram for a three-input summing amplifier.
16. Explain how a summing amplifier can determine the instantaneous average value of its inputs.
17. Draw the schematic diagram for a difference amplifier, and give the equation for determining the output voltage.
18. What is an instrumentation amplifier, and how does it differ from a difference amplifier?

Exercises

1. An operational amplifier has an open-loop gain of 300,000. If the closed-loop gain is 100 using negative feedback, determine the loop gain.
2. An operational amplifier has an open-loop gain of 150,000. If the loop gain is 6000 using negative feedback, determine the closed-loop gain.
3. Negative feedback is applied to an operational amplifier having an open-loop gain of 200,000. If the feedback factor is 0.025, determine the exact closed-loop gain.
4. Repeat Exercise 3, but determine the approximate closed-loop gain.
5. Repeat Exercise 3 for a feedback factor 0.1.

Figure 2–22

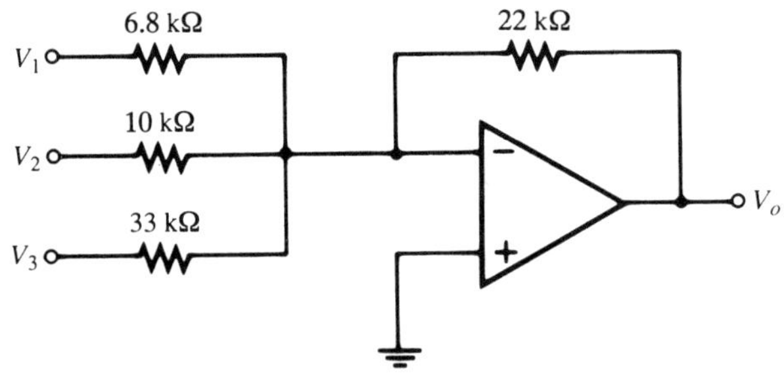

6. A 150-mV RMS voltage is applied to the amplifier of Exercise 3. Determine each value.
   (a) The RMS output voltage
   (b) The RMS difference voltage between the inverting and noninverting input terminals of the operational amplifier
7. Determine the closed-loop gain for the noninverting amplifier circuit of Figure 2–5 for the given values of R_1 and R_2.
   (a) $R_1 = 27\ \text{k}\Omega$ and $R_2 = 150\ \text{k}\Omega$
   (b) $R_1 = 680\ \text{k}\Omega$ and $R_2 = 39\ \text{k}\Omega$
   (c) $R_1 = 47\ \text{k}\Omega$ and $R_2 = 47\ \text{k}\Omega$
   (d) $R_1 = 1\ \text{M}\Omega$ and $R_2 = 330\ \text{k}\Omega$
8. Determine the loop gains for the conditions of Exercise 7 if the operational amplifier used has an open-loop gain of 200,000.
9. Determine the value of R_2 needed for the noninverting amplifier circuit of Figure 2–5 to give a closed-loop gain of 10 for each value of R_1.
   (a) $R_1 = 3.6\ \text{k}\Omega$
   (b) $R_1 = 160\ \text{k}\Omega$
   (c) $R_1 = 5.1\ \text{k}\Omega$
   (d) $R_1 = 11\ \text{k}\Omega$
10. The noninverting amplifier of Figure 2–5 has $R_1 = 39\ \text{k}\Omega$ and $R_2 = 120\ \text{k}\Omega$. Determine the output voltage if the DC input voltage is −1.7 V.
11. The noninverting amplifier of Figure 2–5 has $R_1 = 1.8\ \text{k}\Omega$ and $R_2 = 5.6\ \text{k}\Omega$. Sketch both the input and output voltages if the AC input signal is a sine wave with a peak voltage of 2.5 V.
12. Determine the closed-loop gain for the inverting amplifier circuit of Figure 2–10 for each value of R_i and R_f.
    (a) $R_i = 22\ \text{k}\Omega$ and $R_f = 180\ \text{k}\Omega$
    (b) $R_i = 820\ \text{k}\Omega$ and $R_f = 47\ \text{k}\Omega$
    (c) $R_i = 56\ \text{k}\Omega$ and $R_f = 56\ \text{k}\Omega$
    (d) $R_i = 1\ \text{M}\Omega$ and $R_f = 330\ \text{k}\Omega$
13. Determine the value of R_f needed for the inverting amplifier circuit to give a closed-loop gain of 15 for each vlaue of R_i.
    (a) $R_i = 3.3\ \text{k}\Omega$
    (b) $R_i = 180\ \text{k}\Omega$
    (c) $R_i = 5.6\ \text{k}\Omega$
    (d) $R_i = 10\ \text{k}\Omega$

Figure 2–23

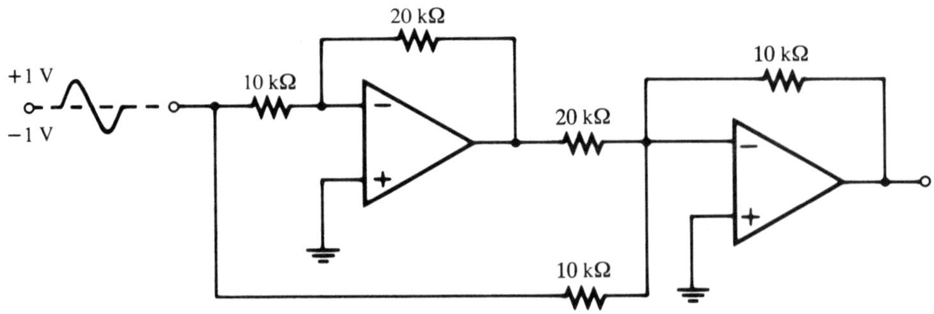

Figure 2–24

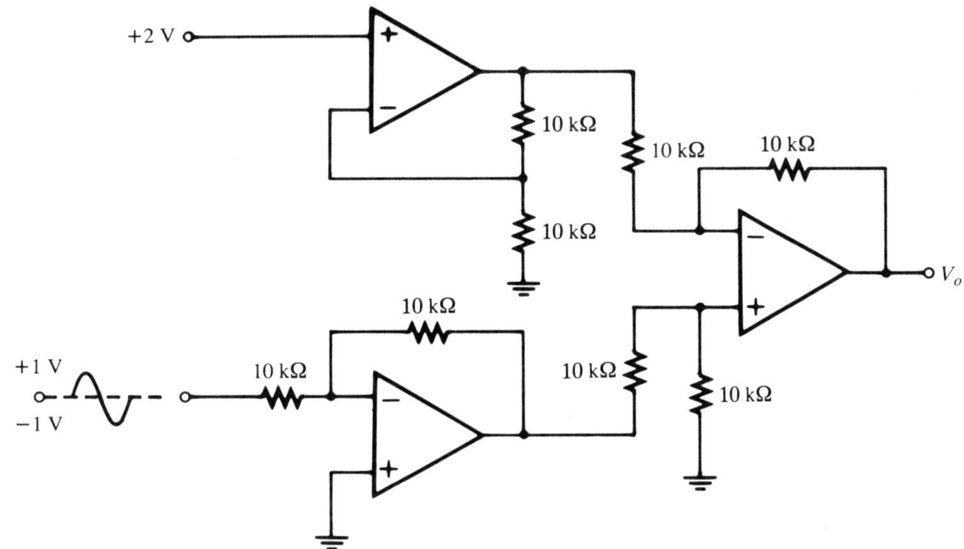

14. For the summing amplifier of Figure 2–22, determine the output voltage if
$V_1 = 0.5$ V
$V_2 = -4.6$ V
$V_3 = 2.1$ V

15. Sketch the output voltage for the summing amplifier circuit of Figure 2–23.

16. Sketch the output voltage for the circuit of Figure 2–24.

Figure 2–25

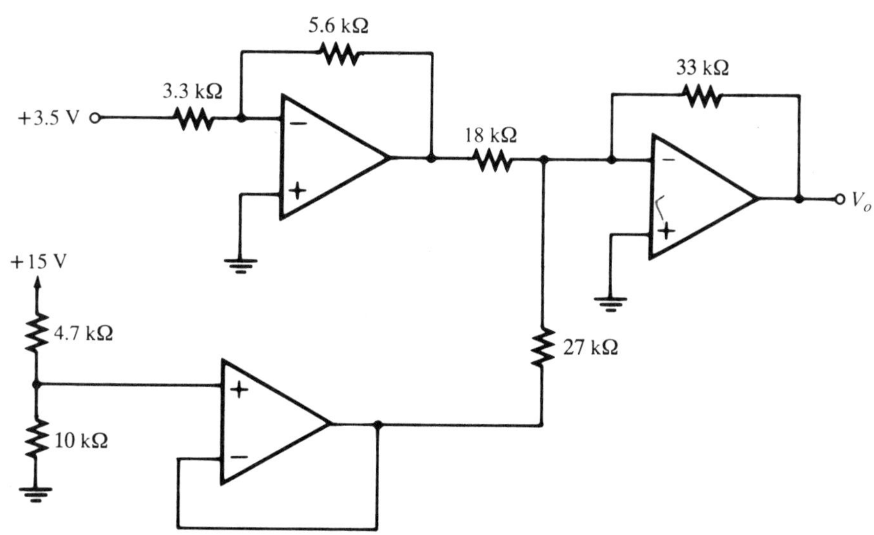

17. A summing amplifier is used to determine the average of four input signals. If the inverting summing amplifier circuit of Figure 2–12 is used, determine the required value for R_f if all four input resistors are made equal to 120 kΩ.

18. Repeat Exercise 17 using the noninverting summing amplifier of Figure 2–16.

19. Determine the output voltage for the difference amplifier of Figure 2–18 if $R_1 = 10$ kΩ, $R_2 = 33$ kΩ, $R_3 = 27$ kΩ, $R_4 = 6.8$ kΩ, $V_1 = 1.3$ V, and $V_2 = 2.5$ V.

20. Determine the output voltage for the difference amplifier of Figure 2–18 if $R_1 = R_4 = 12$ kΩ, $R_2 = R_3 = 56$ kΩ, $V_1 = -0.5$ V, and $V_2 = 1.5$ V.

21. Determine the DC output voltage for the circuit of Figure 2–25.

22. For the instrumentation amplifier of Figure 2–20, determine the value for R_G required to give a differential gain of 1000 if R_1 is 470 Ω.

3 The Practical Operational Amplifier and Circuit Performance

Instructional Objectives

Upon completing this chapter, you will be able to

- Read and understand a data sheet of a typical operational amplifier.
- Determine the magnitude of output offset voltage as a result of an operational amplifier's input offset voltage, input bias current, and input offset current.
- Explain how to compensate externally for output offset voltages for a voltage follower and inverting, noninverting, summing, and difference amplifiers.
- Determine the input and output resistances of amplifier stages.
- Explain how the frequency response of most operational amplifier circuits is affected by the gain-bandwidth product, rise time, and slew rate.
- Calculate the effect of the common-mode rejection on the performance of a difference amplifier.
- Calculate the change in output offset voltage due to the operational amplifier's power-supply rejection ratio.
- Discuss the difference between compensated and uncompensated operational amplifiers.

Introduction

The concept of the ideal operational amplifier is often used in designing operational amplifier circuits. In noncritical applications where the performance requirements are not very stringent, this approach is quite satisfactory. For more demanding applications, however, many of the actual characteristics of the operational amplifier to be used must be taken into consideration if the circuit is to meet given specifications.

In this chapter we will study the following concepts:

1. The electrical characteristics of real operational amplifiers.
2. How to obtain the electrical characteristics from a manufacturer's data sheet.

3. How the electrical characteristics affect actual circuit performance.
4. How to compensate for some of the undesirable operational amplifier characteristics.

Many of these parameters are even more important in audio applications. These are treated further in Chapter 6.

3–1 The Data Sheet

Perhaps the best way to understand the factors that may affect the performance of an operational amplifier is to examine its data sheet. Figure 3–1 shows the data sheet of the type-741 operational amplifier, which is perhaps the most popular general-purpose type in use. Although this particular data sheet is for the unit made by Signetics, type-741 operational amplifiers made by other manufacturers—such as RCA and National Semiconductor—will have nearly the same specifications.

The data sheet for the operational amplifier (as well as most other types of integrated circuits) usually contains the following information:

1. A general description of the operational amplifier
2. An internal equivalent circuit schematic
3. Pin configuration and packaging styles
4. Absolute maximum ratings
5. Electrical characteristics
6. Typical performance curves
7. Typical application circuits and design information

Most of the important parameters are discussed in this section, using the type-741 operational amplifier as a representative example. Later sections in this chapter discuss how these parameters affect circuit performance. For most operational amplifiers, as well as other linear integrated-circuit devices, there will be several versions listed, such as a military-grade ("mil-spec") version and a commercial-grade version. Signetics, for example, manufactures both a 741, or military-grade, and a 741C, or commercial-grade, version. Because of its improved characteristics, the cost of the military-grade version may average three to five times the cost of the commercial version. In this case, you get what you pay for. Unless specified otherwise, the commercial-grade (741C) is assumed whenever the parameters of the 741-type device are discussed.

General Description

For the average user, the general description is of little value. However, there may be an indication as to whether or not the type of operational amplifier has FET input stages, is internally compensated, or is chopper-stabilized; or there may be information about those characteristics that may lead to improved performance.

Figure 3–1
Manufacturer's data sheet for the μA741 operational amplifier. (Courtesy Signetics Corporation.)

signetics

μA741

LINEAR INTEGRATED CIRCUITS

The μA741 is a high performance operational amplifier with high open loop gain, internal compensation, high common mode range and exceptional temperature stability. The μA741 is short-circuit protected and allows for nulling of offset voltage.

- INTERNAL FREQUENCY COMPENSATION
- SHORT CIRCUIT PROTECTION
- OFFSET VOLTAGE NULL CAPABILITY
- EXCELLENT TEMPERATURE STABILITY
- HIGH INPUT VOLTAGE RANGE
- NO LATCH-UP

	μA741C	μA741
Supply Voltage	±18V	±22V
Internal Power Dissipation (Note 1)	500mW	500mW
Differential Input Voltage	±30V	±30V
Input Voltage (Note 2)	±15V	±15V
Voltage between Offset Null and V⁻	±0.5V	±0.5V
Operating Temperature Range	0°C to +70°C	-55°C to +125°C
Storage Temperature Range	−65°C to +150°C	−65°C to +150°C
Lead Temperature (Solder, 60 sec)	300°C	300°C
Output Short Circuit Duration (Note 3)	Indefinite	Indefinite

Notes
1. Rating applies for case temperatures to 125°C; derate linearly at 6.5mW/°C for ambient temperatures above +75°C.
2. For supply voltages less than ±15V, the absolute maximum input voltage is equal to the supply voltage.
3. Short circuit may be to ground or either supply. Rating applies to +125°C case temperature or +75°C ambient temperature.

A PACKAGE
(Top View)

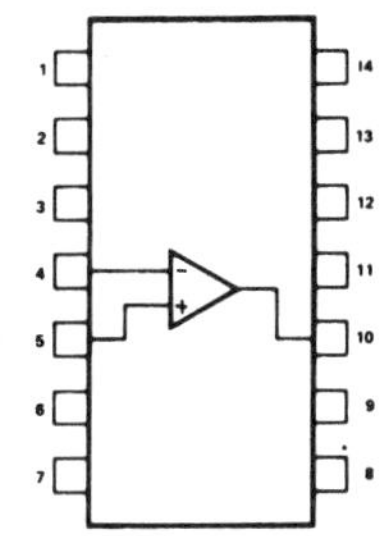

1. NC
2. NC
3. Offset Null
4. Inv. Input
5. Non-Inv. Input
6. V⁻
7. NC
8. NC
9. Offset Null
10. Output
11. V⁺
12. NC
13. NC
14. NC

ORDER PART NO. μA741CA

T PACKAGE

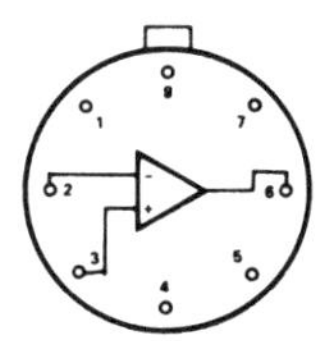

1. Offset Null
2. Inverting Input
3. Non-Inverting Input
4. V⁻
5. Offset Null
6. Output
7. V⁺
8. NC

ORDER PART NOS. μA741T/μA741CT

V PACKAGE

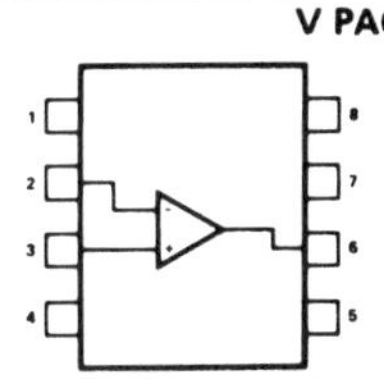

1. Offset Null
2. Inv. Input
3. Non-Inv. Input
4. V⁻
5. Offset Null
6. Output
7. V⁺
8. NC

ORDER PART NO. μA741CV

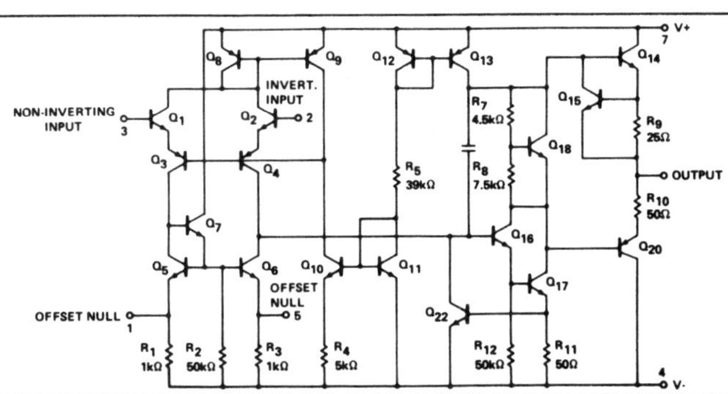

Figure 3–1
continued

($V_S = \pm 15V$, $T_A = 25°C$ unless otherwise specified)

PARAMETER	MIN.	TYP.	MAX.	UNITS	TEST CONDITIONS
μA741C					
Input Offset Voltage		2.0	6.0	mV	$R_S \leq 10k\Omega$
Input Offset Current		20	200	nA	
Input Bias Current		80	500	nA	
Input Resistance	0.3	2.0		MΩ	
Input Capacitance		1.4		pF	
Offset Voltage Adjustment Range		±15		mV	
Input Voltage Range	±12	±13		V	
Common Mode Rejection Ratio	70	90		dB	$R_S \leq 10k\Omega$
Supply Voltage Rejection Ratio		10	150	μV/V	$R_S \leq 10k\Omega$
Large-Signal Voltage Gain	20,000	200,000			$R_L \geq 2k\Omega$, $V_{out} = \pm 10V$
Output Voltage Swing	±12	±14		V	$R_L \geq 10k\Omega$
	±10	±13		V	$R_L \geq 2k\Omega$
Output Resistance		75		Ω	
Output Short-Circuit Current		25		mA	
Supply Current		1.4	2.8	mA	
Power Consumption		50	85	mW	
Transient Response (unity gain)					$V_{in} = 20mV$, $R_L = 2k\Omega$, $C_L \leq 100pF$
Risetime		0.3		μs	
Overshoot		5.0		%	
Slew Rate		0.5		V/μs	$R_L \geq 2k\Omega$
The following specifications apply for $0°C \leq T_A \leq +70°C$					
Input Offset Voltage			7.5	mV	
Input Offset Current			300	nA	
Input Bias Current			800	nA	
Large-Signal Voltage Gain	15,000				$R_L \geq 2k\Omega$, $V_{out} = \pm 10V$
Output Voltage Swing	±10	±13		V	$R_L \geq 2k\Omega$
μA741					
Input Offset Voltage		1.0	5.0	mV	$R_S \leq 10k\Omega$
Input Offset Current		10	200	nA	
Input Bias Current		80	500	nA	
Input Resistance	0.3	2.0		MΩ	
Input Capacitance		1.4		pF	
Offset Voltage Adjustment Range		±15		mV	
Large-Signal Voltage Gain	50,000	200,000			$R_L \geq 2k\Omega$, $V_{out} = \pm 10V$
Output Resistance		75		Ω	
Output Short Circuit Current		25		mA	
Supply Current		1.4	2.8	mA	
Power Consumption		50	85	mW	
Transient Response (unity gain)					$V_{in} = 20mV$, $R_L = 2k\Omega$, $C_L \leq 100pF$
Risetime		0.3		μs	
Overshoot		5.0		%	
Slew Rate		0.5		V/μs	$R_L \geq 2k\Omega$
The following specifications apply for $-55°C \leq T_A \leq +125°C$					
Input Offset Voltage		1.0	6.0	mV	$R_S \leq 10k\Omega$
Input Offset Current		7.0	200	nA	$T_A = +125°C$
		20	500	nA	$T_A = -55°C$
Input Bias Current		0.03	0.5	μA	$T_A = +125°C$
		0.3	1.5	μA	$T_A = -55°C$
Input Voltage Range	±12	±13		V	
Common Mode Rejection Ratio	70	90		dB	$R_S \leq 10k\Omega$
Supply Voltage Refection Ratio		10	150	μV/V	$R_S \leq 10k\Omega$
Large-Signal Voltage Gain	25,000				$R_L \geq 2k\Omega$, $V_{out} = \pm 10V$
Output Voltage Swing	±12	±14		V	$R_L \geq 10k\Omega$
	±10	±13		V	$R_L \geq 2k\Omega$
Supply Current		1.5	2.5	mA	$T_A = +125°C$
		2.0	3.3	mA	$T_A = -55°C$
Power Consumption		45	75	mW	$T_A = +125°C$
		45	100	mW	$T_A = -55°C$

Figure 3–1
continued

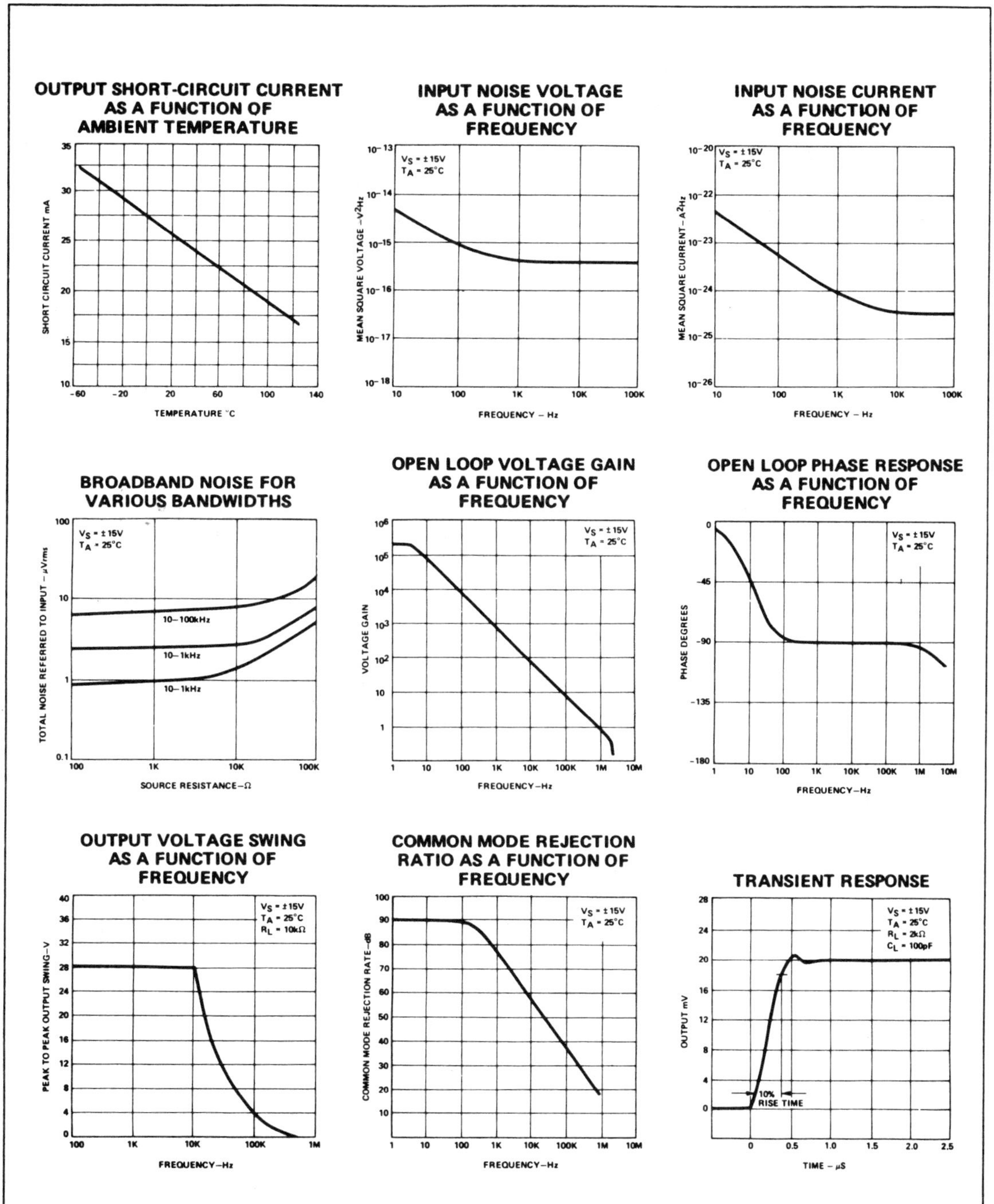

Figure 3–1
continued

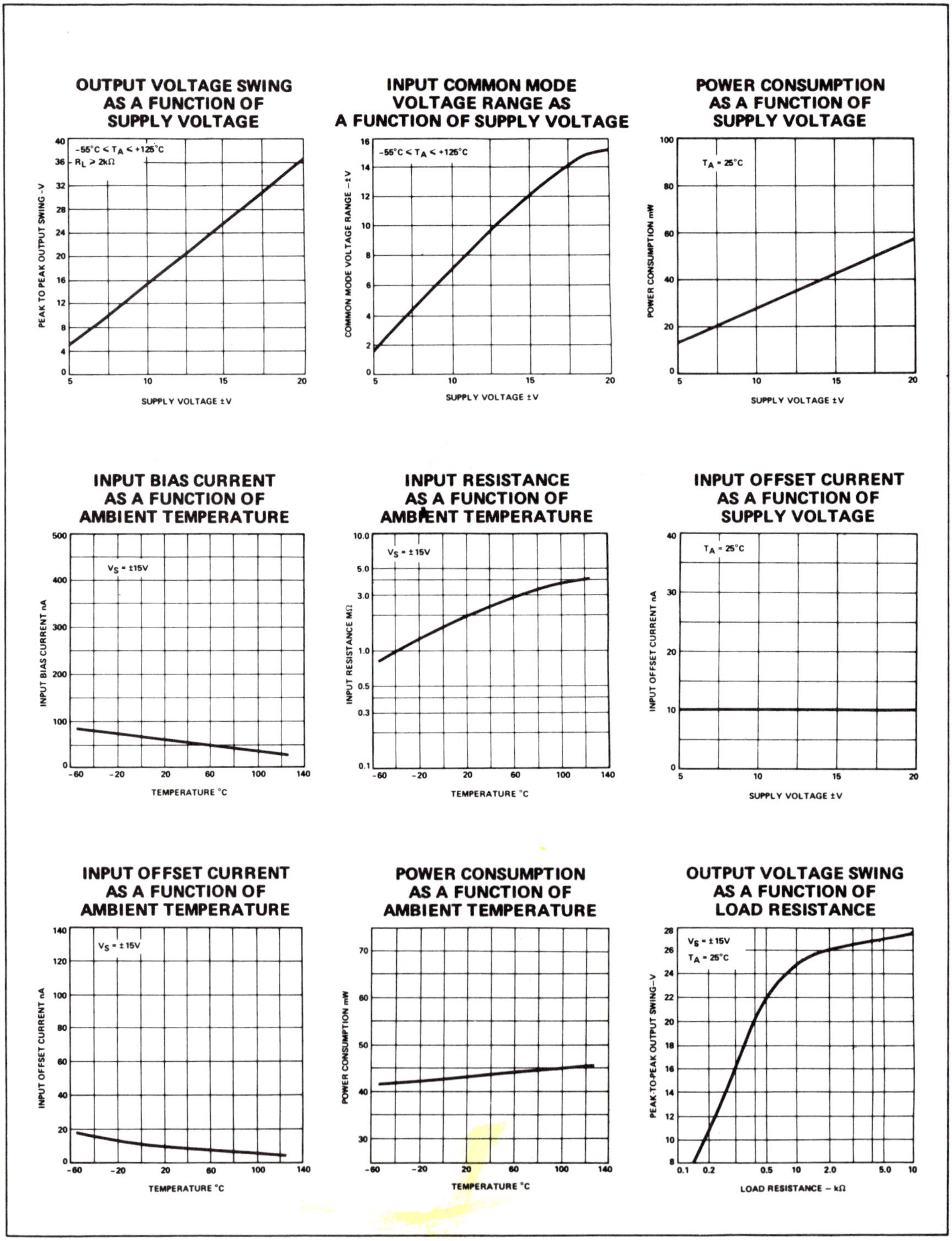

Internal Equivalent Circuit

Like the general description, the internal equivalent circuit has little practical value for the user, especially the technician. Since the actual integrated circuit is inside a sealed package, there is nothing that can be done to change it. An operational amplifier that has 40 equivalent transistors works essentially the same as one having 20 transistors.

Maximum Ratings

The maximum ratings listed are the maximum values that the operational amplifier can safely handle without the possibility of destruction. Under no circumstances should they be equaled or exceeded. For proper circuit operation, none of these maximum ratings affect the overall performance of the circuit involved.

- *Supply voltage* ($+V_{CC}$, $-V_{EE}$) is the maximum positive and negative supply voltage that can be used to power the operational amplifier.
- **Internal power dissipation** (P_D) is the maximum power that the operational amplifier is capable of dissipating at a specified ambient temperature. For higher temperatures, this must be derated, or reduced, accordingly.
- **Differential input voltage** (V_{id}) is the maximum voltage that can be applied between the inverting and noninverting inputs.
- *Input voltage* (V_{icm}) is also called the *common-mode voltage,* and it is the maximum input voltage that can be simultaneously applied between both inputs and ground. In general, this is equal to the positive supply voltage.
- **Operating temperature** (T_A) is the ambient temperature for which the operational amplifier will operate within the manufacturer's specifications. The military-grade version usually has a wider temperature range than the commercial version.
- **Output short-circuit duration** is the amount of time for which the operational amplifier's output terminal can be short-circuited to ground or either supply voltage without damage. For the 741, the output can be short-circuited indefinitely.

Electrical Characteristics

Electrical characteristics are those parameters that may severely affect or limit the performance of many operational amplifier circuits. Those cited are usually specified for a given supply voltage and operating temperature unless specified otherwise. For the 741 and 741C versions listed in the data sheet, the specifications are given for supply voltages of $V_{CC} = +15$ V and $V_{EE} = -15$ V and an ambient operating temperature of 25°C. In addition, other conditions may apply, such as a specified load resistance. Each parameter listed may have a minimum, typical, and/or maximum value. Electrical parameters can apply to static (DC) and dynamic (AC) circuit operation. The following gives the electrical characteristics by function. First we consider *static input parameters*.

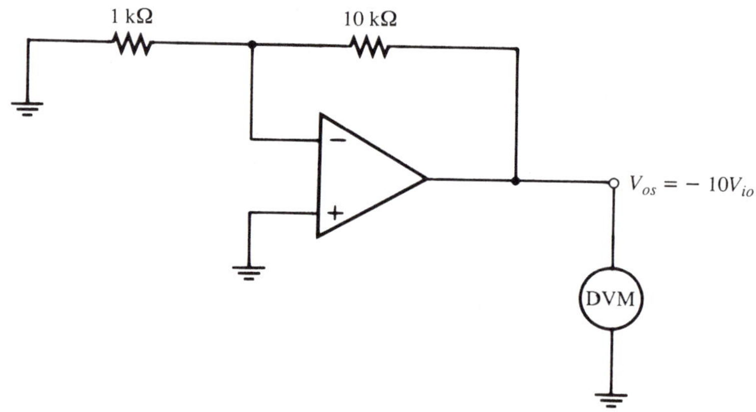

Figure 3–2
Measuring the offset voltage of an inverting amplifier circuit.

1. **Input offset voltage** (V_{io}) is the equivalent DC voltage that must be applied to one of the input terminals to produce a zero output voltage with the other input grounded. Remember, for the ideal operational amplifier, the output offset voltage is zero. For the 741, the input offset voltage is typically 2 mV and can be of either polarity.

   Figure 3–2 shows one method for measuring the offset voltage. An inverting amplifier with a closed-loop gain of −10 has its input grounded. A DMM then measures the DC output offset voltage (V_{os}) with no input. The input offset voltage is then the output offset divided by the closed-loop gain (A_{CL}).

$$V_{io} = -\frac{V_{os}}{A_{CL}} \quad (3\text{–}1)$$

2. **Input bias current** (I_B) is the average of the two DC currents that flow into the inverting and noninverting inputs. In an ideal operational amplifier, input bias current is zero due to the ideal infinite impedance. For the 741, the input bias current is typically 80 nA.

   The inverting amplifier circuit of Figure 3–3 shows how the two input bias currents can be measured. The voltage across each of the 100-kΩ resistors is measured. Using Ohm's law, the current flowing through the resistors is determined:

$$I_{B+} = \frac{V_+}{R_1} \quad (3\text{–}2a)$$

and

$$I_{B-} = \frac{V_-}{R_2} \quad (3\text{–}2b)$$

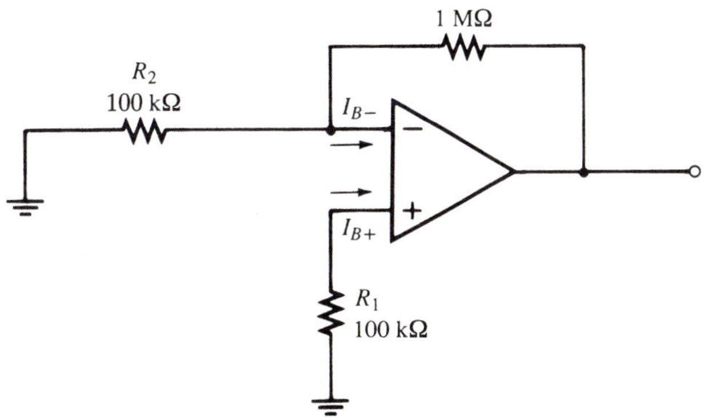

Figure 3–3
Input bias current measurement.

The input bias current is then the average of the two values obtained with Equations 3–2a and 3–2b:

$$I_B = \frac{I_{B+} + I_{B-}}{2} \tag{3–2c}$$

3. **Input offset current** (I_{os}) is the difference of the two input bias currents:

$$I_{os} = I_{B+} - I_{B-} \tag{3–3}$$

I_{os} can be either positive or negative. For the 741, the input offset current is typically 20 nA.

Example 3–1

The circuit of Figure 3–3 is used to measure the input offset voltage and input bias current of a given 741 operational amplifier. Determine V_{io}, I_B, and I_{os} from the following measurements:

$$V_+ = 15.2 \text{ mV}$$
$$V_- = 12.7 \text{ mV}$$
$$V_{os} = -25 \text{ mV}$$

Solution:

(a) Since the closed-loop voltage gain is -1 MΩ/100 kΩ, or -10, the input offset voltage, from Equation 3–1, is

$$V_{io} = \frac{-25 \text{ mV}}{-10} = \mathbf{2.5\ mV}$$

(b) Using Equations 3–2a and 3–2b, the individual bias currents are

$$I_{B+} = \frac{15.2\text{ mV}}{100\text{ k}\Omega} = 152\text{ nA}$$

and

$$I_{B-} = \frac{12.7\text{ mV}}{100\text{ k}\Omega} = 127\text{ nA}$$

so that the measured input bias current is the average of 152 nA and 127 nA, or **139.5 nA.**

(c) Since the input offset current is the difference of the two input bias currents, from Equation 3–3,

$$I_{os} = 152\text{ nA} - 127\text{ nA} = \mathbf{25\text{ nA}}$$

4. **Input resistance** (R_i) is the internal input impedance seen at either the inverting or the noninverting input to ground while the remaining input terminal is grounded. In an ideal device, it is infinite. For the 741, it is typically 2 MΩ.

Next, we consider *static output parameters.*

1. **Output resistance** (R_{oi}) is the internal output resistance as seen from the output terminal to ground. For an ideal operational amplifier, it is zero. For the 741, it is typically 75 Ω.
2. **Output short-circuit current** (I_{osc}) is the maximum DC output current that can be supplied to a load (e.g., 25 mA).
3. **Output voltage swing** ($V_{+\text{SAT}}$ and $V_{-\text{SAT}}$) is also called the *saturation voltage;* it is the maximum peak output voltage that the operational amplifier can produce without saturation or clipping. This is typically about two diode voltage drops less than the corresponding supply voltages.

The *dynamic parameters* are as follows:

1. **Open-loop gain** (A_{OL}) is sometimes called the *large-signal gain* and is the voltage gain of the operational amplifier without any external feedback. As briefly mentioned in Chapter 2, the DC open-loop gain for the 741 is typically 200,000 (+106 dB).
2. **Slew rate** (SR) is the time rate of change of the output voltage with the operational amplifier circuit having a closed-loop gain of 1. Ideally, the output voltage should exactly follow the input signal without distortion. For a rapidly changing input signal such as a square wave, the output should also resemble a square wave. However, the slew rate of an operational amplifier, as a result of the inability of its internal circuitry to drive capacitive loads, causes the output signal to increase (or decrease) at a slower rate than the corresponding

Figure 3–4
Slew rate in operational amplifiers. (a) Effect on square waves. (b) Effect on sine waves.

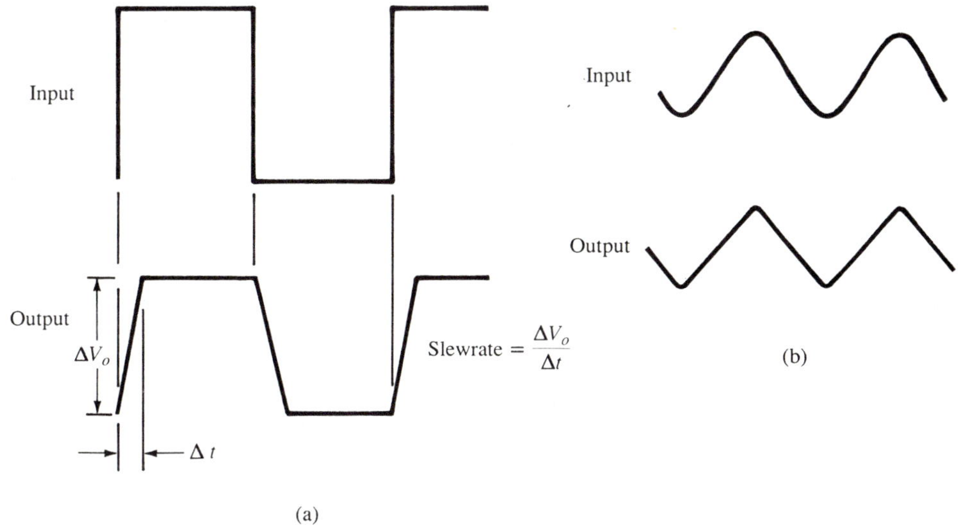

input signal, as shown in Figure 3–4a. The slew rate is calculated by dividing the output voltage swing (ΔV_o) by the corresponding time interval (Δt):

$$SR = \frac{\Delta V_o}{\Delta t} \qquad (3\text{–}4)$$

The effect of the slew rate on sine waves is shown in Figure 3–4b. It is measured in volts per microsecond. For an ideal operational amplifier, it is infinite. For the 741, it is typically a relatively slow 0.5 V/μs, whereas the faster LM318 is 70 V/μs.

3. *Unity-gain bandwidth* (f_T) is the frequency at which the open-loop voltage gain decreases to 1 (0 dB). As is discussed later in this chapter, it is related to the operational amplifier's gain-bandwidth product.
4. **Common-mode rejection ratio (CMRR)** is a measure of the ability of the operational amplifier to reject *common-mode* signals, i.e., signals that are present simultaneously on both inputs. Although the difference input is zero, a finite output will exist as a result of the nonideal common-mode rejection. The common-mode rejection ratio is a dimensionless ratio usually expressed in decibels, which is then referred to as the *common-mode rejection* (CMR). The relation between CMRR and CMR is

$$\text{CMR (dB)} = 20 \log_{10}(\text{CMRR}) \qquad (3\text{–}5)$$

For the 741, the CMR is typically 90 dB.

5. **Channel separation** is a measure of how well cross talk is minimized in packages having more than one internal operational amplifier. That is, a signal applied to the input of one operational amplifier section will develop a small but finite output signal in the remaining section(s), *even though there is no input signal applied to the unused section(s)*. For the 747 and 5558 dual-operational amplifiers, the channel separation is typically 120 dB. For the 4156 quad- (four-) operational amplifier, it is 108 dB. This means that if one operational amplifier section of a 747 has an output of 1 V, there will be a signal that is 120 dB less (i.e., 1 μV) at the output of the unused section. For most applications however, channel separation is not a problem.
6. The 10 to 90% *transient response rise time* (t_r) is specified under closed-loop conditions. This is the amount of time it takes for the output signal, in response to an input pulse, to rise from 10 to 90% of its steady-state value. For the 741, it is typically 0.3 μs.

The following sections now discuss how many of these parameters affect the performance of those linear amplifier circuits discussed in Chapter 2. In addition, techniques are given to effectively eliminate or minimize these limitations.

3–2 Input and Output Resistances

For any operational amplifier circuit to be free from loading effects, the input resistance (or impedance) of the circuit should be high enough that it does not load down the output of the signal source, whereas the output resistance should be low enough that any load that is connected to it does not load down the output of the circuit.

For an ideal operational amplifier, the input resistance is infinite, and the output resistance is zero. From the data sheet of a typical operational amplifier, we know that this is not the case in practice. However, negative feedback has several other advantages besides stabilizing the gain of the operational amplifier circuit. For voltage followers and inverting, noninverting, and difference amplifiers, negative feedback lowers the output resistance of the circuit. If the closed-loop gain is substantially smaller than the open-loop gain, which is generally the case, then the loop gain is much larger than 1. As a consequence, the closed-loop output resistance is essentially equal to the internal (or intrinsic) output resistance of the operational amplifier (R_o) divided by the loop gain:

$$R_o(\text{CL}) \simeq \frac{R_o}{A_L} \qquad \textbf{(3–6a)}$$

Since the loop gain is dependent on the closed-loop gain and the operational amplifier's open-loop gain, Equation 3–6a can be rewritten as

$$R_o(\text{CL}) \simeq \frac{A_{CL}}{A_{OL}} R_o \qquad \textbf{(3–6b)}$$

For the case of a voltage follower with closed-loop gain of 1, the circuit output resistance then reduces to

$$R_o(\text{CL}) \simeq \frac{R_o}{A_{OL}} \tag{3–6c}$$

If the loop gain is sufficiently large compared with the closed-loop gain, then the closed-loop output resistance of a voltage follower or inverting or noninverting amplifier circuit will then be substantially less than 1 Ω.

Example 3–2

Determine the closed-loop output resistance of a noninverting amplifier having a closed-loop gain of 20 using a 741 operational amplifier with the following typical values:

$$A_{OL} = 200{,}000$$
$$R_o = 75\ \Omega$$

Solution: From Equation 3–6b, the closed-loop output resistance of the noninverting amplifier is

$$R_o(\text{CL}) = \frac{20}{200{,}000}(75\ \Omega) = \mathbf{0.0075\ \Omega}$$

For most circuits having negative feedback, the output resistance calculation is not much to be concerned about. Even for open-loop circuits (such as a comparator, discussed in Chapter 8), the worst that the circuit's output resistance would be is the operational amplifier's intrinsic output resistance, R_o.

The closed-loop input resistance of an operational amplifier circuit must be determined separately for inverting and noninverting amplifiers. For the inverting amplifier of Figure 3–5a, the inverting input terminal is a virtual ground. All the input signal source sees is R_1 effectively shunted to ground. For all practical purposes, then, the closed-loop input resistance of an inverting amplifier equals the value of the input resistor, R_1:

$$R_i(\text{CL}) \simeq R_1 \tag{3–7}$$

Because R_1 effectively determines the inverting amplifier's input resistance, the feedback resistor R_2 is the one that is varied to change the circuit's closed-loop gain. As shown in Figure 3–5b, the series combination of the potentiometer (R_{2A}) and the fixed resistor (R_{2B}) is the total feedback resistance. Changing the potentiometer resistance then varies the closed-loop gain while keeping the amplifier's input resistance constant, equal to R_1.

Figure 3–5
Amplifier input resistance. (a) Inverting amplifier. (b) Constant input impedance with variable gain.

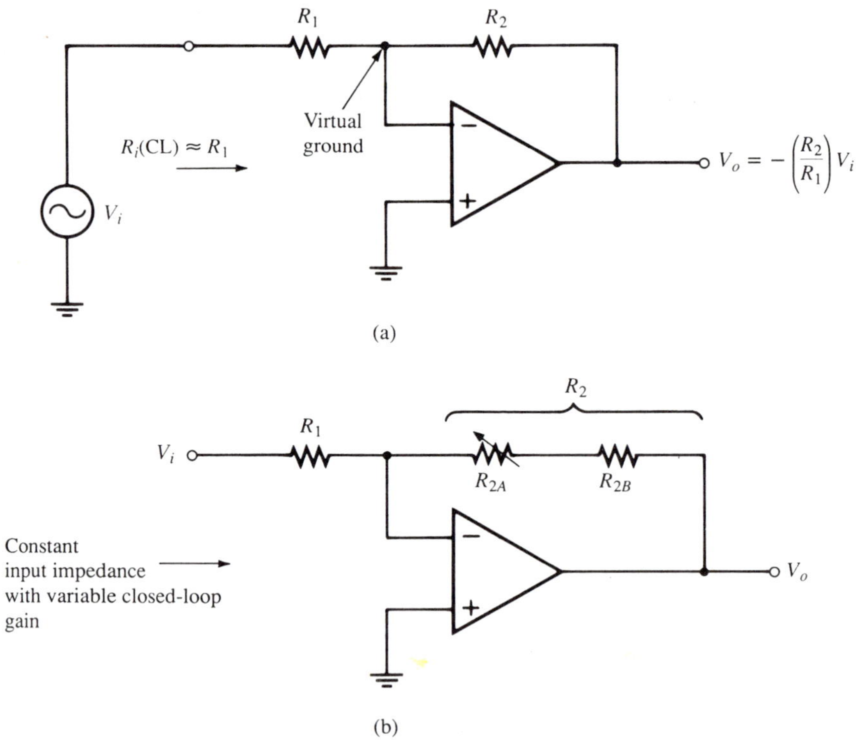

Very often, the design of an inverting amplifier circuit requires that it have a specific input impedance. For a required closed-loop gain, the feedback resistor is then immediately determined, as illustrated by the following example.

Example 3–3
Determine the required resistor values for the inverting amplifier of Figure 3–5a that calls for an input resistance of 100 kΩ and a closed-loop gain of 10 (+20 dB).

Solution: Since the amplifier circuit input resistance is specified as 100 kΩ, this immediately sets $\boldsymbol{R_1}$ **= 100 kΩ.** For a closed-loop gain of 10, the feedback resistor (R_2) is then

$$R_2 = (10)(100\ \text{k}\Omega) = \mathbf{1\ M\Omega}$$

The completed circuit is shown in Figure 3–6.

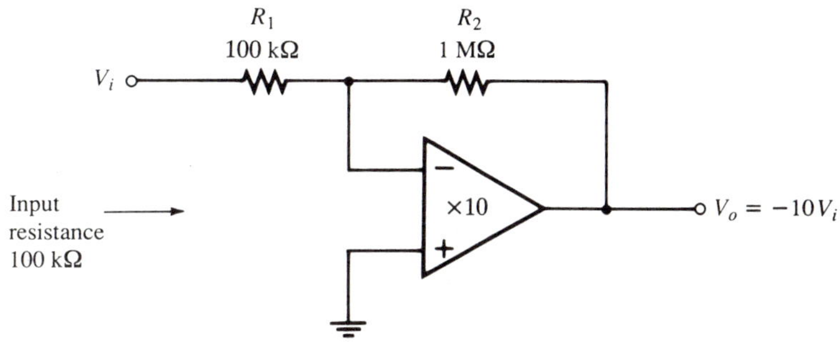

Figure 3–6
Completed inverting amplifier circuit for Example 3–3.

For the noninverting amplifier, the closed-loop input impedance is significantly increased by the amplifier's loop gain, such that

$$R_i(\text{CL}) = (1 + A_L)R_i \qquad \textbf{(3–8a)}$$

Since the loop gain is usually much larger than 1, the circuit input resistance can be rewritten approximately in terms of the open- and closed-loop gains as

$$R_i(\text{CL}) \simeq \frac{A_{OL}}{A_{CL}}(R_i) \qquad \textbf{(3–8b)}$$

This is usually well in excess of the operational amplifier's input impedance, so it generally does not have to be determined as a practical matter in most cases. Unlike the inverting amplifier circuit, however, it is not possible independently to set the input resistance of a noninverting amplifier circuit.

Example 3–4

non-inverting

Determine the input and output resistances for an ~~inverting~~ amplifier circuit using a 741 operational amplifier. The circuit has a closed-loop voltage gain of 10, while the 741 has the following typical values:

$$A_{OL} = 200{,}000$$
$$R_i = 2\ \text{M}\Omega$$
$$R_o = 75\ \Omega$$

Solution:

(a) From Equation 3–8b, the circuit input resistance is

$$R_i(\text{CL}) \simeq \frac{200{,}000}{10}(2\ \text{M}\Omega)$$
$$= \mathbf{4 \times 10^{10}\ \Omega} \qquad \mathbf{(40\ G\Omega)}$$

(b) From Equation 3–6b, the circuit output resistance is

$$R_o(\text{CL}) \simeq \frac{10}{200{,}000}(75\ \Omega)$$
$$= \mathbf{0.00375\ \Omega}$$

This is about as close as you can get to a buffer amplifier having an ideal infinite input resistance and zero output resistance.

3–3 Contributions to Output Offset Voltage

In Chapter 2 it was stated that the output voltage of an ideal operational amplifier is zero if the voltage difference between both inputs is zero. This can occur when (1) both inputs are connected to ground or (2) both inputs are connected to the same voltage source. With an actual operational amplifier, however, it is not uncommon to find a small, but finite, DC output voltage present in either of these two cases. This undesired output voltage level is called the *output offset voltage* (V_{os}) and can result from one or more of the following sources:

1. Input offset voltage
2. Input bias current
3. Input offset current

Although the output offset voltage is usually very small, it nevertheless can contribute significant errors to the performance of the voltage follower, inverting, noninverting, summing, and difference amplifier circuits. We will look at all these sources, the calculation of their relative contributions to the total output offset voltage, and practical methods for their elimination.

Input Offset Voltage Contribution

The input offset voltage of an operational amplifier is represented by a DC voltage source in series with one of its two input terminals. This offset voltage results from the slight mismatches in its internal components. Although it really makes no difference with which input terminal this equivalent voltage is in series, it is conveniently placed in series with the *noninverting input* so that the input offset and resulting output offset have the same polarity. If the output offset voltage due solely to the input offset voltage is positive with respect to ground, then the DC voltage source representing this input offset voltage has its positive terminal connected to the noninverting input terminal of the operational amplifier, as shown in Figure 3–7a. On the other hand, if the output offset voltage is negative with respect to ground, then the DC voltage source representing this input offset voltage has its negative terminal connected to the noninverting input terminal of the operational amplifier, as shown in Figure 3–7b.

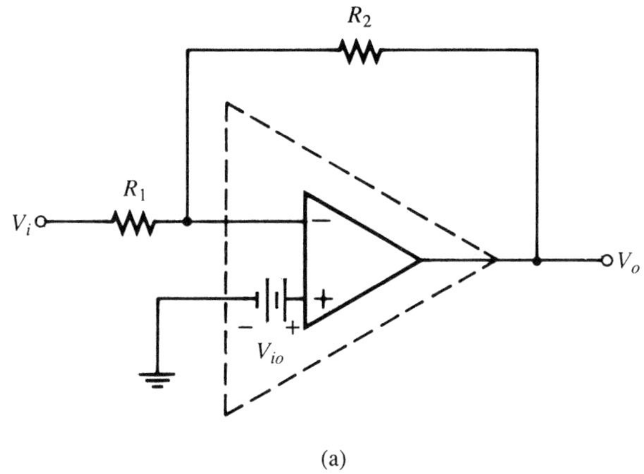

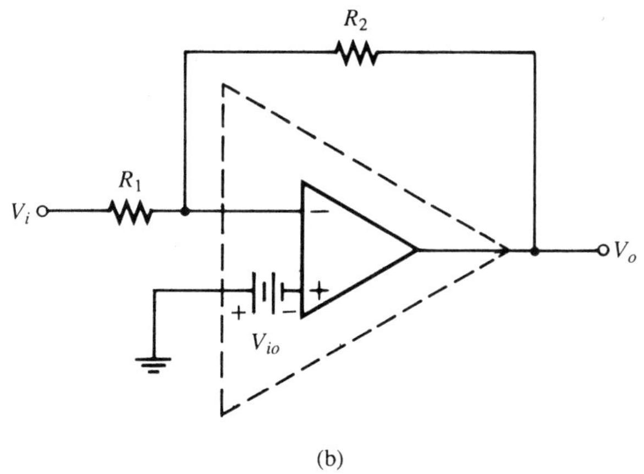

Figure 3–7
Contribution of DC input offset voltage. (a) Schematic representation of inverting amplifier with the positive terminal of the input offset voltage source connected to the noninverting input of the operational amplifier.
(b) Negative terminal of the input offset voltage source connected to the noninverting input of the operational amplifier.

The equivalent circuit shown in Figure 3–8 results from either an inverting or a noninverting amplifier having its input signal grounded. For the inverting amplifier, R_1—which is normally connected to the input signal—is now tied to ground. For the noninverting amplifier, its noninverting input—which normally receives the input signal—is now tied to ground. In either case, the equivalent circuit is the

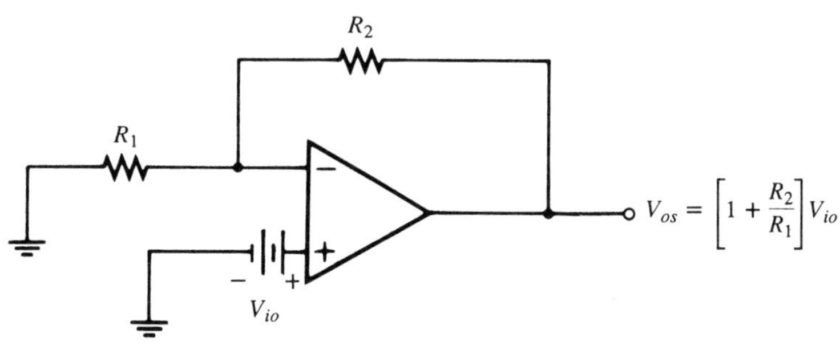

Figure 3–8
Equivalent circuit with input offset voltage source that results from either an inverting or noninverting amplifier having its input signal grounded.

same as a nonideal *noninverting* amplifier whose input signal is now the input offset voltage (V_{io}). Using the closed-loop gain for a noninverting amplifier, the DC output offset voltage of the circuit due solely to this DC input offset voltage of the nonideal operational amplifier is

$$V_{os} = A_{CL} V_{io} = \left[1 + \frac{R_2}{R_1}\right] V_{io} \tag{3–9}$$

If a voltage follower is used instead of a noninverting or inverting amplifier, then the output offset voltage will be the same as the input offset voltage. For inverting and noninverting amplifiers, the effect of the input offset voltage may be easily minimized by either (1) using the lowest possible DC closed-loop gain compatible with system requirements or (2) using an operational amplifier having a very low value for V_{io}. For a 741 inverting amplifier normally having a closed-loop voltage gain of 15, the typical output offset voltage magnitude would then be 16 × 2 mV, or 32 mV. The polarity of the resultant output offset voltage is dependent on the polarity of the input offset voltage. For some 741 devices, V_{os} may be positive. For others, it may be negative.

Input Bias Current Contribution

In an ideal operational amplifier, the input bias currents are zero due to the infinite input resistance. However, input bias currents must be supplied to the two inputs of a real operational amplifier to provide the proper bias for the differential input stage transistors of the amplifier. This current, though usually quite small, results in a small deviation from zero in the output voltage. Because of slight mismatches of the input stages of an operational amplifier, the two bias currents flowing into the two inputs of the operational amplifier are not equal. Consequently, the input bias offset current is not zero.

Regardless of whether we have an inverting or noninverting amplifier, the inverting input is still at a virtual ground. With both input terminals grounded, the

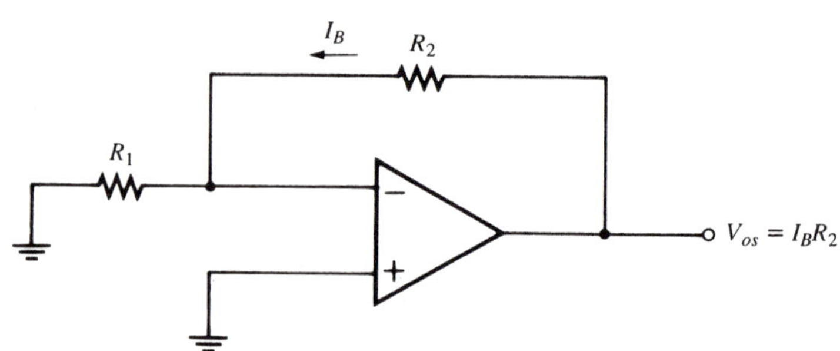

Figure 3–9
Output voltage offset due to input bias current error.

bias current (I_B) flowing into the inverting input (Figure 3–9) is also the current flowing through resistor R_2 that is connected between the operational amplifier's output and inverting input terminals. From Ohm's law, the output offset voltage due to the input bias current is then

$$V_{os} = I_B R_2 \tag{3–10}$$

The output offset voltage due to the input bias current can then be minimized by (1) keeping the feedback resistor R_2 as small as possible for a given closed-loop gain and/or (2) using an operational amplifier having low values for input bias current, such as those having FET-input stages (Chapter 6).

Reducing R_2 too much, however, increases the loading on the operational amplifier's output and lowers the input impedance of the stage. Furthermore, R_2, in the case of inverting amplifiers, may already be fixed at some value by a required input resistance and closed-loop gain. If both the input resistance and

Figure 3–10
Addition of a resistor to compensate for input bias current error. (a) Inverting amplifier. (b) Noninverting amplifier.

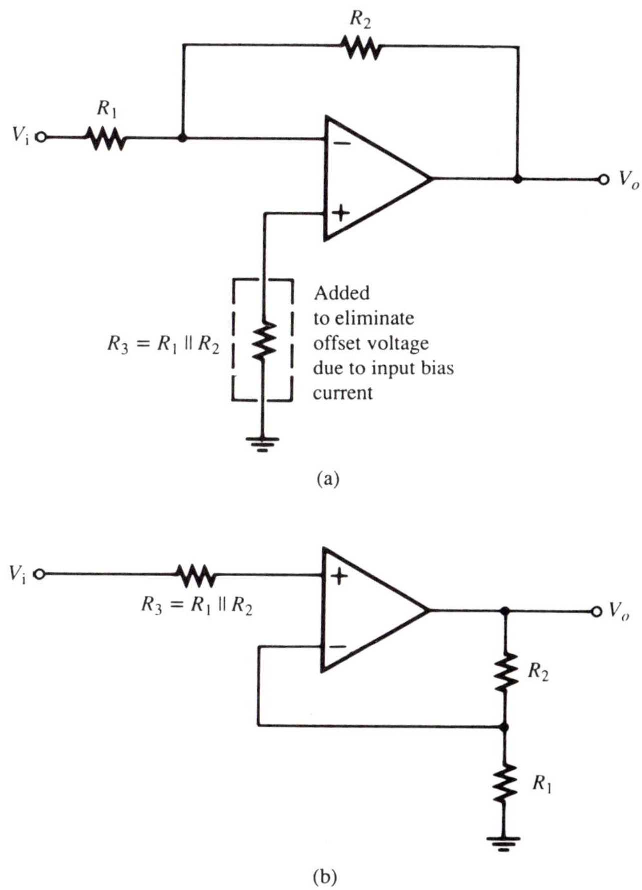

closed-loop gain of an inverting amplifier circuit are required to be high, then the output offset voltage due to the input bias current will also be high.

For the inverting amplifier shown in Figure 3–10a, the output offset voltage due solely to the input bias current is minimized by adding a resistor from the amplifier's noninverting terminal to ground. The value of this resistor (R_3) is chosen so that the current flowing into the noninverting input will exactly offset the current flowing into the inverting input terminal. This required resistance is the parallel combination of R_1 and R_2:

$$R_3 = \frac{R_1 R_2}{R_1 + R_2} \tag{3–11}$$

For a noninverting amplifier, R_3 is placed in series with the noninverting input and the actual signal input (Figure 3–10b).

Input Offset Current Contribution

The use of the external resistor (R_3) reduces the output offset voltage due to the input bias current. However, such use assumes that, at best, the two input bias currents are equal. This then would result in the input offset current being zero. In virtually all real operational amplifiers, however, the input offset current is not zero, which produces a DC offset output voltage equal to

$$V_{os} = I_{os} R_2 \tag{3–12}$$

Since any difference in input bias currents is usually much smaller than the bias currents themselves, the addition of R_3, though not perfect, does reduce the output offset voltage.

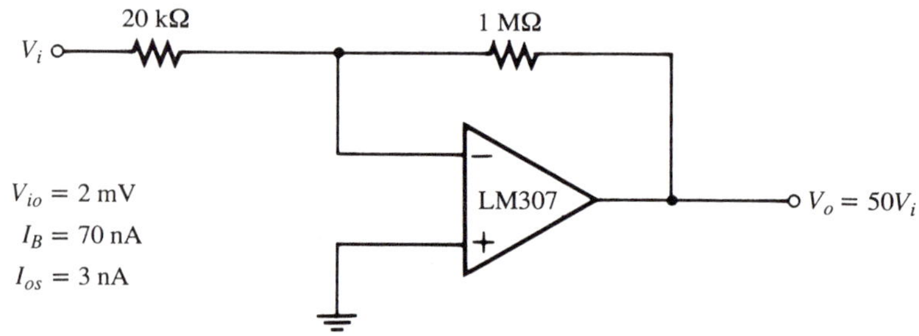

Figure 3–11
Inverting amplifier circuit for Example 3–5.

Example 3–5

For the inverting amplifier circuit of Figure 3–11, determine the output offset voltage if an LM307 is used having the following typical values:

$$V_{io} = 2 \text{ mV}$$
$$I_B = 70 \text{ nA}$$
$$I_{os} = 3 \text{ nA}$$

Solution:

(a) Using Equation 3–9, the output offset voltage due solely to the input offset voltage is

$$V_{os} = \left[1 + \frac{1 \text{ M}\Omega}{20 \text{ k}\Omega}\right](2 \text{ mV})$$
$$= \mathbf{102 \text{ mV}}$$

(b) From Equation 3–10, the output offset voltage due solely to the input bias current is

$$V_{os} = (70 \text{ nA})(1 \text{ M}\Omega)$$
$$= \mathbf{70 \text{ mV}}$$

(c) From Equation 3–12, the output offset voltage due solely to the input offset current is

$$V_{os} = (3 \text{ nA})(1 \text{ M}\Omega)$$
$$= \mathbf{3 \text{ mV}}$$

(d) Since it is not known beforehand whether these individual offset voltage components will be positive or negative voltages, the worst-case values must be computed. If all the components are of the same polarity, then the maximum DC output offset voltage possible from all three sources is

$$V_{os}(\text{max}) = \pm(102 \text{ mV} + 70 \text{ mV} + 3 \text{ mV})$$
$$= \mathbf{\pm 175 \text{ mV}}$$

On the other hand, some components might be of one polarity, whereas the remaining would be of the other polarity. Therefore, the minimum DC output offset voltage would be

$$V_{os}(\text{min}) = \pm[102 \text{ mV} - (70 \text{ mV} + 3 \text{ mV})]$$
$$= \mathbf{\pm 29 \text{ mV}}$$

so that the possible range of the magnitude of the DC output offset is **29 to 175 mV.** Using the same operational amplifier, the output offset could be reduced somewhat by adding a 19.6-kΩ resistor from the noninverting input to ground.

3–4 Output Offset Null Adjustments

If the DC output offset voltage that occurs in an operational amplifier requiring DC coupling to a load is a critical factor in the circuit's performance, one then has several options:

1. Select an operational amplifier with low values of input offset current and input offset voltage.
2. Use minimum resistance values compatible with other circuit requirements.
3. Use the lowest permissible value of closed-loop DC voltage gain.
4. Use a resistor of the proper value in series with the operational amplifier's noninverting input to cancel the output offset voltage due to the input bias current.

Figure 3–12
Output offset voltage nulling with operational amplifiers having internal design provisions requiring a single external potentiometer. (a) Connections required for a 741 wired as an inverting amplifier. (b) Connections required for an LM318 wired as a noninverting amplifier.

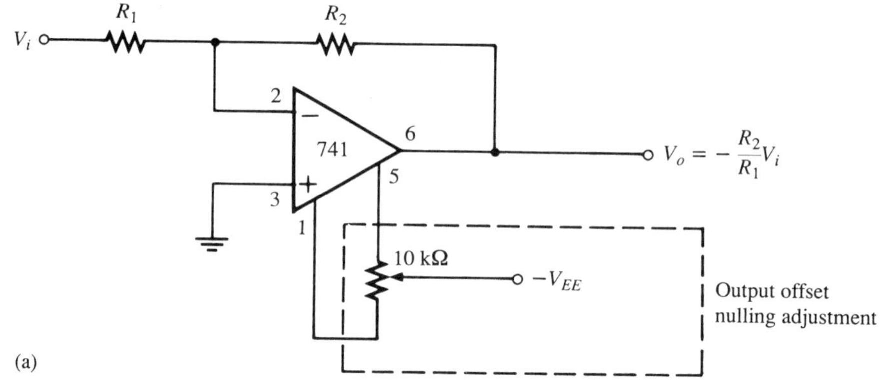

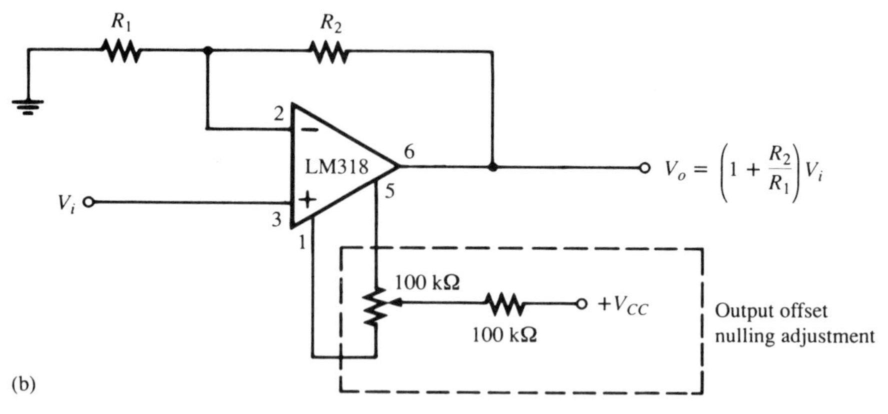

For those cases where the preceding steps still do not reduce the output offset sufficiently, some form of external circuitry must then be used to cancel the total DC output offset. As part of their internal design, some operational amplifier devices have provisions for connecting a *nulling,* or *zeroing,* multiturn potentiometer to trim any DC output offset voltage to zero when the input signal is zero. The manufacturer's data sheet will show the proper potentiometer value to use and the required connections for a particular type operational amplifier.

Figure 3–13
Output offset voltage nulling with operational amplifiers not having internal design provisions. (a) Inverting amplifier. (b) Voltage follower. (c) Noninverting amplifier. (d) Alternate method for voltage followers.

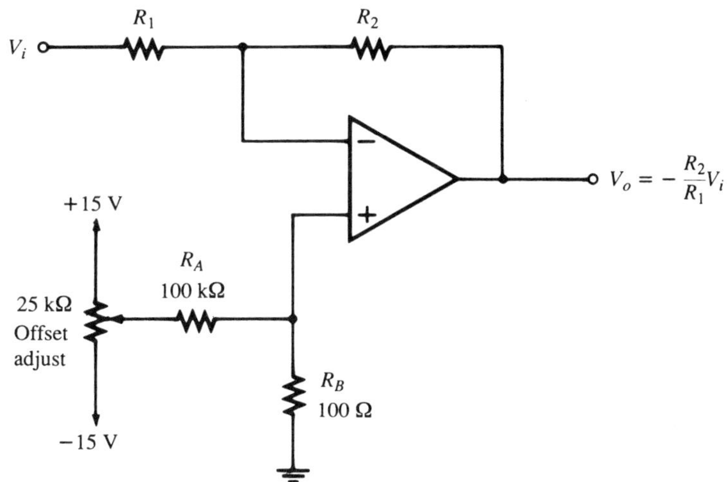

(a)

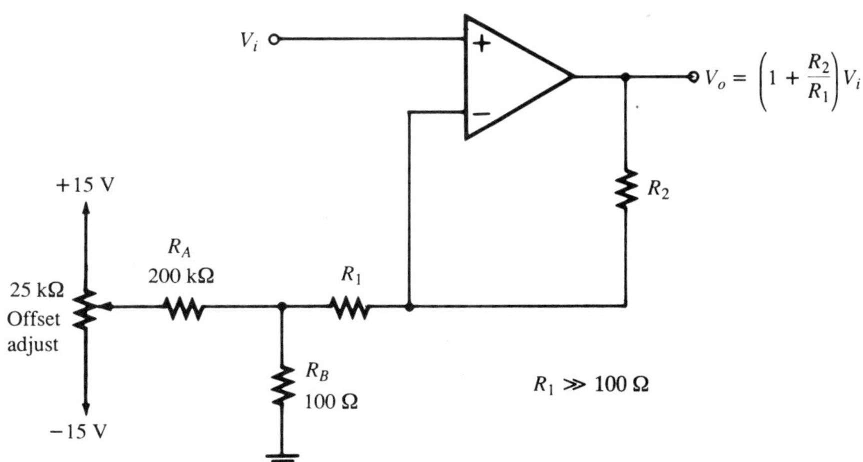

(b)

(continued)

Figure 3–13
continued

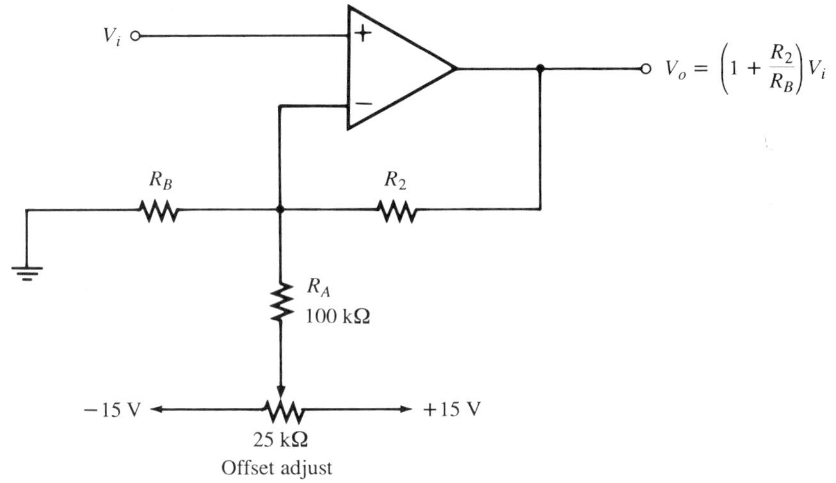

(c)

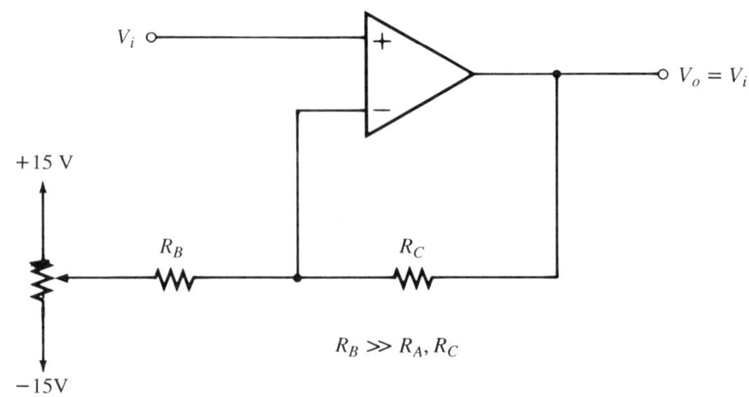

(d)

Figure 3–12a illustrates the required connections recommended by most manufacturers for the type-741 operational amplifier used in an inverting amplifier circuit. Figure 3–12b is for an LM318 device. To adjust the potentiometer for no offset, a digital voltmeter, because of its high resolution (typically ±1 mV), is connected to the circuit's output while grounding the circuit's input. The potentiometer is then adjusted until the voltmeter reads zero.

For those operational amplifiers not having this type of provision, the circuits of Figure 3–13 are used for inverting amplifiers, noninverting amplifiers, and voltage followers. In all these circuits, the voltage at the wiper of the potentiometer can vary from $-V_{EE}$ to $+V_{CC}$. When the potentiometer is properly adjusted, resistors R_A and R_B then form a voltage divider network that reduces the potentiometer voltage by an amount equal to the DC output voltage but of the opposite polarity. For the voltage follower of Figure 3–13d, the voltage developed across the feedback resistor R_C cancels the offset. In terms of the three resistances, the output voltage is given by

$$V_o = \left[1 + \frac{R_C}{R_B + (R_A/4)}\right] V_i \qquad (3\text{–}13)$$

For operational amplifiers having internal provisions for zeroing the offset voltage, a digital voltmeter is connected to the circuit's output while grounding the circuit's input. The potentiometer is then adjusted until the voltmeter reads zero.

3–5 Common-Mode Rejection Ratio

For those difference amplifiers built using ideal operational and instrumental amplifiers, any signal that is simultaneously present at both the inverting and noninverting inputs will be eliminated. Figure 3–14 shows the difference amplifier with both inputs tied together. Since $V_1 = V_2$, the input signal, which could be 60-Hz noise, is called the *common-mode input* (V_{icm}). Because the difference amplifier amplifies the voltage difference between its two inputs, ideally the output voltage in this case is zero. This follows from the ideal input-output voltage equation:

$$V_o = \frac{R_2}{R_1}(V_2 - V_1) \qquad (3\text{–}14)$$

Figure 3–14
Illustration of common-mode rejection ratio in a difference amplifier with both inputs tied together.

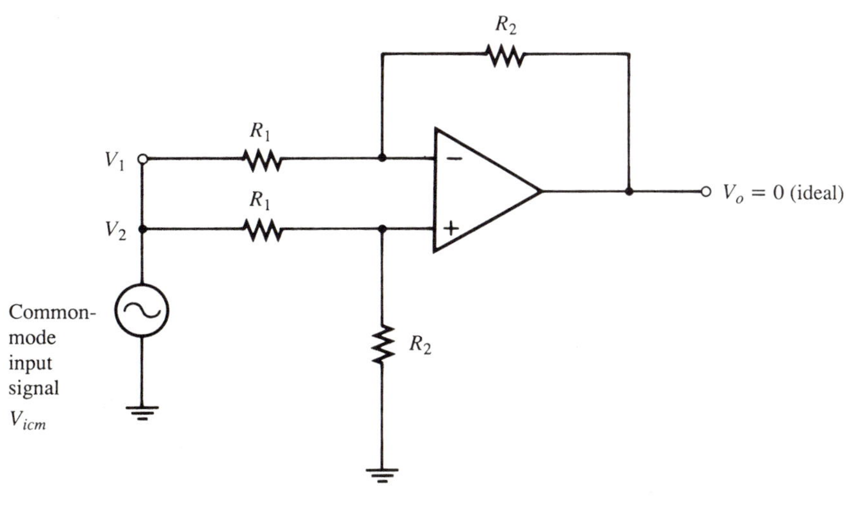

where the ratio R_2/R_1 is the amplifier's *differential gain, A_D*.

Here, the ideal difference amplifier circuit has infinite CMR. For an actual operational amplifier such as the 741, there will nevertheless be a small common-mode output signal component having the same frequency as the common-mode input signal in addition to no phase shift. The **common-mode gain,** A_{cm}, is then defined as the ratio of the common-mode output voltage to the common-mode input voltage:

$$A_{cm} = \frac{V_{ocm}}{V_{icm}} \qquad \textbf{(3–15)}$$

The measure of how well a difference amplifier rejects amplification of a common-mode input voltage is called the *common-mode rejection ratio (CMRR)* and is based on both the common-mode and differential gain; by definition,

$$\text{CMRR} = \frac{A_D}{A_{\text{cm}}} \qquad \textbf{(3–16)}$$

By then substituting Equation 3–16 into Equation 3–15, the common-mode output voltage is

$$V_{ocm} = \frac{V_{icm}}{\text{CMRR}} A_D \qquad \textbf{(3–17a)}$$

and, since $A_D = R_2/R_1$,

$$V_{ocm} = \left(\frac{R_2}{R_1}\right) \frac{V_{icm}}{\text{CMRR}} \qquad \textbf{(3–17b)}$$

Equation 3–17b can then be used to measure experimentally the CMRR of a given operational amplifier. In addition, since Equation 3–17 also gives the output voltage due solely to a common-mode input signal, the total output voltage of a nonideal difference amplifier is the sum of the output of the ideal amplifier (Equation 3–14) and the common-mode, or error, output component (Equation 3–17b):

$$V_o = \frac{R_2}{R_1}(V_2 - V_1) + \frac{R_2}{R_1}\left[\frac{V_{icm}}{\text{CMRR}}\right] \qquad \textbf{(3–18)}$$

The higher the value of CMRR, the smaller will be the common-mode error voltage (the second right-hand term of Equation 3–18). Operational amplifiers used as difference amplifiers should then have the highest CMRR possible. However, most data sheets express CMRR in terms of decibels, which is the CMR. The relation between CMRR and CMR is

$$\text{CMR} = 20 \log_{10}(\text{CMRR}) \qquad (3\text{–}19)$$

A CMRR of 1,000,000 is the same as a CMR of 120 dB.

In many cases, however, simply selecting an operational amplifier with a high CMRR may not be sufficient to reduce the the common-mode "error" component of Equation 3–18 to a level where it does not severely affect the amplifier's performance. This is particularly true if the differential input signal levels are very small, such as in the millivolt range for the unbalanced DC or AC bridge circuit of Figure 3–15, in comparison with a common-mode input signal level that might be as high as several volts RMS. Here, the difference amplifier converts the differential output voltage ($V_2 - V_1$) of an unbalanced bridge circuit to a single-ended output (V_o).

In order to amplify the desired differential input to a level that would exceed the unwanted common-mode output level, a high differential gain is required. From Equation 3–18, however, high differential gain also increases the common-mode component. Trying to zero the common-mode output with any of the offset

Figure 3–15
Using a difference amplifier with a Wheatstone resistance bridge to convert a differential floating input into a single-ended output.

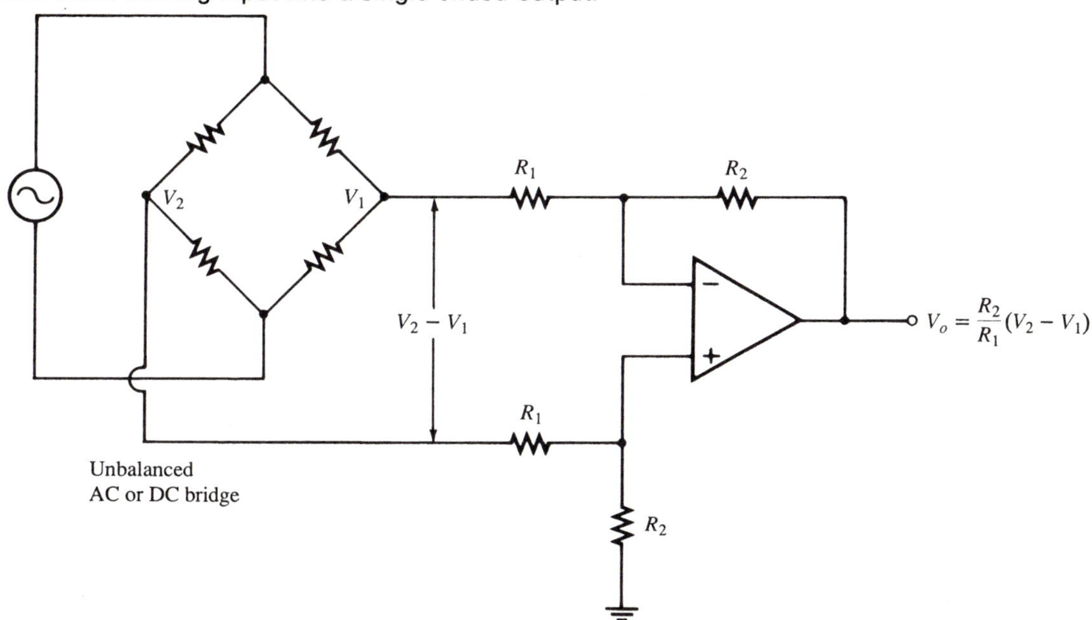

methods discussed earlier in this chapter will not work. Output voltage offset is a *DC* phenomenon, whereas the common-mode error is primarily an *AC* problem.

One way to increase the difference amplifier's common-mode rejection is to use a potentiometer, as shown in Figure 3–16a. Even though the resistance ratio R_2/R_1 for both the inverting and noninverting inputs may be exactly equal, slight tolerances in manufacturing still cause an inbalance in the input currents of the operational amplifier. If both inputs are tied together to an AC source, such as 60 Hz, as shown in Figure 3–16b, the potentiometer is then adjusted so that the resultant common-mode output signal is zero or a minimum.

Figure 3–16
Increasing the common-mode rejection of an operational amplifier. (a) Using a potentiometer. (b) Adjustment of potentiometer for minimum output voltage with both inputs tied together to a common-mode input voltage.

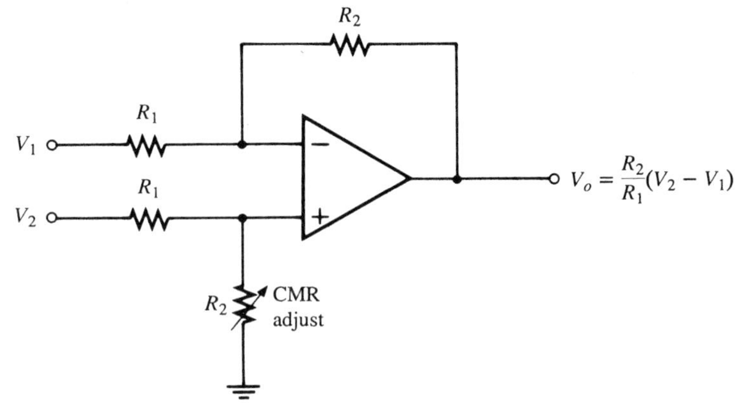

(a)

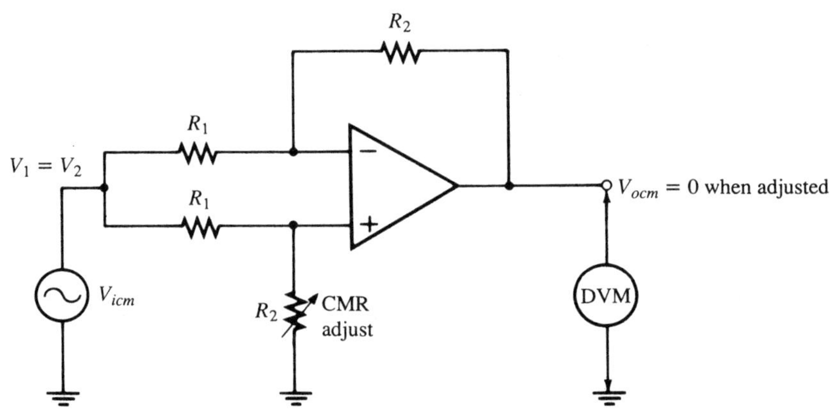

(b)

Example 3–6
The difference amplifier of Figure 3–17 is used to determine the common-mode rejection of a given 741 operational amplifier. Determine the CMR based on the following measurements taken:

$$V_{icm} = 3.52 \text{ V RMS}$$
$$V_{ocm} = 0.245 \text{ V RMS}$$

Solution:
(a) From Equation 3–15, the common-mode gain is

$$A_{\text{cm}} = \frac{0.245 \text{ V}}{3.52 \text{ V}} = 0.0696$$

(b) The differential gain of the amplifier circuit is 100 kΩ/100 Ω, or 1000. Using Equation 3–16,

$$\text{CMRR} = \frac{1000}{0.0696} = 14{,}368$$

and, when expressed in decibels,

$$\text{CMR (dB)} = 20 \log_{10}(14{,}368) = \mathbf{83.1 \text{ dB}}$$

Power-Supply Rejection Ratio

So far, neither the positive nor negative supply voltages are factors in any of the equations. Ideally then, a given amplifier circuit functions the same for supply

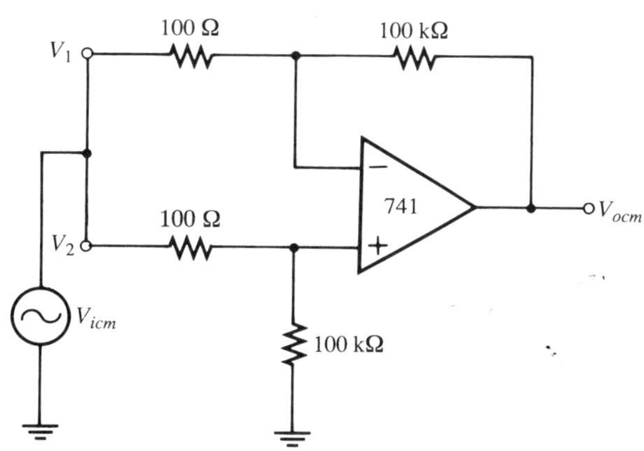

Figure 3–17
Difference amplifier circuit for Example 3–6.

voltages of ±6 V as for those of ±15 V. What happens, however, when one or both supply voltages either slowly drift from one voltage to another or vary back and forth between two other values? For the nonideal operational amplifier, the parameter that must be considered in these situations is the **power-supply rejection ratio (PSRR),** also called the *supply-voltage rejection ratio* or *power-supply sensitivity*. It is defined as the ratio of the change in input offset voltage to a corresponding change in the supply voltage:

$$\text{PSRR} = \frac{\Delta V_{io}}{\Delta V_{CC}} \tag{3–20a}$$

$$= \frac{\Delta V_{io}}{\Delta V_{EE}} \tag{3–20b}$$

It is the rate in terms of microvolts per volt. For the 741, it is typically 10 μV/V.

As is the case with the common-mode rejection ratio, PSRR is frequently expressed in decibels; this form is called the **power-supply rejection (PSR),** and

$$\text{PSR (dB)} = 20 \log_{10} \left(\frac{1}{\text{PSRR}}\right) \tag{3–21}$$

For the 741, a PSRR of 10 μV/V is equivalent to 100 dB.

Since the PSRR represents an input offset voltage, the DC output offset voltage for a circuit due solely to this DC input offset voltage of a nonideal inverting amplifier is

$$\Delta V_{os} = (1 + A_{CL})\Delta V_{io} \tag{3–22a}$$

By substituting Equation 3–20a or 3–20b,

$$\Delta V_{os} = (1 + A_{CL})\text{PSRR}\Delta V_{CC} \tag{3–22b}$$

$$= (1 + A_{CL})\text{PSRR}\Delta V_{EE} \tag{3–22c}$$

For a noninverting amplifier, Equations 3–22a–3–22c reduce to

$$\Delta V_{os} = A_{CL}\Delta V_{io} \tag{3–23a}$$

$$= A_{CL}\text{PSRR}\Delta V_{CC} \tag{3–23b}$$

$$= A_{CL}\text{PSRR}\Delta V_{EE} \tag{3–23c}$$

Example 3–7
An inverting amplifier using an LM318 has a closed-loop gain of 100. If the positive power supply varies from +13 to +15 V, determine the change in the DC output offset voltage if the data sheet specifies PSRR = 30 μV/V.

Solution: From Equation 3–22a, the input offset voltage change is

$$\Delta V_{io} = (1 + 100)(30\ \mu\text{V/V})(2\ \text{V}) = \mathbf{6.1\ mV}$$

Consequently, a power-supply voltage change of 2 V has the same effect as an input offset voltage of 60 μV would have caused.

3–6 The Effect of Frequency on Circuit Performance

One characteristic of the ideal operational amplifier discussed in Chapter 2 is that its bandwidth is infinite. That is, it has no upper frequency limit. Unfortunately, the bandwidth is limited, and there are several concepts that must be kept in mind to ensure that the particular operational amplifier circuit works as intended at all frequencies of interest.

Open-Loop Gain and Gain-Bandwidth Product

Figure 3–18 shows the frequency response of the open-loop voltage gain of a typical operational amplifier. At DC to frequencies below approximately 10 Hz, A_{OL} is constant at 100 dB, equivalent to an ordinary voltage gain of 100,000. Above 10 Hz, the response drops off linearly at a rate of −20 dB/decade, or −6

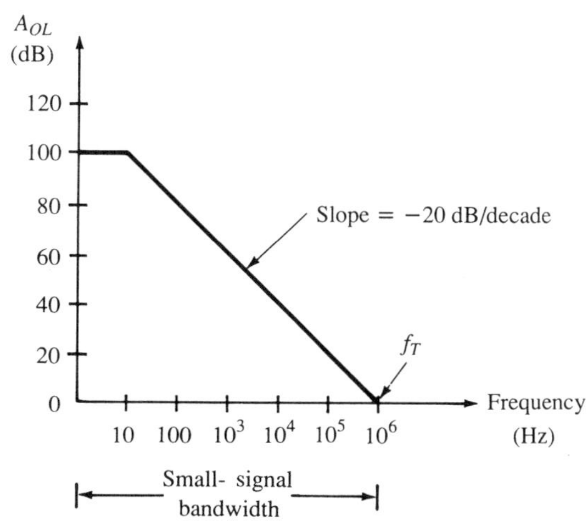

Figure 3–18
Open-loop frequency response of a typical operational amplifier.

dB/octave. This decrease continues until the open-loop gain equals 1. The frequency at which this occurs is the **unity-gain frequency** (f_T), also called either the *unity-gain crossover frequency* or the *small-signal unity-gain bandwidth.*

Since the open-loop gain varies with frequency, those parameters that are dependent on the open-loop gain are also dependent on frequency. Those parameters discussed so far include the amplifier's closed-loop input and output resistances and, to a certain extent, the common-mode rejection. In most cases, it is not necessary to have a plot of the open-loop gain response. Instead, a more useful parameter is the **gain-bandwidth product** (GBP), which is a constant for each operational amplifier. For the 741, it is typically 1 MHz. It is 15 MHz for the LM318. Numerically, the GBP is equal to the unity-gain frequency. Mathematically, it is dependent on the operational amplifier's open-loop gain and the required bandwidth:

$$\text{GBP} = A_{OL} \times \text{BW} \qquad (3\text{–}24)$$

Using Equation 3–24, one is then able to determine the value of the operational amplifier's open-loop gain for a given bandwidth. For example, the gain-bandwidth product for a 741 is 1 MHz. If the maximum input frequency expected is 20 kHz, then the maximum bandwidth is also 20 kHz. Consequently, the available open-loop gain at 20 kHz is (1 MHz)/(20 kHz), or 50. Since the open-loop and closed-loop gains are interrelated, the GBP also specifies the maximum closed-loop gain possible.

As shown in Figure 3–19, the bandwidth graphically is the point at which the closed-loop gain curve intersects the open-loop gain curve. Here is plotted a family of closed-loop gain curves for a 741 operational amplifier. For a closed-loop

Figure 3–19
Illustration of gain-bandwidth product (GBP) of a 741.

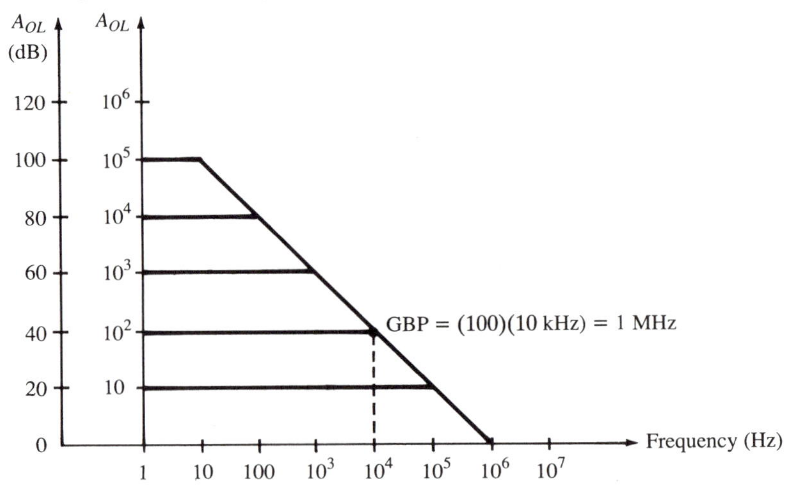

gain of 100 (40 dB), the maximum bandwidth possible is then 10 kHz. However, if the circuit's closed-loop gain is reduced to 10 (20 dB), then the maximum bandwidth possible is increased to 100 kHz. This inverse relationship between gain and bandwidth obtained from Figures 3–18 and 3–19 is the gain-bandwidth product equation of Equation 3–24. For both cases,

1. $100 \times 10 \text{ kHz} = 1 \text{ MHz}$
2. $10 \times 100 \text{ kHz} = 1 \text{ MHz}$

Consequently, the trade-off of lower gain allows for a larger bandwidth, and vice versa.

To be consistent with good engineering practices, it is never a good idea to operate any circuit or device at its limits. Consequently, the maximum closed-loop gain should be limited to about one-half the available open-loop gain to give a reasonable degree of safety margin. If the GBP of a given operational amplifier is 15 MHz, the maximum open-loop gain at 30 kHz then (15 MHz)/(30 kHz), or 500. The maximum allowable closed-loop gain for problem-free performance should then be approximately $\frac{1}{2} \times 500$, or 250. Of course, the closed-loop gain can be less than 250, but it should not exceed 250.

Rise Time

Some data sheets may not give either the open-loop gain response curve, the gain-bandwidth product, or the unity-gain frequency. However, a parameter called the *unity-gain transient response rise time* (t_r) is usually given, which is related to the operational amplifier's unity-gain bandwidth (or GBP) as

$$\text{GBP} = \frac{0.35}{t_r} \qquad (3\text{–}25)$$

For the 741, t_r is typically 0.3 μs, giving a GBP of 0.35/0.3 μs, or 1.16 MHz. Although there may be a slight difference between calculating the GBP using the rise time and the value given in the data sheet, the two can be considered equal.

Slew-Rate Limiting

In addition to the open-loop gain response curve and GBP, the operational amplifier's slew rate is another factor limiting the frequency response. For high-frequency, large-output voltage swings, the slew rate is a greater factor than the GBP.

The maximum time rate of change of which the output signal is capable is dependent on both its peak amplitude and frequency. For a sine wave, the maximum rate of change occurs at the zero crossing. If this rate of change is greater than the slew rate of the operational amplifier, then the output signal will be distorted somewhat. For a sine-wave input, the distorted output appears like a

Figure 3–20
Slew rate distortion of a sine wave.

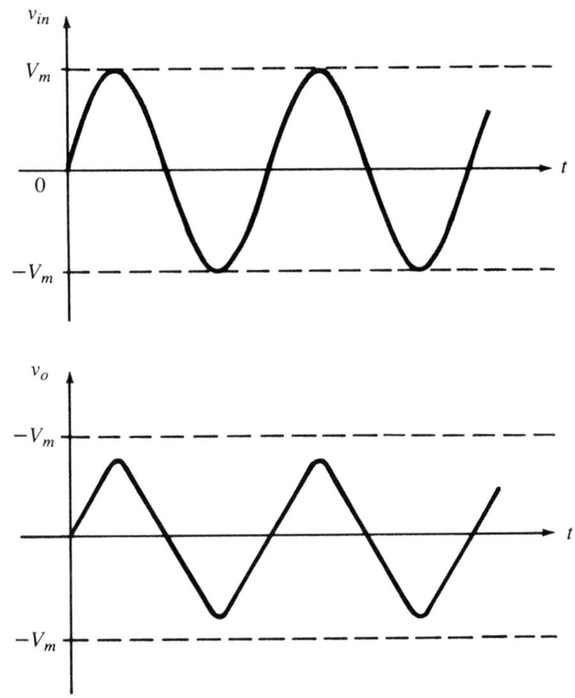

Figure 3–21
Graph of undistorted peak output voltage for a given maximum sine wave frequency and slew rate (V_{CC} = 15 V, V_{+SAT} = 13 V).

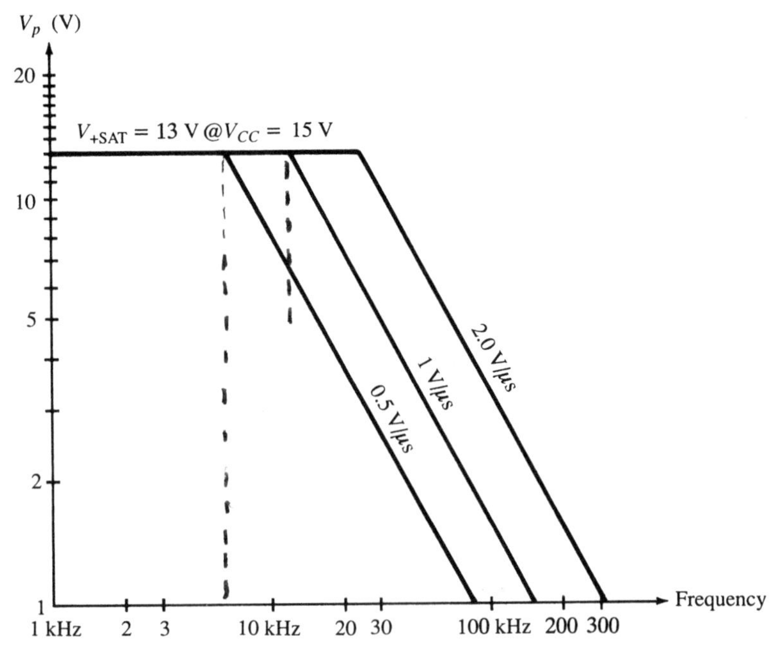

triangle waveform (Figure 3–20). Here, in effect, the slew rate limits how fast the output sine-wave signal can change with time.

For a given slew rate (SR) and peak output voltage (V_P), the maximum input sine-wave frequency that can occur without causing the output signal to take on a triangular shape is given by

$$f_{max} = \frac{SR}{2\pi V_P} \tag{3–26}$$

This equation shows that the borderline between "small-signal" response and the slew-rate-limited response is not just a function of the peak output voltage level, but that by trading off either frequency or peak amplitude, one can continue to have a distortion-free output. A graphical representation of Equation 3–26 is shown in Figure 3–21 such that the area below any V_P line then represents an *undistorted* sine-wave response.

As a matter of convenience, many manufacturers often use the terms *full-power bandwidth* and *large-signal response* in their data sheets to represent f_{max}. For example, a 741 has a slew rate of 0.5 V/μs and, when using a ±15-V supply, the maximum output voltage swing capable is about ±13 V. From Equation 3–26, the full-power bandwidth for the 741 is then

$$f_{max} = \frac{0.5\ \text{V}/\mu\text{s}}{(2\pi)(13\ \text{V})}$$
$$= \mathbf{6.1\ kHz}$$

3–7 Frequency Compensation

Despite some of their nonideal characteristics, which present some minor drawbacks, general-purpose operational amplifiers like the 741 nevertheless are very easy to work with. As long as we keep the gain-bandwidth product rule in mind, not much will go wrong. However, except for most noncritical audio applications having frequencies below 15 kHz to 20 kHz, the 741's 1-MHz unity-gain frequency is not much to brag about. For a simple 60-dB (×1000) microphone preamplifier having a bandwidth of 10 kHz, a single operational amplifier would need a GBP of at least 10 MHz, which rules out the 741. However, two or three 741 operational amplifier stages in cascade, each with gains of 20 to 30 dB at 10 kHz, would work.

The 741 is an example of an *internally compensated* operational amplifier. This means that the manufacturer has already optimized its design to give a controlled roll-off characteristic to prevent the device from oscillating. Inside the 741 is a "compensating" capacitor, which controls the frequency stability and obviously cannot be removed or disconnected. In exchange for this hassle-free device, one usually settles for the following trade-offs:

1. Reduced gain-bandwidth product
2. Slower slew rate
3. Reduced full-power bandwidth
4. Higher rise time

To achieve a higher performance level, one either has to use a compensated operational amplifier with higher ratings or use an uncompensated operational amplifier and provide some amount of external compensation. No operational amplifier can be left entirely uncompensated because there would be a very strong likelihood that the amplifier would become unstable with high gains at higher frequencies and oscillate. In an uncompensated device, therefore, the terminals for the required frequency compensating capacitor are brought out to allow the user to select the optimum condition of stability by the connection of one or more external capacitors.

Figure 3–22 shows a typical response curve of an uncompensated amplifier. Internally, each stage of the operational amplifier's circuit is similar to an RC network, each having a roll-off of −20 dB/decade. When these RC networks are combined, the combined roll-off is additive. For example, the one-decade region from 1 kHz to 10 kHz has a roll-off of −20 dB/decade. From 10 kHz to 40 kHz, a second RC network becomes active, and the overall roll-off becomes −(20 dB + 20 dB), or −40 dB/decade. Above 40 kHz, because of a third internal RC network, the roll-off now becomes −60 dB/decade. Each time the roll-off of the open-loop response changes, the response curve "breaks," and the internal phase shift increases by a maximum of 90°. At the frequency when the roll-off changes from −40 dB/decade to −60 dB/decade, the maximum internal phase shift is 180°, a

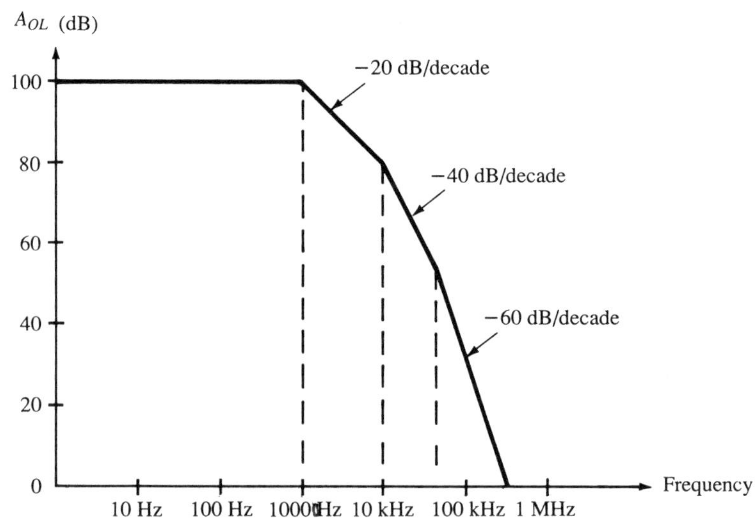

Figure 3–22
Frequency response of a typical uncompensated operational amplifier.

situation known as *conditionally stable*. Circuit operation with this combination of gain and frequency should be avoided. In fact, for those frequencies where the roll-off is −60 dB/decade, the maximum internal phase shift is 270°. Since the phase shift is 180° or more, the circuit will definitely be unstable and will oscillate. To prevent oscillation in an uncompensated operational amplifier, some form of external compensation is required.

Figure 3–23
Frequency compensation of an LM301 operational amplifier. (a) External connections. (b) Open-loop frequency response.

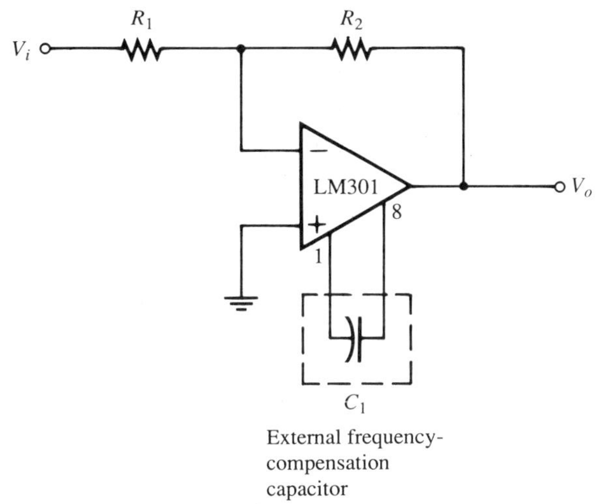

(a)

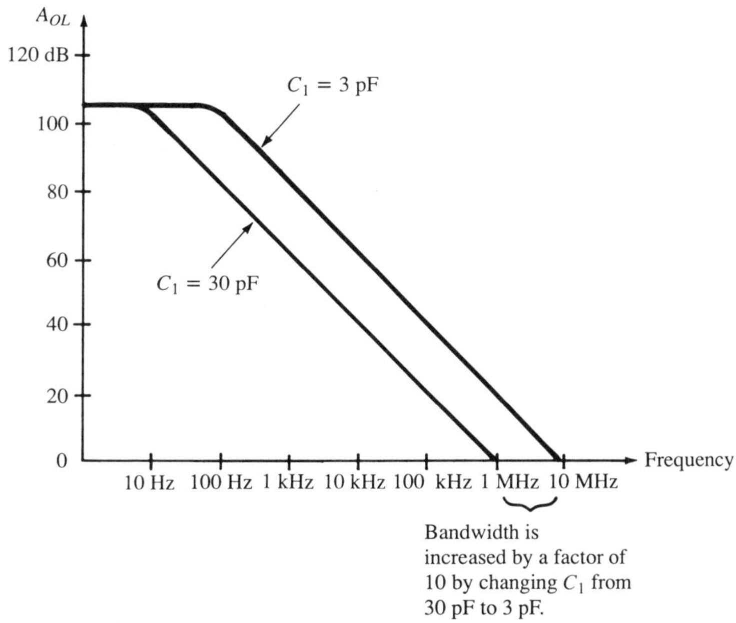

(b)

As a practical matter, the data sheets of uncompensated operational amplifiers give the details of how to determine the amount of compensation for a required gain and frequency response. As an example, the LM301 is a typical uncompensated, general-purpose operational amplifier. As shown in Figure 3–23a, one form of external compensation is achieved by connecting a single capacitor (C_1) between pins 1 and 8. If $C_1 = 3$ pF, the bandwidth is approximately 10 MHz. Increasing the value of C_1 decreases the bandwidth as well as full-power bandwidth. For example, if $C_1 = 30$ pF, then the bandwidth decreases to approximately 1 MHz, similar to the internally compensated 741.

Because the types of compensation networks and the component values required are given in the data sheets and vary from one type device to another, they will not be expanded upon here.

Summary

In noncritical applications where performance requirements are not very stringent, circuit design based on the concept of the ideal amplifier is quite satisfactory. For more demanding applications, however, many of the actual characteristics of the operational amplifier to be used must be taken into consideration if the circuit is to meet given specifications.

This chapter examined the practical operational amplifier and its limitations by looking at those parameters given on a typical data sheet. Information contained on the data sheet includes device description, pin diagram, and absolute ratings as well as static and dynamic characteristics. Particular attention was paid to the magnitude of the output voltage offset as a result of an operational amplifier's input offset voltage, input bias current, and input offset current; compensation methods for voltage follower, inverting, noninverting, summing, and difference amplifier circuits were also illustrated.

Besides the DC output offset voltage with no input signal, difference amplifiers were shown to have additional problems due to their inability to completely eliminate common-mode input signals. How well the difference amplifier does this is given by its common-mode rejection ratio. In addition, those parameters that can significantly affect the frequency response of most operational amplifier circuits were discussed. These include the gain-bandwidth product, rise time, and slew rate.

Finally, the performance difference between compensated and uncompensated operational amplifiers was illustrated. External compensation of operational amplifiers allows the user to tailor the amplifier performance in terms of improved gain-bandwidth product, higher slew rate, and a reduction in rise-time values.

Glossary

channel separation A measure of how well the signal present at the input of one operational amplifier section will be rejected or attenuated at the output in the remaining section(s) of a dual or quad operational amplifier, though no input signal is applied to the unused section(s).

common-mode gain In a difference amplifier, the ratio of the common-mode output voltage to the common-mode input voltage.

common-mode rejection ratio (CMRR) The measure of how well a difference amplifier rejects amplification of a common-mode input voltage, which is based on both the common-mode and differential gains.

differential input voltage The maximum voltage that can be applied between the inverting and noninverting inputs.

gain-bandwidth product (GBP) The product of the closed-loop voltage gain of an operational amplifier and its corresponding bandwidth.

input bias current The average of the two DC currents that flow into the inverting and noninverting inputs of an operational amplifier.

input offset current The difference of the two input bias currents of an operational amplifier.

input offset voltage The equivalent DC voltage that must be applied to one of the input terminals of an operational amplifier to produce a zero output voltage with the other input grounded.

input resistance The internal input impedance of an operational amplifier as seen at either the inverting or the noninverting input to ground while the remaining input terminal is grounded.

internal power dissipation The maximum power that the operational amplifier is capable of dissipating at a specified ambient temperature. For higher temperatures, this must be derated or reduced accordingly.

open-loop gain The voltage gain of the operational amplifier without any external feedback.

operating temperature The ambient temperature for which the operational amplifier will operate within the manufacturer's specifications.

output resistance The internal output resistance as seen from the output terminal to ground.

output short-circuit current The maximum DC output current that can be supplied to a load.

output short-circuit duration The amount of time that the operational amplifier's output terminal can be short-circuited to ground or supply voltage without damage.

output voltage swing The maximum peak output voltage that the operational amplifier can produce without saturation or clipping; it is typically about two diode voltage drops less than the corresponding supply voltages.

power-supply rejection (PSR) The power-supply rejection ratio expressed in decibels.

power-supply rejection ratio (PSRR) The ratio of the change in input offset voltage of an operational amplifier to a corresponding change in the supply voltage. Also called the *power-supply sensitivity*.

slew rate (SR) The time rate of change of the output voltage with the operational amplifier circuit having a unity closed-loop gain.

unity-gain frequency The frequency at which the operational amplifier's open-loop gain is unity. It is also called either the *unity-gain crossover frequency* or the *small-signal, unity-gain bandwidth*.

Review Questions

1. What seven categories of information are contained on the data sheet of a typical operational amplifier?
2. Define and explain how to measure the following parameters of a given operational amplifier: input offset voltage, input bias current, input offset current, and slew rate.
3. Explain the effect of the open-loop, closed-loop, and loop gains on both the input and output impedances of an operational amplifier.
4. Explain how the input offset voltage, input bias current, and input offset current contribute to output offset voltage of an operational amplifier.
5. Describe several techniques used to cancel an output offset voltage.
6. What is the common-mode rejection ratio? How does it differ from the common-mode rejection?
7. How does the common-mode rejection ratio affect the output of a difference amplifier?

8. Describe what adjustments may be made to improve the common-mode rejection characteristics of a difference amplifier.
9. How does frequency affect the performance of an operational amplifier?
10. What is the gain-bandwidth product, and how does it relate to the unity-gain frequency of an operational amplifier?
11. What is an amplifier with internal frequency compensation? How does its performance compare with an operational amplifier having external compensation?

Exercises

1. Determine the bandwidth of an operational amplifier circuit that has a unity-gain bandwidth of 1 MHz and a closed-loop gain of 1000.
2. Determine the maximum closed-loop voltage gain possible for an operational amplifier circuit having a bandwidth of 5 kHz and a unity-gain frequency of 1 MHz.
3. If the operational amplifier of Exercise 2 is now used as an inverting amplifier having a closed-loop gain of 10, determine the circuit's bandwidth.
4. Determine the gain-bandwidth product of an operational circuit if the rise time is 1 μs.
5. What is the rise time of a given operational amplifier if its bandwidth is 25 kHz with a closed-loop gain of 80?
6. A given operational amplifier circuit has a closed-loop gain of 40 and a unity-gain frequency of 1 MHz. Determine the rise time of the operational amplifier.
7. A high-frequency square wave is applied to a voltage follower. Determine the slew rate if the output voltage changes from −4.9 V to +4.8 V in 0.75 μs.
8. The slew rate of an operational amplifier is 50 V/μs. How long will it take for a square-wave output signal to change from +3 V to −3 V?
9. What is the maximum frequency of which an operational amplifier circuit is capable without slew-rate limiting if its slew rate is 10 V/μs and it has a peak output voltage of 2V?
10. What is the maximum peak output voltage an operational amplifier circuit can handle without slew-rate limiting if its slew rate is 2.5 V/μs at 45 kHz?
11. What is the common-mode gain of a difference amplifier whose CMR is 85 dB and differential gain is 2?
12. The difference amplifier of Figure 3–14, having a differential gain of 2500, is used to determine the CMR of a given operational amplifier. Determine the CMR in decibels if the applied common-mode input of 5.3 V RMS gives a corresponding common-mode output of 134 mV RMS.
13. An operational amplifier having a CMR of 93 dB is used as a difference amplifier with a differential gain of 50. Determine the common-mode output voltage that would result if the common-mode input voltage is 9.7 V RMS. Assume that the circuit had been previously nulled.
14. An operational amplifier has the following parameters, taken from a data sheet:

$$A_{OL} = 250{,}000$$
$$R_i = 2\ \text{M}\Omega$$
$$R_o = 75\ \Omega$$

Determine the input and output impedances of a noninverting amplifier having a closed-loop gain of 25.

15. Repeat Exercise 14 using the same operational amplifier in a voltage follower.
16. An operational amplifier has the following parameters, taken from a data sheet:

$$A_{OL} = 300{,}000$$
$$R_o = 75\ \Omega$$

Determine the output impedance of an inverting amplifier if its closed-loop gain is 100.

Figure 3–24

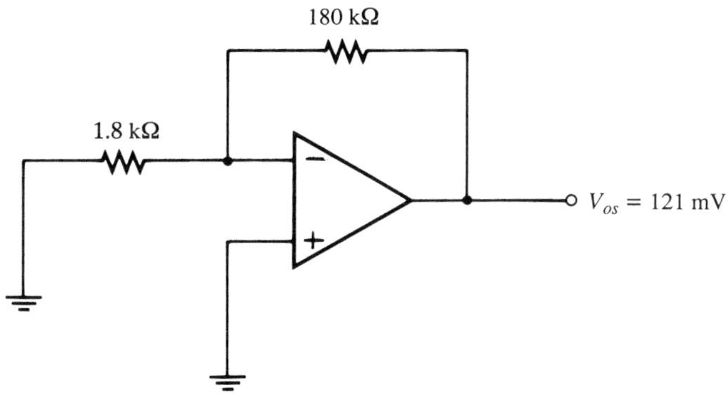

17. The circuit of Figure 3–24 is used to measure the input offset voltage of the operational amplifier. What is the input offset voltage if the measured output voltage is 121 mV?

18. Neglecting the input offset voltage, determine the input bias current for the operational amplifier of Figure 3–25 if the output offset voltage is 83 mV across the ~~10-kΩ~~ resistor. [100 kΩ]

Figure 3–25

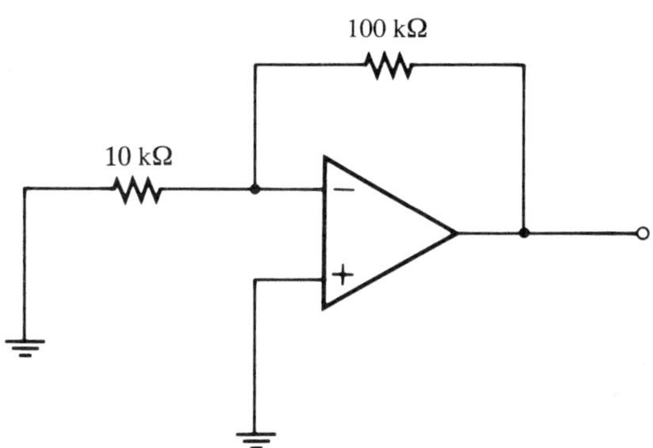

19. Determine the value of the resistor R_3 for the inverting amplifier of Figure 3–10a needed to minimize the output offset voltage due to the operational amplifier's input bias current if R_1 = 22 kΩ and R_2 = 47 kΩ.

20. Repeat Exercise 19 for the noninverting amplifier circuit of Figure 3–10b. What now is the circuit's input impedance?

21. An operational amplifier specifies a power-supply rejection ratio of 25 μV/V. Express this in terms of decibel power-supply rejection.

22. The power-supply rejection of an operational amplifier is 75 dB. What is the power-supply rejection ratio in microvolts per volt?

23. A noninverting amplifier circuit has a closed-loop gain of 20. If the operational amplifier used has a PSRR of 25 μV/V and the positive power supply periodically varies between 12.1 and 12.4 V, determine the change in the output voltage of the amplifier circuit.

24. What would be the equivalent change in the input offset voltage of the operational amplifier of Exercise 23?

4 Differentiators and Integrators

Instructional Objectives

Upon completing this chapter, you will be able to

- Explain the process of differentiation in terms of the slope of the line.
- Explain the process of integration as computing the area underneath the curve.
- List several possible uses for differentiators and integrators.
- Determine the output voltage of the basic differentiator circuit built around an operational amplifier.
- Explain why it is often necessary to stabilize the basic differentiator circuit and how this is accomplished.
- Draw the schematic diagrams for the following differentiator circuits:
  - Basic differentiator
  - Compensated differentiator
  - Summing differentiator
  - Difference differentiator
- Determine the output voltage of the basic integrator circuit built around an operational amplifier.
- Explain why it is often necessary to stabilize the basic integrator circuit and how this is accomplished.
- Draw the schematic diagrams for the following integrator circuits:
  - Basic integrator
  - Compensated integrator
- For several common input waveforms, show what the output waveforms from both a differentiator and integrator would look like.

Introduction

Up to this point, the following linear amplifier circuits discussed in earlier chapters by themselves limited the analog computer to solving only simultaneous algebraic equations:

- Noninverting amplifier: multiplies a value by a constant
- Inverting amplifier: changes algebraic sign
- Summing amplifier: performs addition
- Difference amplifier: performs subtraction

In order for the analog computer to be capable of solving complex equations in calculus that have wide-ranging applications in many technological fields, two additional circuits are required. These are the operations of *differentiation* and *integration;* they are the mathematical inverses of each other, just as multiplication is the inverse of division. The electronic circuits that perform these operations are called *differentiators* and *integrators,* respectively.

Besides increasing the computational power of the analog computer, differentiators and integrators serve other useful purposes. Very often these circuits are used to process further certain signals such as a detector in some FM modulators and to change a triangle wave into a square wave, a process often referred to as *waveshaping*. Elsewhere, integrators are fundamental building blocks of several types of waveform generators (Chapter 10), analog-to-digital converters (Chapter 12), and several active filters (Chapter 11).

4–1 Differentiators

The **differentiator,** not to be confused with a differential or difference amplifier, is a circuit whose output voltage is proportional to the instantaneous time rate of change of its input signal:

$$v_o(t) = k\,\frac{\Delta v_i}{\Delta t} \tag{4–1}$$

where k is a constant of proportionality. The operation of a differentiator is the same as taking the slope of the line at any point on the input signal.* For input signals not having straight-line segments, such as a sine wave, the slope at any point on that curve is the slope of the line that is *tangent* to that point (Figure 4–1).

If the amplitude of an input signal is not changing with time, such as a DC voltage, then its slope is zero, and the differentiator produces no output signal. For the most part, it is not the amplitude of the input signal that is important but rather the *time rate of change*. Differentiators have many uses in electronic circuits, such as measuring the rate of change of a signal, waveshaping, and producing control signals that coincide with rapid changes in signal level.

*Students of calculus will recognize this operation as taking the first derivative of the input voltage with respect to time, dv_i/dt.

Figure 4–1
Graphical representation of differentiation as the slope of the line that is tangent to the curve at time t_1.

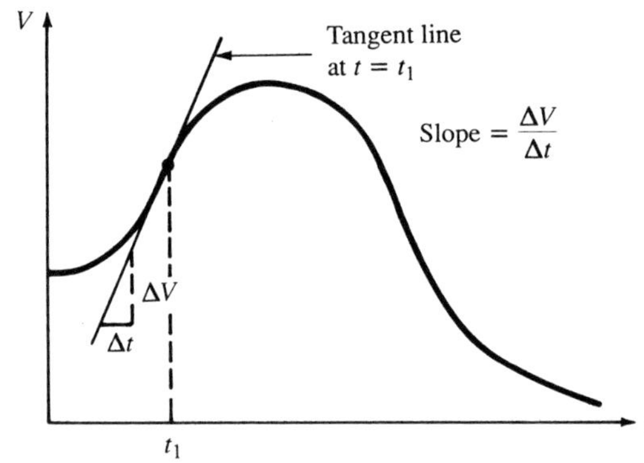

The Basic Circuit

Figure 4–2a shows the basic operational amplifier differentiator. Note that it is similar in form to the inverting amplifier, except that the input element is a capacitor instead of a resistor. From basic electricity, the current flowing in a capacitor (C) is given by

$$i_{in} = C\frac{\Delta v_{in}}{\Delta t} \quad \textbf{(4–2)}$$

Since the current (i_F) flowing through R_F is the same as that through C because of the virtual ground at the inverting input, the circuit's output voltage v_o is

$$v_o = -i_{in}R_F \quad \textbf{(4–3)}$$

Figure 4–2
Basic differentiator. (a) Operational amplifier active circuit. (b) Equivalent passive RC high-pass filter circuit.

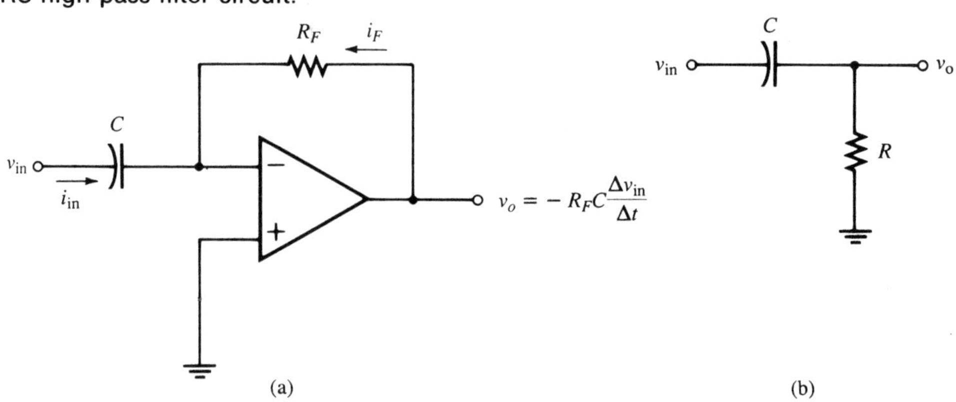

When Equation 4–3 is substituted into Equation 4–2, solving for v_o yields

$$v_o = -R_F C \frac{\Delta v_{in}}{\Delta t} \tag{4–4}$$

where the quantity $\Delta v_{in}/\Delta t$ represents the rate of change, or slope, of the input signal over any small time change. In the notation of differential calculus, Equation 4–4 is written as

$$v_o = -R_F C \frac{dv_{in}}{dt} \tag{4–5}$$

The quantity $R_F C$ is simply a scaling factor, as represented by the proportionality constant k in Equation 4–1. This quantity is often referred to as the *differentiator gain*. The minus sign in Equation 4–5 implies that the polarity of the output signal is the opposite of what it normally would be. If the input signal is increasing with time (i.e., a positive slope), the resulting differentiated output signal will be negative. If the input signal is decreasing with time, the resulting output signal will be positive.

The operational amplifier differentiator is identical in function to the passive RC circuit of Figure 4–2b provided that the RC time constant of the passive circuit is much shorter than the time period of the input signal. This circuit is often used as a simple passive high-pass filter to reject input signals having frequencies less than $1/2\pi RC$.

The constant slopes of simple straight-line (linear) segments, such as those of ramps and triangle waves, are easily calculated here by simply dividing the change in voltage by the corresponding time change, $\Delta v/\Delta t$. For example, if the voltage ramp of Figure 4–3 decreases from +15 V to +10 V in 2.5 ms, the slope over this time period is

$$\text{Slope} = \frac{10\text{ V} - 15\text{ V}}{2.5\text{ ms}}$$
$$= -2000\text{ V/s}$$

Calculation of the instantaneous slope of the more complex, time-varying waveforms such as sine waves requires the methods of calculus because the instantaneous slope of those signals is (1) different at each point on the curve and (2) not as easily described as was the ramp voltage of Figure 4–3. Differentiated waveforms of common AC signals are discussed in Section 4–3.

One serious problem with the basic differentiator circuit is that it is very susceptible to high-frequency electrical noise. This is because the reactance of the capacitor decreases with frequency, causing a corresponding increase in the closed-loop voltage gain at the rate of +6 dB/octave (Figure 4–4). Like the inverting amplifier, the magnitude of the closed-loop gain is the ratio of the feedback to input element impedances, so that

Figure 4–3
Example of slope calculation.

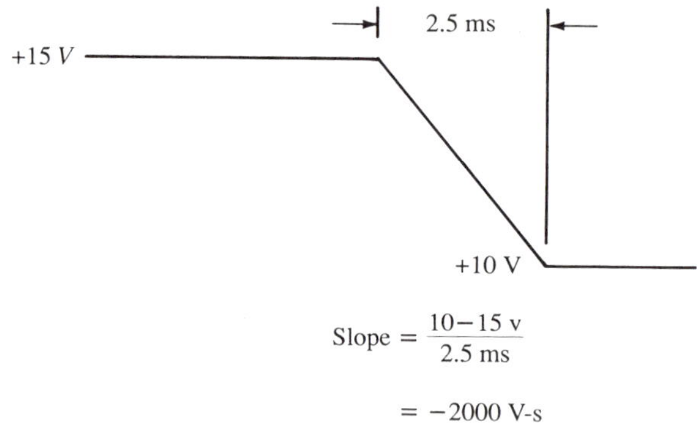

$$A_{CL} = \frac{R_F}{X_C} = 2\pi f R_F C \qquad (4\text{–}6)$$

Although the basic differentiator's closed-loop gain increases with frequency, it is nevertheless limited at the high end by the operational amplifier's open-loop response curve. For those frequencies above where the response intersects the open-loop curve, the circuit stops acting as a differentiator.

To put a limit on the closed-loop gain at high frequencies before being limited by the operational amplifier's open-loop response curve, resistor R_S is added in series with the input capacitor, as shown in the stabilized, compensated circuit of Figure 4–5a.

Figure 4–4
Frequency response of an operational amplifier differentiator.

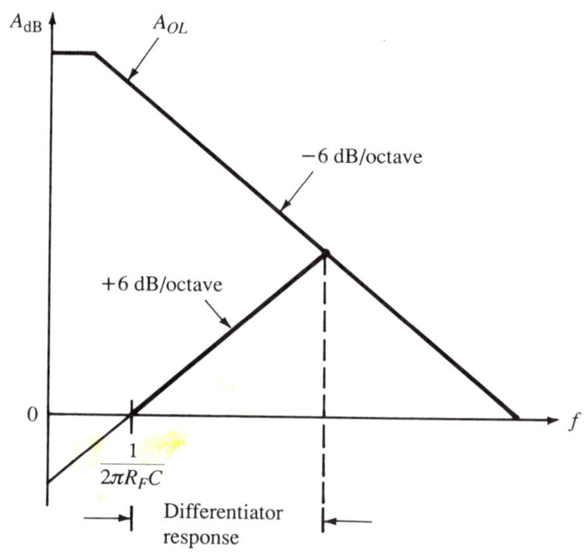

Figure 4–5
Compensated differentiator. (a) Basic circuit. (b) Equivalent high-frequency circuit. (c) Frequency response graph showing high-frequency limiting.

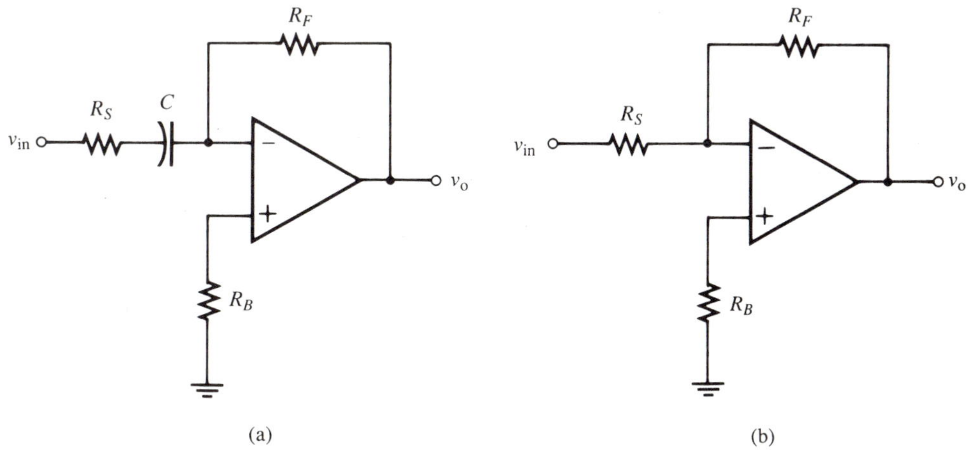

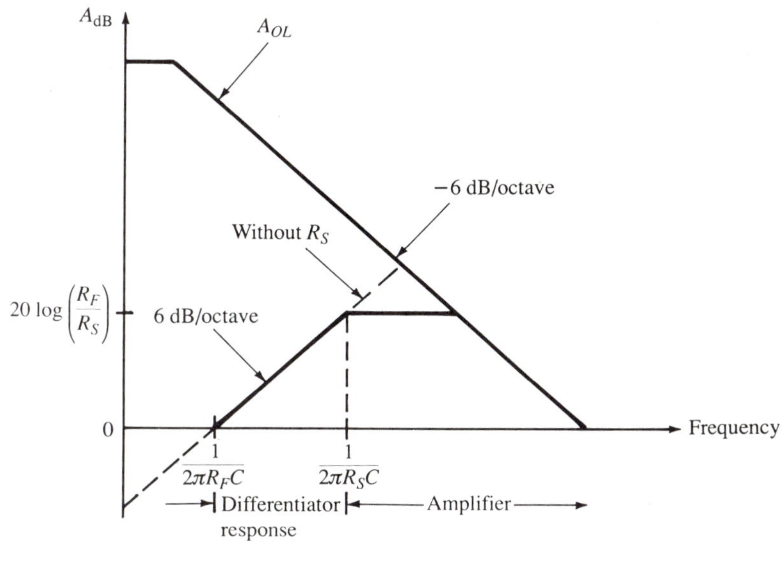

This type of compensated-differentiator circuit has a much-improved noise-handling ability, but the maximum usable input frequency over which the improved circuit acts as a differentiator is now limited to input frequencies less than

$$f_1 = \frac{1}{2\pi R_S C} \tag{4–7}$$

and should be greater than the frequency at which the differentiator's low-frequency closed-loop gain is unity (0 dB), such that

$$f_1 > \frac{1}{2\pi R_F C} \tag{4–8}$$

At frequencies above f_1, the capacitor's reactance is very small compared with R_S, so that the circuit essentially looks like an inverting amplifier (Figure

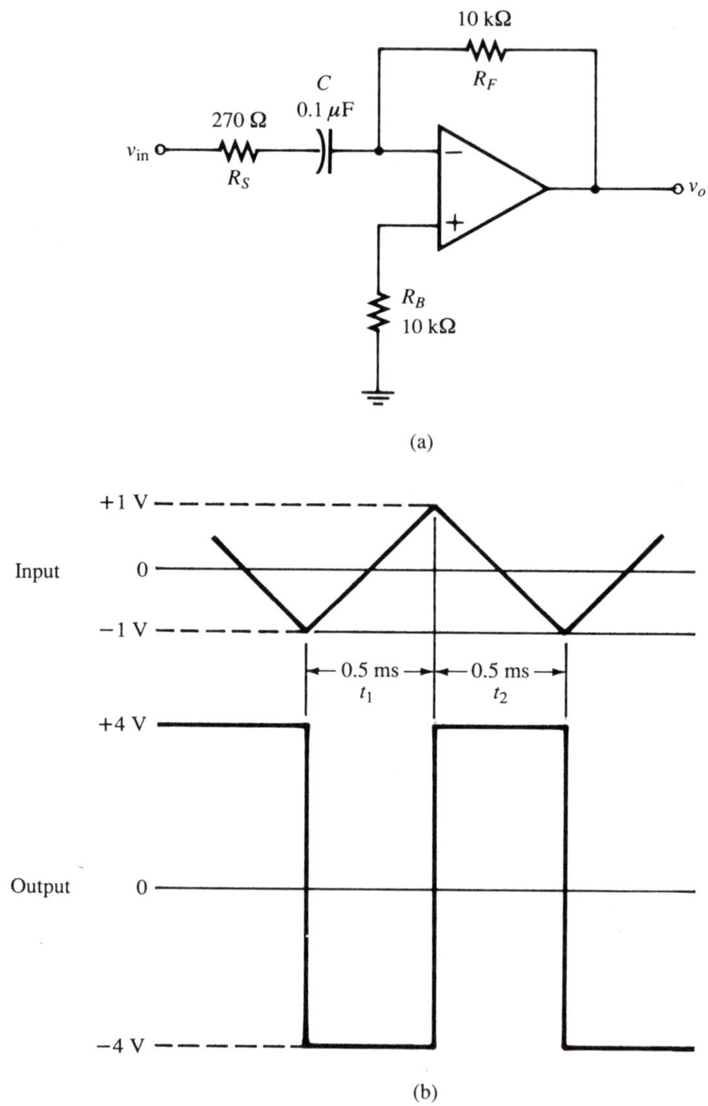

Figure 4–6
Example 4–1. (a) Circuit. (b) Input and output waveforms.

4–5b) having a closed-loop voltage gain equal to $-R_F/R_S$. The resultant frequency response of the differentiator in relation to the open-loop voltage gain of the operational amplifier is graphed in Figure 4–5c. This constant gain continues until it meets with the open-loop response. From the gain-bandwidth product, the frequency at which this intersection occurs is unity-gain frequency (f_T) divided by the closed-loop gain, R_F/R_S.

For input frequencies lower than f_1, the improved circuit works properly as a differentiator. Frequency f_1 is set approximately one decade above the highest frequency at which the circuit can act properly as a differentiator. For example, if the highest input-signal frequency is anticipated to be 500 Hz, then f_1 should be set to approximately 10 × 500 Hz, or 5 kHz, for proper operation.

As a general rule, the design of any circuit with capacitors is first accomplished by selecting a given standard value for the capacitor(s) and calculating the resulting value(s) for the resistors. This is because the range of standard capacitor values is not as extensive as it is for those resistors having 5 or 10 percent tolerances. As with many linear amplifier circuits, resistor R_B is sometimes used to reduce the effects of the operational amplifier's input bias current. Since the capacitor blocks DC current, R_B in practice is made equal to R_F. In many circuits, however, R_B can be omitted, with the noninverting input then being connected directly to ground.

For differentiators, the operational amplifier should have a high slew rate in order to respond properly to very quickly changing signals as well as a high gain-bandwidth product. The use of high-quality capacitors, such as polystyrene and Teflon, is preferred.

Example 4–1

A 1-kHz triangle wave having a peak amplitude of 1 V is the input signal to the differentiator circuit of Figure 4–6a. Determine the shape and amplitude of the output signal.

Solution: The input waveform must be analyzed in two sections, covering time periods t_1 and t_2. Since the period is 1/1 kHz, or 1 ms, and the triangle wave is symmetrical, both t_1 and t_2 are 0.5 ms.

For time period t_1, the positive-going slope is

$$\text{Slope }(t_1) = \frac{2\text{ V}}{0.5\text{ ms}} = 4000\text{ V/s}$$

In similar fashion, the slope of the negative-going portion during time period t_2 is

$$\text{Slope }(t_2) = \frac{-2\text{ V}}{0.5\text{ ms}} = -4000\text{ V/s}$$

Using Equation 4–4, the output voltage of the differentiator for time period t_1 is the slope times the differentiator gain, or

$$v_o = -(10\ \text{k}\Omega)(0.1\ \mu\text{F})(4000\ \text{V/s})$$
$$= \mathbf{-4\ V}$$

whereas for time period t_2,

$$v_o = -(10\ \text{k}\Omega)(0.1\ \mu\text{F})(-4000\ \text{V/s})$$
$$= \mathbf{4\ V}$$

Figure 4–6b shows the corresponding input and output waveforms for the differentiator circuit. Note that the positive-going slope is constant for period t_1, and the output signal is a constant negative voltage. Remember, there is a polarity reversal. For period t_2, the negative-going constant slope produces a constant positive voltage of 4 V.

Example 4–2

Determine the component values required for the differentiator circuit of Figure 4–5a to properly differentiate input signals up to 1.5 kHz. Assume that the operational amplifier's gain-bandwidth product is 1 MHz.

Solution: For proper differentiation of signals up to 1.5 kHz, f_1 should be set approximately one decade higher, or 15 kHz. Letting $\boldsymbol{C} = \mathbf{0.047\ \mu F}$ in Equation 4–7 and solving for R_S,

$$R_S = \frac{1}{(2\pi)(15\ \text{kHz})(0.047\ \mu\text{F})}$$
$$= \mathbf{226\ \Omega} \qquad \text{(use a 220-}\Omega\text{ standard value)}$$

Above 15 kHz, the closed-loop voltage gain should be limited to some convenient value—for example, 50. Therefore,

$$R_F = (50)(226\ \Omega)$$
$$= \mathbf{11.3\ k\Omega} \qquad \text{(use an 11-k}\Omega\text{ standard value)}$$

Checking to make sure that—based on the gain-bandwidth product of the operational amplifier—it is possible to have a closed-loop gain of 50 at 15 kHz:

$$f_{max} = \frac{1\ \text{MHz}}{50}$$
$$= \mathbf{20\ kHz}$$

Thus the completed circuit of Figure 4–7 behaves as an inverting amplifier for input frequencies between 15 kHz and 20 kHz. Below 15 kHz, the circuit acts as a differentiator.

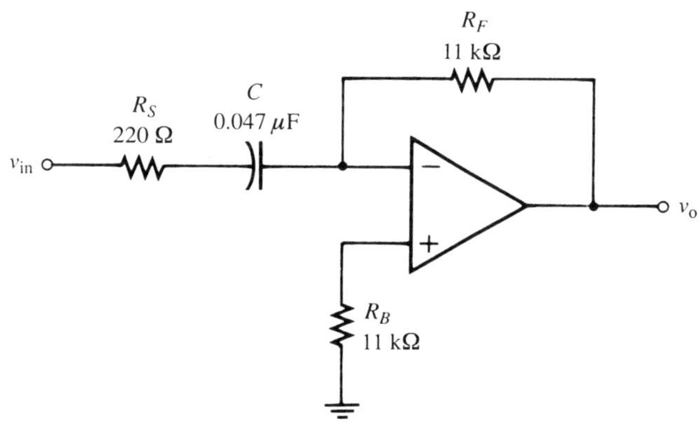

Figure 4–7
Completed differentiator circuit for Example 4–2.

Summing Differentiator

Like the standard linear amplifier circuits built around operational amplifiers, the basic differentiator circuit can be similarly expanded to allow for the differentiation of the sum of several input signals. Figure 4–8 shows the basic circuit of a two-input uncompensated summing differentiator. This is the same form as the linear summing amplifier discussed in Chapter 2, except that the input resistors are replaced by capacitors to provide the differentiating action. Taking each input one at a time, the total output voltage for N inputs can be shown to be equal to

$$v_o = -R_F C_1 \frac{\Delta v_1}{\Delta t} - R_F C_2 \frac{\Delta v_2}{\Delta t} - \cdots - R_F C_N \frac{\Delta v_N}{\Delta t} \tag{4–9}$$

Difference Differentiator

The difference amplifier discussed in Chapter 2 can be modified in a similar manner as the standard inverting amplifier—the input resistors are replaced by capaci-

Figure 4–8
Summing differentiator.

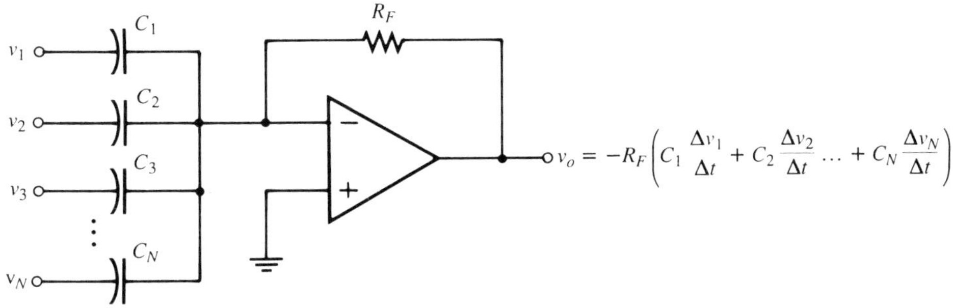

Figure 4–9
Difference differentiator.
(a) Basic circuit.
(b) Compensated circuit.

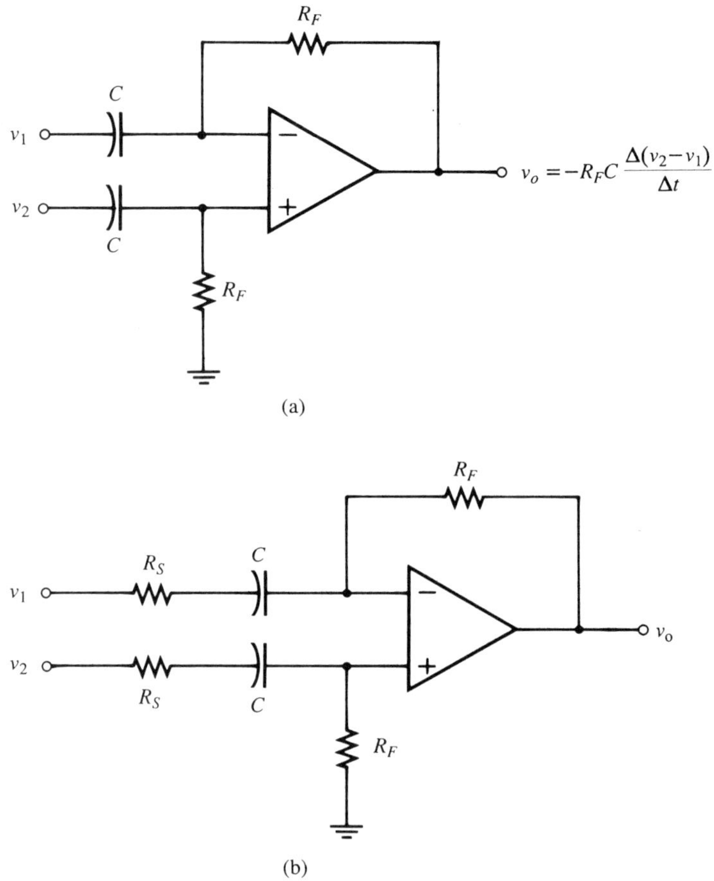

tors, as shown in Figure 4–9a. What results is an uncompensated differentiator circuit that operates on the instantaneous difference of two input signals, $v_2 - v_1$:

$$v_o = -R_F C \frac{\Delta(v_2 - v_1)}{\Delta t} \quad \textbf{(4–10)}$$

Figure 4–9b shows the high-frequency stabilized version. The closed-loop, high-frequency differentiator gain is limited to R_F/R_S. As with the difference amplifier, the components must be closely matched to minimize output errors.

4–2 Integrators

The mathematical inverse of differentiation is *integration,* and the circuit that performs this is the **integrator.** Integration is the mathematical process of determining the cumulative area underneath the curve or upper boundary of a wave-

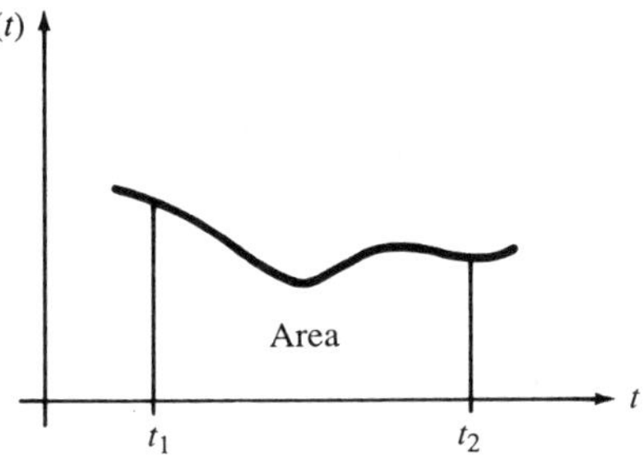

Figure 4–10
Graphical representation of integration as the process of determining the accumulative area underneath the curve over a given time period.

form over a given period of time (Figure 4–10)*. The electronic integrator that is built around an operational amplifier produces an output voltage signal proportional to the area underneath the boundary formed by the input voltage signal.

As an example of what integration means, let's consider the DC voltage shown in Figure 4–11a. Its value remains at a constant +2.5-V level indefinitely. Since integration is the computation of the area underneath the curve, the time axis is first divided into convenient intervals, as shown in Figure 4–11b. However, these intervals do not necessarily have to be equal. In this case we have chosen equally spaced intervals of 1 s. Notice now that each 1-s segment resembles a rectangle with a height of 2.5 V and a base of 1 s. When time equals zero, there is no area. The area (A_1) of the first rectangle is 2.5 V × 1 s, or 2.5 V-s (volt-seconds). As shown in Figure 4–11c, the area of the DC signal is graphed as a function of time. After 1 s, the area is 2.5 V-s.

Now compute the area of the second segment, which is also 2.5 V-s. However, since integration is a *cumulative total,* the total area after 2 s is 2.5 + 2.5 V-s, or 5 V-s. This process then can be repeated for as long as there is an input signal. As shown in Figure 4–11d, integration of a DC (constant) voltage results in a straight line having a constant slope. Its value at any point in time will then depend on the cumulative area underneath the curve of the input signal.

As mentioned previously, integration of a signal represents the area under the curve as measured from a given starting time up to a time for which the integral is computed. For many periodic signals, the integral is usually measured over either a half- or a full cycle. As another example, the integral over the positive half-cycle of the 1-kHz square wave of Figure 4–12a is equal to the area under that part of the curve (A_1)—the product of its height times its width. For a peak-positive amplitude of 10 V, the integral over the 0.5-ms positive half-cycle is 10 V × 0.5 ms, or 0.005 V-s (Figure 4–12b). The integral over the 0.5-ms negative half-cycle is −10 V × 0.5 ms, or −0.005 V-s. At the end of each cycle, the integral is the sum of the two half-cycle integrals, or zero. The zero sum is obtained because the two integrals cover equal areas having opposite polarities.

---

*In the notation of integral calculus, the integral, or accumulated area of a function of time $f(t)$ over the time period from t_1 to t_2, is written as

$$\text{Accumulated area} = \int_{t_1}^{t_2} f(t)\, dt$$

Figure 4–11
Example of integrating a +2.5-V DC voltage. (a) Input signal versus time. (b) Time axis divided into equally spaced 1-s time intervals. (c) Graph of integrator output after 1 s. (d) Graph of output after 3 s.

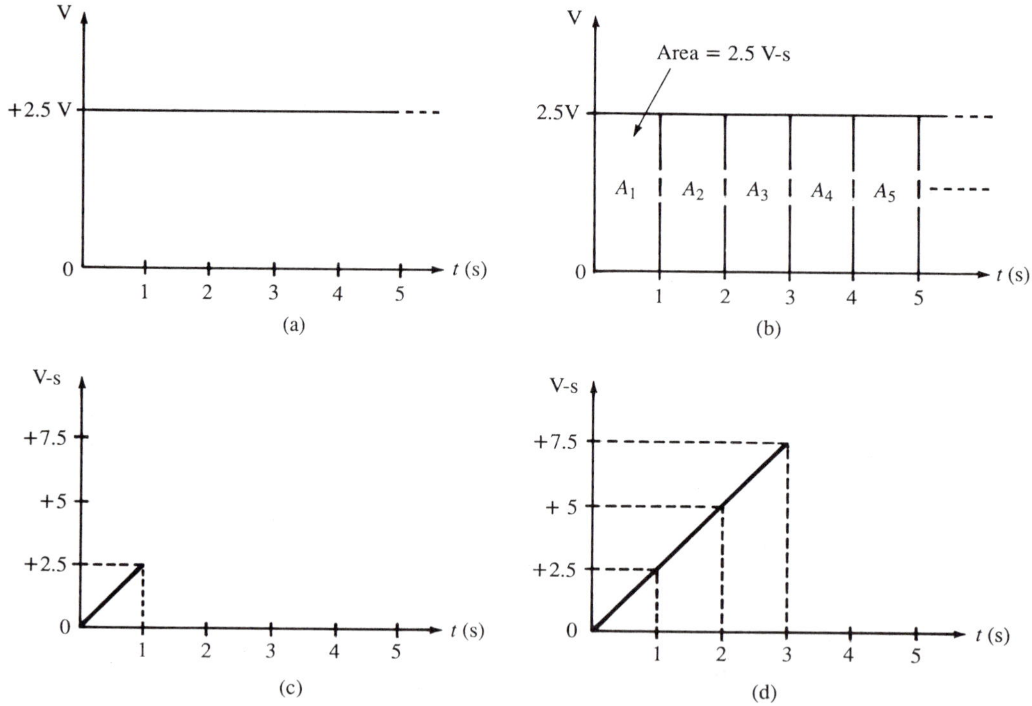

These examples serve to illustrate how the integral of signals having simple geometric shapes composed of straight-line segments such as a square wave are computed. To calculate the integral of more complex wave shapes such as sine waves and rectified sine waves requires calculus. Nevertheless, waveforms representing the integral of common AC signals are discussed in Section 4–3. Since integration is the opposite process of differentiation, if a given signal is differentiated and then integrated, we obtain the original signal.

Owing to this inverse relationship, we have the basic operational amplifier integrator of Figure 4–13a by simply interchanging the positions of the resistor and capacitor of the basic differentiator circuit (Figure 4–2a). Here the resistor is now the input element, and the capacitor becomes the feedback element.

The analysis of the basic integrator is performed in a similar manner as for the differentiator. The input current flowing through the input resistor is

$$i_{in} = \frac{v_{in}}{R} \qquad \textbf{(4–11)}$$

Figure 4–12
Integration of a square wave. (a) Input waveform. (b) Output for positive areas, negative areas, and composite output.

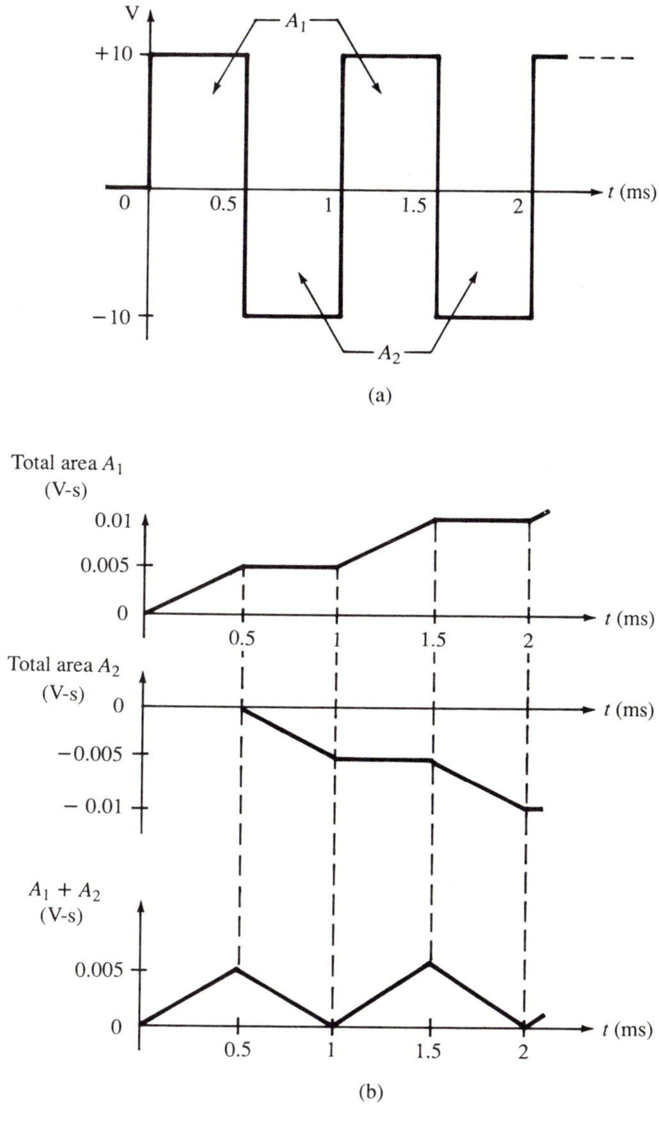

whereas the feedback current flowing in the capacitor is given by

$$i_F = -C_F \frac{\Delta v_o}{\Delta t} \qquad \textbf{(4–12)}$$

Since the feedback current (i_F) flowing through C_F is the same as the input current (i_{in}) through R, setting Equations 4–11 and 4–12 equal to each other and solving for Δv_o gives

Figure 4–13
Basic integrator. (a) Operational amplifier active circuit. (b) Equivalent passive RC low-pass filter circuit.

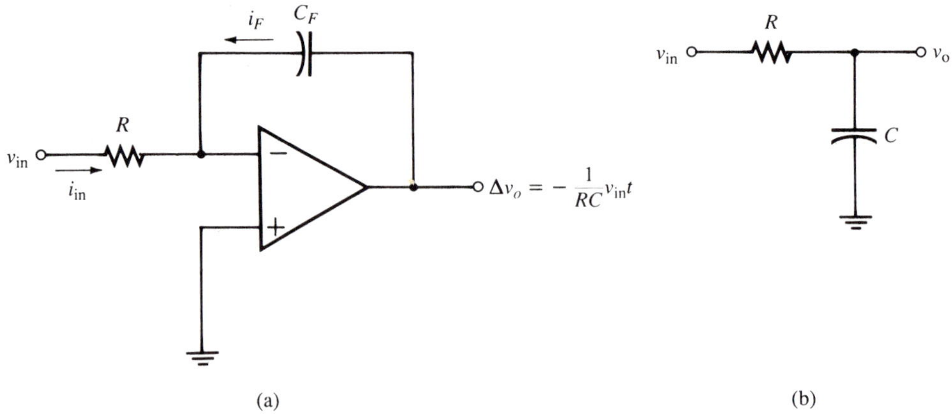

$$\Delta v_o = -\frac{1}{RC_F} v_{in}\,\Delta t \qquad \textbf{(4–13)}$$

The $1/RC_F$ term in Equation 4–13 is a scaling factor, commonly called the *integrator gain*. The minus sign tells us that the polarity of the output voltage representing the integral, or accumulated area underneath the curve, of the input signal is of opposite polarity from the normal sense. The operational amplifier integrator is identical in function to the passive RC circuit of Figure 4–13b provided that the RC time constant of the passive circuit is much longer than the time period of the input signal. This circuit is often used as a simple passive low-pass filter to reject input signals having frequencies greater than $1/2\pi RC_F$.

As with the differentiator, this basic integrator circuit presents its own problems under certain conditions. First, any of the nonideal characteristics of the

Figure 4–14
Operational amplifier integrator frequency response.

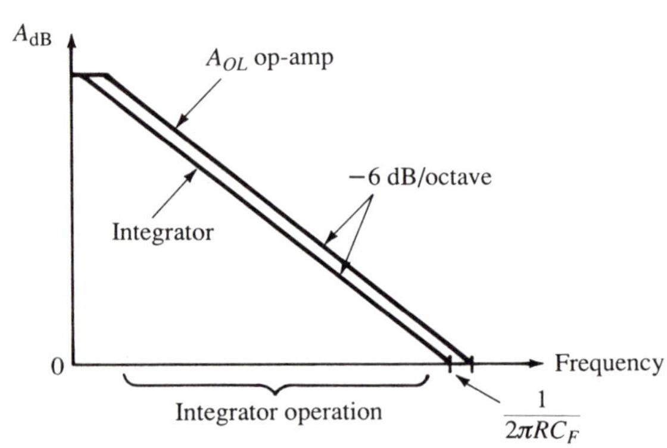

operational amplifier that produce an output offset voltage with no applied input signal will also be integrated over the same period of time. If left unchecked, however, the output voltage will either slowly rise from zero toward the positive supply voltage or decrease toward the negative supply voltage. In either case, this action eventually saturates the operational amplifier. Secondly, the reactance of the feedback capacitor varies with frequency. At very low frequencies, the closed-loop gain becomes very large and approaches the operational amplifier's open-loop gain. With increasing frequency, the capacitor's reactance decreases, so that the closed-loop gain drops off by −6 dB/octave (Figure 4–14). Like that of the inverting amplifier, the magnitude of the closed-loop gain is the ratio of the feedback to input element impedances, so that

$$A_{CL} = \frac{X_C}{R} = \frac{1}{2\pi f R C_F} \qquad (4\text{–}14)$$

A stabilized, compensated integrator circuit that overcomes—or at least minimizes—these problems is shown in Figure 4–15a. To limit the DC closed-loop gain to a reasonable value (typically 10 to 100), resistor R_F is placed in parallel with the integrating capacitor. At DC, the equivalent circuit shown in Figure 4–15b looks like an inverting amplifier with a closed-loop voltage gain of $-R_F/R$. In addition to limiting the closed-loop gain, R_F also reduces the output offset voltage due to the operational amplifier's input offset voltage. However, a price is paid for these improvements. Resistor R_F now limits the use of the integrator to those input frequencies above

$$f_1 = \frac{1}{2\pi R_F C_F} \qquad (4\text{–}15)$$

Practically, f_1 should be set approximately one decade below the lowest anticipated input frequency. For input frequencies below f_1, the circuit looks like an inverting amplifier with a closed-loop gain equal to $-R_F/R$ (Figure 4–15b). For input frequencies greater than f_1, the compensated circuit works properly as an integrator if the RC time constant is made smaller than the period of the input signal. Furthermore, the input impedance of the integrator is set by the value of R.

To keep the effects of the input bias current from eventually saturating the operational amplifier, resistor R_B is connected between the noninverting input and ground. The value for R_B is equal to the parallel combination of R_F and R. For best overall performance, the feedback capacitor should be a high-quality type having very low leakage, such as polystyrene, Teflon, or mica. For relatively short integration times, Mylar capacitors are satisfactory.

Figure 4–15
Compensated integrator. (a) Basic circuit. (b) Equivalent low-frequency circuit. (c) Frequency response graph showing low-frequency limiting.

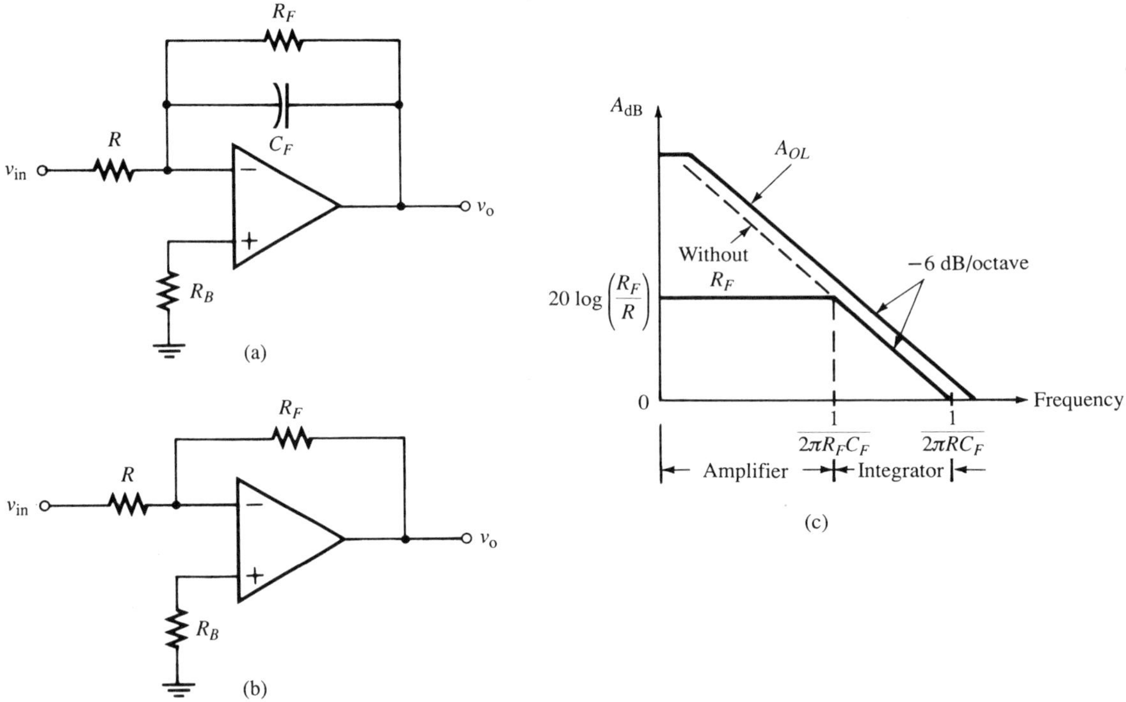

The choice of the operational amplifier will affect the accuracy of operation. Chopper-stabilized operational amplifiers, which serve to reduce DC drift, are preferred for long-term integrators. FET-input devices are used for medium-term periods because of their low-input bias current. The standard bipolar-type device is used for very short integration times, such as those in certain active filters and most audio applications.

Figure 4–16
Completed integrator circuit for Example 4–3.

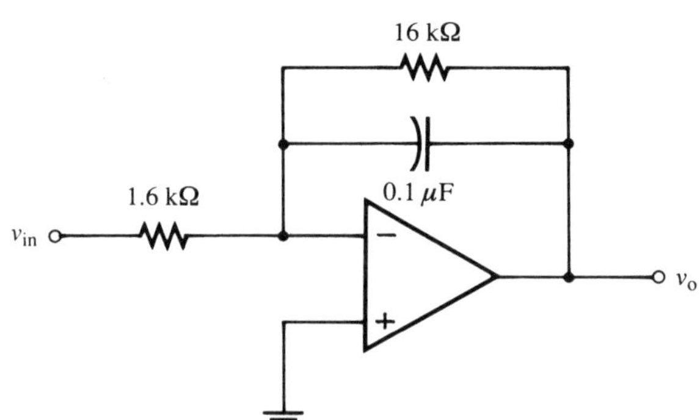

Example 4–3
Using the compensated integrator circuit of Figure 4–15a, determine the component values necessary to integrate fully any waveform greater than 1 kHz while keeping the DC closed-loop gain limited to 10 for frequencies below 100 Hz.

Solution: For a minimum input signal frequency of 1 kHz, f_1 should be set approximately one decade lower, or 100 Hz. Assuming $C = \mathbf{0.1\ \mu F}$ and using Equation 4–15, solving for R_F gives

$$R_F = \frac{1}{(2\pi)(100\text{ Hz})(0.1\ \mu\text{F})}$$

$$= \mathbf{15.9\ k\Omega} \qquad \text{(use a 16-k}\Omega\text{ standard value)}$$

For a DC closed-loop gain of 10,

$$R = \frac{15.9\text{ k}\Omega}{10}$$

$$= \mathbf{1.59\ k\Omega} \qquad \text{(use a 1.6-k}\Omega\text{ standard value)}$$

The final circuit is shown in Figure 4–16.

4–3 Differentiation and Integration of Some Common Waveforms

Figure 4–17 shows the effects of integration and differentiation on some common waveforms. Differentiating a sine wave using calculus results in another sinusoi-

Figure 4–17
Differentiation and integration of common waveforms.

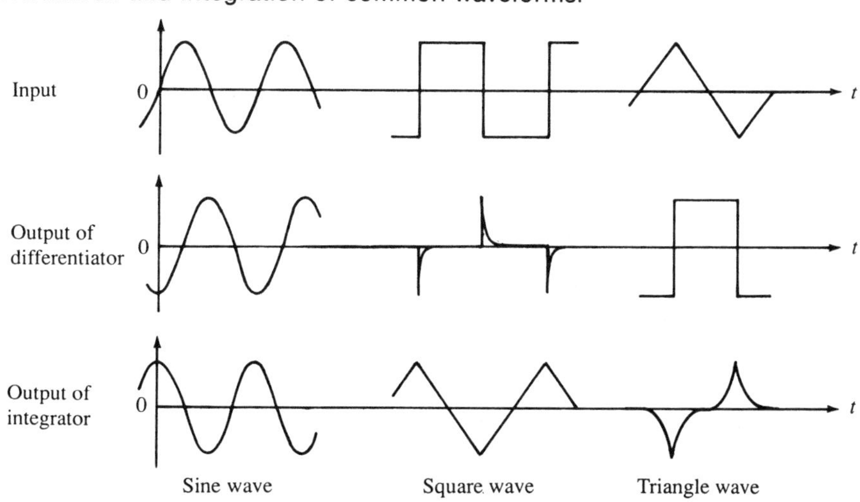

dal waveform advanced in time by 90°, or a cosine wave. Integration of a sine wave produces another sinusoidal waveform retarded in time by 90°, again a cosine. Since the operational amplifier causes a polarity inversion, both waveforms then show an additional 180° phase shift.

Summary

Besides the common linear amplifier circuits, the integrator and differentiator allow for other types of signal processing. The differentiator is a circuit whose output is proportional to the instantaneous slope of the line of its input signal. With it, one is able to change triangular waves into square waves. On the other hand, an integrator is a circuit whose output is proportional to the instantaneous accumulated area underneath the curve of its input signal. With it, one is able to convert square waves into triangle waveforms. In addition, it also serves as a fundamental building block for the formation of certain types of active filters and waveform generators.

Glossary

differentiator A circuit whose output is proportional to the instantaneous slope of the line of its input signal.

integrator A circuit whose output is proportional to the instantaneous accumulated area underneath the curve of its input signal.

Review Questions

1. What is a differentiator? How does its output voltage compare with its input voltage?
2. Draw the schematic diagram of a basic differentiator. How does it compare with the circuit of an inverting amplifier?
3. What can be done to the basic differentiator circuit to compensate it against high-frequency noise?
4. Explain how to determine at what frequencies a compensated differentiator functions as an inverting amplifier and at what frequencies it functions properly as a differentiator.
5. What is an integrator? How does its output voltage compare with its input voltage?
6. Draw the schematic diagram of a basic integrator. How does it compare with the circuit of an inverting amplifier?
7. What can be done to the basic integrator circuit to compensate for DC drift and low-frequency noise?
8. Explain how to determine at what frequencies a stabilized, compensated integrator circuit functions as an inverting amplifier and at what frequencies it functions properly as an integrator.

Exercises

1. For the basic differentiator circuit of Figure 4–2a, determine the required value for R if $C = 0.022\ \mu\text{F}$ and $v_o = -0.01\ \Delta v_{in}/\Delta t$.
2. For the compensated differentiator of Figure 4–5a with $C = 0.01\ \mu\text{F}$, $R_B = 4.7\ \text{k}\Omega$, $R_F = 5.6\ \text{k}\Omega$, and $R_S = 1\ \text{k}\Omega$, determine the fre-

quency below which it will act as a differentiator. Assume that the operational amplifier's gain-bandwidth product is 10 MHz.

3. For the compensated differentiator of Exercise 2, what is the high-frequency gain at which it stops acting as a differentiator?

4. For the compensated differentiator of Figure 4–5a with $C = 0.01\ \mu F$, determine suitable values for R_B, R_F, and R_S so that it will properly differentiate all signals below 20 kHz while limiting its high-frequency gain to 10. Assume that the operational amplifier's gain-bandwidth product is 10 MHz.

5. The compensated differentiator circuit of Figure 4–5a is for an input signal frequency of 1 kHz. Determine the required values suitable for R_B, R_F, and R_S if $C = 0.1\ \mu F$ and the high-frequency gain is limited to 10. Assume that the operational amplifier's gain-bandwidth product is 10 MHz.

6. A 500-Hz triangle waveform having a peak-to-peak amplitude of 0.15 V is to be differentiated using the circuit of Figure 4–5a with $C = 0.047\ \mu F$, $R_B = 10\ k\Omega$, $R_F = 10\ k\Omega$, and $R_S = 1\ k\Omega$.

Determine each of the following.

(a) The shape and amplitude of the output signal

(b) The frequency above which the circuit stops acting as a differentiator

Assume that the operational amplifier's gain-bandwidth product is 10 MHz.

7. For the basic integrator circuit of Figure 4–13a, determine the required value for R if $C_F = 0.01\ \mu F$ and $\Delta v_o = -1000\ v_{in}\Delta t$.

8. For the compensated integrator of Figure 4–15a with $C_F = 0.01\ \mu F$, $R_B = 820\ \Omega$, $R_F = 5.6\ k\Omega$, and $R = 1\ k\Omega$, determine the frequency above which it will act as an integrator. What is the low-frequency gain below this frequency?

9. For the compensated integrator of Figure 4–15a with $C_F = 0.047\ \mu F$, determine the values for R_B, R_F, and R so that it will properly integrate all signals above 2 kHz while limiting its low-frequency gain to 10.

5 Single-Supply Biasing and Norton Amplifiers

Instructional Objectives

Upon completing this chapter, you will be able to

- Design and predict the operation of a bipolar operational amplifier using a single-supply voltage in the following AC-coupled circuits:
  - Voltage follower
  - Inverting and noninverting amplifiers
  - Summing amplifier
  - Difference amplifier
- Design and predict the operation of a Norton amplifier, such as the LM3900, using a single-supply voltage in the following AC-coupled circuits:
  - Voltage follower
  - Inverting and noninverting amplifiers
  - Inverting and noninverting summing amplifiers
  - Difference amplifier

Introduction

Most operational amplifiers are designed to be used with a bipolar, or split, power supply such that the magnitude of both the positive and negative supply voltages with respect to ground are equal. Ignoring any small output offset voltage caused by the nonideal characteristics of the operational amplifier, the DC (and AC) output voltage is zero when the input signal is zero. The DC reference, or baseline, value of the resulting output signal is also zero and the maximum output voltage possible is slightly less than the positive or negative supply voltage. As an example, if the positive supply pin of an operational amplifier is connected to +12 V and the negative supply pin is connected to −12 V, there is a 24-V potential difference between the two supply pins. Taking the operational amplifier's saturation voltage into account, the output could then swing from almost −12 V to almost +12 V.

The cost of implementing a given operational amplifier circuit using two power supplies is therefore higher than one that could be built using a single-ended power supply. For battery-powered portable circuits, the extra battery required for a conventional split-supply operation not only adds to the cost of the circuit, it also adds some weight. This chapter discusses how many of the bipolar linear operational amplifier circuits discussed in earlier chapters can be externally biased to function using a single-supply voltage. Because these circuits use capacitor coupling to remove unwanted DC levels, single-supply biased circuits are designed only for AC operation. In addition, a special type of operational amplifier, the Norton operational amplifier, whose design lends itself specifically to single-supply operations, is discussed.

5–1 Single-Supply Biasing

The operational amplifier, using a single-supply voltage, must be able to produce both negative-going and positive-going signals. Similar to a class-A transistor amplifier, the best approach is then to set the operational amplifier's quiescent DC output voltage equal to one-half the positive supply voltage, or $\frac{1}{2}V_{CC}$.

When an input signal such as a sine wave is applied to a single-supply biased circuit, the output voltage will then vary about the DC, or quiescent, voltage level of $\frac{1}{2} \times V_{CC}$ when no signal is applied (Figure 5–1). The resultant sine-wave signal is now a superposition of both the quiescent DC component and the amplified AC signal. For proper amplifier operation, this DC component should then be removed from the output signal before the desired AC output is coupled to a load.

Voltage Follower

The basic bias, or DC, circuit for the voltage follower powered by a single-supply voltage is shown in Figure 5–2a. Resistors R_1 and R_2, whose values are made equal, form a voltage divider for which the DC voltage across R_2 is one-half the

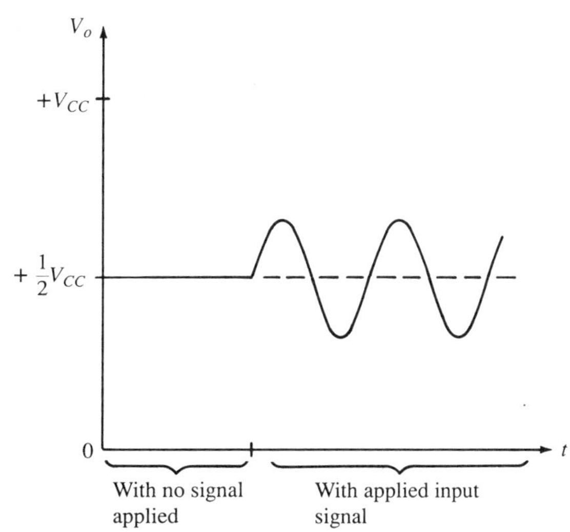

Figure 5–1
The quiescent DC output is biased to one-half the supply voltage similar to a class-A amplifier.

Figure 5–2
Single-supply voltage follower. (a) DC bias circuit. (b) Complete circuit.

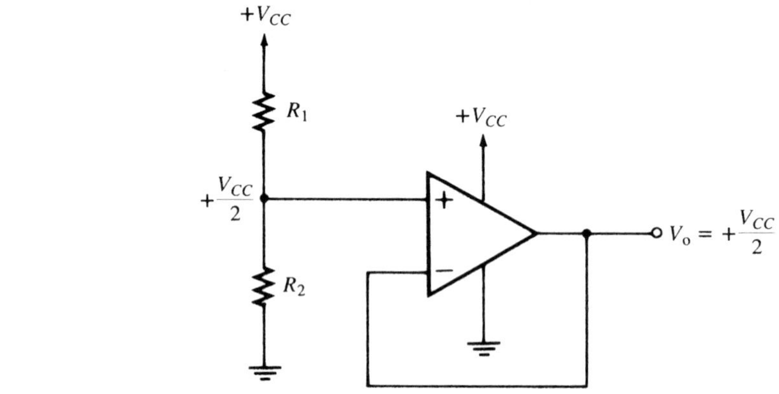

(a)

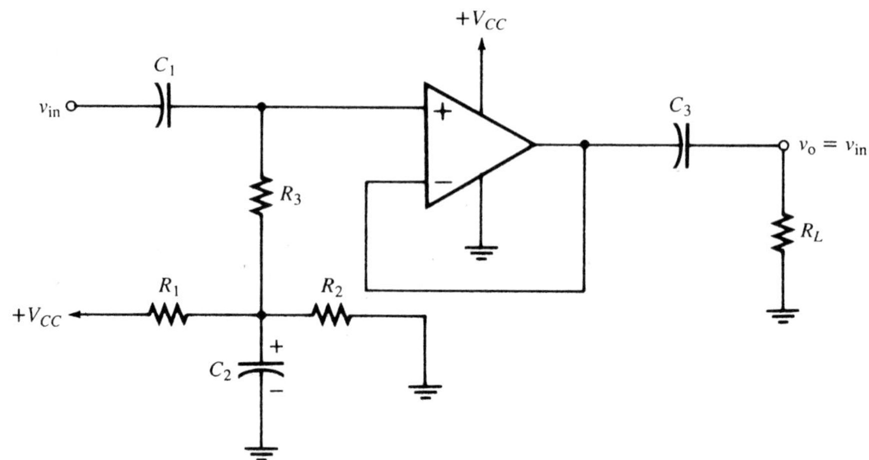

(b)

supply voltage V_{CC}. Note now that the operational amplifier's negative supply pin is grounded, and, contrary to the convention for split supplies, the power-supply connections for single-supply biasing are not omitted from schematic diagrams.

Since the closed-loop voltage gain for the voltage follower is 1 and the DC input voltage at the operational amplifier's noninverting input is one-half the supply voltage, the DC output voltage also equals one-half the supply voltage.

Figure 5–2b shows the completed circuit for the single-supply biased AC voltage follower. As in the basic bias circuit, resistors R_1 and R_2 form the voltage divider that puts the operational amplifier's noninverting input at one-half the supply voltage. In general, R_1 and R_2 can be any value, as long as they are equal. In practice, however, they are conveniently chosen to be in the 10-kΩ to 30-kΩ range.

Provided R_3 is made very large compared with R_1 or R_2, resistor R_3 has almost no DC voltage across it, since only a very small DC bias current flows through R_3. Capacitor C_2 is a bypass capacitor to ensure that the junction of R_1, R_2, and R_3 is always at AC ground. In this case, the reactance of C_2 must be very small compared with the parallel combination of R_1 and R_2. This capacitor is also referred to as a *decoupling capacitor* as it acts as a low-pass filter to prevent low-frequency interfering signals (such as 60 Hz) from reaching the operational amplifier's input via the power supply. On the other hand, any AC noise signal on the supply voltage reaching the operational amplifier's positive supply terminal would be effectively eliminated by the rejection ratio of the operational amplifier's supply voltage. For the 741 operational amplifier, this is typically a 31,600 : 1 (90-dB) reduction.

For AC signals, C_2 is in parallel with R_1 and R_2, so the cutoff frequency above which this RC network sets the junction of R_1, R_2, R_3, and C_2 at AC ground is

$$f_2 = \frac{1}{2\pi(R_1 \| R_2)C_2} \qquad \textbf{(5–1)}$$

For frequencies where C_2's reactance is less than the parallel combination of R_1 and R_2, the equivalent AC circuit is shown in Figure 5–3. Resistor R_3 is not really necessary as it sets the input impedance of the circuit. If it is omitted—along with C_2—the voltage follower's input impedance is essentially equal to R_1 in parallel with R_2.

For any AC input signal to reach the operational amplifier's input, C_1 must provide an AC short circuit. Any AC input signal will then be superimposed on the DC bias voltage, producing an identical output signal that is in phase with the AC input signal. For AC signals resistors, R_1 and R_2 of the original amplifier circuit of Figure 5–2b are shorted to ground by C_2, so the input circuit as seen by the AC input signal is the series capacitor C_1 and the shunt resistor R_3 to ground. The

Figure 5–3
Voltage follower equivalent AC circuit.

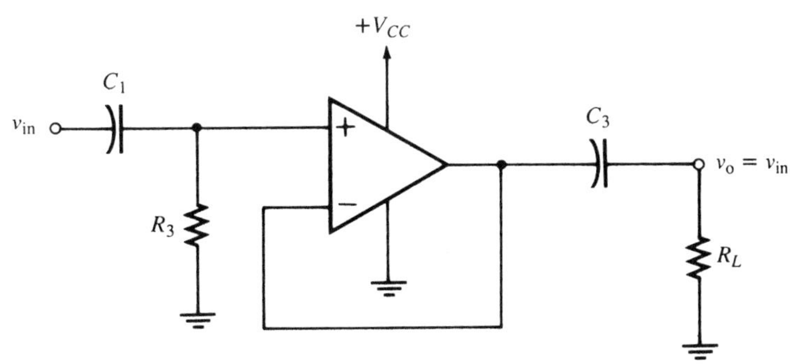

input impedance of the voltage follower is then effectively determined by R_3. The cutoff frequency (f_1) for the input RC network is found from

$$f_1 = \frac{1}{2\pi R_3 C_1} \tag{5–2}$$

Finally, C_3 is used to remove the quiescent DC component from the output signal as well as to produce an AC signal that is now referenced to ground. The cutoff frequency (f_3) of the output RC network is determined by C_3 and either the load or input impedance of the next stage (R_L):

$$f_3 = \frac{1}{2\pi R_L C_3} \tag{5–3}$$

As a general rule for simplicity, all three cutoff frequencies are made approximately equal and are set to approximately one-tenth the lowest input frequency to be amplified.

Although operational amplifiers are capable of amplifying both DC and AC signals, single-supply biasing now requires that coupling capacitors C_1 and C_3 effectively block the passage of DC levels. The technique for building other AC amplifier circuits, both inverting and noninverting types, that use a single-ended

Figure 5–4
Summary of the single-supply biased voltage follower.

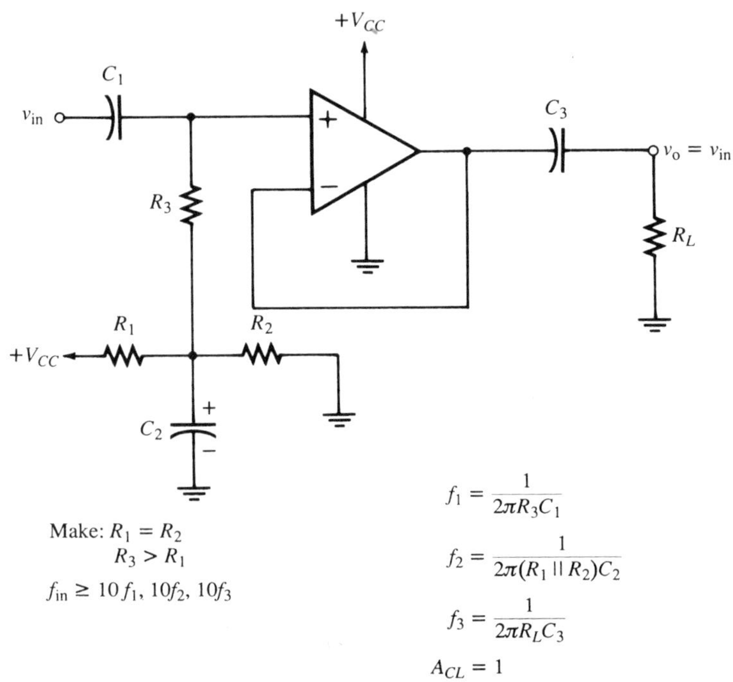

power supply is very similar. Figure 5–4 summarizes the characteristics of the single-supply biased voltage follower.

Example 5–1

Using the voltage follower circuit of Figure 5–5, determine each of the following.

(a) The quiescent DC output voltage
(b) The voltage follower's input impedance
(c) The low-frequency response

Solution:

(a) Based on the two 10-kΩ resistors and the +15-V supply, the quiescent DC voltage at the operational amplifier's noninverting input is one-half the supply voltage, or +7.5 V. Since the voltage follower's closed-loop voltage gain is 1, the DC quiescent output voltage is also **+7.5 V.**

(b) The input impedance of the voltage follower circuit is equal to the **180-kΩ** resistor.

(c) Using Equations 5–1, 5–2, and 5–3, the frequencies above which each capacitor essentially looks like an AC short circuit are as follows.

For C_1:

$$f_1 = \frac{1}{(2\pi)(180 \text{ k}\Omega)(0.22\ \mu\text{F})} = \mathbf{4\ Hz}$$

For C_2:

$$f_2 = \frac{1}{(2\pi)(10 \text{ k}\Omega \| 10 \text{ k}\Omega)(5\ \mu\text{F})} = \mathbf{6.4\ Hz}$$

For C_3:

$$f_3 = \frac{1}{(2\pi)(1.5 \text{ k}\Omega)(25\ \mu\text{F})} = \mathbf{4.2\ Hz}$$

Of these three cutoff frequencies, the one associated with the voltage divider network (6.4 Hz) is the dominant one, so that the circuit operates properly as a single-supply biased voltage follower for AC input signals above 64 Hz.

Noninverting Amplifier

As shown in Figure 5–6, the technique for biasing a noninverting amplifier with a single supply is identical to that used for the voltage follower. Resistors R_1 and R_2 form the required voltage divider, which puts the noninverting input at a DC level

Figure 5–5
Single-supply biased voltage follower circuit for Example 5–1.

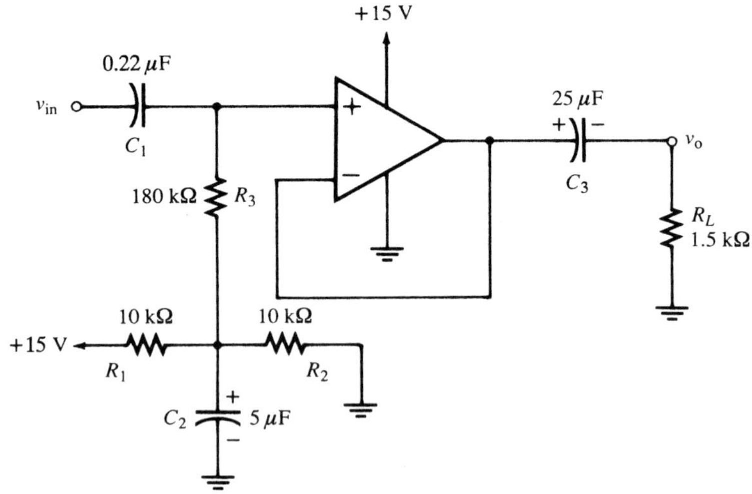

equal to one-half the power supply voltage, whereas capacitor C_2 provides power-supply noise decoupling. The input impedance is equal to R_3 if both R_1 and R_2 are made very small compared to R_3. The cutoff frequency for the input RC network is

$$f_1 = \frac{1}{2\pi R_3 C_1} \qquad (5\text{–}4)$$

Figure 5–6
Single-supply biased noninverting amplifier.

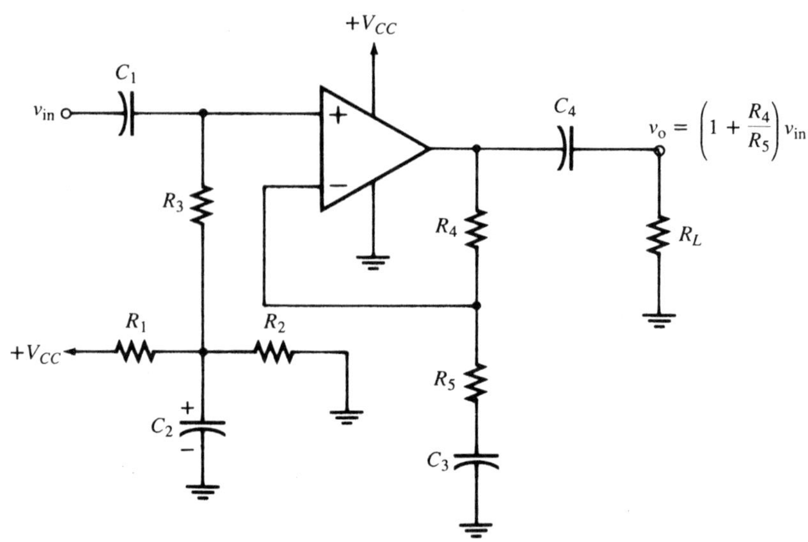

Figure 5–7
Single-supply biased noninverting amplifier. (a) DC equivalent circuit. (b) High-frequency equivalent circuit.

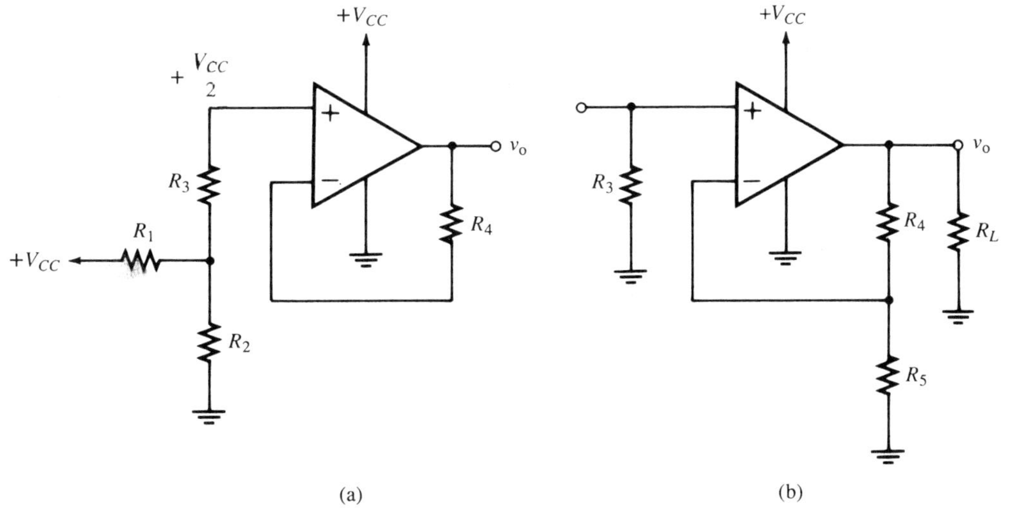

Capacitor C_3 serves to block DC, so that the DC equivalent circuit (Figure 5–7a) looks essentially like a voltage follower. Thus, the DC quiescent output voltage is one-half the supply voltage, as is the DC voltage at the amplifier's noninverting input. For AC input signals, however, capacitor C_3 effectively grounds the lower end of R_5 so that the AC closed-loop voltage gain is the same as for noninverting amplifiers powered by a bipolar supply,

$$A_{CL} = 1 + \frac{R_4}{R_5} \qquad \textbf{(5–5)}$$

and the output signal will be in phase with the input. Therefore, the DC closed-loop gain ($A_{CL} = 1$) is different than for AC signals (Equation 5–5). The reactance of C_3 then must be small compared to the resistance of R_5 at the lowest frequency (f_2) to be amplified with a closed-loop voltage gain of $1 + R_1/R_2$, so that

$$f_2 = \frac{1}{2\pi R_5 C_3} \qquad \textbf{(5–6)}$$

As a general rule, f_2 is made about five to ten times larger than f_1.

Using Equation 5–3, the output coupling capacitor (C_3) is selected in a similar manner as that of the voltage follower circuit so that it effectively blocks the quiescent DC output voltage from being passed onto the load. The cutoff frequency for the output RC network is made the same as the input RC network

Figure 5–8
Summary of the single-supply biased noninverting amplifier.

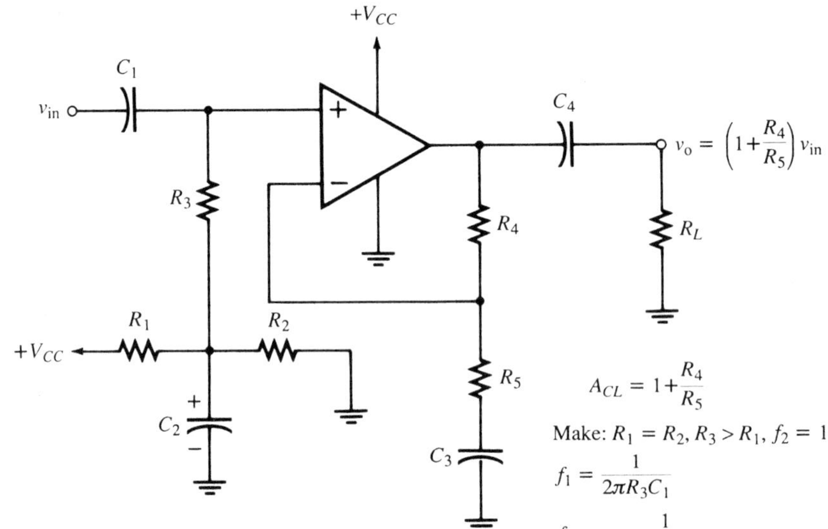

$$A_{CL} = 1+\frac{R_4}{R_5}$$

Make: $R_1 = R_2, R_3 > R_1, f_2 = 10f_1, f_3 = f_1$

$$f_1 = \frac{1}{2\pi R_3 C_1}$$

$$f_2 = \frac{1}{2\pi (R_1 \parallel R_2) C_2}$$

$$f_3 = \frac{1}{2\pi R_5 C_3}$$

$$f_4 = \frac{1}{2\pi R_L C_4}$$

(Equation 5–4). Figure 5–8 summarizes the characteristics of the single-supply biased noninverting amplifier.

Inverting Amplifier

Figure 5–9a shows the circuit of an AC inverting amplifier that uses a single-ended power supply. Resistors R_3 and R_4, when made equal to each other, act as the required voltage divider to hold the noninverting input at one-half the supply voltage. As with the single-supply noninverting amplifier and voltage follower circuits, the voltage dividing resistors are generally 10 kΩ to 30 kΩ, and the decoupling capacitor C_2 merely acts as a filter to remove any high-frequency noise from reaching the operational amplifier's input via the power supply. In many situations, however, C_2 could be eliminated.

Input coupling capacitor C_1 is used to block any DC voltage from reaching R_2. For DC signals, the equivalent circuit of Figure 5–9b acts as a voltage follower, and the quiescent DC output signal is the same as the DC voltage on the noninverting input (i.e., $V_{CC}/2$). In practice, the reactance of C_1 is made very small compared to the value of R_2, so that the cutoff frequency for the input RC network is

$$f_1 = \frac{1}{2\pi R_2 C_1} \qquad (5\text{–}7)$$

For AC input signals, the equivalent circuit of Figure 5–9c shows a simple inverting amplifier having a closed-loop voltage gain the same as for inverting amplifiers powered by a bipolar supply:

$$A_{CL} = \frac{R_1}{R_2} \qquad (5\text{–}8)$$

Figure 5–9
Single-supply biased inverting amplifier. (a) Basic circuit. (b) DC equivalent circuit. (c) High-frequency equivalent circuit.

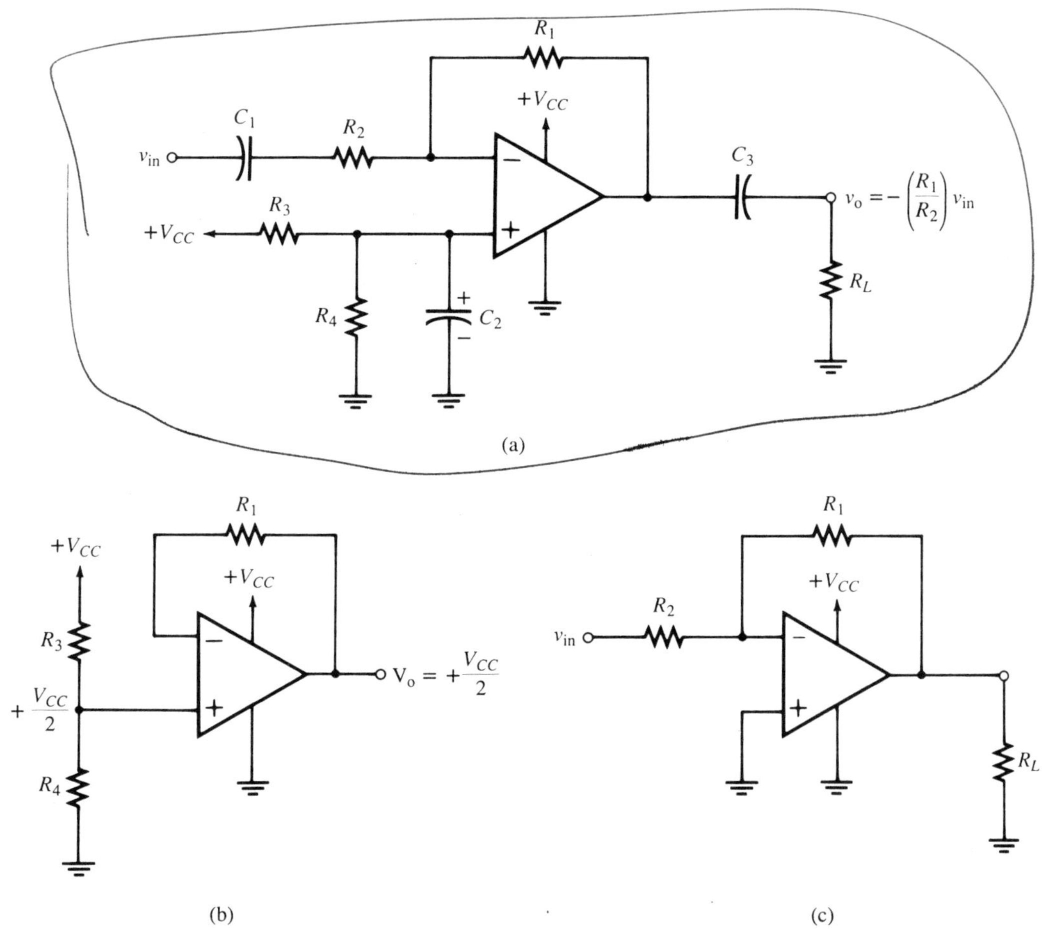

Thus the output signal will be 180° out of phase with the input. In addition, the AC input impedance is effectively equal to R_2. The output coupling capacitor, C_3, is selected using Equation 5–3 so that it effectively blocks the quiescent DC output voltage from being passed onto the load. In general, the cutoff frequencies of the input and output RC networks are made the same.

Inverting Summing Amplifier

As discussed in Chapters 2 and 3, the inverting summing amplifier, or adder, is a variation of the basic inverting amplifier. As illustrated by the three-input circuit of Figure 5–10, each input is AC-coupled, the output is AC-coupled to the load, and the amplifier's noninverting input is held at one-half the supply voltage. As a general rule, the cutoff frequencies of the input, bypass, and output RC networks have approximately the same cutoff frequency.

Difference Amplifier

The difference amplifier powered by a single positive supply voltage is shown in Figure 5–11. When R_3 equals R_4, the noninverting input is one-half the supply voltage. As with the conventional split-supply difference amplifier discussed in Chapter 2, the design is greatly simplified by making $R_1 = R_2$, with both equal to the desired input impedance of the amplifier. In addition, the AC equivalent circuit places R_3 in parallel with R_4. If the closed-loop differential gain is

$$A_d = \frac{R_5}{R_1} \qquad \textbf{(5–9a)}$$

Figure 5–10
Single-supply biased summing amplifier.

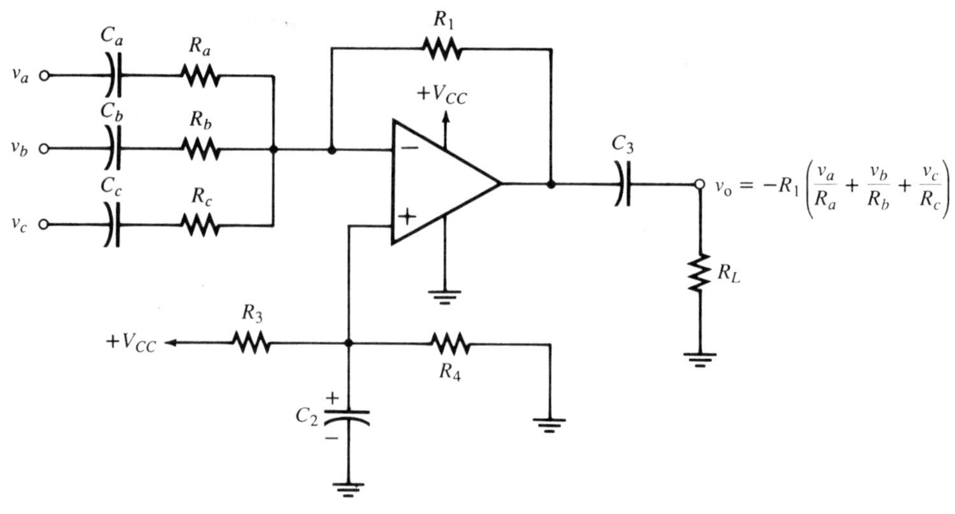

Figure 5–11
Single-supply biased difference amplifier.

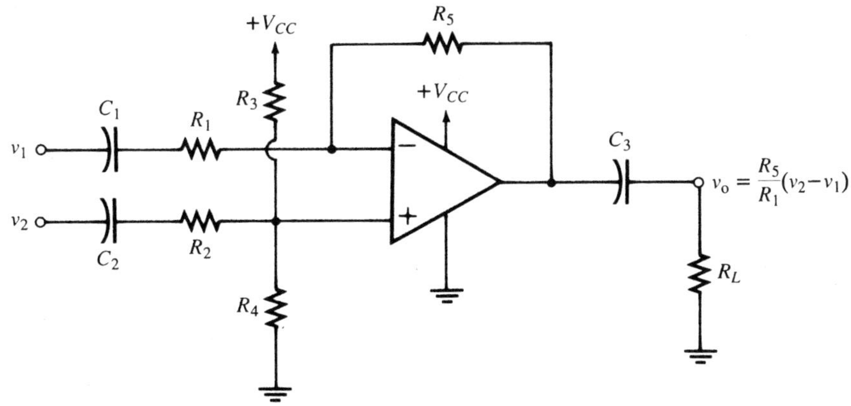

then the following must hold:

$$A_d = \frac{R_3}{2R_2} \tag{5–9b}$$

In turn, Equations 5–9a and 5–9b require that

$$R_3 = 2R_5 \tag{5–9c}$$

With these simplifications, the AC output voltage of the difference amplifier is given by

$$v_o = \frac{R_5}{R_1}(v_2 - v_1) \tag{5–10}$$

The cutoff frequencies of the two input RC networks are given as

$$f_1 = \frac{1}{2\pi R_2 C_1} \tag{5–11}$$

and

$$f_1 = \frac{1}{2\pi(R_2 + R_3 \| R_4)C_2} \tag{5–12}$$

The output coupling capacitor (C_3) is selected using Equation 5–3, so that it effectively blocks the quiescent DC output voltage from being passed onto the load. In practice, all three cutoff frequencies are made equal in order to provide the best common-mode rejection.

Example 5–2

For the difference amplifier of Figure 5–11 driving a 1-kΩ load, determine the component values required for an input impedance of 10 kΩ and a differential gain of 10 for frequencies above 75 Hz.

Solution:

(a) Since the input impedance is specified at 10 kΩ, then $R_1 = R_2 =$ **10 kΩ.**

(b) Resistors R_5 and R_3 are then determined from Equations 5–9a and 5–9b, so that:

$$R_5 = (10)(10\text{ k}\Omega) = \mathbf{100\text{ k}\Omega}$$

and

$$R_3 = (2)(10)(10\text{ k}\Omega) = \mathbf{200\text{ k}\Omega}$$

This also requires that

$$R_4 = \mathbf{200\text{ k}\Omega}$$

(c) Since the closed-loop gain is to be constant above 75 Hz, then the cutoff frequencies of all three RC networks are made equal for the best common-mode rejection, equal to approximately one-tenth of 75 Hz, or 7.5 Hz. From Equations 5–11, 5–12, and 5–3,

$$C_1 = \frac{1}{(2\pi)(10\text{ k}\Omega)(7.5\text{ Hz})} = \mathbf{2.12\ \mu F} \quad \text{(use a 2.2-}\mu\text{F standard value)}$$

$$C_2 = \frac{1}{(2\pi)(10\text{ k}\Omega + 200\text{ k}\Omega \| 200\text{ k}\Omega)(7.5\text{ Hz})} = \mathbf{0.19\ \mu F} \quad \text{(use a 0.22-}\mu\text{F standard value)}$$

and

$$C_3 = \frac{1}{(2\pi)(1\text{ k}\Omega)(7.5\text{ Hz})} = \mathbf{21.2\ \mu F} \quad \text{(use a 25-}\mu\text{F standard value)}$$

The completed circuit is shown in Figure 5–12.

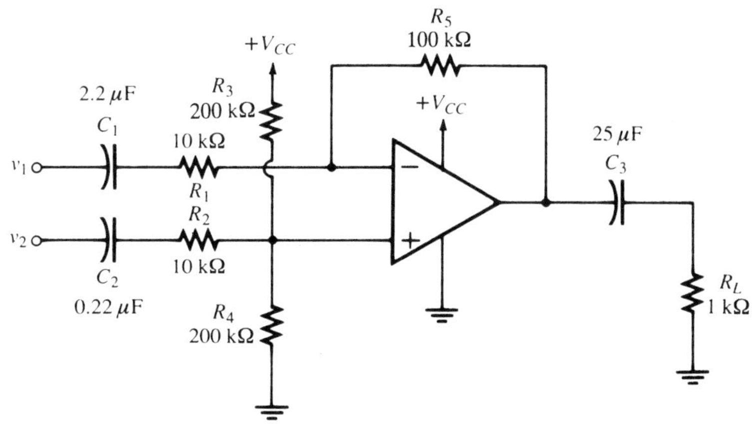

Figure 5–12
Completed single-supply biased difference amplifier circuit for Example 5–2.

5–2 Norton Amplifiers

A type of operational amplifier that is designed specifically for single-supply operation is the **Norton amplifier**, or *current-differencing amplifier (CDA)*. This type of device is very different from the conventional operational amplifier, as it was developed to provide a low-cost amplifier of modest performance that would work well from a single-ended power supply over a wide range of voltages with little change in its open-loop characteristics. As a trade-off, some of the parameters are somewhat inferior to many general-purpose operational amplifiers such as the 741. Table 5–1 compares the LM3900 and MC3401 Norton amplifiers with that of the 741 operational amplifier. For many applications, however, the Norton amplifier more than compensates by requiring only a single power supply, costing less, and performing well at low-supply voltages.

Unlike a conventional operational amplifier, which amplifies the difference in voltage between its inputs, the Norton amplifier produces an output voltage that is proportional to the difference in the currents flowing into its input terminals. To

Table 5–1
Comparison of typical parameters of the LM3900 and MC3401 Norton amplifiers with the 741 operational amplifier

Parameter	741	LM3900	MC3401
Supply voltage range	±15 V	+4 to +36 V	+5 to +18 V
Open-loop gain	200,000	2,800	2,000
Unity-gain frequency	1.0 MHz	2.0 MHz	5 MHz
Input bias current	80 nA	30 nA	50 nA
Input impedance	2 MΩ	1 MΩ	1 MΩ
Output impedance	75 Ω	8 kΩ	8 kΩ
Slew rate	0.5 V/μs	0.5 V/μs	0.6 V/μs
PSRR	90 dB	70 dB	55 dB

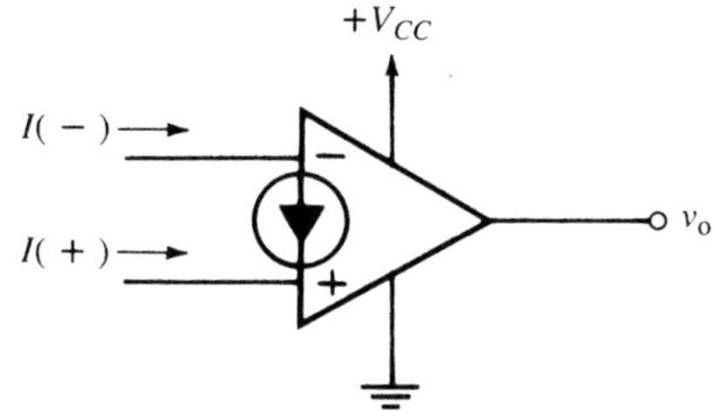

Figure 5–13
Norton amplifier schematic symbol.

distinguish itself from the conventional operational amplifier, the schematic symbol for the Norton amplifier includes a current source symbol, as shown in Figure 5–13.

In an ideal Norton operational amplifier, the input terminals would be considered to be at zero volts—a true current sink. Actually, the input voltage on both input terminals is about one diode drop (0.7 V) above ground potential. Just as it might be considered that a conventional operational amplifier has almost zero input current, it might also be considered that a Norton amplifier has almost zero input voltage. A conventional operational amplifier is voltage-driven, whereas a Norton operational amplifier is current-driven.

The Norton operational amplifier is driven by the difference in currents flowing into its inputs. With the addition of negative feedback, the output voltage will swing in such a direction as to minimize the difference between these two input currents, which must flow through the feedback resistor. This is somewhat analogous to a conventional operational amplifier with negative feedback, where the output swings to minimize the *voltage difference* between the inputs.

When analyzing the operation of a Norton amplifier with negative feedback, it is helpful to remember that the two input currents must be approximately equal, just as the two input voltages of a conventional operational amplifier must be approximately equal when negative feedback is applied. Since the two inputs terminals appear as diodes going to ground, they must never be connected directly to a voltage source without a current-limiting resistor.

Figure 5–14 shows the internal circuit of an LM3900 Norton amplifier. Diode D_1 and transistor Q_1 form a current mirror such that Q_1 extracts a current from the inverting input that is equal to the current flowing in the noninverting input. The input current to the base of transistor Q_2 is then approximately equal to the difference between the current flowing into the inverting input and the current flowing into the noninverting input—hence the term *current-differencing*. When $I(-)$ is larger than $I(+)$, the output voltage swings in the negative direction. When $I(+)$ is larger than $I(-)$, the output swings positive.

Biasing the Norton Amplifier

Since the Norton amplifier is normally used to amplify AC signals, its output is usually biased to a DC level equal to one-half the power supply voltage. As with the bipolar operational amplifier, this class-A quiescent output level permits an AC output signal to swing equally in either direction before clipping.

Figure 5–14
Equivalent internal circuit of an LM3900 Norton amplifier.

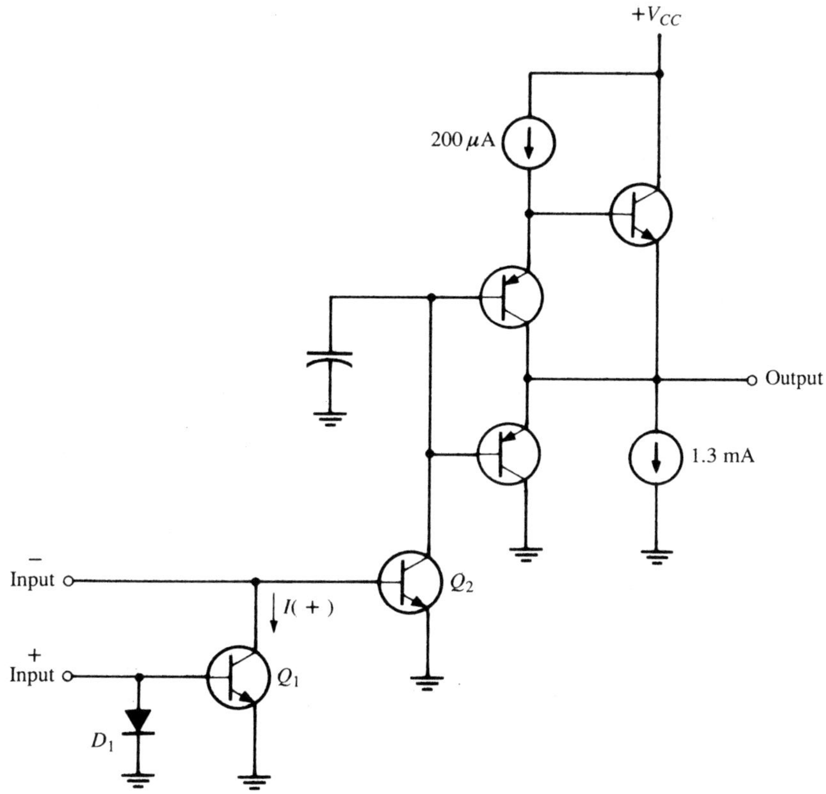

Figure 5–15 shows the most common basic biasing arrangement for the Norton amplifier, known as *current-mirror biasing*. Neglecting the amplifier's internal diode drop (D_1 of Figure 5–14), the current flowing into the noninverting input is equal to

$$I(+) = \frac{V_{CC}}{R_1} \qquad \textbf{(5–13)}$$

If the base-emitter diode drop of Q_2 (Figure 5–14) is also neglected, the current flowing into the amplifier's inverting input is

$$I(-) = \frac{V_o}{R_2} \qquad \textbf{(5–14)}$$

Figure 5–15
Current mirror biasing.

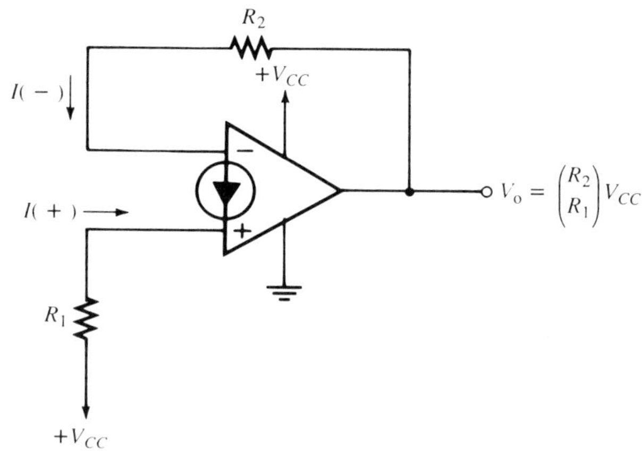

where V_o is the DC quiescent output voltage. With this arrangement, the currents flowing into both inputs are equal. Setting Equations 5–13 and 5–14 equal to each other gives

$$\frac{V_{CC}}{R_1} = \frac{V_o}{R_2} \tag{5–15}$$

If $V_o = V_{CC}/2$, then $R_2 = R_1/2$.

Example 5–3
The Norton amplifier of Figure 5–15 is biased with $R_1 = 470\ \text{k}\Omega$ and $R_2 = 220\ \text{k}\Omega$. Determine the DC quiescent output voltage for a supply voltage of 12 V.

Solution: From Equation 5–15, after rearranging,

$$\begin{aligned} V_o &= \frac{220\ \text{k}\Omega}{470\ \text{k}\Omega} \times (12\ \text{V}) \\ &= \mathbf{5.62\ V} \end{aligned}$$

Biasing with Low-Supply Voltages

If the supply voltage is less than 7 V, the forward bias voltage drops of internal diode junctions of the Norton amplifier cannot be neglected. Taking the diode's forward voltage drop (typically 0.7 V) into account, Equations 5–13 and 5–14, respectively, are rewritten as

$$I(+) = \frac{V_{CC} - 0.7}{R_1} \tag{5–16}$$

and

$$I(-) = \frac{V_o - 0.7}{R_2} \tag{5–17}$$

Taking the two input currents to be equal as before,

$$R_1 = \frac{V_{CC} - 0.7}{V_o - 0.7} R_2 \tag{5–18}$$

Since Norton amplifiers are designed to operate from a single-ended power supply, the following subsections illustrate those amplifier circuits that are AC-coupled.

Norton Inverting Amplifier

Figure 5–16 shows a Norton inverting amplifier using current-mirror biasing so that the quiescent DC output is one-half the supply voltage. The AC input signal (v_{in}) will cause an AC current to flow through input resistor R_3 just as in a conventional inverting amplifier. The amplifier's output will then swing (1) to prevent any difference in input currents and (2) just enough so that any difference in the

Figure 5–16
Norton inverting amplifier.

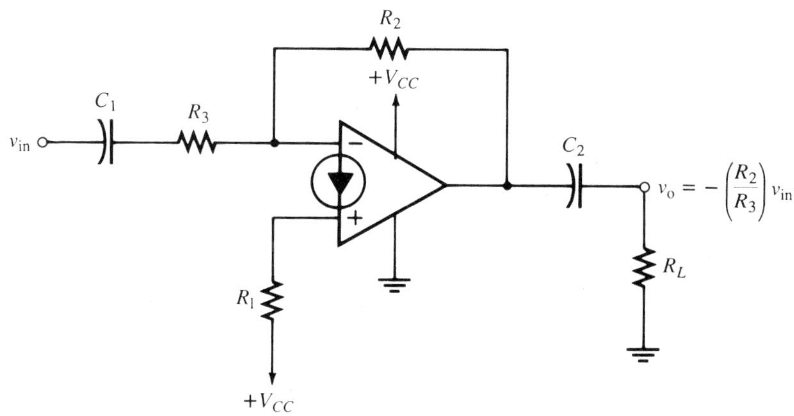

currents flowing through R_2 and R_3 will pass through R_1. For example, if the input signal swings positive, the output signal will swing negative enough to cause any increase in the current through R_3 to flow directly through R_2 onto the output, and vice versa. This action keeps the Norton amplifier's two input currents equal during input-signal excursions.

Because the inverting input is essentially at ground, the AC input current $i(-)$ flowing through R_3 is approximately equal to v_{in}/R_3. Since the current through the feedback resistor R_2 is the same as through R_3 due to the virtual ground, the AC output voltage is approximately $i(-) \times R_2$. The Norton inverting amplifier's closed-loop voltage gain (v_o/v_{in}) is then written as

$$A_{CL} = -\frac{R_2}{R_3} \qquad \textbf{(5–19)}$$

For the Norton circuit, the closed-loop AC voltage gain is simply the ratio of two resistors (R_2/R_3), whereas the output signal is 180° output of phase with the input.

The cutoff frequency of the input RC network,

$$f_1 = \frac{1}{2\pi R_3 C_1} \qquad \textbf{(5–20)}$$

and that of the output RC network,

$$f_2 = \frac{1}{2\pi R_L C_2} \qquad \textbf{(5–21)}$$

Figure 5–17
Summary of design equations for the Norton inverting amplifier.

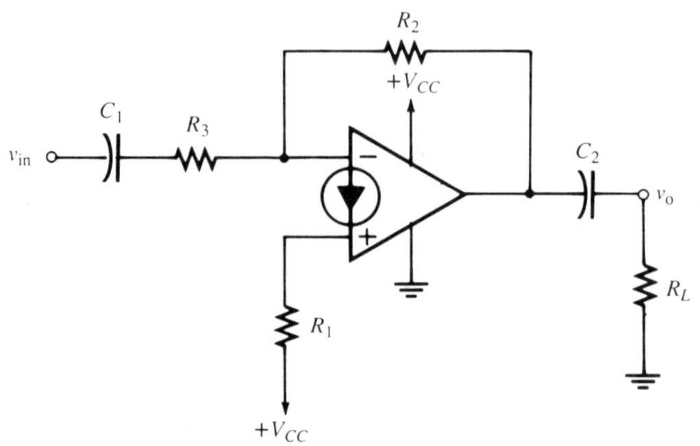

Make: $R_1 = 2R_2$, $f_{in} \geq 10f_1, 10f_2$

$Z_{in} = R_3$

$A_{CL} = -\frac{R_2}{R_3}$

$f_1 = \frac{1}{2\pi R_3 C_1}$

$f_2 = \frac{1}{2\pi R_L C_2}$

are made equal and are set approximately one-tenth times the lowest frequency of the signal to be amplified. The input impedance of this stage is approximately equal to the value of R_3, whereas the output impedance is very low, like the conventional inverting operational amplifier stage. Figure 5–17 summarizes the characteristics of the Norton inverting amplifier.

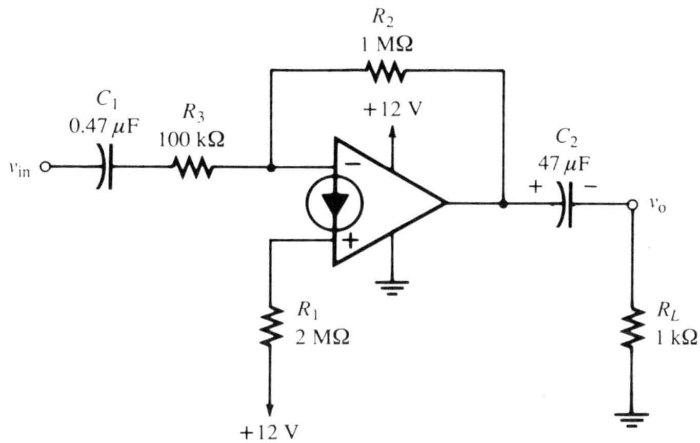

Figure 5–18
Norton amplifier circuit for Example 5–4.

Example 5–4

For the Norton inverting amplifier circuit of Figure 5–18, determine each of the following.

(a) The quiescent DC output voltage
(b) The closed-loop AC voltage gain
(c) The low-frequency response
(d) The input impedance of the stage

Solution:

(a) From Equation 5–15, the quiescent DC output voltage is

$$V_o = \frac{1\ \text{M}\Omega}{2\ \text{M}\Omega} \times (12\ \text{V})$$
$$= \mathbf{6\ V}$$

(b) From Equation 5–19, the closed-loop AC voltage gain is

$$A_{CL} = -\frac{1\ \text{M}\Omega}{100\ \text{k}\Omega}$$
$$= \mathbf{-10}$$

(c) From Equations 5–20 and 5–21, the cutoff frequencies for the input and output RC networks, respectively, are

$$f_1 = \frac{1}{(2\pi)(100\ \text{k}\Omega)(0.47\ \mu\text{F})} = \mathbf{3.4\ Hz}$$

and

$$f_2 = \frac{1}{(2\pi)(1\ \text{k}\Omega)(47\ \mu\text{F})} = \mathbf{3.4\ Hz}$$

so that the circuit will have a closed-loop voltage gain of 10 for input frequencies above approximately 10 × 3.4 Hz, or **34 Hz**.

(d) The input impedance is approximately equal to R_3, or **100 kΩ**.

Norton Noninverting Amplifier

By moving R_3 and C_1 from the inverting input to the noninverting input, as shown in Figure 5–19, we have a noninverting amplifier. For all practical purposes, the closed-loop AC voltage gain and input impedance are the same as for the inverting amplifier circuit of Figure 5–16. The closed-loop AC voltage gain is approximately

$$A_{CL} = \frac{R_2}{R_3} \tag{5–22}$$

whereas the AC input impedance is approximately R_3.

Notice now that unlike the conventional bipolar noninverting amplifier, the closed-loop voltage gain of the Norton noninverting amplifier can be less than unity, equal to unity, or greater than unity. Like the Norton inverting amplifier,

Figure 5–19
Norton noninverting amplifier.

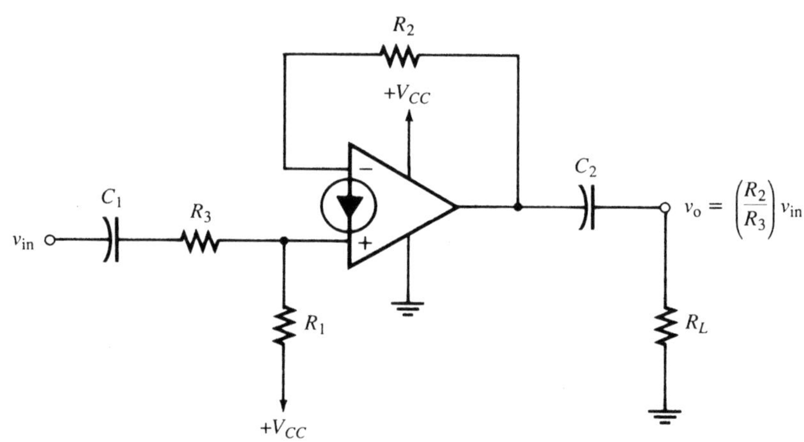

resistor R_1 still determines the DC quiescent output voltage, and its value is determined in the same manner (Equation 5–15) as well as the cutoff frequencies for the input and output RC networks (Equations 5–20 and 5–21).

Norton Voltage Follower

By making $R_2 = R_3$ in the noninverting amplifier circuit of Figure 5–19, we have a unity-gain noninverting amplifier, or voltage follower, as shown in Figure 5–20. In order to achieve a high input impedance, R_3 should then be chosen as high as possible.

Norton Summing Amplifiers

Summing amplifiers using Norton amplifiers are available in both inverting and noninverting configurations. Figure 5–21 shows a noninverting summing amplifier with three inputs. The AC output is easily obtained by considering each input as the input to a separate Norton noninverting amplifier and then combining the results. For the three-input circuit shown, the AC output voltage is

$$v_o = R_F\left(\frac{v_1}{R_1} + \frac{v_2}{R_2} + \frac{v_3}{R_3}\right) \qquad \textbf{(5–23a)}$$

If the circuit is allowed to be simplified such that all three input resistors are made equal to R_1, the output voltage is then

$$v_o = \frac{R_F}{R_1}(v_1 + v_2 + v_3) \qquad \textbf{(5–23b)}$$

Figure 5–20
Norton voltage follower.

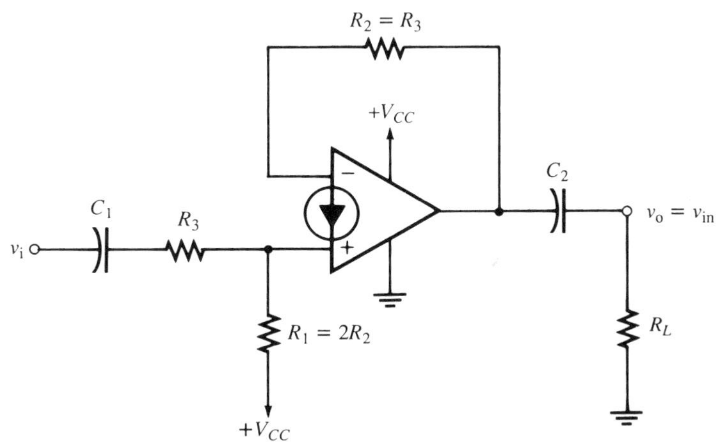

Figure 5–21
Norton noninverting summing amplifier.

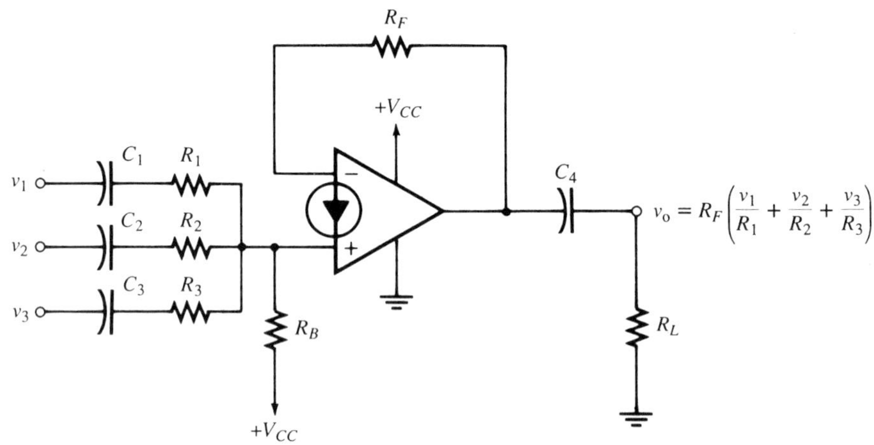

In either case, the DC bias level at the Norton amplifier's output equals

$$\frac{V_{CC}}{R_B} = \frac{V_o}{R_F} \tag{5–24}$$

Whereas the input impedance seen by each input signal source is approximately equal to the value of its corresponding series resistor:

$$Z_1(\text{in}) = R_1$$
$$Z_2(\text{in}) = R_2$$
$$Z_3(\text{in}) = R_3$$

If we now move the input circuit to the Norton amplifier's inverting input, as shown in Figure 5–22, we have the corresponding three-input Norton inverting summing amplifier. The analysis and output voltage equation are identical to that for the noninverting configuration, except that there is now a minus sign to reflect the change in output polarity:

$$v_o = -R_F\left(\frac{v_1}{R_1} + \frac{v_2}{R_2} + \frac{v_3}{R_3}\right) \tag{5–25a}$$

If the circuit is allowed to be simplified such that all three input resistors are made equal to R_1, the output voltage is then

$$v_o = -\frac{R_F}{R_1}(v_1 + v_2 + v_3) \tag{5–25b}$$

Figure 5–22
Norton inverting summing amplifier.

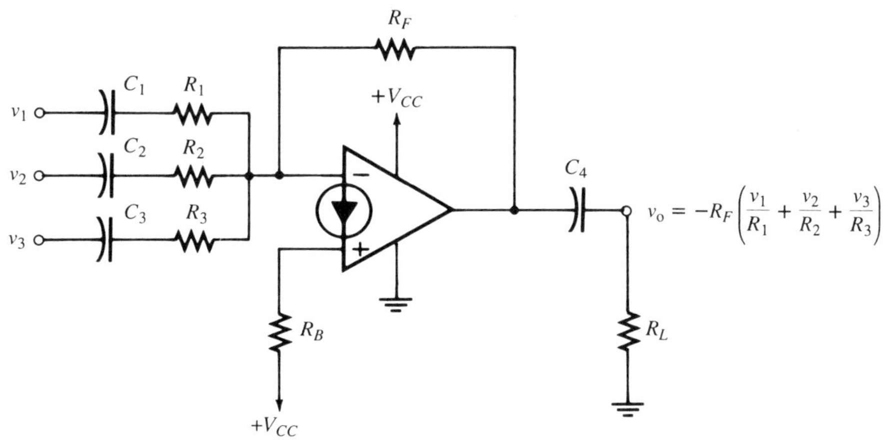

As for the noninverting summing amplifier, the DC bias level at the Norton amplifier's output is given by Equation 5–24, whereas the input impedance seen by each input signal source is approximately equal to the value of its corresponding series resistor.

Norton Difference Amplifier

The Norton difference amplifier, unlike its bipolar operational amplifier counterpart, does not require two pairs of matched resistors to set the closed-loop differential gain. Like the operational amplifier circuit, however, the Norton circuit can

Figure 5–23
Norton difference amplifier.

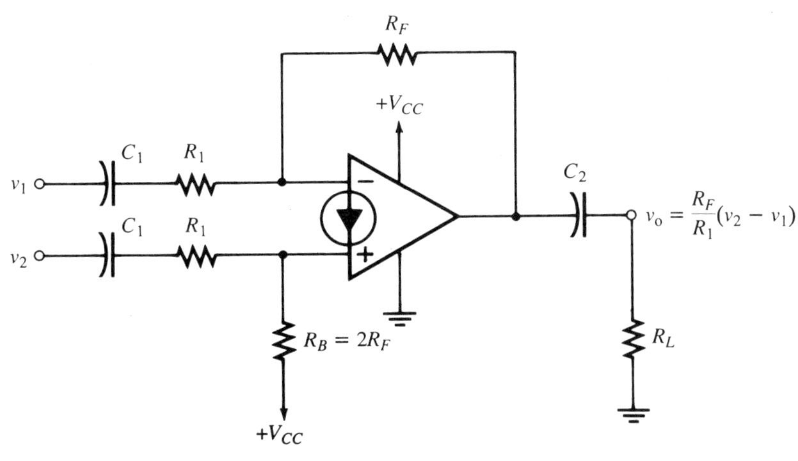

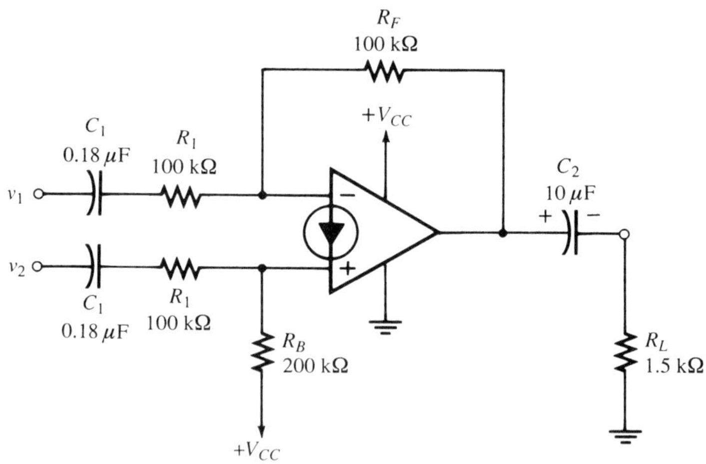

Figure 5–24
Completed Norton difference amplifier circuit for Example 5–5.

be analyzed easily by first determining the output to each input separately and then combining the two outputs to give the total output.

Figure 5–23 shows the Norton difference amplifier using current-mirror biasing. The DC quiescent output voltage when both input voltages are equal is given by Equation 5–24. The differential gain is simply the ratio R_F/R_1, with an AC output voltage equal to

$$v_o = \frac{R_F}{R_1}(v_2 - v_1) \qquad \textbf{(5–26)}$$

Example 5–5

Determine the necessary component values for the difference amplifier circuit of Figure 5–23 having a unity closed-loop voltage gain for frequencies above 100 Hz, an input impedance of 100 kΩ for both inputs, and a load impedance of 1.5 kΩ.

Solution:

(a) For an input impedance of 100 kΩ, then $R_1 = R_2 =$ **100 kΩ.**

(b) For unity differential gain, $R_F = R_1 =$ **100 kΩ,** so that bias resistor R_B = (2)(100 kΩ), or **200 kΩ.**

(c) For a low frequency of 100 Hz, the cutoff frequencies for all the RC networks should be approximately one-tenth of 100 Hz, or 10 Hz. Therefore,

$$C_1 = \frac{1}{(2\pi)(100\ \text{k}\Omega)(10\ \text{Hz})}$$
$$= \mathbf{0.16\ \mu F} \qquad \text{(use an 0.18-}\mu\text{F standard value)}$$

and

$$C_2 = \frac{1}{(2\pi)(1.5\ \text{k}\Omega)(10\ \text{Hz})}$$
$$= \mathbf{10.6\ \mu F} \qquad \text{(use a 10-}\mu\text{F standard value)}$$

The completed circuit is shown in Figure 5–24.

Besides the AC amplifiers just discussed, other circuits such as active filters and waveform generators are possible with operational amplifiers and Norton amplifiers powered from a single-ended supply.

Summary

Although operational amplifiers are traditionally designed to be used with a bipolar power supply, this chapter discusses how the basic amplifier configurations can be biased to operate from single-supply voltages. This is similar to biasing transistor amplifiers for class-A operation such that the DC output voltage is one-half the supply voltage with no input signal applied. The cost of implementing a given operational amplifier circuit using two power supplies is therefore higher than one that could be built using a single-ended power supply. Because these circuits use capacitor coupling to remove unwanted DC levels, single-supply biased circuits are designed only for AC operation.

Besides the standard bipolar operational amplifier, this chapter also discussed the Norton, or current-differencing, amplifier whose design lends itself specifically to single-supply operation with current-mirror biasing. Unlike a conventional operational amplifier, which amplifies the difference in voltage between its inputs, the Norton amplifier produces an output voltage that is proportional to the difference in the currents flowing into its input terminals. Furthermore, in order to distinguish itself from the conventional operational amplifier, the schematic symbol for the Norton amplifier includes a current source symbol.

Glossary

Norton amplifier An integrated-circuit device that amplifies the difference in the currents flowing into its input terminals. Also called a *current-differencing amplifier (CDA)*.

Review Questions

1. Explain how an operational amplifier should be biased to operate properly from a single-supply voltage.
2. Why are decoupling capacitors often used?
3. Why are capacitors used in the input and output connections to amplifier circuits biased with a single-supply voltage?
4. Draw the schematic diagram of a single-

supply biased voltage follower using a bipolar operational amplifier. Include the equations necessary for the input and output coupling capacitors.

5. Draw the schematic diagram and write the equation for the closed-loop gain of a single-supply biased inverting amplifier using a bipolar operational amplifier. Include the equations necessary for the input and output coupling capacitors.
6. Draw the schematic diagram and write the equation for the closed-loop gain of a single-supply biased noninverting amplifier using a bipolar operational amplifier. Include the equations necessary for the input and output coupling capacitors.
7. Draw the schematic diagram and write the equation for the output voltage of a single-supply biased, three-input inverting summing amplifier using a bipolar operational amplifier. Include the equations necessary for the input and output coupling capacitors.
8. Draw the schematic diagram and write the equation for the output voltage of a single-supply biased difference amplifier using a bipolar operational amplifier. Include the equations necessary for the input and output coupling capacitors.
9. Draw the schematic symbol for a current-differencing, or Norton, amplifier.
10. What is the fundamental difference between a Norton amplifier and a conventional bipolar operational amplifier?
11. What type of biasing is used in a Norton amplifier?
12. Draw the schematic diagram of a Norton-type inverting amplifier. Include the equations necessary for the input and output coupling capacitors.
12. Draw the schematic diagram and write the equation for the closed-loop gain of a Norton-type inverting amplifier. Include the equations necessary for the input and output coupling capacitors.
13. Draw the schematic diagram and write the equation for the closed-loop gain of a Norton-type noninverting amplifier. Include the equations necessary for the input and output coupling capacitors.
14. Draw the schematic diagram and write the equation for the output voltage of a Norton-type two-input noninverting summing amplifier. Include the equations necessary for the input and output coupling capacitors.
15. Draw the schematic diagram and write the equation for the output voltage of a Norton-type difference amplifier. Include the equations necessary for the input and output coupling capacitors.

Exercises

1. For the voltage follower circuit of Figure 5–25, determine:
   (a) The input impedance
   (b) The low-frequency response
2. For the voltage follower circuit of Figure 5–4, determine suitable component values required for the buffering of all signals above 40 Hz when connected to a 5-kΩ load and operated from a 9-V supply. Assume that the voltage follower is to have an input impedance of 100 kΩ.
3. For the noninverting amplifier circuit of Figure 5–26, determine:
   (a) The input impedance
   (b) The closed-loop gain
   (c) The low-frequency response
4. For the noninverting amplifier circuit of Figure 5–8, determine suitable component values required for a closed-loop gain of 19 for all signals above 50 Hz when connected to a 1-kΩ load and operated from a 9-V supply. Assume that the circuit is to have an input impedance of 100 kΩ.
5. For the inverting amplifier circuit of Figure 5–27, determine:
   (a) The input impedance

Figure 5–25

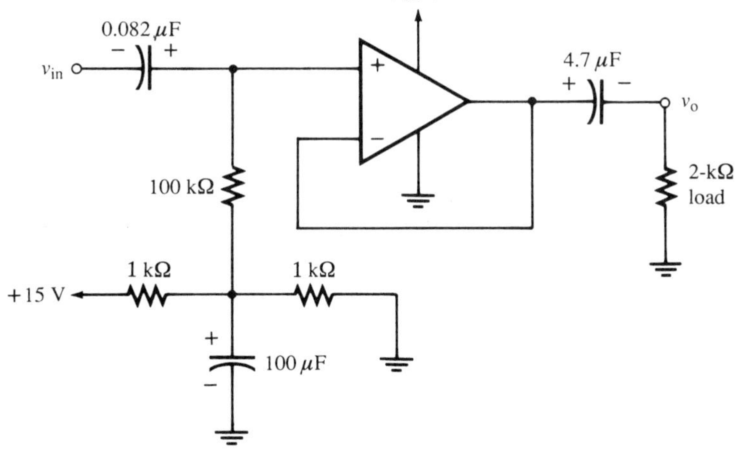

Figure 5–26

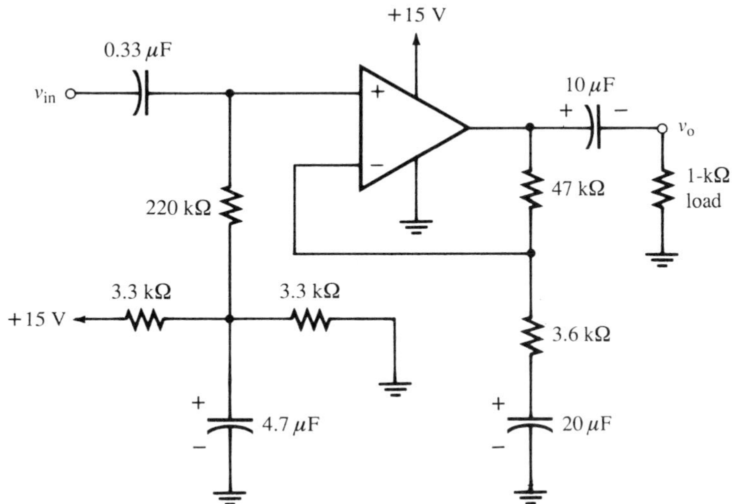

(b) The closed-loop gain
(c) The low-frequency response

6. For the inverting amplifier circuit of Figure 5–9a, determine suitable component values required for a closed-loop gain of 30 for all signals above 25 Hz when connected to a 3-kΩ load and operated from a 12-V supply. Assume that the circuit is to have an input impedance of 10 kΩ.
7. For the summing amplifier circuit of Figure 5–28, determine:
   (a) The input impedance
   (b) The closed-loop gain
   (c) The low-frequency response
8. For the three-input summing amplifier circuit of Figure 5–10 to be used as an audio mixer, determine suitable component values required for a closed-loop gain of 10 for all signals above 20 Hz when connected to a 1-kΩ load and operated from a 9-V supply. Assume that the circuit is to have an input impedance of 10 kΩ for two inputs and 2 kΩ for one input.

Figure 5–27

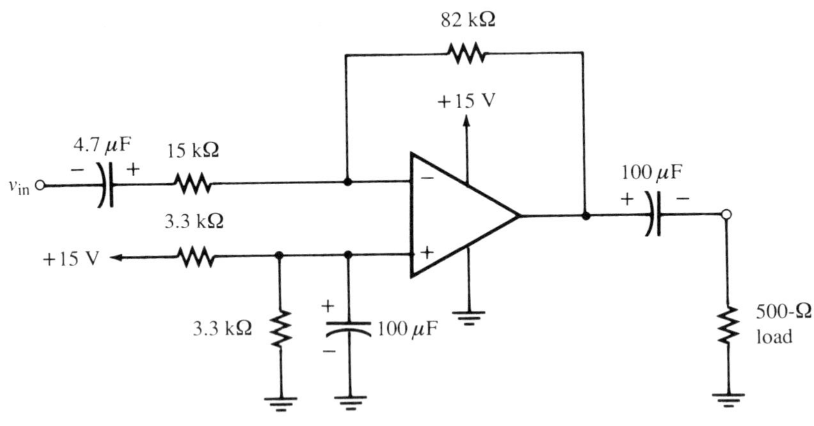

Figure 5–28

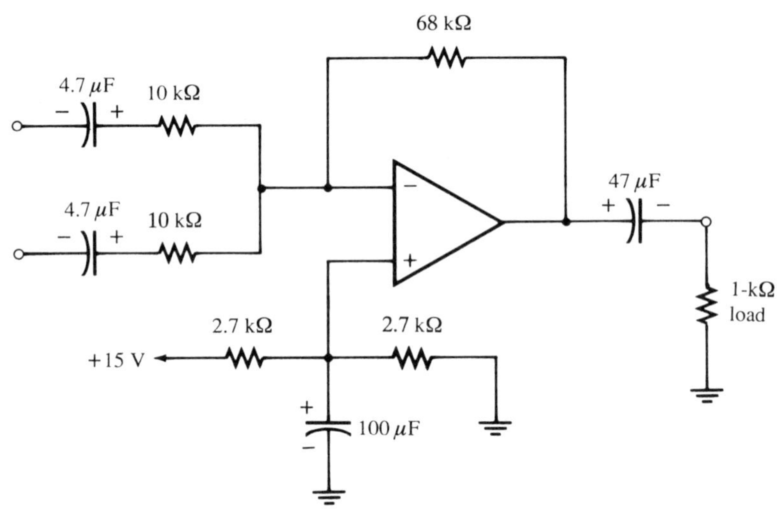

9. For the difference amplifier circuit of Figure 5–29, determine:
   (a) The input impedance
   (b) The differential gain
   (c) The low-frequency response
10. For the difference amplifier circuit of Figure 5–11, determine suitable component values required for a differential gain of 20 for all signals above 30 Hz when connected to a 1-kΩ load and operated from a 15-V supply. Assume that the circuit is to have an input impedance of 100 kΩ.
11. For the Norton inverting amplifier circuit of Figure 5–30, determine:
    (a) The quiescent DC output voltage
    (b) The input impedance
    (c) The closed-loop gain
    (d) The low-frequency response
12. For the Norton inverting amplifier circuit of Figure 5–17, determine suitable component values required for a closed-loop gain of 15 for all signals above 50 Hz when connected to a 5-kΩ load and operated from a 9-V supply.

Figure 5–29

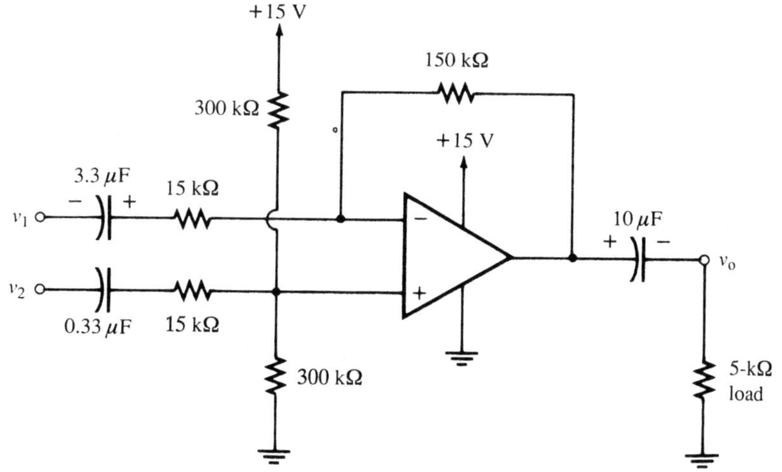

Figure 5–30

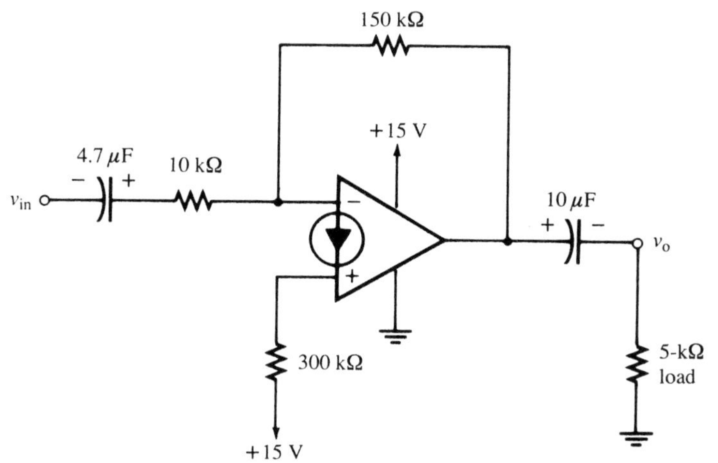

Assume that the circuit is to have an input impedance of 100 kΩ.

13. For the Norton noninverting amplifier circuit of Figure 5–31, determine:
    (a) The quiescent DC output voltage
    (b) The input impedance
    (c) The closed-loop gain
    (d) The low-frequency response

14. For the Norton noninverting amplifier circuit of Figure 5–19, determine suitable component values required for a closed-loop gain of 30 for all signals above 60 Hz when connected to a 2-kΩ load and operated from a 15-V supply. Assume that the circuit is to have an input impedance of 10 kΩ.

15. For the Norton voltage follower circuit of Figure 5–20, determine suitable component values required for the buffering of all signals above 40 Hz when connected to a 1-kΩ load and operated from a 9-V supply. Assume that the voltage follower is to have an input impedance of 100 kΩ.

Figure 5–31

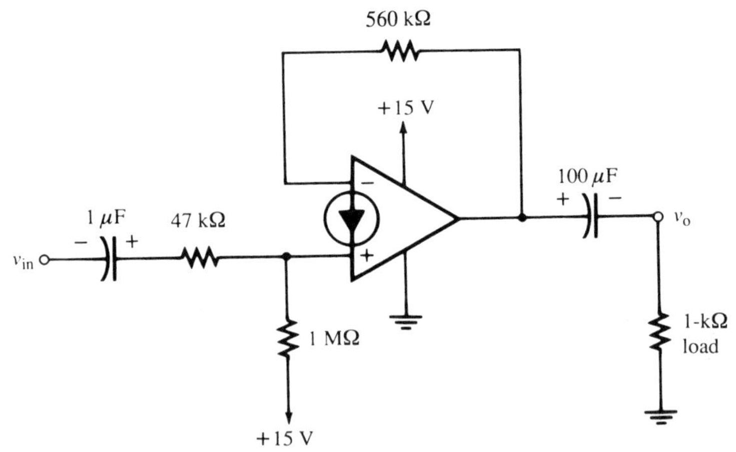

16. For the Norton summing amplifier circuit of Figure 5–32, determine:
    (a) The input impedance
    (b) The closed-loop gain
    (c) The low-frequency response
17. For the Norton three-input noninverting summing amplifier circuit of Figure 5–21 to be used as an audio mixer, determine suitable component values required for a closed-loop gain of 10 for all signals above 20 Hz when connected to a 1-kΩ load and operated from a 9-V supply. Assume that the circuit is to have an input impedance of 10 kΩ for all inputs.
18. For the difference amplifier circuit of Figure 5–23, determine suitable component values required for a differential gain of 20 for all signals above 30 Hz when connected to a 1-kΩ load and operated from a 15-V supply. Assume that the circuit is to have an input impedance of 10 kΩ.

Figure 5–32

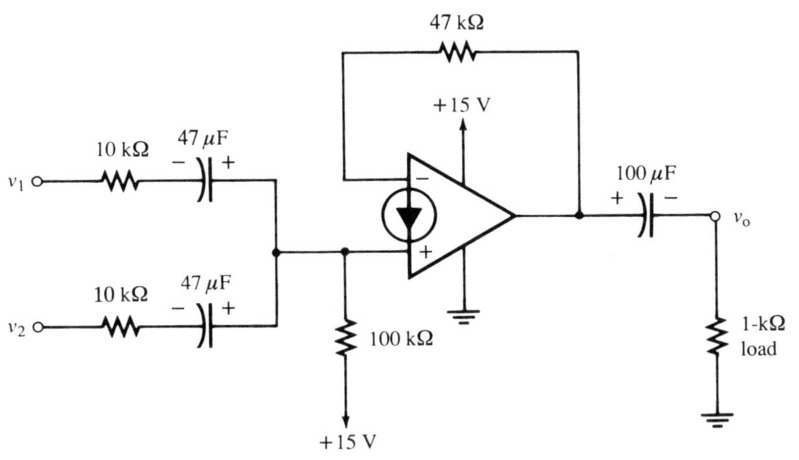

6 Different Types of Operational Amplifiers

Instructional Objectives

Upon completing this chapter, you will be able to

- Discuss the characteristics, operation, and possible advantages of the following devices:
  - An operational transconductance amplifier (OTA), such as a CA3080
  - A programmable operational amplifier, such as an LM4250
  - A BIFET operational amplifier, such as an LF351
  - A CMOS operational amplifier, such as a CA3130
- Properly bias a CA3080 OTA.
- Use a CA3080 as a voltage-controlled amplifier and an amplitude modulator.

Introduction

Besides the bipolar operational amplifier and the current-differencing, or Norton, amplifier, which is discussed in Chapter 5, there are several other types of operational amplifier devices that are sometimes used for specialized applications. This chapter discusses the characteristics and operation of these devices that set themselves apart from the bipolar and Norton amplifier devices. Specifically, these include the following:

- Operational transconductance amplifiers
- Programmable operational amplifiers
- BIFET operational amplifiers
- CMOS operational amplifiers

Unlike the standard bipolar operational amplifier, the operational transconductance amplifier is a voltage-to-current device whose gain is controlled by an externally applied bias current. Because of this feature, it is often used in amplitude modulators and analog multipliers. The programmable operational amplifier and its operating current may be controlled by an externally supplied bias current, which affects the gain-bandwidth product, power consumption, slew rate, input noise, input bias current, and input offset current.

The advantages of using a BIFET device are its extremely high input impedance and virtually negligible input bias and offset current levels. On the other hand CMOS devices are generally preferred for portable operation because of their low power consumption and, at the same time, their reasonably high slew rates.

6–1 The Operational Transconductance Amplifier

The **operational transconductance amplifier (OTA)** differs from a conventional operational amplifier in that it is a *voltage-to-current* amplifier rather than a straight voltage amplifier. Just as the Norton operational amplifier, or current-differencing amplifier (Chapter 5), has a schematic symbol that distinguishes it from the conventional bipolar operational amplifier, so does the OTA. Figure 6–1 shows the comparison of the schematic symbol for a conventional operational amplifier with that of an OTA. Here the OTA symbol includes a current source with an external bias current (I_{BIAS}) near the output to reflect the voltage-to-current nature of the amplification process.

Figure 6–1
Comparison of amplifier schematic symbols. (a) Bipolar operational amplifier. (b) OTA.

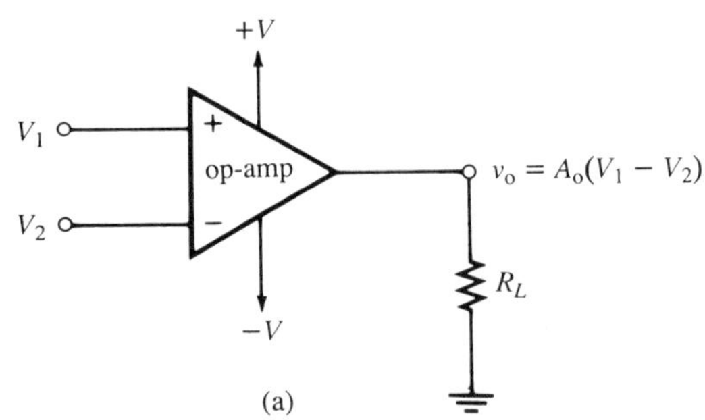

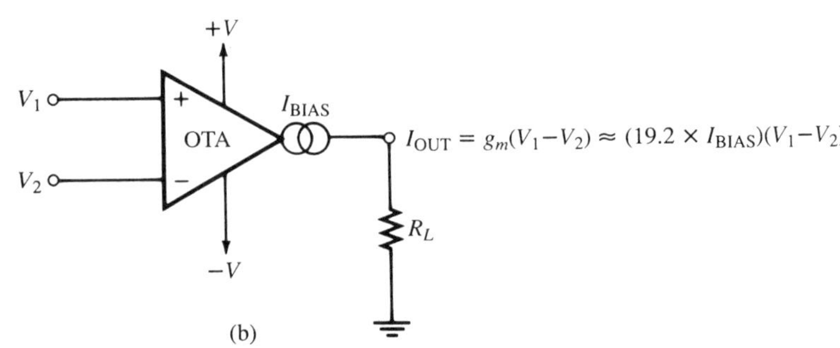

A conventional operational amplifier has differential input terminals and produces an output voltage equal to its open-loop gain times the input voltage difference, or

$$V_o = A_o(V_1 - V_2) \qquad (6\text{–}1)$$

In addition, the conventional operational amplifier normally provides a high input impedance and a low output impedance. In contrast, the OTA also has differential input terminals, but the differential input voltage controls an output current source. The ratio of a change in output current to a corresponding change in input voltage is called the *transconductance* (g_m):

$$g_m = \frac{I_{\text{out}}}{V_{\text{in}}} \qquad (6\text{–}2)$$

which is measured in siemens (S)*. In effect, transconductance is a measure of the voltage-to-current gain of the device. As a consequence, this type of operational amplifier has a high output impedance as well as a high input impedance. Unlike the fixed gain of a conventional operational amplifier, the transconductance of an OTA is directly proportional to an external bias current (I_{BIAS}) fed into the OTA's bias input terminal. For a typical OTA, this bias current may be varied from 0.1 μA to 1 mA, providing a 10,000 : 1 gain-control range.

An OTA will operate like a conventional voltage-amplifying operational amplifier if a suitable load resistor (R_L) is connected to its output terminal. This converts its output current,

$$I_{\text{out}} = g_m(V_1 - V_2) \qquad (6\text{–}3)$$

to a proportional voltage,

$$V_{\text{out}} = I_{\text{out}} R_L \qquad (6\text{–}4)$$

The total current drawn by an OTA is equal to twice the value of I_{BIAS}. Since I_{BIAS} may be as low as 0.1 μA, the OTA may be used in true micropower applications. The value of I_{BIAS} is easily set by using the proper size of resistor between an external voltage source and the bias terminal.

---

*The SI unit of transconductance (g_m), the reciprocal of resistance, is officially the *siemen*. This replaced the *mho*, which is *ohm* spelled backwards. As a practical matter, g_m is expressed either in millisiemens (mS) or microsiemens (μS). Owing to tradition, however, many manufacturers still express transconductance in terms of mhos.

Because of its variable-gain nature, the OTA can be easily adapted to construct voltage-controlled amplifiers (VCA), voltage-controlled oscillators (VCO), and voltage-controlled filters (VCF).

The CA3080 OTA

Perhaps the most popular OTA is the CA3080, developed by RCA, whose pin connections and equivalent internal circuit are shown in Figure 6–2.*

Basic Operation

The CA3080 consists of a single differential amplifier and four current mirrors (CM). A current mirror is a three-terminal circuit that will produce an in-phase output current having the same magnitude as the current flowing at its input terminal. Figure 6–3 shows the representation of two types of current mirrors. One is labeled a current source (Figure 6–3a), whereas the other is a current sink (Figure 6–3b). The basic circuit of a sink-type current mirror is illustrated in Figure 6–3c.

When both a current-mirror source and a current-mirror sink are connected as shown in Figure 6–4 and powered from a split power supply, a differential current, $I_{\text{SOURCE}} - I_{\text{SINK}}$, then flows through a load connected between their common point and ground.

Figure 6–5 shows the circuit of the differential amplifier section of the CA3080. The emitter current (I_C)† of the amplifier is equal to the sum of the two collector currents ($I_A + I_B$). When the differential input voltage (V_{in}) is zero, $I_A = I_B$. On the other hand, when V_{in} is not zero (± 25 mV maximum), I_A and I_B are different, so that

$$I_B - I_A = g_m V_{\text{in}} \qquad \textbf{(6–5)}$$

Also, I_B is proportional to I_C. At 25°C, g_m is approximately equal to $19.2 I_C$, so that Equations 6–3 and 6–5 may be rewritten as

$$I_{\text{out}} \simeq 19.2 I_{\text{BIAS}}(V_1 - V_2) \qquad \textbf{(6–6)}$$

and

$$I_B - I_A \simeq 19.2 I_C V_{\text{in}} \qquad \textbf{(6–7)}$$

If the current source of Figure 6–5 is replaced with a current mirror (C in Figure 6–6), the amplifier's transconductance is then controlled by an external current source (I_{BIAS}). The addition of three more current mirrors (A, B, and D in

*The data sheet for the CA3080 is given in Appendix A.

†This assumes that, for high current-gain transistors, the emitter and collector currents are equal to within 0.5%.

Figure 6–2
CA3080 OTA. (a) Pin diagram. (b) Internal equivalent circuit.

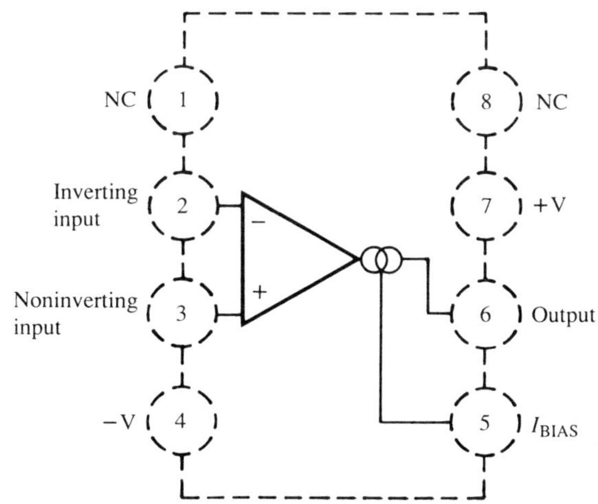

(a)

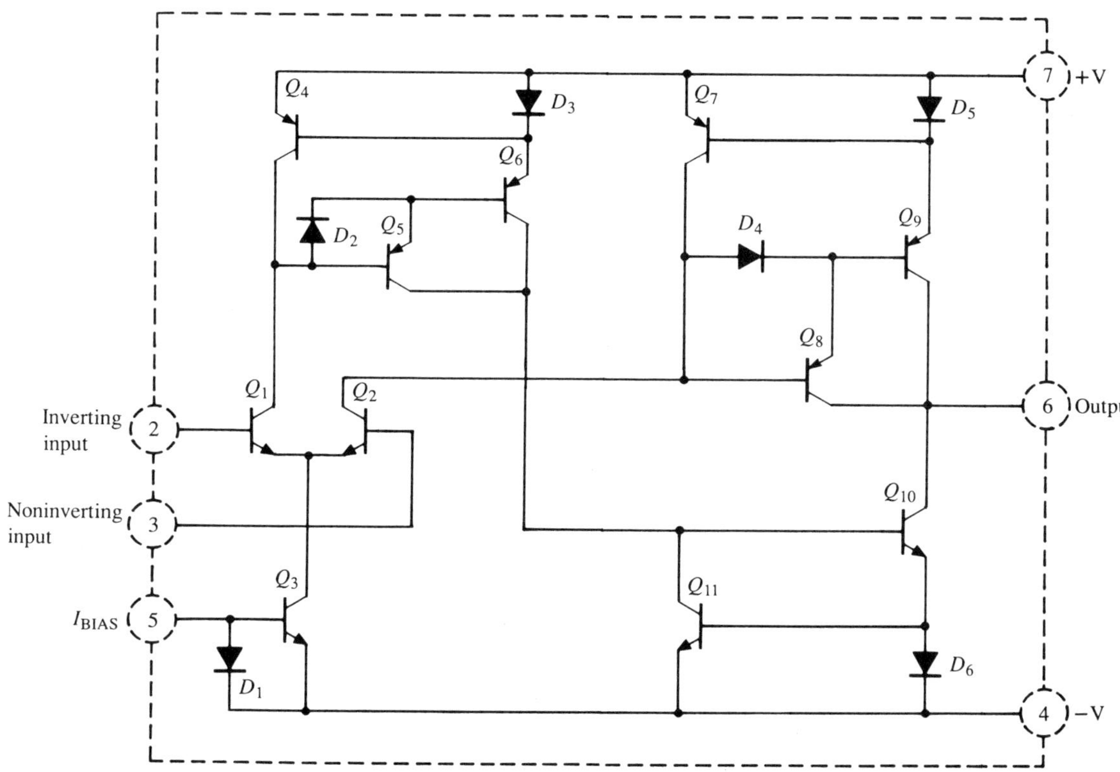

(b)

Figure 6–3
Current mirrors. (a) Schematic symbol of a current-mirror source. (b) Schematic symbol of a current-mirror sink. (c) Internal equivalent circuit of a sink-type current mirror.

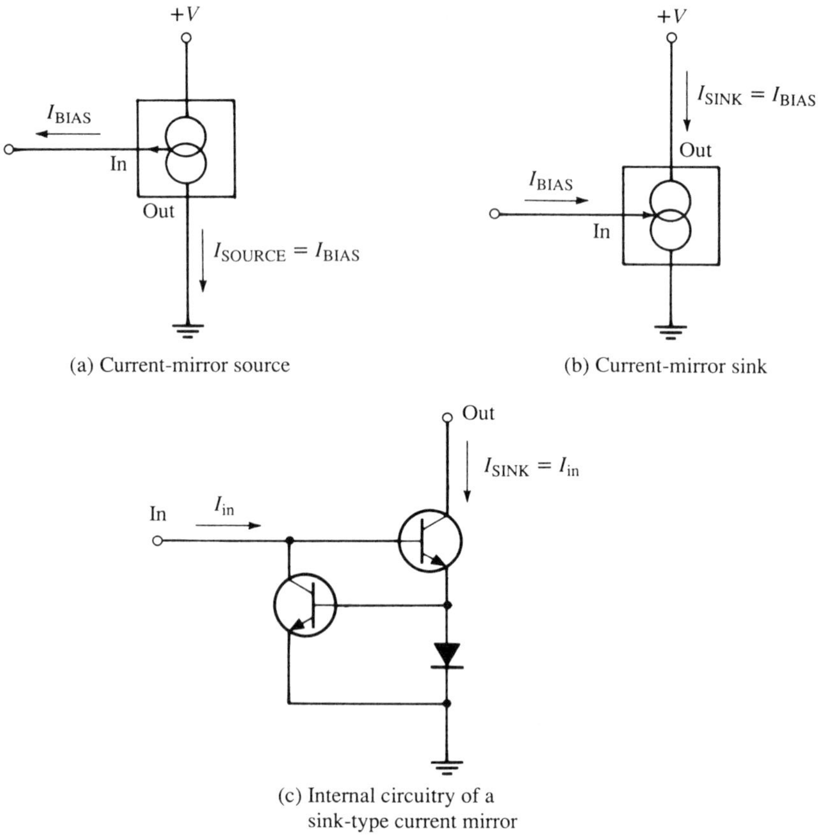

Figure 6–6) causes the difference between I_A and I_B to be delivered as an unbalanced output current to an external load. Since I_{BIAS} controls the emitter current, it in turn controls the transconductance of the differential amplifier (Q_1, Q_2). The collector current of Q_1 is reproduced by current mirror *A* and fed to the bias terminal of current mirror *D*. In a similar manner, Q_2's collector current is reproduced by current mirror *B* and fed to the sink terminal of current mirror *D*. The output current is then equal to $I_B - I_A$.

Referring back to the equivalent circuit of the CA3080 in Figure 6–2b, diode D_1 and transistor Q_3 form a unity-gain current mirror with the I_{BIAS} input, which is the function of current mirror *C* in Figure 6–6. Since the current-voltage transfer (transconductance) characteristics of D_1 and Q_3 are matched, I_{BIAS} is then duplicated in the collector current of Q_3. Therefore, the sum of the emitter currents of the Q_1-Q_2 differential pair is equal to I_{BIAS}. In this manner, the input bias current, the input resistance, and transconductance of the differential pair are directly

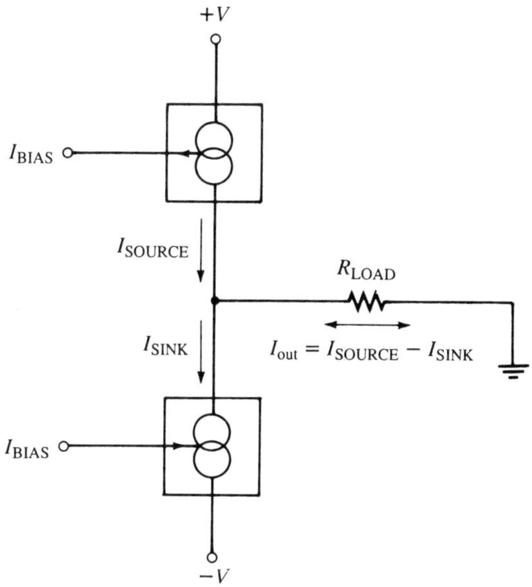

Figure 6–4
A current-mirror source and sink combined to produce a differential current in an external load.

controlled by I_{BIAS}. The push-pull current outputs of the Q_1-Q_2 differential pair are in turn applied to a pair of PNP current-mirror stages (Q_4-Q_6, and Q_7-Q_9), which are, respectively, represented by current mirrors *A* and *B* in Figure 6–6. These stages also have a unity current-transfer ratio but provide signal inversion. The output of Q_4-Q_6 is applied to a third current mirror, Q_{10}-Q_{11}, which corresponds to current mirror *D* in Figure 6–6. This NPN transistor stage inverts the signal from Q_2, after which the signal appears at the output in parallel with the output from Q_7-Q_9.

Both current mirror A (Q_4-Q_6, D_2, D_3) and current mirror B (Q_7-Q_9, D_4, D_5) stages use Darlington transistor pairs and speed-up diodes to improve their performance. The outputs of these current-mirror stages have a very high impedance, which in turn causes the output of the OTA to have a very high impedance. The entire circuit operates as a class-A amplifier. The peak output current is equal to

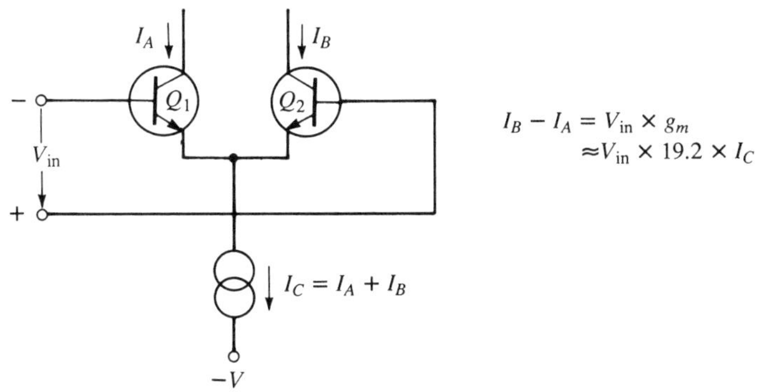

Figure 6–5
The differential amplifier section of a CA3080 OTA.

Figure 6–6
Equivalent circuit of a CA3080 OTA showing a differential amplifier with four current mirrors.

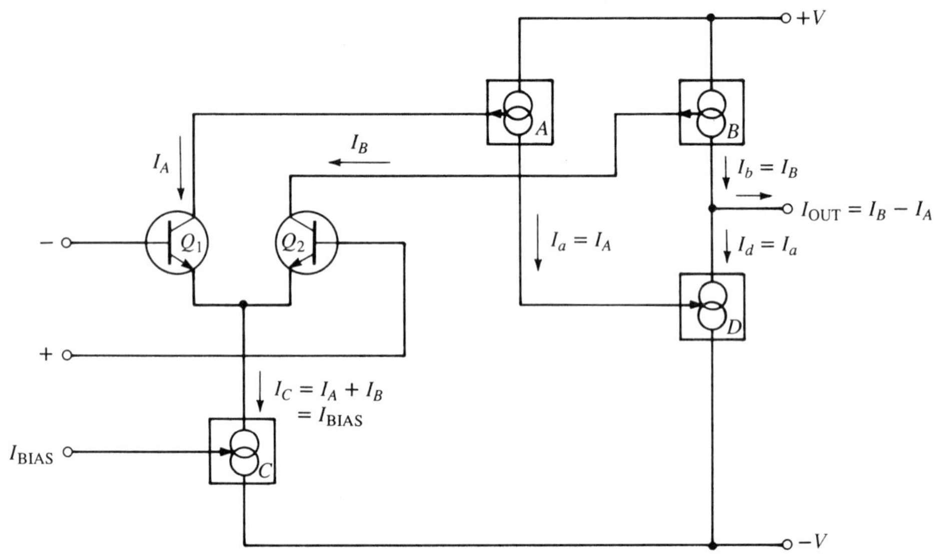

$\pm I_{BIAS}$, and the total operating current is equal to $2I_{BIAS}$. The input bias currents drawn by pins 2 and 3 when the OTA is operating in the linear mode are each equal to approximately $I_{BIAS}/200$, the actual values depending on the current gains of Q_1 and Q_2.

Figure 6–7 shows the operating characteristic curves for the CA3080 operating at 25°C and powered from a ±15-V supply. Figure 6–7a shows how bias

Figure 6–7
CA3080 operating characteristics. (a) Transconductance versus bias current. (b) Input and output resistance versus bias current.

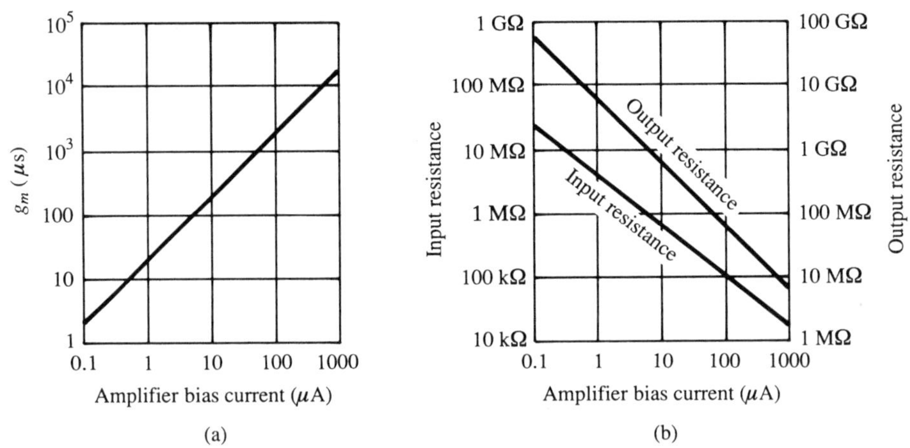

current affects the value of g_m, and Figure 6–7b shows how the input and output resistances are affected by bias current. For example, a CA3080 operating with a bias current of 10 μA would have a g_m of approximately 192 μS (μmhos), an input resistance of 800 kΩ, and an output resistance of 700 MΩ. The same amplifier operating at a bias current of 1 mA would have a g_m of about 19.2 mS, an input resistance of 15 kΩ, and an output resistance of 7 MΩ.

The peak output voltage swing is a function of I_{BIAS} and the size of the external load resistor connected to pin 6. The peak output voltage swing is given by $I_{BIAS}R_L$. Thus a 10-μA bias current with a 100-kΩ load resistor would permit a peak output swing of 10 μA $\times$ 100 kΩ, or 1 V. With no external capacitance connected to pin 6, the maximum slew rate of the CA3080 is typically 50 V/μs. When an external loading capacitor (C_L) is present, however, the slew rate is given by

$$\text{Slew rate} = \frac{I_{BIAS}}{C_L} \qquad \textbf{(6–8)}$$

where I_{BIAS} is measured in microamperes (μA) and C_L in picofarads (pF).

Biasing the CA3080

The external bias current input (pin 5) is internally connected to the negative supply (pin 4) through a base-emitter junction (Q_3 in Figure 6–2b), so that the voltage on pin 5 is normally about 0.6 V above the voltage on pin 4. The proper value of the external bias current input may be established by either connecting pin 5 to ground or to the positive supply through a properly sized resistor, as illustrated in Figure 6–8. If the positive supply is used as the source of bias

Figure 6–8
CA3080 OTA biasing. (a) External bias resistor tied to positive voltage. (b) External bias resistor grounded.

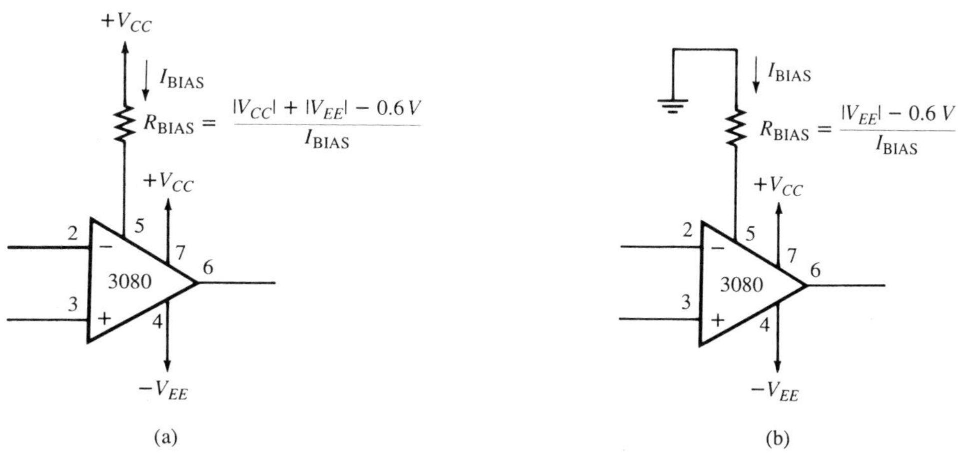

current, the resistor size is given by:

$$R_{\text{BIAS}} = \frac{|V_{CC}| + |V_{EE}| - 0.6}{I_{\text{BIAS}}} \qquad \textbf{(6–9)}$$

If, on the other hand, the bias resistor is connected between pin 5 and ground, V_{CC} in Equation 6–9 is set to zero, so that

$$R_{\text{BIAS}} = \frac{|V_{EE}| - 0.6}{I_{\text{BIAS}}} \qquad \textbf{(6–10)}$$

Example 6–1

Determine the size of the external resistor necessary to set the bias current of a CA3080 OTA at 50 μA when operating from a $\pm$9-V supply.

Solution:

(a) If the resistor is connected to the $\pm$9-V supply, as in Figure 6–8a, then from Equation 6–9,

$$R_{\text{BIAS}} = \frac{9\text{ V} + 9\text{ V} - 0.6\text{ V}}{50\ \mu\text{A}}$$
$$= \mathbf{348\ k\Omega} \qquad \text{(use a 330-k}\Omega\text{ standard value)}$$

(b) If the resistor is connected to ground (Figure 6–8b), then from Equation 6–10,

$$R_{\text{BIAS}} = \frac{9\text{ V} - 0.6\text{ V}}{50\ \mu\text{A}}$$
$$= \mathbf{168\ k\Omega} \qquad \text{(use a 160-k}\Omega\text{ standard value)}$$

OTA Applications

Figure 6–9 shows the circuit of a voltage-controlled amplifier (VCA) using a CA3080 OTA. In this circuit, a variable voltage at the V_C (control) input controls the gain by varying the OTA's bias current. The gain is at its maximum when V_C is zero and is reduced as the control voltage approaches -15 V. At $V_C = 0$, the bias current is approximately 500 μA and g_m is about 9.3 mS, producing the maximum gain value. Since the voltage gain of an OTA is $g_m R_L$, the voltage gain for the circuit of Figure 6–9 as measured between the input terminals of the OTA and the output is

$$A_V = (9.3\text{ mS})(10\text{ k}\Omega)$$
$$= 93$$

Figure 6–9
CA3080 OTA voltage-controlled amplifier with optional buffer for a low-impedance output.

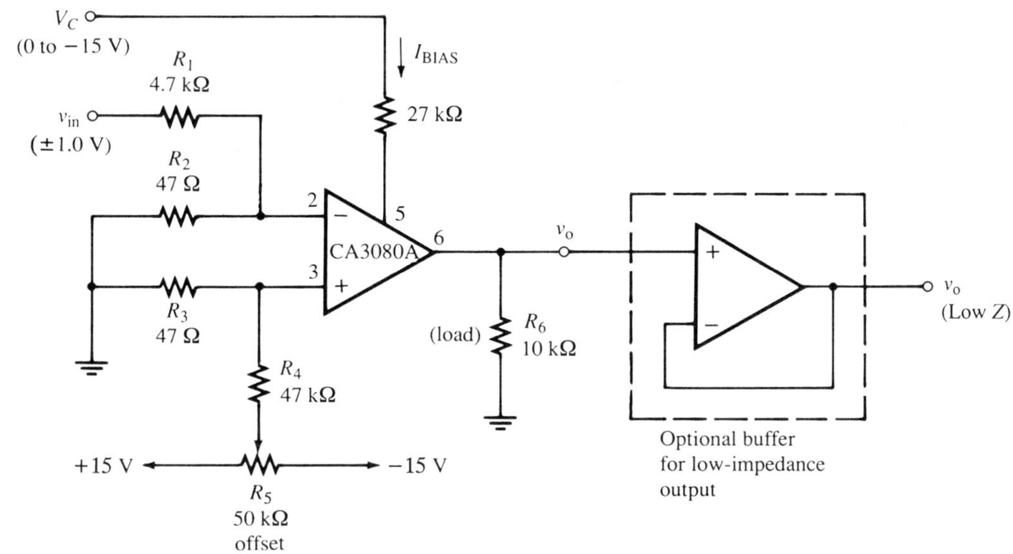

The input voltage applied to the OTA is reduced by a factor of approximately 100 by voltage divider R_1-R_2 for two reasons: (1) to keep the maximum stage gain to unity and (2) to keep the input voltage applied to the OTA within its most linear range. The CA3080A has an output distortion of less than 1% only if the input amplitude is less than ±10 mV. With the 100 : 1 reduction by R_1 and R_2, the input signal can then be as large as ±1 V. Larger input amplitudes can be applied if greater nonlinearity can be tolerated. In this case, the upper limit on the input signal is equal to $\pm I_{BIAS}R_6$. Using Equation 6–9 with a ±15-V supply, the bias current is set at approximately 1.1 mA, which in turn limits the input signal to about 11 V.

Input signals in the range of 10 mV are not much greater than the possible input offset voltage of the OTA, so that some kind of offset voltage adjustment would be required. In Figure 6–9, potentiometer R_5, through a 100 : 1 voltage divider (R_4-R_3), is adjusted for zero volts DC output at maximum voltage gain. This circuit also includes a conventional operational amplifier output buffer (voltage follower) stage, which provides a low output impedance for the OTA circuit.

Figure 6–10 shows a CA3080 OTA similarly connected as an AC-coupled inverting amplifier that would be used with a conventional operational amplifier. Here, the closed-loop voltage gain is determined primarily by the ratio R_2/R_3 ($A_{CL} = 10$) but is true only if the load resistance is infinite. With the values shown, the output impedance of this circuit is equal to R_2/A_{CL}, or 10 kΩ. Any external load will then lessen the output impedance further and also reduce the available output swing. With the values shown, I_{BIAS} is about 50 μA, whereas the total current draw is about 150 μA.

Figure 6–10
OTA inverting AC amplifier using a CA3080 having a closed loop gain of 10.

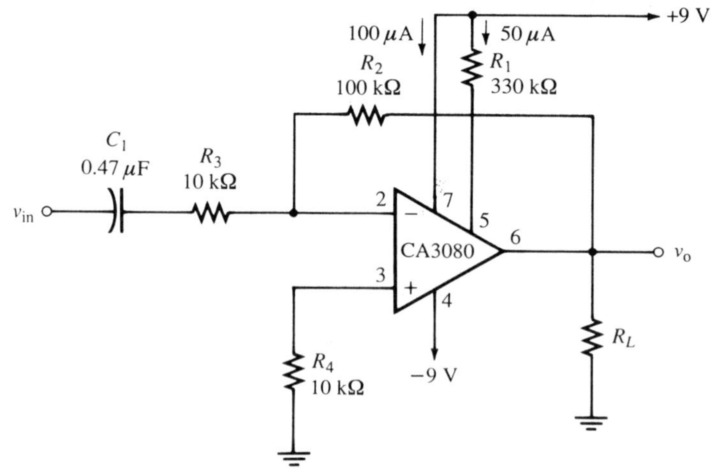

The OTA has many useful applications but is particularly well suited for use as a voltage-controlled amplifier (VCA), or amplitude modulator (Figure 6–11). As a modulator, a high-frequency carrier sine-wave signal is fed to the input of the amplifier, and sine-wave output is modulated by a lower-frequency signal connected to the I_{BIAS} terminal (pin 5) through a series resistor. The shape of either the positive or negative section of the output signal, called the *envelope,* is exactly the same as the signal feeding the I_{BIAS} input.

An amplitude modulator of this type is actually a special case of what is known as a *two-quadrant multiplier*. The modulating signal adds to and subtracts

Figure 6–11
CA3080 OTA amplitude modulator.

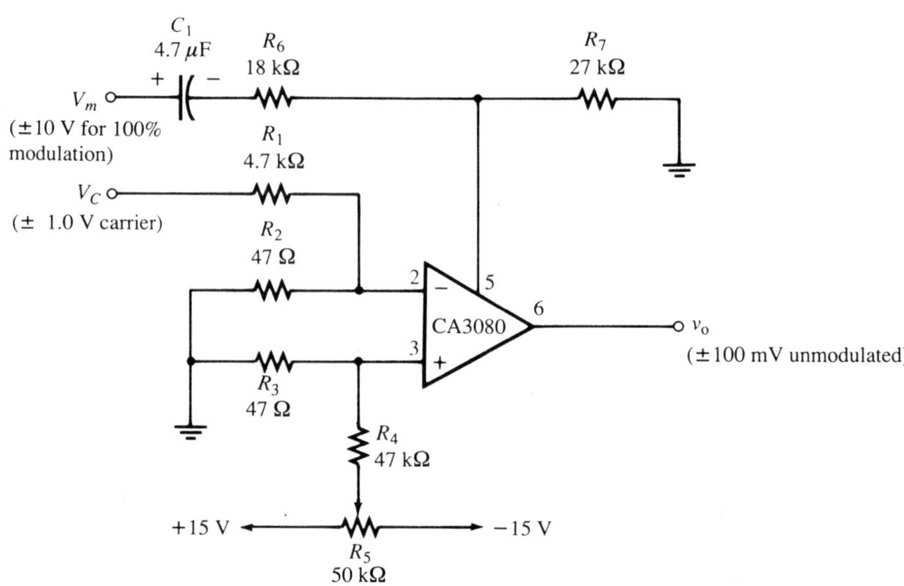

Figure 6–12
LM13600 dual OTA. (a) Pin diagram. (b) Internal equivalent circuit. (c) Effect of internal linearizing diodes on distortion levels.

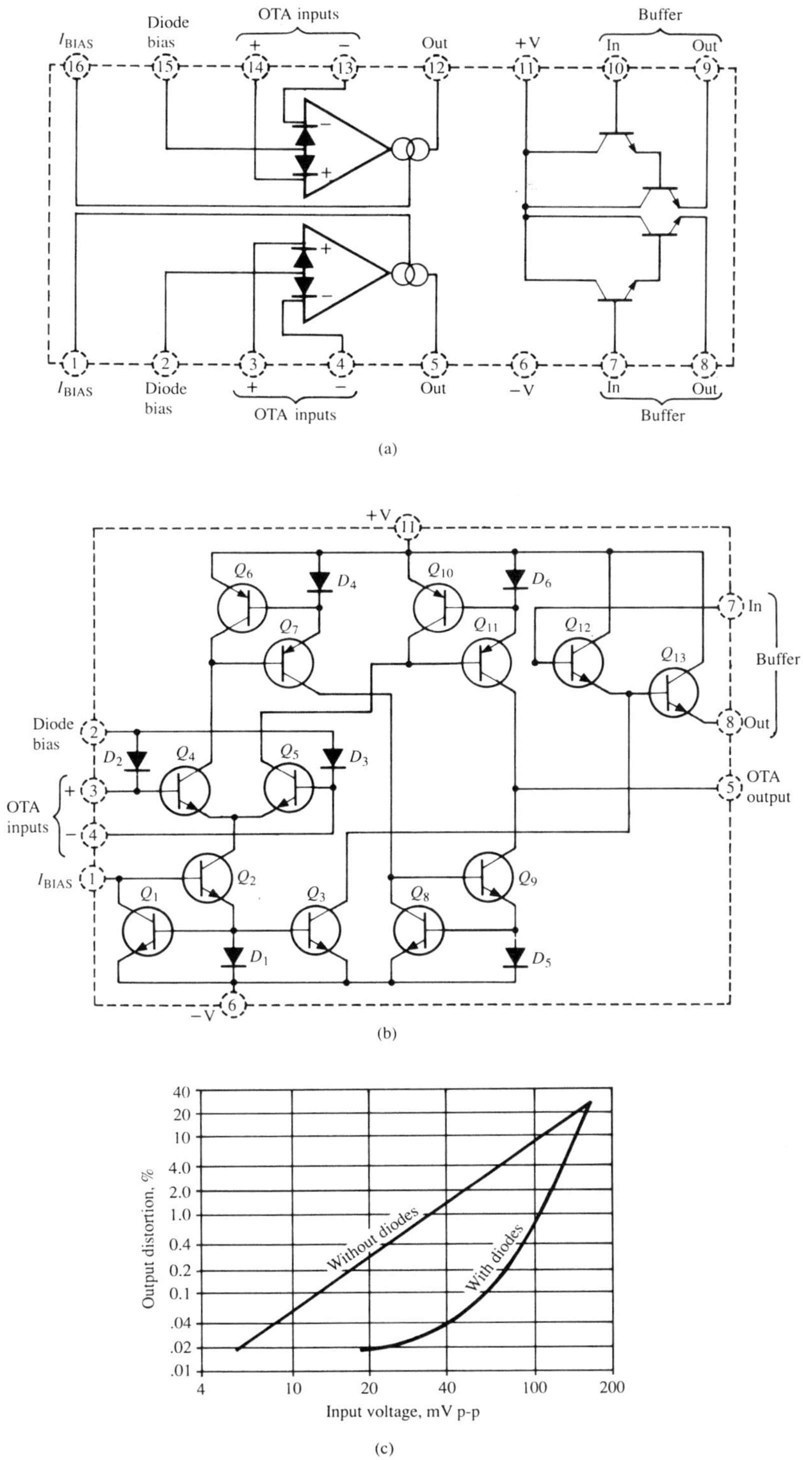

from a constant DC level to establish an average carrier level. In the circuit of Figure 6–11, R_7 sets the quiescent value of I_{BIAS} flowing into pin 5 at about 500 μA. This then establishes a constant ±100-mV output carrier level with ±1.0-V carrier input level. The C_1-R_6 network converts the AC modulating voltage (V_M) to a corresponding current that is superimposed on the quiescent DC current established by R_7. With the values shown, a modulation input level of ±10 V gives 100% modulation with a ±1-V carrier level. To increase modulation sensitivity, it is only necessary to decrease the value of R_6.

Although potentiometer R_5 controls the output offset voltage, as in Figure 6–9, it has the effect of controlling the modulation symmetry. This is easily accomplished by reducing the carrier input (V_C) to zero and adjusting R_5 until the OTA's output level is zero.

Second-Generation OTAs

Even though the popular CA3080 has found many uses, it nevertheless suffers from the pitfalls of having (1) high distortion and (2) an unbuffered high-impedance output stage. The LM13600, whose pin diagram is shown in Figure 6–12a, is an improved dual-OTA version. Each OTA section uses linearizing diodes to reduce signal distortion and has an optional output buffer stage that may be used to provide a low-impedance output. Figure 6–12c illustrates the effect of the linearizing diodes on the distortion levels for the LM13600.

Figure 6–13
LM13600 OTA voltage-controlled amplifier.

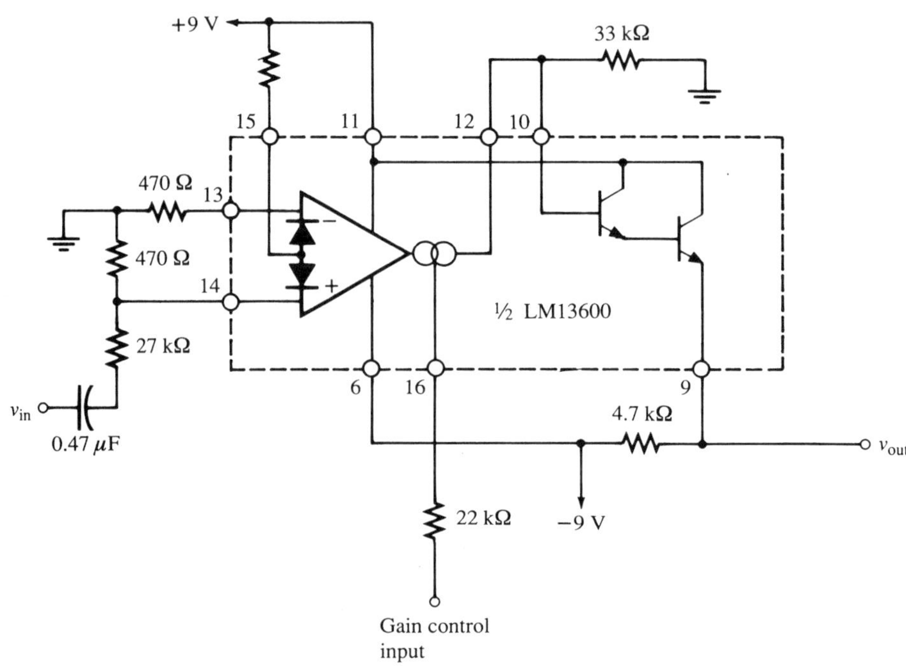

Figure 6–13 shows the LM13600 connected as a voltage-controlled amplifier. Because it has two identical sections, it is particularly well suited for the applications that require two matched amplifiers.

6–2 The Programmable Operational Amplifier

A **programmable operational amplifier** is one designed in such a way that its operating current may be controlled by an externally supplied bias current and is normally set by a single external resistor. This bias current actually controls the quiescent operating currents of the different internal amplifying stages of the operational amplifier. In effect, this arrangement gives the user control over a number of important parameters such as gain-bandwidth product, power consumption, slew rate, input noise, input bias current, and input offset current.

Some of the currently available programmable operational amplifiers include the following:

- LM4250 (National Semiconductor)
- CA3078 (RCA)
- μA776 (Fairchild)
- L144 (Siliconix)

6–3 BIFET Operational Amplifiers

The term *BIFET* is a registered trademark of National Semiconductor that describes a line of wide-bandwidth, JFET-input operational amplifiers that have an internally trimmed input offset voltage. **BIFET operational amplifiers** require a low-supply current and yet maintain a large gain-bandwidth product and a fast slew rate. In addition, well-matched, high-voltage JFET-input devices provide very low input bias and offset currents. In virtually all cases, BIFET devices require no circuit changes when used in place of standard bipolar operational amplifiers.

The LF351, for example, is pin-for-pin compatible with a standard 741 device and uses the same offset voltage adjustment circuitry. This feature then allows designers to upgrade the overall performance of existing designs using 741 operational amplifiers merely by substituting the LF351 for the 741.

6–4 CMOS Operational Amplifiers

Compared with conventional bipolar operation amplifiers, most **CMOS* operational amplifiers** provide almost an infinite input impedance and an output that can swing from one supply rail to the other, a feature not available previously.

---

*CMOS is the acronym for Complementary Metal Oxide Semiconductor and is constructed using both *p*- and *n*-channel (i.e., complementary) MOS transistor pairs whose characteristics are matched.

The TLC277 family of CMOS operational amplifiers, manufactured by Texas Instruments, are low-offset voltage, low-power, dual-operational amplifiers designed to operate from single- or dual-power supplies. They exhibit the following characteristics:

- Stable input offset voltage
- Very high input impedance
- Extremely low input offset and bias currents
- Common-mode input range to the negative supply rail
- Extremely low power consumption

All these features make the CMOS operational amplifier ideally suited for battery-powered or energy-saving applications. Because of their extremely high input impedance, CMOS devices are susceptible to destruction by static discharge during handling as well as by in-circuit discharges. For this reason, the TLC277 family uses electrostatic discharge-protection diode circuits that prevent failures at voltages up to 2000 V for short time periods.

The CA3130 and CA3140 are two very inexpensive CMOS operational amplifiers manufactured by RCA. The CA3130 offers about the same open-loop gain and input offset voltage as the popular 741 bipolar operational amplifier. At room temperature, however, the input impedance is about a million times higher than

Figure 6–14
High-impedance (10 MΩ), low-frequency amplifier using a CA3130 CMOS operational amplifier used ahead of a standard multimeter or VOM to improve its sensitivity and minimize meter loading.

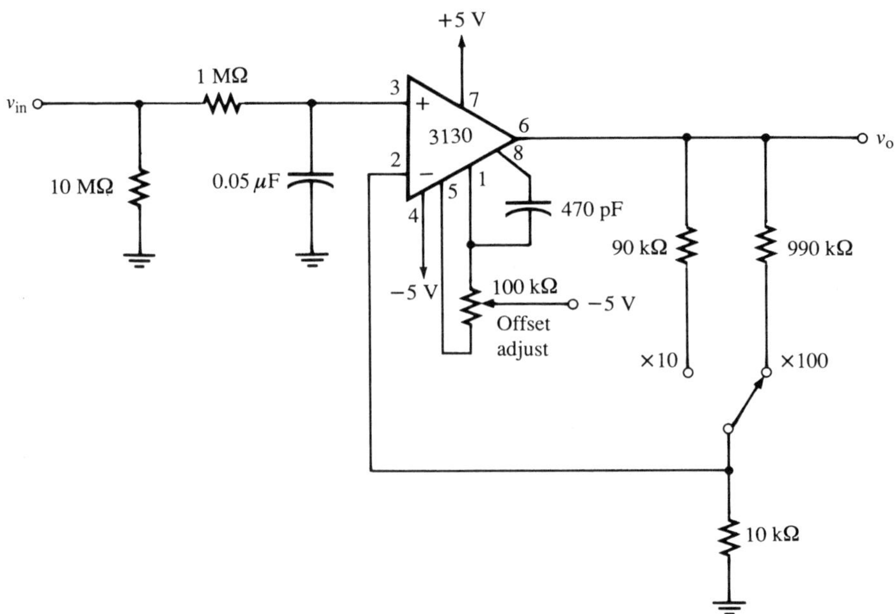

the 741, whereas the input bias and offset currents are proportionately lower. Furthermore, the slew rate of the CA3130 is about 20 times better than that of the 741, and the output can swing from supply rail to supply rail when lightly loaded. The CA3130 also has some disadvantages. It does not drive capacitive loads very well, and although it works well with supply voltages as low as ±5 V, it has an upper limit of ±16 V, which is almost the same as that of the 741. Compared to the 741, the CA3130 also tends to be somewhat noisy.

The CA3140 is a newer CMOS operational amplifier that eliminates some of the problems of the CA3130. The CA3140 can be considered as a "super 741" that does everything that the 741 can do but with a better slew rate and a much higher input impedance. However, the maximum output of the CA3140 will come to only 2 V less than the positive supply voltage. Thus, the CA3140 is not as useful as the CA3130 at low supply voltages or when the entire rail-to-rail output swing is required.

Figure 6–14 shows a CA3130 CMOS operational amplifier connected as a high-impedance, low-frequency amplifier. Such an amplifier could be used ahead of a standard multimeter or VOM to improve its sensitivity and reduce meter loading.

Summary

This chapter presented several other types of operational amplifier devices besides the bipolar and Norton amplifiers that are sometimes used for specialized applications. These included the operational transconductance amplifier, the programmable operational amplifier, and the BIFET and CMOS operational amplifiers.

Unlike the standard bipolar operational amplifier, the operational transconductance amplifier is a voltage-to-current device whose gain is controlled by an externally applied bias current. Because of this feature, it is often used in amplitude modulators and analog multipliers. The programmable operational amplifier and its operating current may be controlled by an externally supplied bias current, which affects the gain-bandwidth product, power consumption, slew rate, input noise, input bias current, and input offset current.

The advantage of using a BIFET device is its extremely high input impedance and virtually negligible input bias and offset current levels, whereas a CMOS device generally is preferred for portable operation because of its low power consumption in exchange for speed.

Glossary

BIFET operational amplifier A type of operational amplifier that has JFET-input stage and is characterized by wide bandwidth, fast slew rate, and high input impedance.

CMOS operational amplifier A type of operational amplifier fabricated from *p*-channel and *n*-channel MOS transistor pairs characterized by low power consumption.

operational transconductance amplifier (OTA) A type of voltage-to-current amplifier.

programmable operational amplifier A type of operational amplifier designed in such a way that its operating current is controlled by an externally supplied bias current, which in turn controls the gain-bandwidth product, power consumption, slew rate, input noise, input bias current, and input offset current.

Review Questions

1. How does an operational transconductance amplifier differ from a conventional operational amplifier?
2. How are CMOS operational amplifiers different from bipolar operational amplifiers?
3. What is a BIFET operational amplifier, and how is it different from a bipolar operational amplifier?
4. What is a programmable operational amplifier?

Exercises

1. A type CA3080 OTA is to be operated from a ±12-V supply. Determine the value of the bias resistor required to set the bias current at 100 μA.
2. Determine the input and output impedances of the type CA3080 OTA biased at 100 μA.
3. A type CA3080 OTA is used to construct a voltage-controlled amplifier, such as the circuit of Figure 6–9. If the OTA is biased at 100 μA, and the output is connected to a 2-kΩ load, determine the resultant voltage gain.

7 Audio Operational Amplifiers: Devices and Applications

Instructional Objectives

Upon completing this chapter, you will be able to

- Translate basic operational amplifier configurations for audio performance.
- Compare the AC performance of inverting and noninverting amplifier configurations.
- Design a preamplifier having an RIAA characteristic.
- Design a microphone preamplifier having unbalanced high impedance as well as balanced differential inputs.
- Explain the use of and the design for a loudness, or tone, control to improve bass/treble performance.
- Explain the use of and the design for a multiband octave equalizer.
- Explain the use of and the design for a two-channel panoramic, or stereo-pan control, circuit.
- Use audio-power-amplifier integrated circuits such as the LM380, 384, and 386.
- Explain the three methods by which heat can be transferred from a heat source.
- Determine the proper size heat sink required to prevent destruction of integrated circuits from overheating.

Introduction

The overwhelming majority of operational amplifier circuits are those involved with audio applications. The term **audio** is generally applied to those complex AC frequencies in the 20-Hz to 20-kHz range. Unlike its DC performance, the operational amplifier behaves differently at higher frequencies and must be corrected and compensated for. To be effective, however, any form of correction and compensation for those frequency effects must be done without adding a significant amount of distortion or noise to the original signal.

Chapter 3 discusses the major parameters that affect operational amplifier performance. Among those that are more significant for audio applications are the slew rate, open- and closed-loop gain, gain-bandwidth product, and noise. Some of these, in turn, ultimately affect other parameters. As an example, the open- and closed-loop gain affect input and output impedances.

Operational amplifier circuits used for audio applications are generally AC-coupled in order to eliminate DC levels from being amplified. The ground rules and considerations for these operational amplifier circuits, whether they be inverting, noninverting, buffer, or differential amplifiers, are virtually identical to those used in the single-supply circuits that are discussed in Chapter 5.

This chapter looks at how the basic operational amplifier configurations of Chapter 2 are wired for audio frequencies. In addition, several specialized audio circuits such as an octave equalizer, stereo pan-pot, tone control, and preamplifier circuits are discussed. This chapter also looks at specialized integrated-circuit power amplifiers such as the LM380, 384, and 386.

7–1 Audio Amplifiers

Operational amplifier circuits used for audio applications can be placed in two broad categories:

- Low-gain amplifiers, such as inverters, buffers, followers, difference amplifiers, and mixers (summing amplifiers).
- High-gain amplifiers, such as preamplifiers and power amplifiers.

The following subsections illustrate how some of the basic operational amplifier configurations are translated to audio performance.

Inverting Amplifier

The circuit for an inverting amplifier as applied to audio is shown in Figure 7–1. In fact, it is similar to a compensated differentiator (see Chapter 4) in that it is a standard DC inverting amplifier with a coupling capacitor (C_1) added in series with the input resistor R_1. This makes the circuit responsive above some low-frequency cutoff, so that

$$f_1 = \frac{1}{2\pi R_1 C_1} \qquad \textbf{(7–1)}$$

Above this frequency, the closed-loop voltage gain is

$$A_{CL} = \frac{R_2}{R_1} \qquad \textbf{(7–2)}$$

Figure 7–1
Audio-type inverting amplifier.

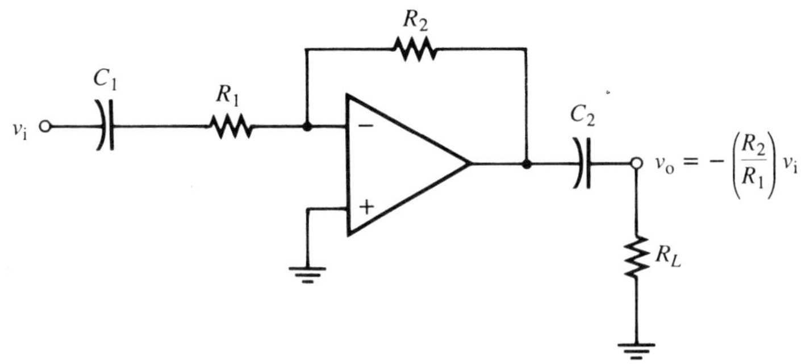

Because of C_1, this amplifier stage does not amplify DC offset voltages or low-frequency noise. The output is capacitively coupled to the load to remove any DC levels. The cutoff frequency for the output RC network is

$$f_2 = \frac{1}{2\pi R_L C_2} \quad \text{(7–3)}$$

and is usually made the same as for the input circuit (Equation 7–1). The circuit for an inverting amplifier powered from a single supply voltage is the same as that of Figure 5–9a.

Noninverting Amplifier

The circuit for a noninverting amplifier as applied to audio is shown in Figure 7–2. As with the standard noninverting amplifier, the closed-loop voltage gain is given by

$$A_{CL} = 1 + \frac{R_2}{R_1} \quad \text{(7–4)}$$

Figure 7–2
Audio-type noninverting amplifier.

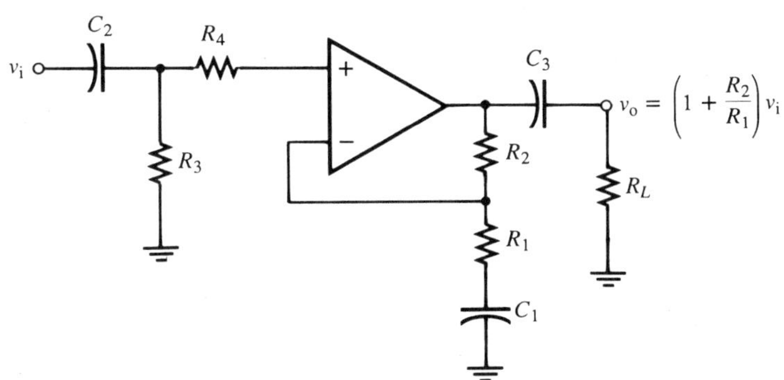

There are three low-frequency cutoff frequencies. The response due solely to the RC network of R_1 and C_1 is, in this case,

$$f_1 = \frac{1}{2\pi R_1 C_1} \tag{7–5}$$

and should be the dominant (i.e., highest) cutoff frequency of the three RC networks to minimize the operational amplifier's low-frequency noise gain. The cutoff frequency due to R_3 and C_2 is

$$f_2 = \frac{1}{2\pi R_3 C_2} \tag{7–6}$$

That due to R_L and C_3 is

$$f_3 = \frac{1}{2\pi R_L C_3} \tag{7–7}$$

In general, $f_2 = f_3$, and f_2 is made approximately five to ten times lower than f_1. Above the highest of these three frequencies, the closed-loop voltage gain is equal to that given by Equation 7–4. The input impedance of the amplifier is equal to the value of R_3, whereas R_4 is made approximately 100 times R_3. The circuit for a noninverting amplifier powered from a single-supply voltage is the same as shown in Figure 5–6.

Example 7–1

For the noninverting amplifier of Figure 7–2, using a 741 operational amplifier determine the components required for a closed-loop voltage gain of 10 above 80 Hz if the load is 1500 Ω and the input impedance is to be 10 kΩ.

Solution:

(a) Assuming an arbitrary value of **1 kΩ** for R_1, then from Equation 7–4,

$$R_2 = (10 - 1)(1\text{ k}\Omega)$$
$$= \mathbf{9\text{ k}\Omega} \quad \text{(use a 9.1-k}\Omega\text{ standard value)}$$

(b) $R_3 = Z_{in}$, so that $R_3 =$ **10 kΩ.**

(c) Since R_4 should be approximately 100 times R_3, then R_4 is 100 × 10 kΩ, or **1 MΩ.**

(d) Using Equation 7–5,

$$C_1 = \frac{1}{(2\pi)(1\ \text{k}\Omega)(80\ \text{Hz})}$$
$$= \mathbf{2\ \mu F} \qquad \text{(use a 2.2-}\mu\text{F standard value)}$$

(e) Since the cutoff frequencies associated with the RC networks for R_3-C_2 (Equation 7–6) and R_L-C_3 (Equation 7–7) should be five to ten times (10 to 20%) less than that found in part (d), choose these frequencies to be 10% of f_1 for convenience. Therefore, $f_2 = f_3 = 80$ Hz/10, or 8 Hz. From Equation 7–6,

$$C_2 = \frac{1}{(2\pi)(10\ \text{k}\Omega)(8\ \text{Hz})}$$
$$= \mathbf{2\ \mu F} \qquad \text{(use a 2.2-}\mu\text{F standard value)}$$

and from Equation 7–7,

$$C_3 = \frac{1}{(2\pi)(1500\ \Omega)(8\ \text{Hz})}$$
$$= \mathbf{13.3\ \mu F} \qquad \text{(use a 15-}\mu\text{F standard value)}$$

The completed circuit is shown in Figure 7–3.

Difference Amplifier

The circuit for a difference, or differential, amplifier as applied to audio is shown in Figure 7–4. It is similar to the standard DC amplifier except that capacitors have been added in series with both inputs. For the best common-mode rejection, the cutoff frequency associated with the input RC networks should be made equal,

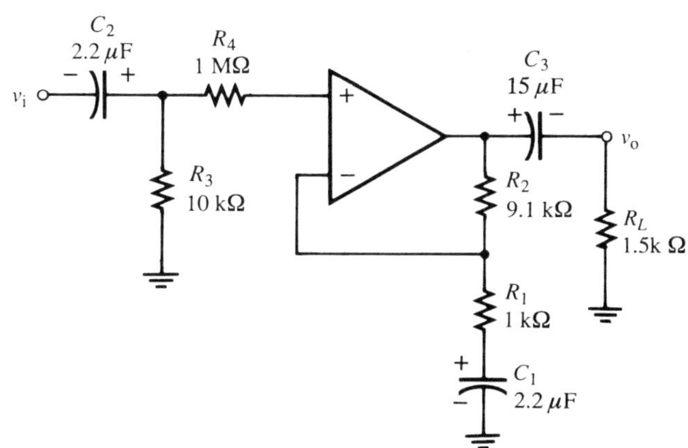

Figure 7–3
Completed noninverting amplifier for Example 7–1.

Figure 7–4
Audio-type difference amplifier.

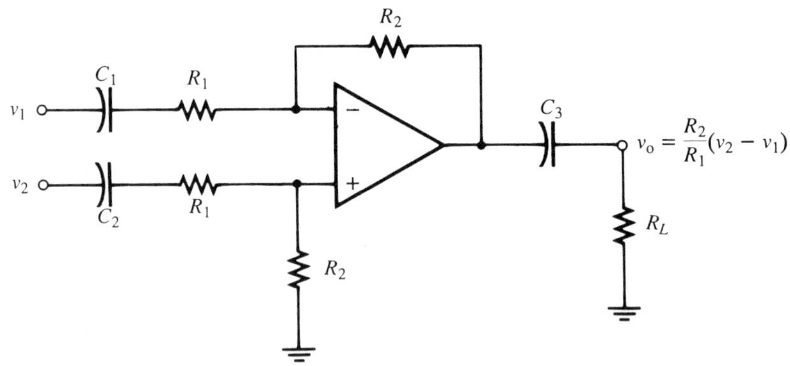

so that

$$f_1 = \frac{1}{2\pi R_1 C_1} \tag{7–8}$$

It is also made equal to the cutoff frequency of the output RC network (R_L and C_3), so that

$$f_3 = \frac{1}{2\pi R_L C_3} \tag{7–9}$$

Above the highest of these three frequencies, the closed-loop differential voltage gain is equal to

$$A_D = \frac{R_2}{R_1} \tag{7–10}$$

The circuit for a difference amplifier powered from a single-supply voltage is nearly the same as shown in Figure 5–11.

Inverting versus Noninverting Amplifiers

The majority of audio amplifier applications use either inverting or noninverting amplifiers. Both circuits employ feedback of a fraction of the output voltage back to the input. To the human ear, there is no difference if the amplified signal is inverted or not—both signals sound the same.

Table 7–1
Summary of loop gain on amplifier parameters

Amplifier Parameter	Noninverting Amplifier	Inverting Amplifier
Input impedance	$(1 + A_{OL})R_{\text{in}}$	R_F/A_{OL}
Output impedance	$R_o/(1 + A_{OL})$	$R_o/(1 + A_{OL})$
Total harmonic distortion	$\text{THD}/(1 + A_{OL})$	$\text{THD}/(1 + A_{OL})$
Noise gain	A_{CL}	$A_{\text{CL}} + 1$
Closed-loop bandwidth	GPB/A_{CL}	$\text{GBP}/(A_{\text{CL}} + 1)$

In order to determine whether the inverting or noninverting amplifier is superior for a given application, we must consider the amplifier's loop gain (A_L), which is the ratio of the operational amplifier's open-loop gain (A_{OL}) to the closed-loop gain (A_{CL}). Table 7–1 summarizes the most important amplifier parameters and the effect of the loop gain upon them. Both the output impedance and total harmonic distortion are affected equally. The input impedance is higher for noninverting amplifiers. The noise gain differs only by unity and becomes significant only for low-gain applications. Here, a unity-gain inverting amplifier has twice the noise gain over a noninverting counterpart. The closed-loop bandwidth is similarly related.

7–2 Preamplifiers and RIAA Equalization

Audio **preamplifiers,** or *preamps,* are amplifiers that are specifically designed to amplify signals at the low end of the dynamic range of practical amplifier circuits. In general, this is approximately less than 10 mV.

Because of low-input signal levels, audio preamplifiers have their own problems. Since it is at the *front end,* or the first stage of a larger amplifier system, its response characteristics ultimately play a significant part in shaping the characteristics of the overall amplifier system. For preamps, noise is of prime consideration. If the internal noise (Johnson, shot, burst, and flicker) of the preamp amounts to 5 μA and the input signal is 1 mV, the signal-to-noise ratio is 1 mV/5 μA, or 200 (46 dB). In preamps, it is the input signal level that is fixed so that the generated noise must be minimized to achieve a high S/N ratio.

Phono preamps are different from other preamps for microphone and tape recorders in that their frequency response is tailored to compensate for, or *equalize,* the characteristics of the recording and playback process. The popular types of phono cartridges—magnetic, crystal, and ceramic—are vastly different. Magnetic cartridges are low-output, constant-velocity devices, so that the response is constant at the rate of +6 dB/octave. Characteristics of constant-velocity recording are that its amplitude varies inversely with frequency and the maximum slope is constant with frequency. On the other hand, ceramic cartridges are high-output (100 mV to 2 V), constant-amplitude devices and do not require equalization.

To assure optimum performance, the RIAA (Record Industry Association of America) has set a standard for the frequency response for phono preamps as a

Figure 7–5
RIAA standard frequency response curve.

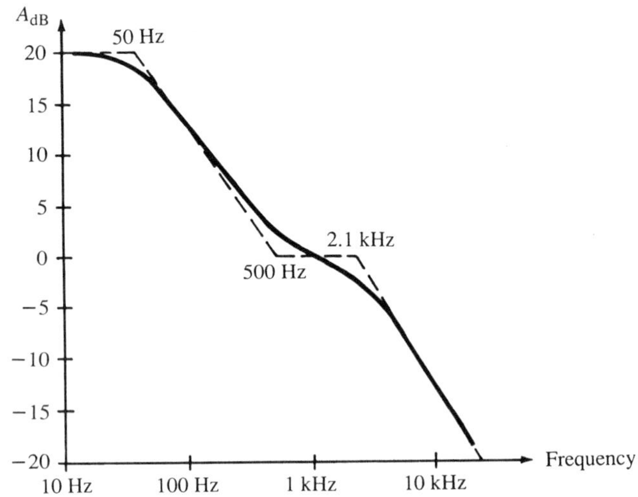

compromise between constant-amplitude and constant-velocity pickups. The RIAA curve, shown in Figure 7–5, has its maximum response below the first cutoff frequency of 50 Hz. Above 50 Hz, the gain rolls off at −6 dB/octave (−20 dB/decade) until the second cutoff frequency of 500 Hz, where the gain is essentially flat. The third cutoff frequency is 2.1 kHz, above which the response again rolls off at −6 dB/octave throughout the rest of the audio region. Table 7–2 summarizes the RIAA curve.

The gain of a preamp that is tailored to the RIAA characteristics is specified in terms of a reference frequency of 1 kHz. In order to examine what effect the standard RIAA response curve has on the loop gain of an operational amplifier, Figure 7–6 graphs the responses of those operational amplifiers having gain-band-

Table 7–2
RIAA standard response specifications

Hz	dB	Hz	dB
20	+19.3	800	+0.7
30	+18.6	1,000	0*
40	+17.8	1,500	−1.4
50	+17.0	2,000	−2.6
60	+16.1	3,000	−4.8
80	+14.5	4,000	−6.6
100	+13.1	5,000	−8.2
150	+10.3	6,000	−9.6
200	+8.2	8,000	−11.9
300	+5.5	10,000	−13.7
400	+3.8	15,000	−17.2
500	+2.6	20,000	−19.6

*Reference frequency.

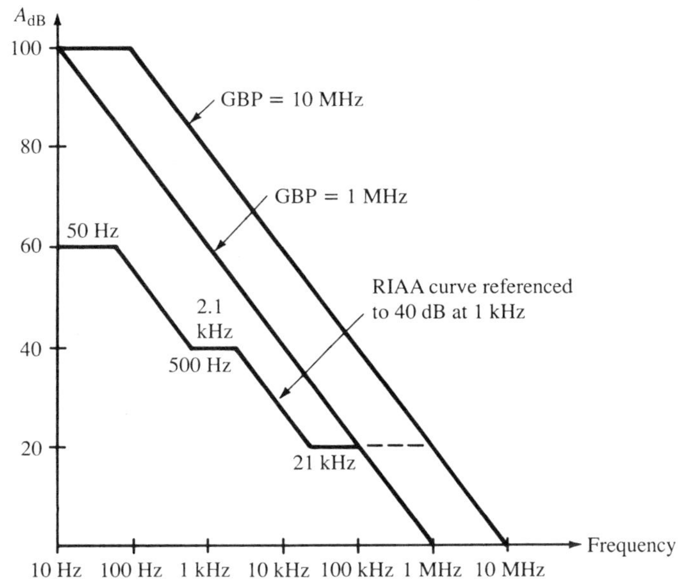

Figure 7–6 Comparison of RIAA curve and open-loop gain responses of operational amplifiers having gain-bandwidth products of 1 MHz and 10 MHz.

width products of 1 MHz (e.g., a 741) and 10 MHz (e.g., an LM318), respectively, along with the standard RIAA curve.

A preamp using an operational amplifier having a GBP of 1 MHz will then have less than 20 dB of feedback at frequencies above 2.1 kHz. Consequently, the only way to increase the feedback is to then lower the gain of the preamp stage and make it up in a later stage. In comparison, a preamp using an operational amplifier with a GBP of 10 MHz will allow approximately a minimum of 34 dB of feedback.

If an RC network having a break frequency above 21 kHz is added, then the RIAA curve can then be made flat at approximately 20 dB. As a result, any operational amplifier that uses external compensation can then be compensated for a gain of 20 dB rather than for unity gain (0 dB), which in turn allows the higher loop gain of a 10-MHz GBP device to be used. As an example, Figure 7–7 shows a magnetic pickup RIAA preamp using a 725 operational amplifier with a closed-loop voltage gain of 100 (40 dB) at the RIAA reference frequency of 1 kHz. The external compensation network of R_7, R_8, C_4, and C_5 allows for compensation for a closed-loop gain of 20 dB to allow a GBP of 10 MHz.

At low frequencies, the closed-loop voltage gain is fixed at 1000 (60 dB) by R_2/R_1. The low-frequency cutoff frequency of approximately 50 Hz is set by C_2 and R_2. Above 50 Hz, the response drops at the rate of −6 dB/octave until the mid-frequency cutoff frequency of 500 Hz, set by R_3 and C_2, and the gain is then flat at 100 (40 dB), set by R_3/R_1. Above the 2.1-kHz high-frequency cutoff, set by R_3 and C_3, the gain again rolls off at −6 dB/octave. This continues until 21 kHz, set by R_4 and C_3, where the closed-loop voltage gain then flattens out at 10 (20 dB), which is set by R_4/R_1.

Figure 7–7
RIAA preamplifier with a magnetic pickup using a 725 externally compensated operational amplifier.

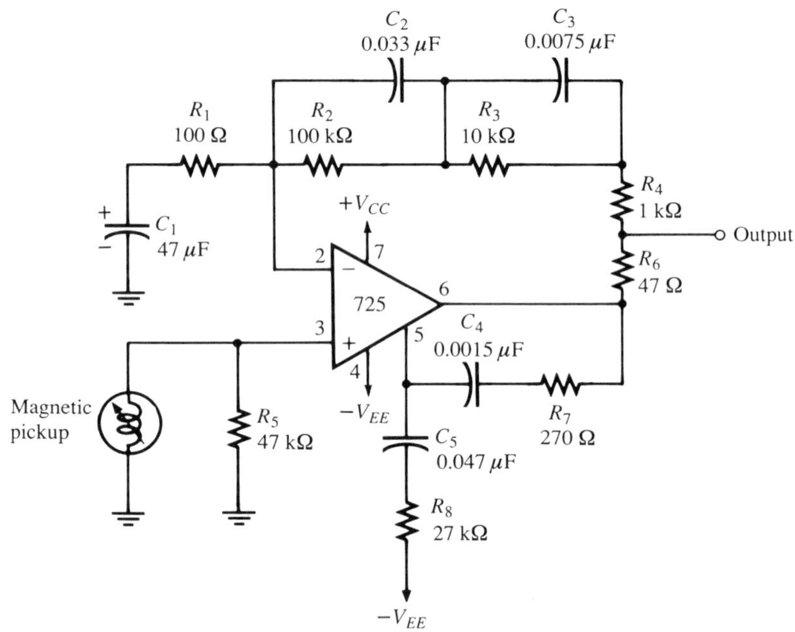

7–3 Microphone Preamplifiers

A microphone preamplifier is another commonly encountered type of preamplifier circuit. Because microphones are available in a wide range of impedances and signal levels, these two parameters greatly influence the type of circuit used. In general, microphones are divided into two groups, as follows:

- High-impedance (~20 kΩ), high-output (~200 mV) microphones
- Low-impedance (~200 Ω), low-output (~2 mV) microphones

The high-impedance, high-output microphone places no special requirements upon the preamp, so that the frequency response is essentially flat, and no equalization is necessary. Because of the high output levels of the microphone, amplifier hum and noise considerations are virtually ignored. However, hum *is* a problem with the microphone itself due to its high impedance because it is susceptible to stray magnetic-field pickup (e.g., 60 Hz). To minimize this effect, microphone cable lengths must be limited to short distances, typically less than 10 ft. Because of this limitation, high-impedance microphones are seldom used.

Low-impedance microphones are available in two forms:

- Unbalanced two-wire output composed of one floating-signal wire and one that is grounded.

- Balanced three-wire output composed of two floating-signal wires and one that is grounded.

Balanced microphone systems are used most frequently, since this arrangement greatly minimizes hum and noise pickup by using differential input connections, either with a grounded tapped transformer primary, single-ended secondary, or differential amplifier. In the three-wire system, the two floating-signal wires are usually twisted tightly together, with a surrounding shield serving as the third wire to be grounded.

Unbalanced, High-Impedance Microphone Preamplifier

A circuit for a simple single-ended (unbalanced), high-impedance ($> 600\ \Omega$) microphone preamplifier built around a 725 externally compensated operational amplifier is shown in Figure 7–8. The noninverting closed-loop voltage gain is determined by

$$A_{CL} = 1 + \frac{R_2}{R_1} \qquad \text{(7–11)}$$

for frequencies above that set by

$$f_C = \frac{1}{2\pi R_1 C_1} \qquad \text{(7–12)}$$

To minimize input noise, the parallel combination of R_1 and R_2 is made less than the output impedance of the microphone (Z_{mic}). Resistor R_5 should be chosen so that

$$R_5 \geq 10 Z_{\text{mic}} \qquad \text{(7–13)}$$

External compensation for the 725 is achieved with R_4 and C_2.

Balanced (Transformerless) Differential-Input Preamplifier

A balanced differential-input preamplifier is shown in Figure 7–9 using an LF357 (FET-input) operational amplifier for low-impedance microphones. Here, the differential gain is fixed by R_2/R_1 at 100 (40 dB), whereas the 10-kΩ potentiometer is used to adjust for the best CMR. As a general rule, make

$$R_1 = 5 Z_{\text{mic}} \qquad \text{(7–14)}$$

Figure 7–8
High-impedance, single-ended microphone preamplifier using a 725 operational amplifier with external compensation.

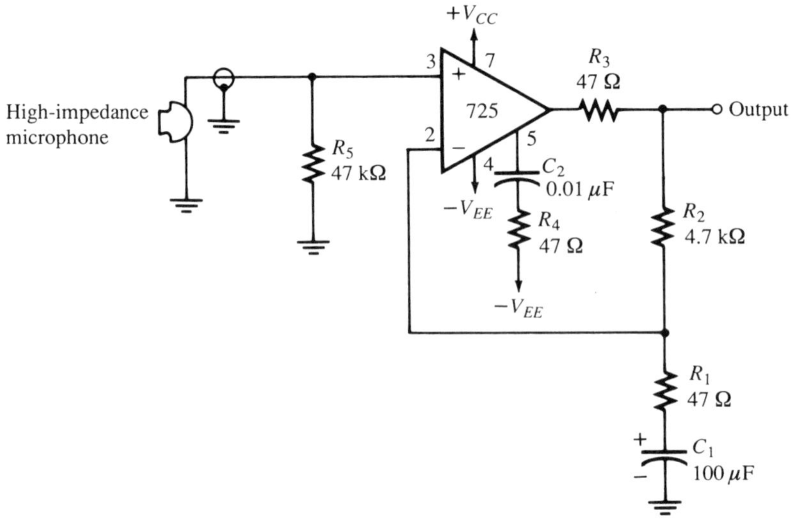

Figure 7–9
Balanced differential-input preamplifier using an LF357 operational amplifier having an FET input.

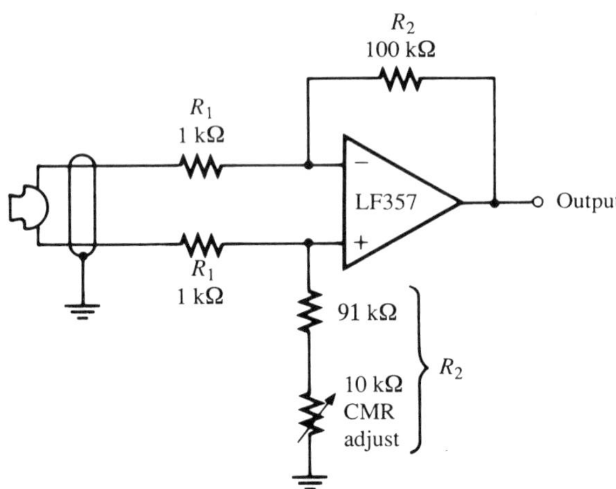

7–4 Loudness Control

A *loudness,* or **tone-control, circuit** compensates for the logarithmic response of the human ear. It provides added bass (low-frequency) and treble (high-frequency) response, which may be adjusted to provide the most pleasing effect to the listener. The added boost afforded by the circuit at these frequencies is inversely proportional to the setting of the amplifier's volume control.

Active tone-control circuits, such as the one shown in Figure 7–10a, have advantages over passive designs in that their response is symmetrical, and they

Figure 7–10
Loudness control. (a) Circuit. (b) Frequency response.

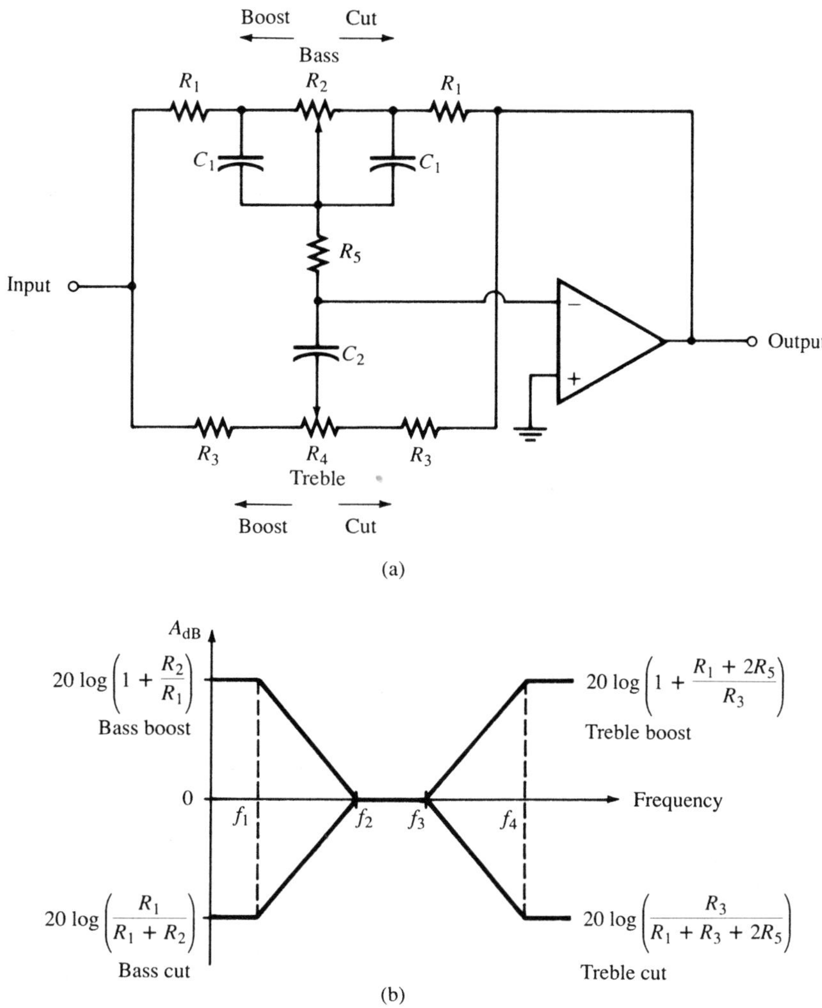

have low total harmonic distortion as a result of the frequency-selective network being placed in the operational amplifier's feedback path. The circuit's frequency response is shown in Figure 7–10b. At very low frequencies, the impedance of the capacitors is large enough that they are considered open circuits, and the bass closed-loop voltage gain is a function of the setting of the potentiometer R_2,

$$A_{CL}(\text{bass}) = 1 + \frac{R_2}{R_1} \tag{7–15}$$

This equation assumes that R_2 is much greater than R_1. The low-frequency bass cutoff frequency f_1 is set by

$$f_1 = \frac{1}{2\pi R_2 C_1} \tag{7–16}$$

after which the response drops off at approximately −6 dB/octave and flattens out to zero decibels at f_2:

$$f_2 = \frac{1}{2\pi R_1 C_1} \tag{7–17}$$

The treble response is such that the gain increases at approximately +6 dB/octave from f_3, set by

$$f_3 = \frac{1}{2\pi(R_1 + R_3 + 2R_5)C_2} \tag{7–18}$$

until the high-frequency treble cutoff frequency f_4,

$$f_4 = \frac{1}{2\pi R_3 C_2} \tag{7–19}$$

where the closed-loop gain flattens out. Since the reactances of the capacitors are now quite small (almost short circuits), the treble closed-loop voltage gain is given by

$$A_{CL}(\text{treble}) = 1 + \frac{R_1 + 2R_5}{R_3} \tag{7–20}$$

which assumes that R_4 is much greater than $R_1 + R_3 + 2R_5$.

The relationship between f_1 and f_2 and between f_3 and f_4 is given by

$$\frac{f_2}{f_1} = \frac{f_4}{f_3} = 10 \tag{7–21}$$

Example 7–2
Determine the component value for a bass-treble tone control using the circuit of Figure 7–10a having a gain of +20 dB up to 45 Hz and above 7 kHz.

Solution:
(a) using a 100-kΩ potentiometer having a linear taper for R_2, R_1 is determined from Equation 7–15 for a voltage gain of 10 (20 dB) so that

$$R_1 = \frac{100 \text{ k}\Omega}{10 - 1}$$
$$= \mathbf{11.1 \text{ k}\Omega} \qquad \text{(use an 11-k}\Omega\text{ standard value)}$$

(b) The design of the bass frequency components starts with Equation 7–16:

$$C_1 = \frac{1}{(2\pi)(45 \text{ Hz})(100 \text{ k}\Omega)}$$
$$= \mathbf{0.0354\ \mu F} \qquad \text{(use a 0.033-}\mu\text{F standard value)}$$

(c) Another arbitrary choice is to set R_5. This is necessary because there are not enough equations to solve for all the unknown components. Here, R_5 is set at **10 kΩ.**

(d) The design of the treble components is begun by determining R_3 from Equation 7–20, knowing R_1 = 11 kΩ, R_5 = 10 kΩ, and A_{CL}(bass) = 10 (+20 dB):

$$R_3 = \frac{(11 \text{ k}\Omega) + (20 \text{ k}\Omega)}{10 - 1}$$
$$= \mathbf{3.4 \text{ k}\Omega} \qquad \text{(use a 3.3-k}\Omega\text{ standard value)}$$

(e) Using Equation 7–19 and solving for C_2,

$$C_2 = \frac{1}{(2\pi)(7 \text{ kHz})(3.3 \text{ k}\Omega)}$$
$$= \mathbf{0.00689\ \mu F} \qquad \text{(use a 0.0068-}\mu\text{F standard value)}$$

(f) Since R_4 should be much greater than $R_1 + R_3 + 2R_5$,

$$R_4 \gg 11\ \Omega + 3.3 \text{ k}\Omega + 20 \text{ k}\Omega$$
$$\gg \mathbf{34.3 \text{ k}\Omega}$$

so that R_4 is then made at least ten times the minimum value of 34.3 kΩ. Using standard components, R_4 is selected as a **500-kΩ** potentiometer. The completed circuit is shown in Figure 7–11.

Figure 7–11
Completed loudness-control circuit for Example 7–2.

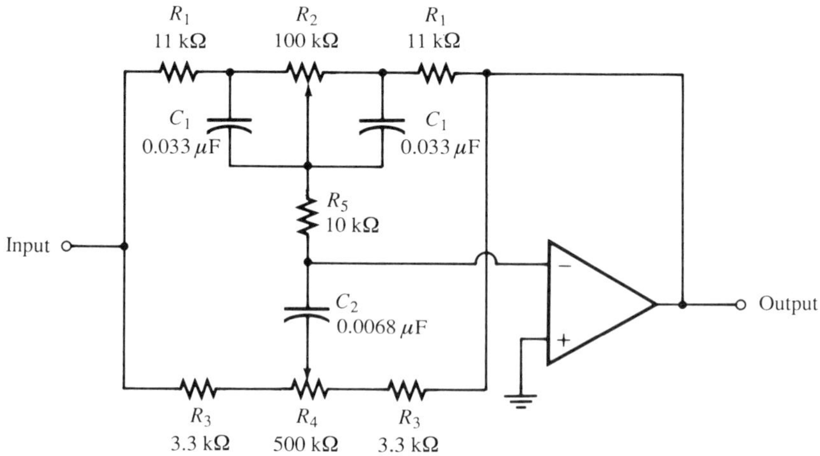

7–5 Octave Equalizers

An octave equalizer allows the user the ability of having independent tone control over several frequency ranges, or *bands*, that are separated an octave apart. Its primary purpose, other than tailoring the frequency response to the likes of the user, is that it compensates for any amplitude-versus-frequency or phase-versus-frequency characteristics of a given audio system.

The circuit of a single-band octave equalizer is shown in Figure 7–12. Like many audio circuits, there are usually more unknown values than there are circuit equations. Consequently, several assumptions or compromises have to be made. First, the closed-loop gain is typically set to approximately 4 (+12 dB) and is more than adequate for most situations. Furthermore, since the circuits are centered one octave apart, the circuit Q doesn't have to be large. A Q of 2 is usually satisfactory.

The center frequency of the equalizer is given by

$$f_o = \frac{1}{20\pi R_2 C_2}\sqrt{2 + \frac{R_2}{R_1}} \qquad \textbf{(7–22)}$$

The center frequency gain is found from

$$A_o = 1 + \frac{R_2}{3R_1} \qquad \textbf{(7–23)}$$

Figure 7–12
Single-band octave equalizer.

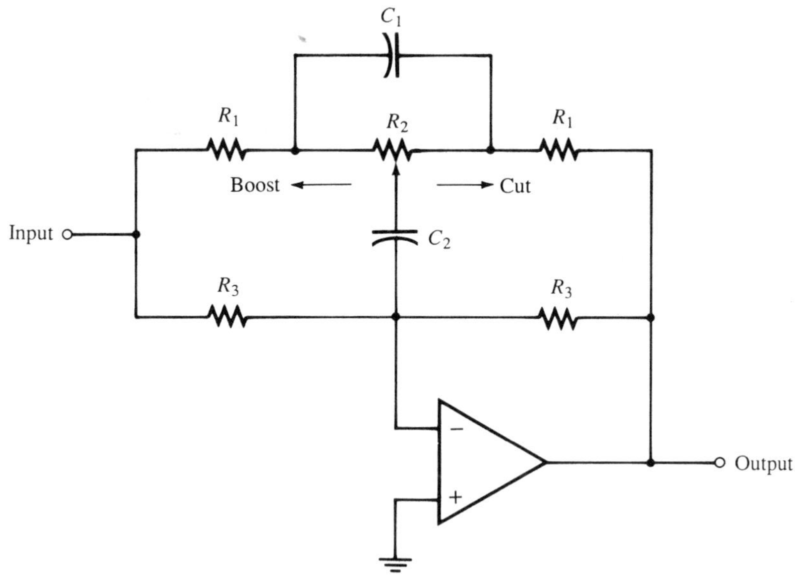

whereas the circuit Q is given by

$$Q = \sqrt{\frac{2R_1 + R_2}{9.61R_1}} \qquad (7\text{–}24)$$

Rewriting Equations 7–22, 7–23, and 7–24 in terms of R_1 and R_2,

$$R_1 = 3(A_o - 1)R_1 \qquad (7\text{–}25)$$

$$R_2 = (9.61Q^2 - 2)R_1 \qquad (7\text{–}26)$$

Potentiometer R_2 is usually of the slide wiper type to allow easy adjusting of the equalizer's cut and boost characteristics. To complete the design, the following general guidelines hold:

$$R_3 = 10R_2 \qquad (7\text{–}27)$$

$$C_1 = 10C_2 \qquad (7\text{–}28)$$

A complete audio equalizer system (for each channel, if stereo) would then be composed of ten such circuits whose center frequencies are spaced one octave apart, from 32 Hz to 16 kHz. Such an arrangement is then referred to as a *ten-band equalizer.*

Example 7–3

Using the circuit of Figure 7–12, determine the component values necessary to construct an equalizer centered on 250 Hz with a center frequency gain of 4 (12 dB).

Solution:

(a) Select R_2 arbitrarily to be equal to **100 kΩ.**

(b) From Equation 7–27,

$$R_3 = (10)(100\ \text{k}\Omega) = \mathbf{1\ M\Omega}$$

(c) From Equation 7–25,

$$R_1 = \frac{100\ \text{k}\Omega}{(3)(4-1)} = \mathbf{11.1\ k\Omega} \quad \text{(use a 10-k}\Omega\text{ standard value)}$$

(d) Using Equation 7–24, we check the value of Q:

$$Q = \sqrt{\frac{20\ \text{k}\Omega + 100\ \text{k}\Omega}{(9.61)(10\ \text{k}\Omega)}} = 1.12$$

which is satisfactory for most purposes.

(e) The value for C_2 is found using Equation 7–22:

$$C_2 = \frac{1}{(20\pi)(250\ \text{Hz})(100\ \text{k}\Omega)}\sqrt{2 + \frac{100\ \text{k}\Omega}{10\ \text{k}\Omega}} = \mathbf{0.00221\ \mu F} \quad \text{(use a 0.0022-}\mu\text{F standard value)}$$

(f) From Equation 7–28,

$$C_1 = (10)(0.0022\ \mu\text{F}) = \mathbf{0.022\ \mu F}$$

The completed circuit is shown in Figure 7–13. Table 7–3 summarizes the other values of C_1 and C_2 necessary to form a complete ten-band equalizer system.

Using ten such equalizer circuits (see Example 7–3 and Table 7–3), we can then form a complete ten-band equalizer system, as illustrated in Figure 7–14. The first operational amplifier stage is a unity-gain inverting amplifier buffer having a high input impedance (100 kΩ). The final stage functions as a summing amplifier, where the 3.3-μF capacitor removes any DC voltage from appearing at the output.

Figure 7–13
Completed 250-Hz, 12-dB octave equalizer circuit for Example 7–3.

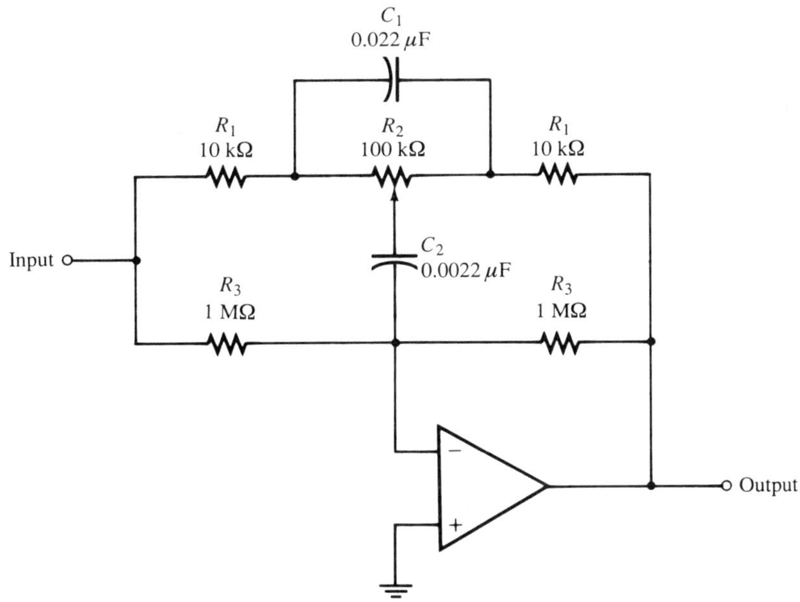

Table 7–3
Octave equalizer values

Frequency	C_1	C_2
32 Hz	0.18 μF	0.018 μF
64 Hz	0.1 μF	0.01 μF
125 Hz	0.047 μF	0.0047 μF
250 Hz	0.022 μF	0.0022 μF
500 Hz	0.012 μF	0.0012 μF
1 kHz	0.0056 μF	560 pF
2 kHz	0.0027 μF	270 pF
4 kHz	0.0015μF	150 pF
8 kHz	680 pF	68 pF
16 kHz	360 pF	36 pF

7–6 Two-Channel Panning Circuit

Often, recording engineers can slowly move the signal from one stereo channel to another. This has the effect of making the apparent position of the recording microphone move about in a panorama from left to right, or vice versa. Such a circuit then is called a *panoramic control, panning,* or *stereo pan-pot circuit.* Its output is required to have unity gain at each extreme of the potentiometer travel. Consequently, 100% of the input signal appears at one output channel while the output of the other channel is zero. With the control potentiometer centered, the output from each channel is −3 dB.

Figure 7–14
Block diagram of a 10-band equalizer from 20 Hz to 20 kHz.

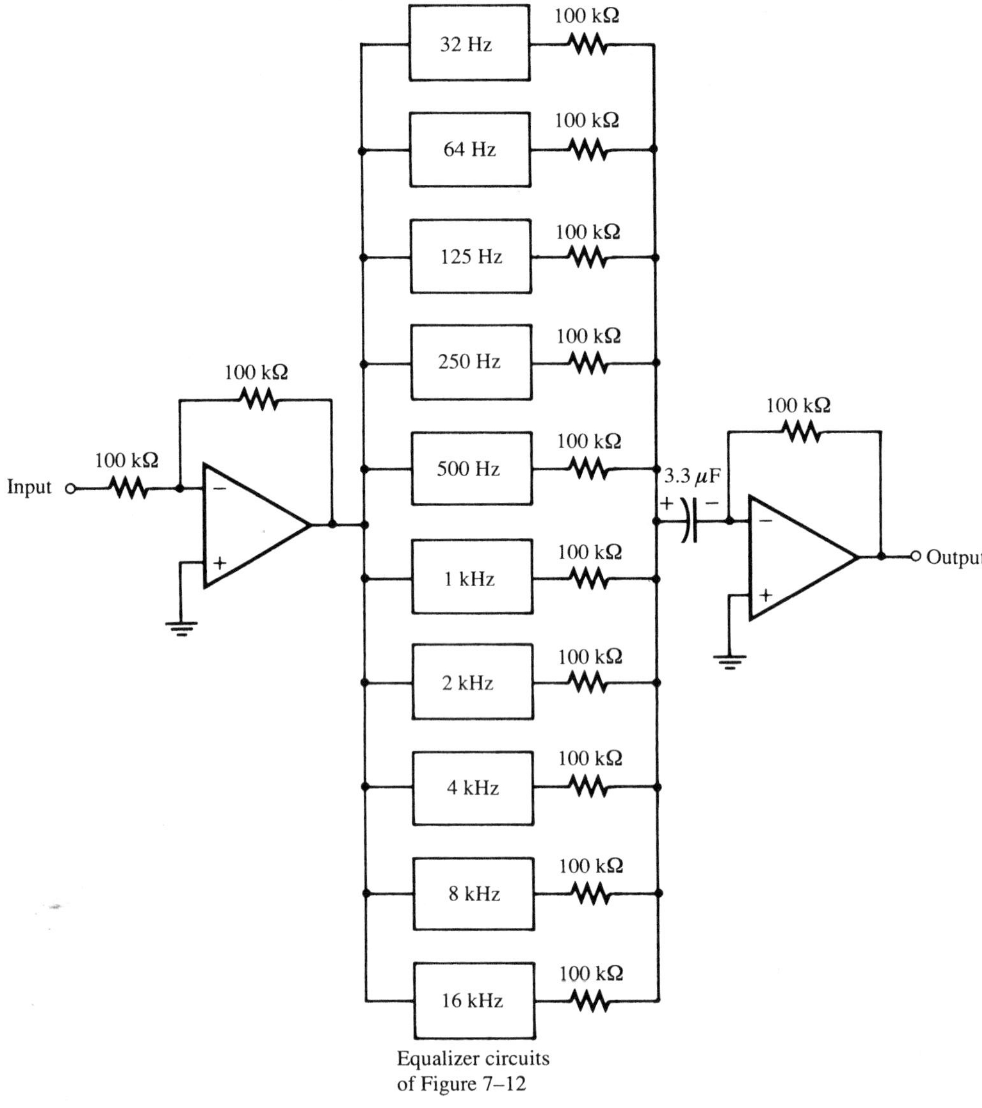

Panning circuits normally require two oppositely wound controls that must be ganged together. However, the panning control circuit of Figure 7–15 uses a single linear-taper potentiometer. The selection of the circuit resistances begins with the arbitrary choice of R_1, so that

$$R_2 = 0.707R_1 \tag{7–29}$$

$$R_3 = 3.41R_1 \tag{7–30}$$

Figure 7–15
Two-channel panoramic control circuit.

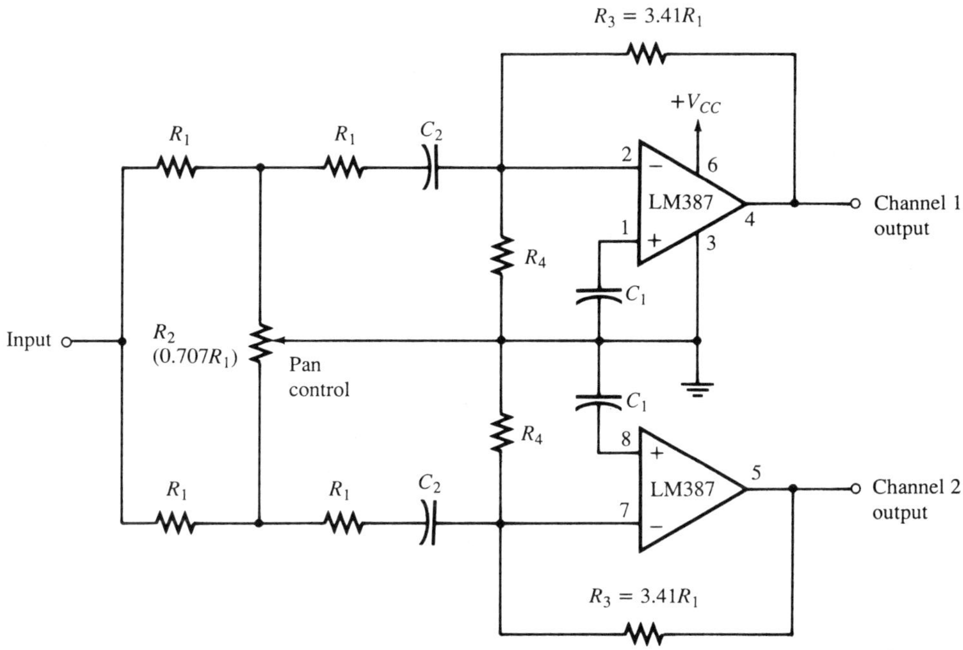

whereas R_4 is a biasing resistor for the LM387 and is chosen from

$$R_4 = \frac{1}{R_3}\left(\frac{V_{CC}}{2.6} - 1\right) \qquad (7\text{–}31)$$

7–7 Integrated-Circuit Audio-Power Amplifiers

As a logical step up from the basic operational amplifier with limited output capabilities, many manufacturers have devices capable of providing high-power output levels from approximately 300 mW to well above 25 W. Although they are generally intended for single-supply operation, these devices can also be operated from split-supply voltages. The following discusses the LM380, 384, and 386 as illustrative examples.

The LM380, whose pin diagram is shown in Figure 7–16, is a fixed-gain power amplifier intended for consumer applications. It has a voltage gain of 50 (34 dB), and the DC quiescent output is automatically fixed at one-half the single-supply voltage level with short-circuit protection. Its output power varies with both the level of the load as well as supply voltage. Table 7–4 summarizes the electrical characteristics of the LM380.

Figure 7–16
LM380 power amplifier pin diagrams. (a) 14-pin DIP. (b) 8-pin DIP.

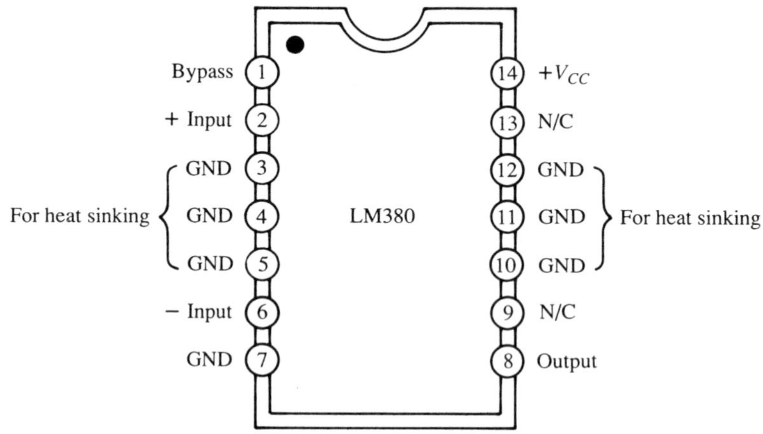

(a)

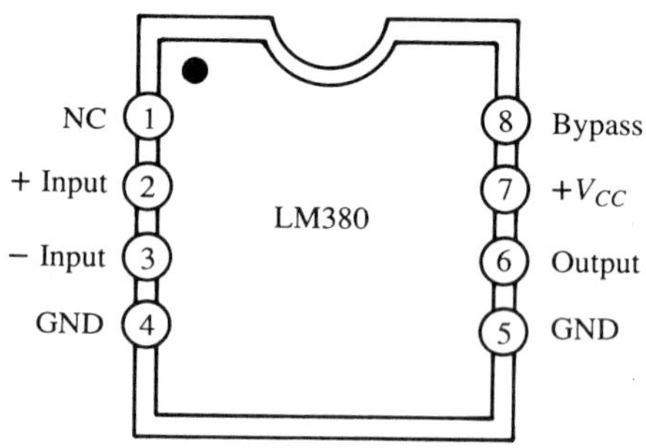

(b)

Table 7–4
LM380 electrical characteristics

Parameter	Conditions*	Minimum	Typical	Maximum	Units
V_{CC}		8		22	V
Bandwidth	$P_o = 2$ W $R_L = 8\ \Omega$		100		kHz
I_{SC}			1.3		A
P_o(RMS)	$R_L = 8\ \Omega$	2.5			W
Gain		40	50	60	V/V
V_o	$R_L = 8\ \Omega$		14		V p-p
R_{in}			150		kΩ

*$V_{CC} = +18$ V

Because of the manner in which it is internally biased, signal sources may be directly coupled to either the inverting or noninverting inputs. As shown in the noninverting amplifier circuit of Figure 7–17a, the inverting input is typically left floating. For the inverting amplifier circuit of Figure 7–17b, the noninverting input can be handled in one of three ways:

1. Connect a capacitor between the noninverting input and ground, which is preferred when using high-impedance signal sources or transducers.
2. Connect a resistor between the noninverting input and ground, which is preferred when using moderate to low-impedance signal sources or transducers. The resistor is made equal to the signal source impedance.
3. Connect the noninverting input directly to ground, which is preferred when using low-impedance DC signal sources or transducers.

For both noninverting and inverting amplifier configurations, all that is needed is an output coupling capacitor to remove the DC quiescent output voltage from the passing onto the load.

When more power is required than can be supplied with one LM380, two such devices can be connected in a "bridge" configuration, as illustrated in Figure 7–18. Here, the maximum output voltage swing across the floating load is now twice that with a single LM380, so that the output power is increased by a factor of four.

Figure 7–17
LM380 audio amplifier circuits. (a) Noninverting amplifier. (b) Inverting amplifier.

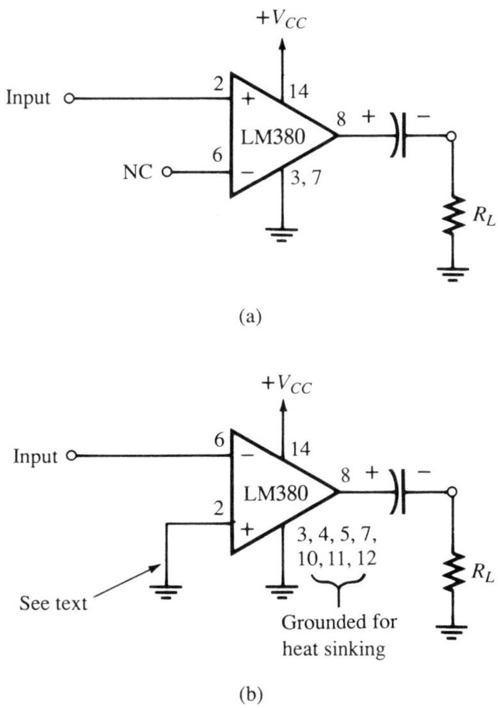

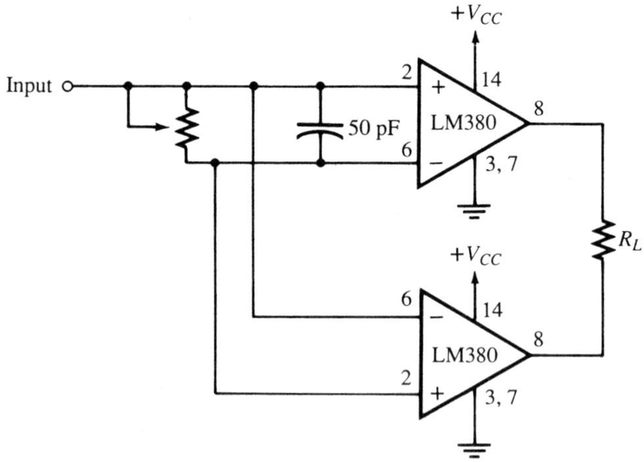

Figure 7–18
Bridge power amplifier using two LM380s to increase output power to load.

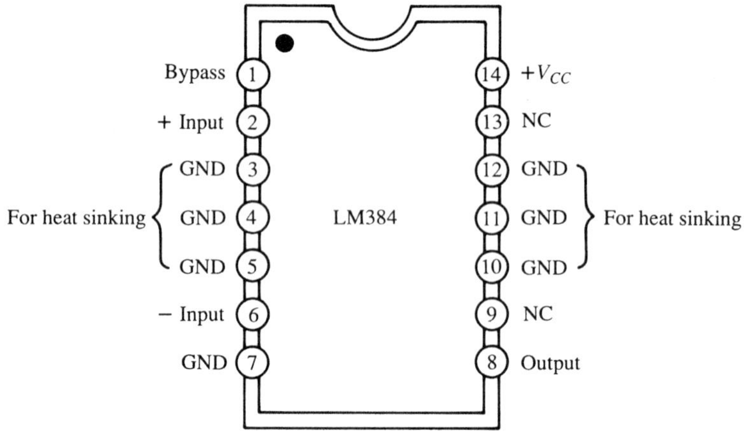

Figure 7–19
LM384 amplifier pin diagram.

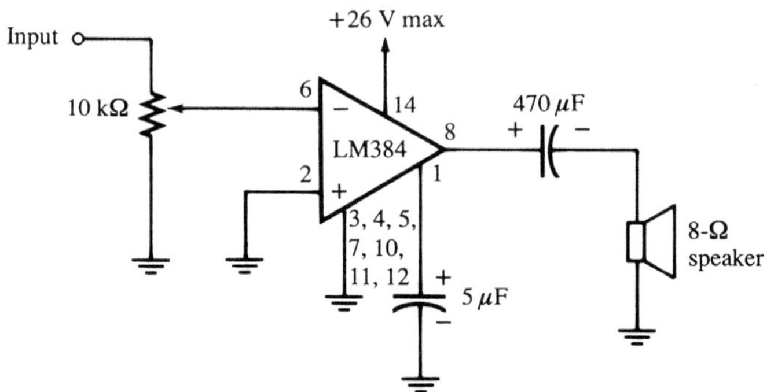

Figure 7–20
A 5-W audio-power amplifier using an LM384.

Figure 7–21
LM386 amplifier pin diagram.

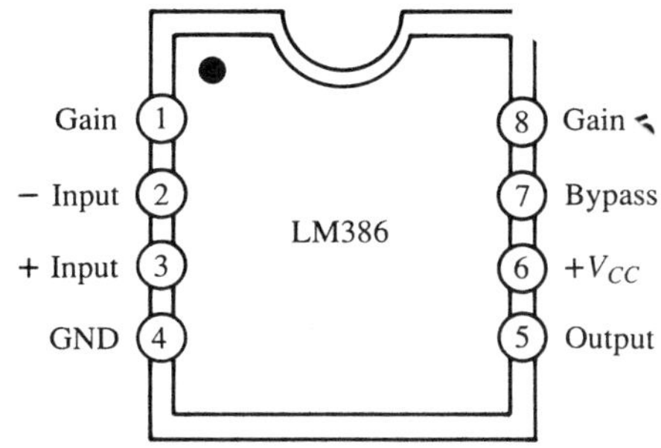

Besides the LM380, some manufacturers, such as National Semiconductor, have LM384 and LM386 devices. The LM384 (Figure 7–19) is similar in virtually every regard to the LM380, but the maximum supply voltage is increased to +26 V, which in turn provides higher output power. A typical power output of 7.5 W with 10% total harmonic distortion can be achieved with an 8-Ω load and a supply voltage of 26 V. As an illustration, Figure 7–20 shows the simple circuit of a 5-W audio power amplifier designed to operate with an 8-Ω speaker load.

The LM386 (Figure 7–21) is designed for low supply voltage (4-V to 12-V) applications; as a result, the output power is limited to about 325 mW with an 8-Ω load and a 6-V supply. In addition, the voltage gain is internally set to 20 (26 dB) over the 300-kHz bandwidth. The circuit shown in Figure 7–22 is a simple fixed-gain amplifier. The 10-kΩ potentiometer at the input serves as a volume control.

Figure 7–22
26-dB audio amplifier using an LM386.

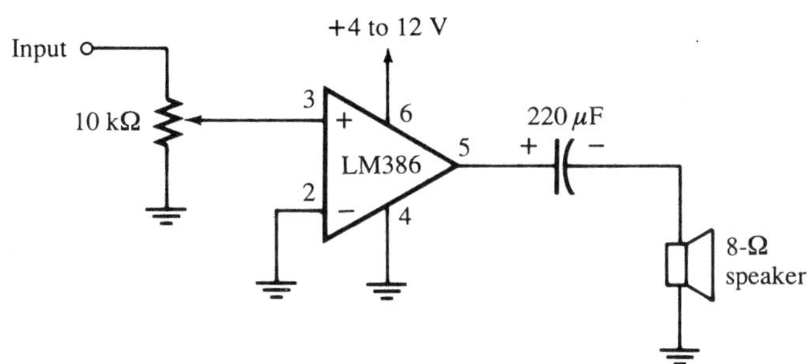

Figure 7–23
Increasing the voltage gain of an LM386 up to 46 dB using an external RC network between pins 1 and 8.

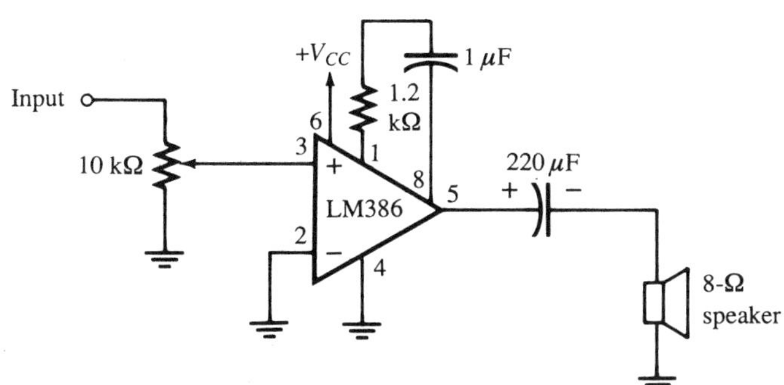

Table 7–5
Comparison of LM380, LM384, and LM386 amplifier characteristics

		Typical Output Power (W) at 10% THD		
Device	**Supply Voltage**	R_L **= 4 Ω**	R_L **= 8 Ω**	R_L **= 16 Ω**
LM380	12	2.5	2.5	0.5
	14	3.3	2.2	1.0
	18	4.2	4.0	2.2
LM384	18	4.2	4.0	2.2
	22	3.5	5.7	3.5
LM386	6	0.32	0.32	0.18
	9		0.5	0.5
	12		0.3	0.9

Although the voltage gain is internally fixed at 20 (26 dB), it can be increased up to 200 (46 dB) by the addition of a 1.2 k-Ω resistor and capacitor (100 μF maximum) between the gain pins 1 and 8 (Figure 7–23). Here, a 1-μF capacitor typically provides a voltage gain of 50 (34 dB).

Table 7–5 compares the output capabilities of LM380, LM384, and LM386 devices.

7–8 Heat Sinks

Many integrated-circuit devices, such as the power amplifiers discussed in this chapter and monolithic voltage regulators (see Chapter 9), will operate at their maximum ratings only if adequate means are taken to prevent the integrated-circuit device from overheating. Since every electronic device internally generates some amount of heat, care should be taken to ensure that the power dissipation ratings are not exceeded, whereby the device then self-destructs.

The most common form of temperature control for semiconductor devices is a **heat sink.** This is nothing more than a piece of metal placed in contact with the case of the device to draw the heat away from it to the ambient environment. The object here is to increase the available surface area for the transfer of heat away from the device. To have a large surface area without a large overall size, heat sinks generally have *fins,* as shown in Figure 7–24.

Heat can be transferred away from an integrated-circuit package by one or more of the following three methods:

- Conduction
- Convection
- Radiation

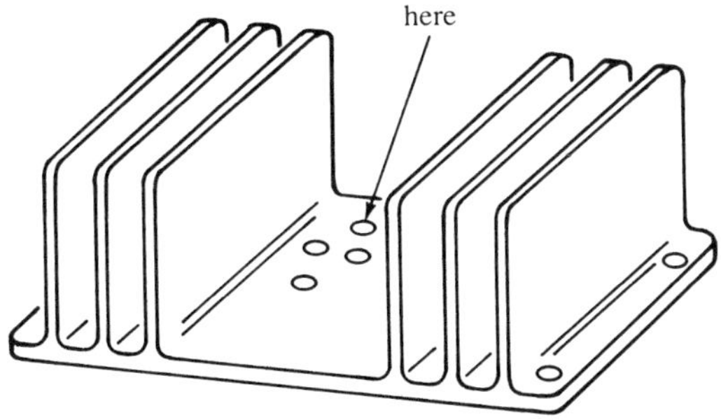

Figure 7–24
Heat sink.

Conduction involves the transfer of heat away from the integrated circuit through the heat sink material. The heat flow is analogous to the current flow through a resistor. The lower the resistance (i.e., the greater the conductance), the larger the current flow. **Convection** involves the removal of heat away from the heat source by air currents. This is similar to having a fan nearby or any other movement of cooler air passing by the heat source. Finally, **radiation** is similar to convection, but it is the transfer of heat away from its source in still air.

Thermal Resistance

The concept of **thermal resistance** is a figure of merit used to relate temperature and power. It is defined simply as the temperature difference between two regions divided by the dissipated power (P_D) and is expressed in units of degrees Celsius per watt (°C/W). Mathematically,

$$\theta_{AB} = \frac{T_A - T_B}{P_D} \tag{7–32}$$

Very often, the study of heat transfer is expressed in terms of electrical analogs so that the modeling of the overall heat-transfer system can be easily accomplished. Here, Equation 7–32 is the thermal equivalent of Ohm's law ($R = V/I$), whose individual terms are related in Table 7–6.

Figure 7–25 illustrates a simplified heat-transfer circuit, or model, for the entire integrated circuit, heat sink, and ambient environment system. Beginning from inside the device, the temperature of the semiconductor junction (T_J) is the driving force. For most integrated circuits, a maximum junction temperature of

Table 7–6
Analogous parameters

Electrical	Thermal
Resistance (R)	Resistance (θ)
Voltage (V)	Temperature (T)
Current (I)	Power dissipation (P_D)

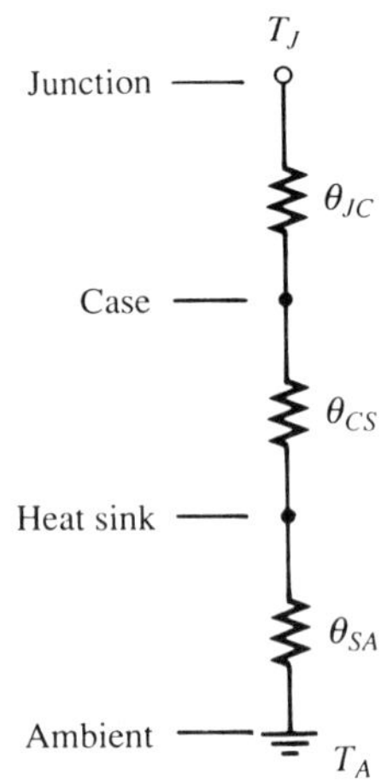

Figure 7–25
Simplified heat transfer model for the integrated circuit, heat sink, and ambient environment.

Table 7–7
Typical case: Heat sink thermal resistances

Mounting	Thermal Resistance
Dry	0.25°C
Silicone grease only	0.1°C
Dry with mica washer	0.8°C
Silicone grease + mica washer	0.4°C

150°C is permitted and is specified on the data sheet. From the junction to the device's case, it encounters a junction-case thermal resistance (θ_{JC}), which is a measure of how many degrees hotter the semiconductor junction will be than the case for each watt of power dissipated. This is generally given on the data sheet.

The thermal resistance from the case to the heat sink (θ_{CS}) depends on how the device is actually mounted to the heat sink. The device may be mounted directly (i.e., dry) to the heat sink, or it may be mounted with some silicone grease to aid the transfer of heat from the case to the heat sink. For those applications where the case must be electrically insulated from the heat sink, which is usually ground, a mica washer is placed between the case and heat sink and insulating standoff washers are used for the mounting holes. For better heat transfer, silicone grease can also be used with the mica washer. Table 7–7 gives typical values for θ_{CS}. Finally, θ_{SA} is the thermal resistance from the heat sink to the ambient environment, which is taken as near room temperature (T_A). As is explained in the next section, this is the value that is determined so that the proper heat sink may be used.

Selecting the Heat Sink

From the simple electrical model shown in Figure 7–25, the total series thermal resistance from the semiconductor junction to the ambient (θ_{JA}) is the sum of the three individual resistances:

$$\theta_{JA} = \theta_{JC} + \theta_{CS} + \theta_{SA} \tag{7–33}$$

Rewriting Equation 7–32 in terms of the proper subscripts,

$$\theta_{JA} = \frac{T_J - T_A}{P_D} \tag{7–34}$$

or, after substitution of Equation 7–33 and solving for θ_{SA},

$$\theta_{SA} = \frac{T_J - T_A}{P_D} - (\theta_{JC} + \theta_{CS}) \tag{7–35}$$

Manufacturers then specify the heat sink in terms of its thermal resistance. Any thermal resistance less than the value calculated from Equation 7–35 will be adequate.

The dissipated (DC) power (P_D) of a device is simply the difference of the power delivered by the DC power supply and the power delivered to the load:

$$P_D = P_{\text{power supply}} - P_{\text{load}} \tag{7–36}$$

On the other hand, data sheets usually provide "power dissipation versus power output" curves. This is particularly true for audio-power amplifier devices, such as that shown in Figure 7–26 for the LM378 for 8-Ω loads and total harmonic distortion levels between 3 and 10%. For other loads not shown by the data sheet

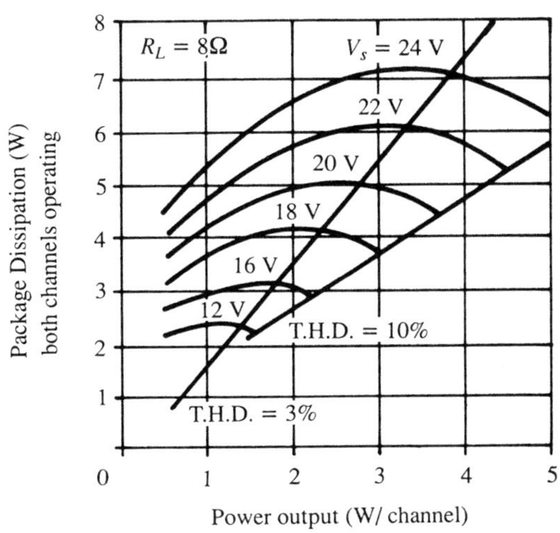

Figure 7–26
Power dissipation versus power output curve for the LM378 for 8-Ω loads.

curves, the dissipated power may be determined from

$$P_D = \frac{V_o(\text{p-p})^2}{20R_L} \qquad \textbf{(7–37)}$$

where V_o(p-p) is the peak-to-peak output signal voltage across the load R_L; it can also be equal to the power supply voltage to arrive at the maximum possible value. Although Equation 7–37 is derived for sine-wave signals, it nevertheless provides a good approximation for any complex waveform except square waves.

Manufacturer's data sheets give the maximum power dissipation of a given device, which is usually specified at a nominal room temperature of 25°C. For ambient temperatures above 25°C, the maximum power dissipation rating must then be decreased, or *derated,* for each degree above 25°C. The value of the derating factor (α) used is specified on the data sheet, or the result is summarized by a derating curve. In terms of the equation of a straight line, the maximum power dissipation that a device can have at any temperature between 25°C and the maximum operating temperature [T_A(max)] is given by

$$P_D = P_D(25°\text{C}) - \alpha(T_A - 25°\text{C}) \qquad \textbf{(7–38)}$$

Example 7–4

The data sheet for an LM380 gives the power dissipation at 25°C at 10 W, which must be derated by a factor of 0.083 W/°C for ambient temperatures over 25°C. Determine the maximum power dissipation allowable if the ambient temperature reaches 35°C.

Solution: From Equation 7–38,

$$\begin{aligned} P_D &= 10\text{ W} - (0.083\text{ W/°C})(35 - 25) \\ &= \mathbf{9.17\ W} \end{aligned}$$

Example 7–5

Determine the thermal resistance of a heat sink needed for an LM378 mounted with a mica washer with silicone grease. The following additional information is provided:

$$\begin{aligned} T_J\text{max} &= 150°\text{C} \\ \theta_{CJ} &= 13.4°\text{C/W} \\ \text{Supply voltage} &= 24\text{ V} \\ \text{Amplifier output voltage} &= 4\text{ W across an } 8\text{-}\Omega\text{ load} \end{aligned}$$

Solution:

(a) From the power output versus power dissipation curve for the LM378 (Figure 7–26, THD = 3%), $P_D = 7$ W.

(b) From Table 7–7, the typical thermal resistance for mounting with a silicone-grease-covered mica washer is 0.4°C/W. Using Equation 7–35,

$$\theta_{SA} = \frac{(150°\text{C} - 25°\text{C})}{7\text{ W}} - (13.4°\text{C/W} + 0.4°\text{C/W})$$
$$= \mathbf{4.06°C/W}$$

Consequently, any heat sink having a thermal resistance of 4.06 °C/W or less can be used.

Summary

This chapter discussed how the basic operational amplifier configurations of Chapter 2 and the operational amplifier parameters discussed in Chapter 3 are translated for audio frequencies. In general, signals are capacitively coupled into and out of the amplifier to remove noise and DC offset voltages. Besides the basic amplifier configurations, designs for specialized audio applications such as RIAA-response preamplifiers, microphone preamplifiers for unbalanced and balanced inputs, octave equalizers, and tone and two-channel panoramic control circuits were examined.

Besides the standard operational amplifier, this chapter also looked at specialized integrated-circuit power amplifiers such as the LM380, 384, and 386, which are capable of providing power levels of up to 10 W. Finally, information was given about how a heat sink is used to permit integrated-circuit devices such as power amplifiers and voltage regulators to operate at or near their maximum ratings without destruction by overheating.

Glossary

audio Referring to those frequencies that can be heard by the human ear and taken to be in the range from 20 Hz to 20 kHz.

conduction A form of heat transfer through or by means of a conductor.

convection A form of heat transfer caused by air currents.

heat sink A metallic device used for dissipating heat away from semiconductor devices.

preamplifier Often called a preamp, a low-noise first-stage amplifier specifically designed to amplify signals at the low end of the dynamic range of practical amplifier circuits.

radiation The transfer of heat away from its source in still air.

thermal resistance A figure of merit defined as the temperature difference between two regions divided by the dissipated power.

tone control circuit A circuit in an audio amplifier that compensates for the logarithmic response of the human ear so that it provides added bass and treble response, which may be adjusted to provide the most pleasing effect to the listener.

Review Questions

1. What is meant by the term *audio?*
2. Explain what considerations should be made when selecting an operational amplifier for audio use.
3. Does it make any difference whether the signal is an inverting or noninverting signal in an audio-amplifier stage?
4. Draw the schematic diagram for a noninverting amplifier used for audio signals. Also give the equation for the closed-loop gain.
5. Draw the schematic diagram for an inverting amplifier used for audio signals. Also give the equation for the closed-loop gain.
6. How does a preamplifier differ from other types of amplifiers?
7. What is the RIAA curve?
8. What is the difference between the use of balanced and unbalanced low-impedance microphones?
9. What is an octave equalizer? Why is one used?
10. What is a panoramic control, or stereo pan-pot circuit?
11. What is the purpose of using a heat sink?
12. Explain the difference between the following methods of heat transfer: conduction, convection, and radiation.
13. Draw the schematic of a simple thermal circuit consisting of a heat sink, integrated circuit (or transistor), and the ambient surroundings.
14. Give the units of thermal resistance.
15. When must the maximum power dissipation rating of an integrated circuit be derated?

Exercises

1. For the noninverting audio amplifier circuit of Figure 7–27, determine each value.
   (a) The input impedance
   (b) The closed-loop gain
   (c) The low-frequency response

Figure 7–27

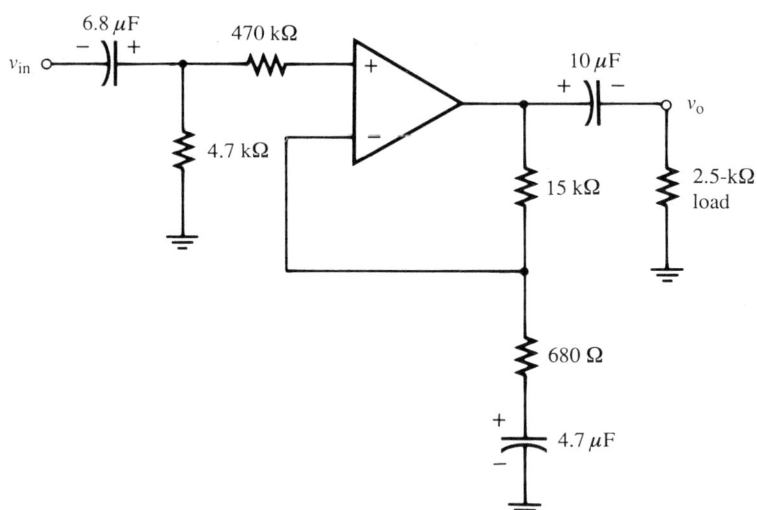

2. Determine suitable component values for the noninverting audio-amplifier circuit of Figure 7–2 having a closed-loop gain of 19 for all signals above 50 Hz when connected to a 1-kΩ load. Assume that the circuit is to have an input impedance of 10 kΩ.

3. Determine suitable component values for the unbalanced high-impedance amplifier circuit of Figure 7–8 having a closed-loop gain of 100 for all signals above 30 Hz. Assume that the circuit is fed by a microphone having an impedance of 2 kΩ. It is not required to determine the values for C_2, R_3, and R_4.

4. Determine suitable component values for the bass-treble tone control circuit of Figure 7–10a having a gain of 40 dB up to 50 Hz and above 3 kHz. Assume that a 50-kΩ potentiometer is used for R_2 and $R_5 = 10$ kΩ.

5. Determine suitable component values using the circuit of Figure 7–12 for an octave equalizer centered on 2 kHz with a center frequency gain of 6 dB.

6. Repeat Exercise 5, this time centered on 64 Hz with a center frequency gain of 10 dB.

7. The power dissipation rating at 25°C for a given integrated circuit is 3 W with a derating factor of 0.12 W/°C for temperatures above 25°C. Determine the maximum power that the integrated circuit can dissipate at an ambient temperature of 40°C.

8. Determine the thermal resistance of the heat sink needed for an LM378 audio amplifier integrated circuit that is mounted with a mica washer with silicone grease. The following additional information is provided:

$$T_J(\text{max}) = 150°\text{C}$$

$$\theta_{CJ} = 11.9°\text{C/W}$$

Supply voltage = 18 V

Amplifier output voltage = 2 W across an 8-Ω load with 3% THD

8 Nonlinear Signal Processing Circuits

Instructional Objectives

Upon completing this chapter, you will be able to

- Describe the operation and advantages of precision half-wave and full-wave rectifiers.
- Describe the operation and limitations of logarithmic-type amplifiers and how they are used to multiply and divide two signals.
- Describe the operation of analog multipliers and how they are used to perform division, squaring, and square roots.
- Describe how an analog multiplier modulates a carrier signal in an AM system.
- Briefly explain the operation and advantage of a compandor.
- Describe the operation of inverting, noninverting, and window comparator circuits, and explain what operational amplifier parameters should be considered.
- Explain the effect of noise on simple comparator circuits and how this may be eliminated.
- Describe the operation of peak detectors, clampers, and sample-and-hold circuits.

Introduction

Very often, signals must be processed in other ways besides amplification. The resultant signal is then often used to control another process, or is transformed into a form that allows easy measurement. This chapter discusses those signal-processing circuits that are nonlinear. Here, the output signal is not a linear function of the input signal, such as precision rectifiers, logarithmic type amplifiers, analog multipliers, compandors, comparators, peak detectors, sample-and-hold amplifiers, and clampers. In many case cases, these circuit functions can be achieved using operational amplifier circuits. For others, specialized integrated-circuit modules offer improved circuit performance and versatility.

8–1 Precision Rectifiers

When a diode is used in either half-wave or full-wave rectifiers, its nonlinear characteristics tend to distort the output waveform at low signal levels. Since silicon diodes must be forward-biased to about 0.7 V and germanium diodes to about 0.3 V before conduction begins, neither of these devices alone is suitable for the rectification of small signal levels below several voltages. The following subsections discuss how precise rectification can be accomplished using operational amplifiers.

Half-Wave Rectifier

Figure 8–1 shows the problems associated with trying to rectify small signals with a simple diode circuit. If a sine wave having a peak-to-peak voltage of 2 V is the input signal to the half-wave rectifier of Figure 8–1a, the peak output voltage is only about 0.3 V. Because of the high diode drop that has to be overcome in relation to the peak input voltage, the rectification action actually takes place for a period lasting substantially less than one-half cycle. On the other hand, a 1-V peak-to-peak sine-wave signal will not allow a typical silicon diode to conduct at all, giving a zero output voltage (Figure 8–1b).

By incorporating conventional diodes as the feedback element in an operational amplifier circuit, as shown in Figure 8–2a, it is now possible to overcome this signal-size limitation. Unlike most operational amplifier circuits, the output signal here is not taken from the output of the operational amplifier itself, but instead it is taken from the junction between R_1 and D_1. When the input signal is

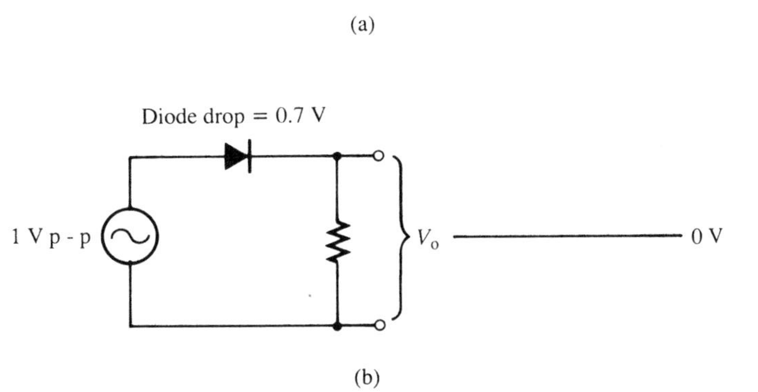

Figure 8–1
Diode half-wave rectifier. (a) Rectification of a signal level slightly larger than the diode drop. (b) Rectification of a signal level smaller than the diode drop.

Figure 8–2
Precision half-wave rectifier. (a) Circuit. (b) Input and output waveforms.

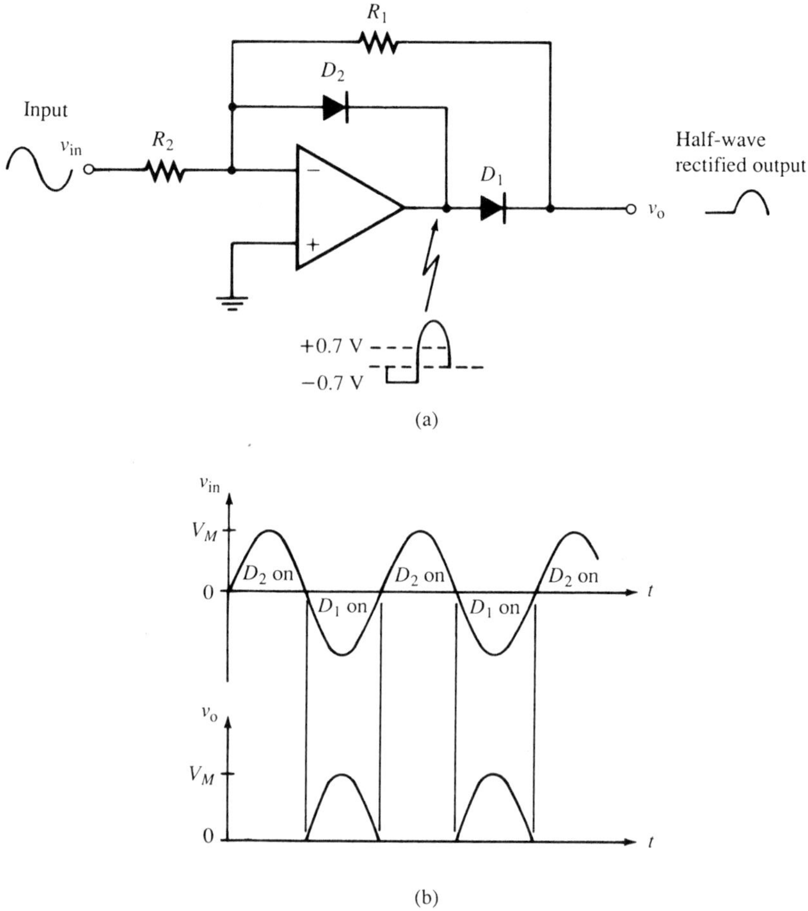

positive, diode D_2 is forward-biased and all the feedback current flows through D_2; the output of the circuit is zero because D_1 is reverse-biased (Figure 8–2b). When the input signal goes negative, D_2 is reverse-biased, and because D_1 is forward-biased, the feedback current flows through both D_1 and R_1 in series.

A signal then appears at the output across R_1. As soon as the input level goes negative, the output of the operational amplifier jumps to the positive level required to cause D_1 to conduct. This effectively eliminates any diode drop from the output signal. Once diode D_1 begins to conduct, the circuit behaves as a conventional inverting amplifier. If $R_1 = R_2$, the closed-loop gain is unity.* Thus, the output signal is nearly an exact half-wave-rectified version of the input without the undesirable effects due to the diode drop. For this reason, this type of circuit is

*The output of the operational amplifier is always one diode drop higher than the output of the circuit for the positive half-cycle of output.

then referred to as a **precision half-wave rectifier.** On the other hand, if desired, the negative-going half-cycle at the output could be achieved merely by reversing the direction of both diodes.

Full-Wave Rectifier

The precision half-wave rectifier circuit of Figure 8–2a forms the basis for the *precision full-wave rectifier* of Figure 8–3a. A close look reveals that it is composed of a half-wave precision rectifier followed by a two-input summing amplifier.

Figure 8–3
Precision full-wave rectifier made from a precision half-wave rectifier and a two-input summing amplifier. (a) Circuit. (b) Waveforms.

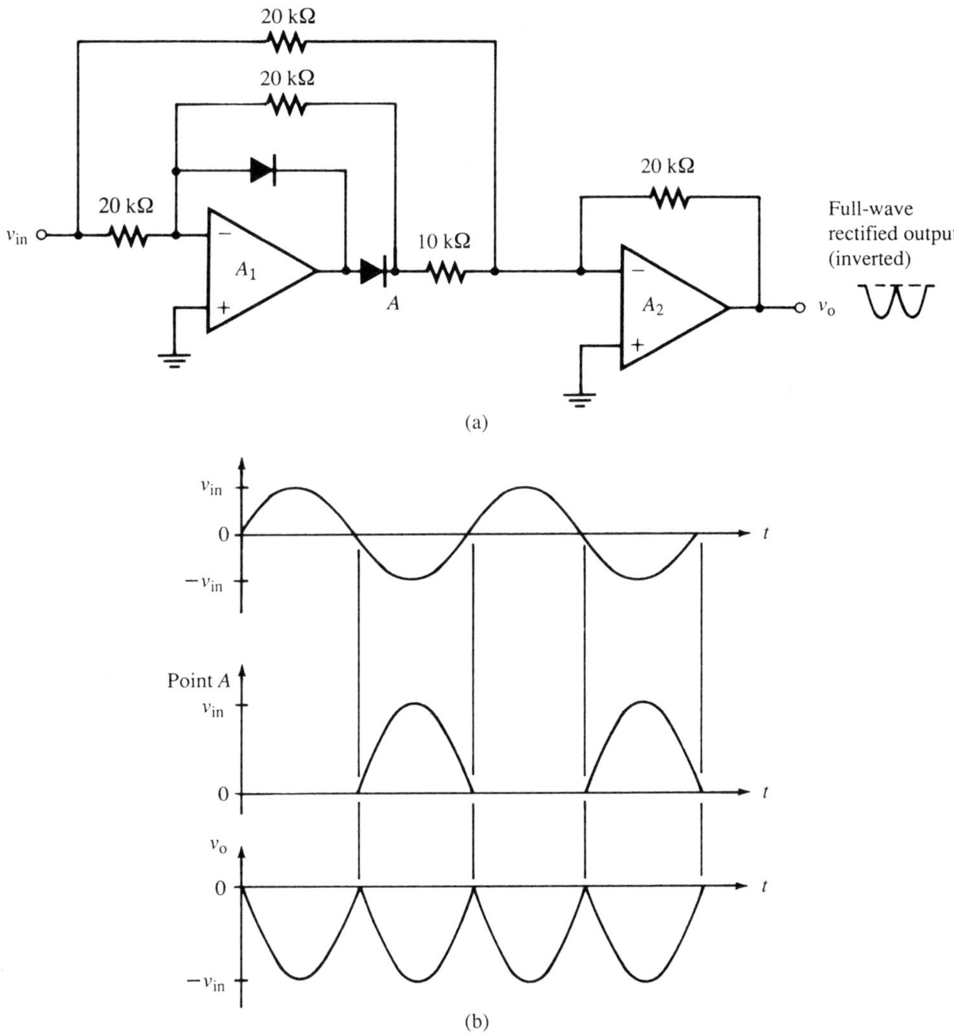

Figure 8–4
Precision half-wave rectifier with complementary outputs.

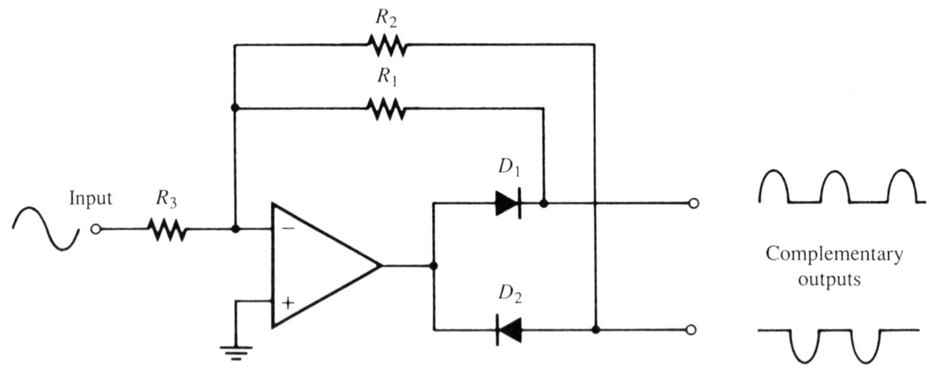

One input to the summing amplifier is the signal to be rectified. The other input to the summing amplifier is the output from the precision half-wave rectifier discussed previously. As a result, the summing amplifier produces an output signal with two components. One component is the same level as the input signal and is inverted. The other component is due to the half-wave rectified output, which is then inverted and amplified by a factor of two. The instantaneous sum of these two signal components is a full-wave rectified version of the input signal. Figure 8–3b shows how these signals combine to form the full-wave rectified sine-wave output. This same analysis can be applied to any complex AC signal.

Another version of a precision full-wave rectifier is given by the building-block circuit of Figure 8–4. Here, this circuit produces two complementary half-

Figure 8–5
Precision full-wave rectifier made from a difference amplifier.

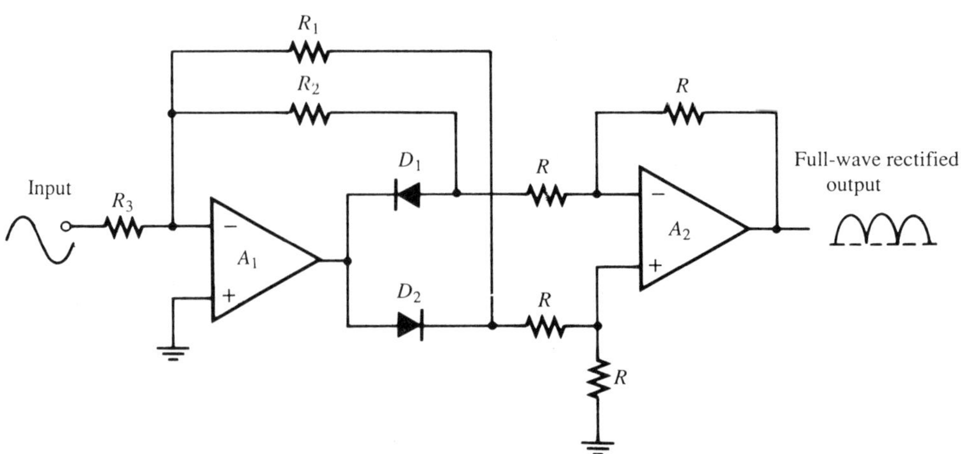

wave outputs. Its function is identical to the circuit of Figure 8–2 except that a resistor has been placed in series with diode D_2 to provide a second output on the other input half-cycle. By connecting the outputs of the circuit of Figure 8–4 to a difference amplifier, we are able to form the full-wave-rectified version of the input signal, as shown in Figure 8–5.

8–2 Peak Detector

A **peak detector** is a circuit that will note and "remember" the peak positive or negative excursion of an input signal for an infinite period of time until it is reset. Such circuits are frequently used to measure the peak values of nonsinusoidal waveforms.

Figure 8–6 shows the circuit for a very simple positive-peak detector. The source signal charges the capacitor through the diode at a rate limited by the output impedance of the source. The capacitor will charge to the highest source voltage, less one diode drop, but cannot discharge except through (1) any leakage of the capacitor itself and (2) the very high input impedance of the operational amplifier. Since the operational amplifier is connected as a voltage follower, its output voltage will be equal to the positive peak value of the applied signal and will remain that way for relatively long periods.

Figure 8–7 shows a more sophisticated peak detector that buffers the signal source from the capacitor. Operational amplifier A_1 presents a high-impedance load to the signal source, whereas A_2 serves as a buffer between the capacitor and the load. With this circuit, the output voltage (V_o) at any given time is equal to the voltage on the capacitor, which in turn equals the peak input voltage that occurred up to that time. This assumes that the capacitor has had sufficient time to charge. The relatively low output impedance of A_1 serves to improve the circuit's response time for high-impedance sources.

Whenever the input signal's level becomes more positive than the current capacitor voltage, the capacitor will now charge toward the higher input level. On the other hand, if the input level now drops below the stored capacitor voltage, diode D_1 is reverse-biased and diode D_2 prevents the output of A_1 from going into negative saturation. This then serves to improve the recovery time of A_1 when the input signal again goes more positive. Resistor R_2 serves as a path for the input

Figure 8–6
Basic peak detector.

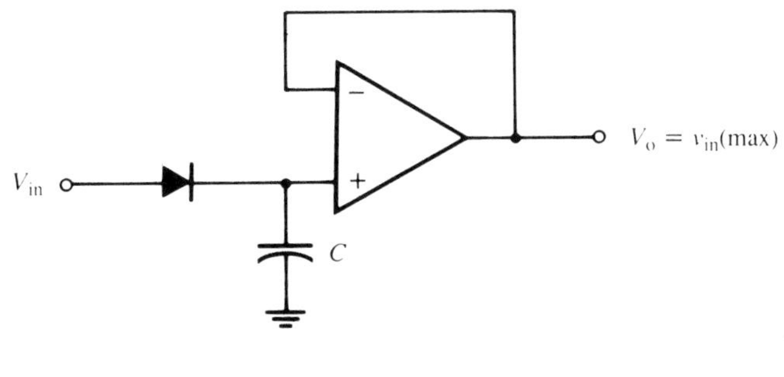

Figure 8–7
Improved peak detector circuit using a buffer amplifier.

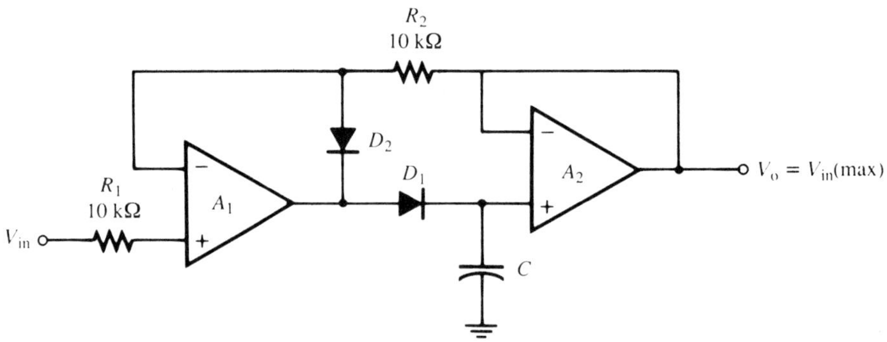

bias current to A_1. Resistor R_1 is made equal to R_2 in order to minimize the effect of the operational amplifier's offset voltage.

Since A_1 is used to drive a capacitive load and is operated with unity gain, it is then important to be sure that any required frequency compensation is provided in order to avoid instability (i.e., oscillations).

8–3 Comparators

When processing analog signals, it is often necessary to know if an instantaneous signal level is more positive or negative than another signal or if a given signal is inside or outside the range of two fixed voltage levels. On the other hand, it is sometimes required to determine when the signal's polarity changes. Circuits that are capable of making these determinations are called **comparators.** Sometimes called a *level detector,* the comparator basically uses the operational amplifier in its open-loop mode to compare the instantaneous input signal level against a fixed reference voltage. The following subsections discuss comparator circuits built around conventional operational amplifiers as well as those using Norton amplifier and precision integrated-circuit comparator modules specifically designed for this purpose.

Basic Operational Amplifier Comparators

Using the open-loop circuit of Figure 8–8a, the input signal V_{in} is connected to the noninverting input and a fixed DC reference voltage V_{REF} is connected to the inverting input. Whenever V_{in} is greater than V_{REF}, even by as little as 1 mV, the high open-loop gain of the operational amplifier forces the output voltage to its positive saturation voltage, V_{+SAT}, which is typically 1.4 V to 2 V less than the corresponding supply voltage. The reference can be either a positive or negative voltage.

Figure 8–8
Noninverting comparator. (a) Circuit. (b) Input and output waveforms.

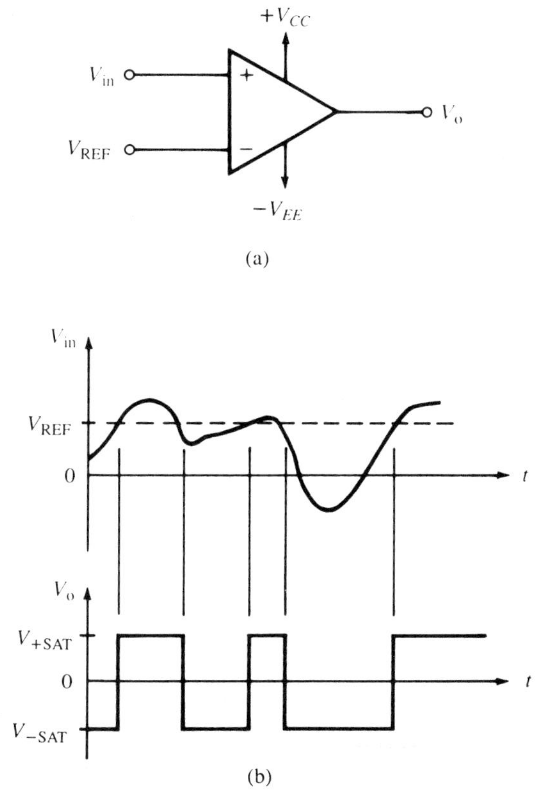

If the DC open-loop voltage gain of the operational amplifier, like the 741, is typically 200,000, then for $V_{in} - V_{REF} = 1$ mV, the output voltage will try to equal 200,000 × 1 mV, or 200 V. However, if the positive supply voltage is +9 V, then the output voltage will equal approximately +7 V when $V_{in} > V_{REF}$. On the other hand, whenever $V_{in} < V_{REF}$, the output voltage then swings to its negative saturation voltage, about −7 V.

The ideal level for switching occurs when $V_{in} - V_{REF} = 0$. However, this action is affected somewhat by the operational amplifier's input offset voltage, which should be compensated for when exact switching is required. The circuit of Figure 8–8a is often called a **noninverting comparator** because the output goes positive when the input signal is more positive than the voltage reference (Figure 8–8b). In summary, we have

$$V_o = V_{+SAT}, \quad \text{when } V_{in} > V_{REF} \qquad \textbf{(8–1a)}$$
$$= V_{-SAT}, \quad \text{when } V_{in} < V_{REF} \qquad \textbf{(8–1b)}$$
$$= 0, \quad \text{when } V_{in} = V_{REF} \qquad \textbf{(8–1c)}$$

Figure 8–9
Common methods for establishing a reference voltage. (a) Voltage divider. (b) Zener diode.

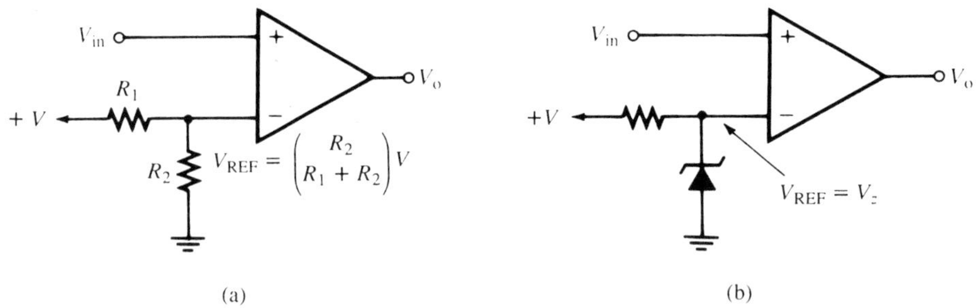

The reference voltage can be generated in variety of ways, as shown in Figure 8–9. The reference voltage can be obtained from a simple fixed resistive voltage divider or made variable using a potentiometer whose source voltage is obtained from the same regulated power supply that is used to power the operational amplifier. On the other hand, a reference voltage can be obtained by using a Zener diode voltage regulator. Here the reference voltage equals the Zener voltage of the diode used.

For the special case when the reference voltage is made zero by grounding the operational amplifier's inverting input, the resultant circuit of Figure 8–10a is called a **zero crossing detector.** The output then switches from one extreme to another when the input signal level crosses zero volts. Furthermore, this same

Figure 8–10
Zero crossing detector. (a) Circuit. (b) Input and output waveforms.

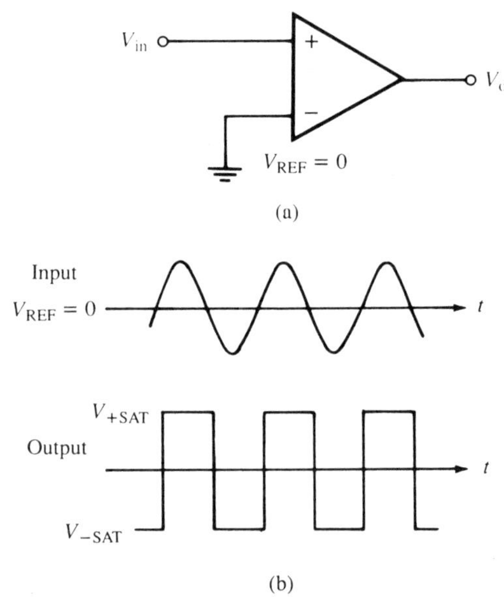

circuit can also be used as a **polarity detector,** since the output is positive when the input is positive and negative when the input is negative with respect to ground. One popular use for this circuit is to convert a sine wave into a square wave, as shown by the waveforms in Figure 8–10b.

By reversing the two inputs as shown in Figure 8–11a, an **inverting comparator** is formed that has the opposite action as the noninverting type. Whenever V_{in} at the inverting input is greater than V_{REF}, the output swings toward V_{-SAT}. Whenever V_{in} is less than V_{REF}, the output voltage then swings toward V_{+SAT} (Figure 8–11b). In summary,

$$V_o = V_{+SAT}, \quad \text{when } V_{in} < V_{REF} \qquad \textbf{(8–2a)}$$

$$= V_{-SAT}, \quad \text{when } V_{in} > V_{REF} \qquad \textbf{(8–2b)}$$

$$= 0, \quad \text{when } V_{in} = V_{REF} \qquad \textbf{(8–2c)}$$

To be useful, comparator circuits must be interfaced with other circuits. This interface may be as simple as the LED driver of Figure 8–12a, which indicates the

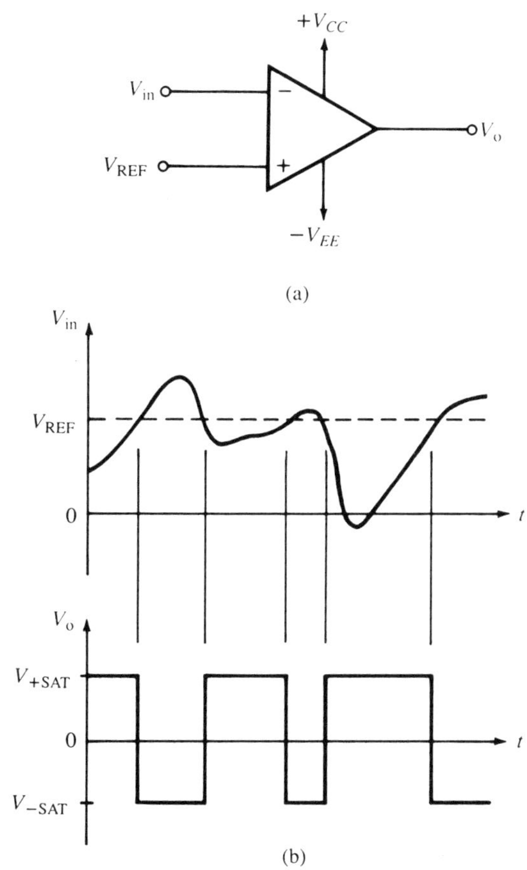

Figure 8–11
Inverting comparator.
(a) Circuit. (b) Input and output waveforms.

Figure 8–12
Comparator interfacing circuits. (a) LED output using an NPN transistor. (b) Limiting output to 5.1 V for TTL devices by placing a Zener diode in the feedback loop of an operational amplifier. (c) Alternative output limiting using a simple Zener diode regulator.

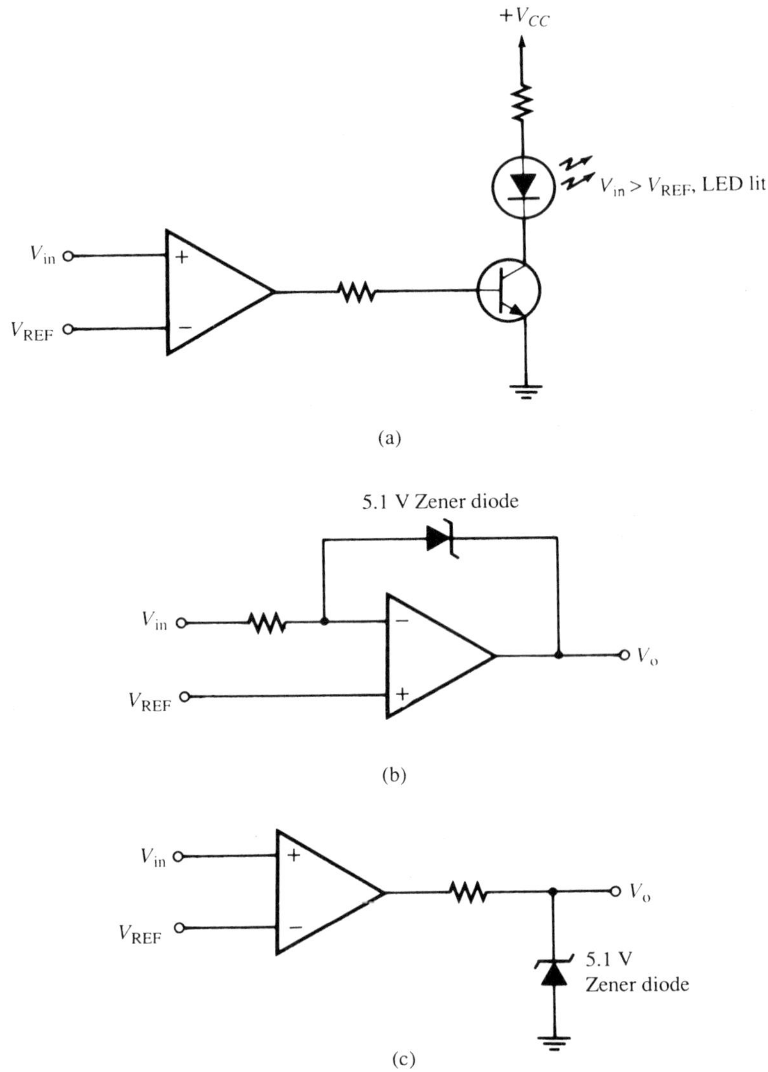

state of the output level. When interfaced with a noninverting comparator as shown, the LED is lit when $V_{in} > V_{REF}$.

On the other hand, comparators frequently drive TTL and CMOS digital circuits where the logic 0 and 1 levels are well defined, and if these levels are exceeded, the digital device may be destroyed. Consequently, it is undesirable for comparators to produce large positive and negative output voltage levels. For example, TTL devices typically require that the maximum positive voltage representing a logic 1 be +5 V (±5%) while logic 0 is at ground. If a comparator operating from a ±12-V supply is used, then the output will swing from −12 V to +12 V, depending on the relationship of V_{in} to V_{REF}.

Figure 8–12b illustrates how a Zener diode is used as a feedback element to limit the positive output voltage to the Zener voltage. If a 5.1-V Zener diode is used for TTL voltage levels, the diode conducts heavily beyond −0.7 V if the output tries to go negative. The negative output level is then limited to −0.7 V. Conversely, the output is limited at +5.1 V for positive output swings. Figure 8–12c shows an alternative circuit to limit the output voltage from −0.7 V to a positive voltage equal to the Zener voltage.

Noise Effects and Comparators with Hysteresis

Suppose that the levels of the two comparator inputs are very close to each other and there is also a small noise level on one of these inputs. If the noise level is large enough, as shown in Figure 8–13, the output of the comparator will switch back and forth between V_{+SAT} and V_{-SAT} each and every time the noisy input signal exceeds or drops below the reference voltage. These rapid changes due to noise are unwanted transitions sometimes called *chatter*. Furthermore, some comparators may oscillate during switching due to stray capacitive feedback.

Figure 8–13
Effect of input noise on the output of a comparator.

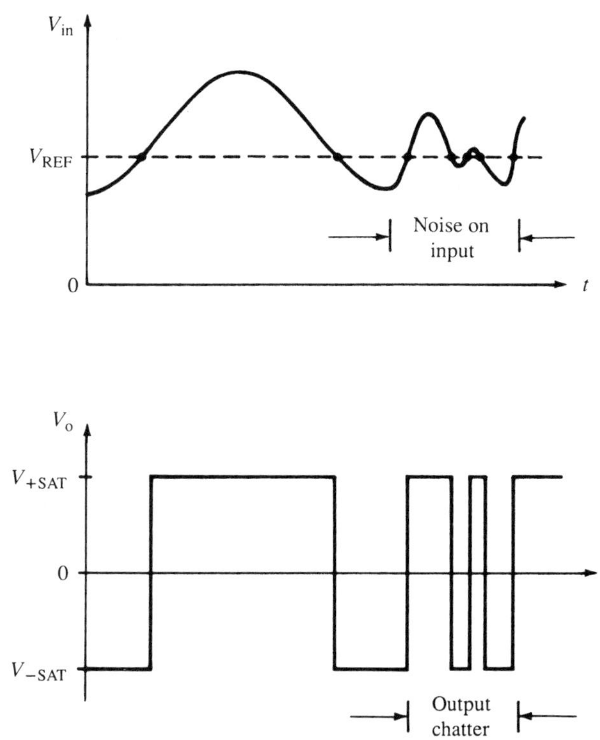

Figure 8–14
Inverting comparator with hysteresis. (a) Circuit. (b) Transfer characteristic. (c) Input and output waveforms.

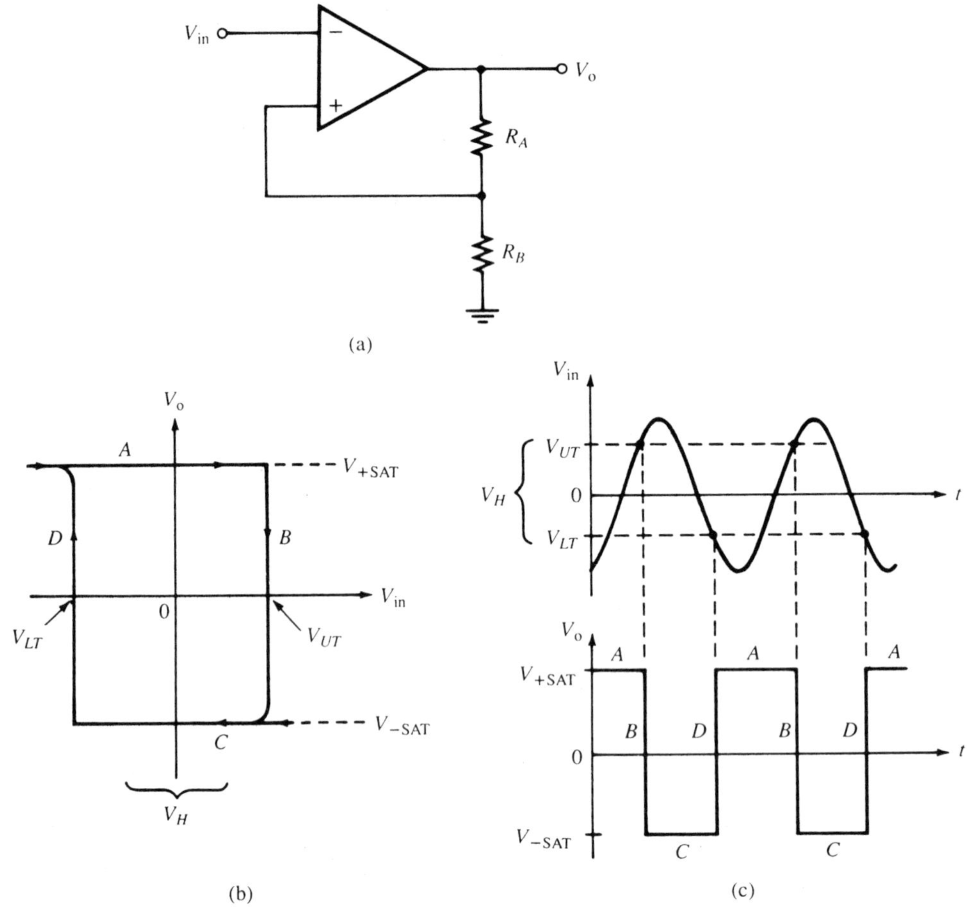

To prevent noise from causing false comparator chattering, positive feedback is often added to provide a "dead zone" over which the comparator will not respond. In the zero reference comparator circuit of Figure 8–14a, which is sometimes called a **Schmitt trigger,** a resistive voltage divider provides a positive feedback path from the junction of R_A and R_B to the operational amplifier's noninverting input. The voltage at the noninverting input determines the positive-going level at which V_{in} will cause the comparator's output to switch to approximately V_{-SAT}. This positive-going level is called the *upper threshold voltage* (V_{UT}) and is found from

$$V_{UT} = \frac{R_B}{R_A + R_B}(V_{+SAT}) \tag{8–3}$$

When V_{in} is less than V_{UT}, the voltage on the noninverting input is more positive than V_{in} and the output equals V_{+SAT}. When V_{in} exceeds V_{UT}, the output switches to V_{-SAT}.

When the output voltage equals V_{-SAT}, the voltage across R_B is called the *lower threshold voltage*, V_{LT}. The input voltage then drops below V_{LT} before the output will switch to V_{+SAT}, which occurs when

$$V_{LT} = \frac{R_B}{R_A + R_B}(V_{-SAT}) \qquad \textbf{(8–4)}$$

The amount of dead zone, or *hysteresis* (V_H), is the difference between the two threshold levels:

$$V_H = V_{UT} - V_{LT} \qquad \textbf{(8–5)}$$

In most cases, $V_{+SAT} = V_{-SAT}$, so that the upper and lower threshold voltages are equal except for polarity.

Graphically, the effect of hysteresis is shown in Figure 8–14b. Assuming that the output is already at V_{+SAT} and that the input voltage level is rising toward the upper threshold voltage V_{UT} on the graph, this effect is line *A*. When $V_{in} = V_{UT}$, the output level then snaps from V_{+SAT} to V_{-SAT} (line *B*). As V_{in} decreases toward V_{LT} (line *C*), the output voltage remains at V_{-SAT} until $V_{in} = V_{LT}$, at which point the output voltage snaps back to V_{+SAT} (line *D*). Figure 8–14c illustrates the effect of hysteresis on a sine-wave signal. For voltage levels within the dead zone from V_{LT} to V_{UT}, the comparator does not respond. Only when the input levels exceed these thresholds does the output change.

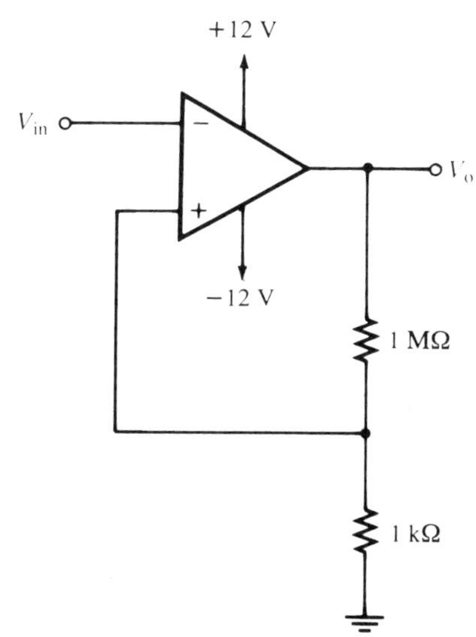

Figure 8–15
Inverting comparator circuit for Example 8–1.

Example 8–1

Find the upper threshold, lower threshold, and hysteresis voltages for the comparator circuit of Figure 8–15. Assume that the positive and negative saturation voltages for the operational amplifier using a ±12-V supply are ±10.5 V.

Solution:

(a) Using Equation 8–3,

$$V_{UT} = \frac{1\ \text{k}\Omega}{(1\ \text{M}\Omega + 1\ \text{k}\Omega)}(10.5\ \text{V})$$
$$= \mathbf{10.5\ mV}$$

(b) Using Equation 8–4,

$$V_{LT} = \frac{1\ \text{k}\Omega}{(1\ \text{M}\Omega + 1\ \text{k}\Omega)}(-10.5\ \text{V})$$
$$= \mathbf{-10.5\ mV}$$

(c) The hysteresis voltage is then

$$V_H = 10.5\ \text{mV} - (-10.5\ \text{mV})$$
$$= \mathbf{21\ mV}$$

When the input level exceeds the upper threshold of 10.5 mV, then output of the comparator switches to −10.5 V. When the input level then drops below −10.5 mV, the comparator's output switches to +10.5 V.

Upper and Lower Switching Levels for Nonzero References

Grounding resistor R_B in the inverting comparator with positive feedback (Figure 8–14a) sets the comparator's reference voltage at zero volts. The hysteresis, or dead zone, will then be centered about zero. In many situations, however, it is desired that the reference be some value other than zero while still having some amount of noise immunity. If the lower end of R_B is tied to a DC reference voltage, as shown by the inverting comparator circuit of Figure 8–16a, the upper and lower threshold voltages, respectively, are now

$$V_{UT} = \frac{R_B(V_{+\text{SAT}} - V_{\text{REF}})}{R_A + R_B} + V_{\text{REF}} \qquad (8\text{–}6)$$

and

$$V_{LT} = \frac{R_B(V_{-\text{SAT}} - V_{\text{REF}})}{R_A + R_B} + V_{\text{REF}} \qquad (8\text{–}7)$$

Figure 8–16
Inverting comparator with hysteresis and nonzero voltage reference. (a) Circuit. (b) Transfer characteristic. (c) Input and output waveforms.

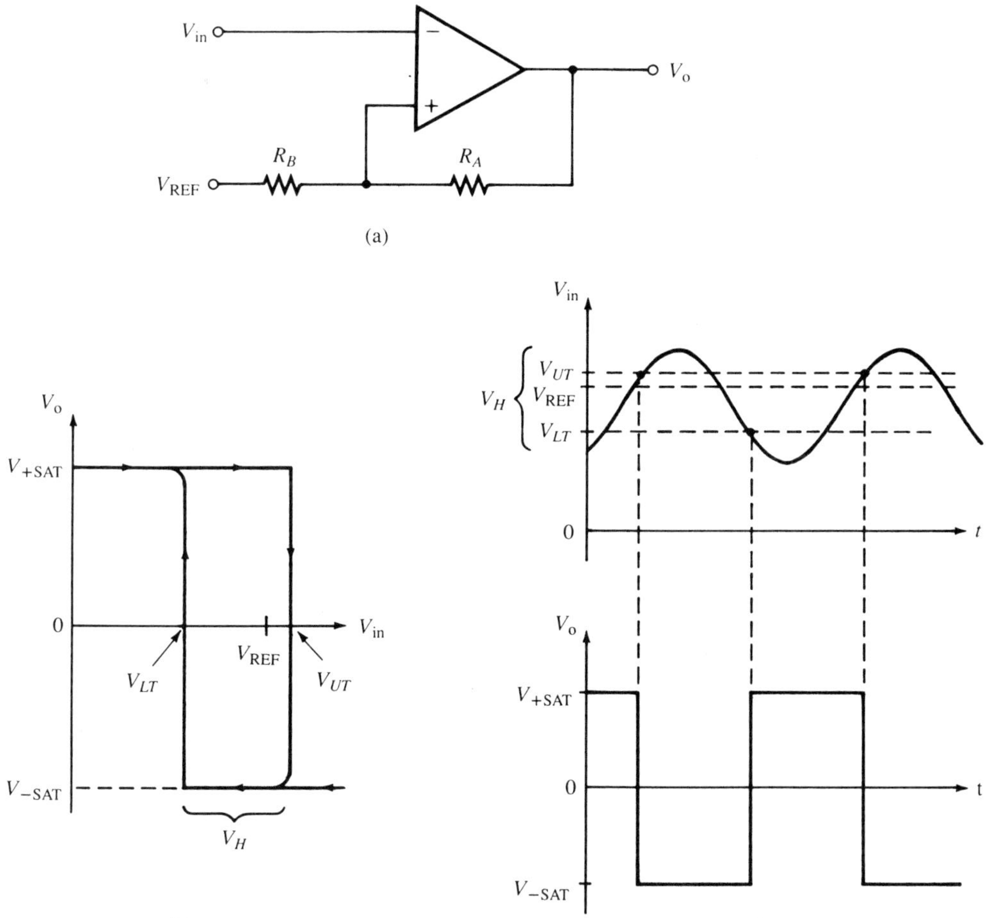

Example 8–2

A comparator circuit of the type shown in Figure 8–16a has the following parameters: $R_A = 9\ \text{k}\Omega$, $R_B = 1\ \text{k}\Omega$, and $V_{REF} = -2$ V. Determine (a) V_{LT}, (b) V_{UT}, and (c) V_H if the saturation voltages are ± 13 V.

Solution:

(a) Using Equation 8–7,

$$V_{LT} = \frac{(1\ \text{k}\Omega)[-13\ \text{V} - (-2\ \text{V})]}{(10\ \text{k}\Omega + 1\ \text{k}\Omega)} + (-2\ \text{V})$$

$$= \mathbf{-3\ V}$$

(b) Using Equation 8–6,

$$V_{UT} = \frac{(1\ \text{k}\Omega)[13\ \text{V} - (-2\ \text{V})]}{(10\ \text{k}\Omega + 1\ \text{k}\Omega)} + (-2\ \text{V})$$
$$= \mathbf{-0.64\ V}$$

(c) The hysteresis voltage is then

$$V_H = -0.64\ \text{V} - (-3\ \text{V})$$
$$= \mathbf{2.36\ V}$$

Figure 8–17a shows the corresponding circuit for a noninverting comparator with hysteresis. Here, the inputs to the operational amplifier are reversed from those of Figure 8–16a. For this circuit, the switching levels are given by:

$$V_{UT} = \frac{R_A + R_B}{R_A}(V_{\text{REF}}) - \frac{R_B}{R_A}(V_{+\text{SAT}}) \quad \textbf{(8–8)}$$

and

$$V_{LT} = \frac{R_A + R_B}{R_A}(V_{\text{REF}}) - \frac{R_B}{R_A}(V_{-\text{SAT}}) \quad \textbf{(8–9)}$$

Window Comparators

A **window comparator,** or *double-ended comparator,* is a circuit designed to detect the presence of a voltage between two specified voltage limits—that is, within a voltage "window." This is normally accomplished by logically combining the outputs from an inverting comparator and a noninverting comparator, as shown in Figure 8–18a. When the input level is greater than the upper reference voltage (V_{UL}) of the window, the output of A_1 is at $V_{+\text{SAT}}$. Since V_{in} is also greater than the lower reference voltage (V_{LL}), the output of the inverting comparator (A_2) is $V_{-\text{SAT}}$. Diode D_1 is then forward-biased, whereas D_2 is reverse-biased. The voltage at the junction of the two diodes and R_A is approximately $V_{+\text{SAT}}$.

If the input level is now in the window between V_{LL} and V_{UL}, the outputs of both A_1 and A_2 are at $V_{-\text{SAT}}$. Both diodes are then reverse-biased, so the voltage at their junction is zero. For input levels less than V_{LL}, the output of A_1 is at $V_{-\text{SAT}}$, whereas the output of A_2 is $V_{+\text{SAT}}$. Diode D_2 is forward-biased, but D_1 is reverse-biased. The voltage at the junction of the two diodes and R_A is approximately $V_{+\text{SAT}}$. Consequently, when the input voltage is between V_{LL} and V_{UL}, the output voltage at D_1 and D_2 is zero. For input levels outside this window, the output voltage equals $+V_{\text{SAT}}$:

$$V_o = 0, \quad \text{for } V_{LL} < V_{\text{in}} < V_{UL} \quad \textbf{(8–10a)}$$
$$= V_{+\text{SAT}}, \quad \text{for } V_{\text{in}} < V_{LL} \text{ or } V_{\text{in}} > V_{UL} \quad \textbf{(8–10b)}$$

Figure 8–17
Noninverting comparator with hysteresis and nonzero voltage reference. (a) Circuit. (b) Transfer characteristic. (c) Input and output waveforms.

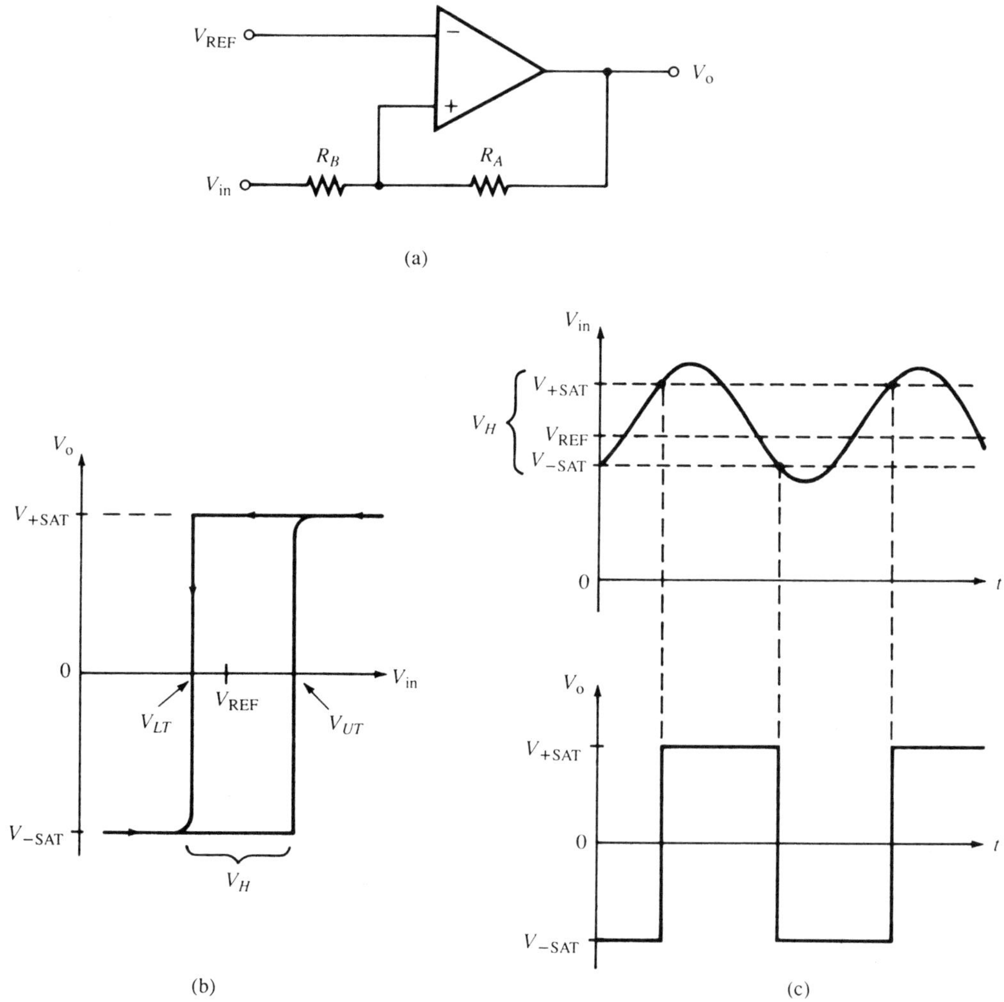

The transistor output driver allows for output-level conversion. As shown, the transistor supply is +5 V. If the input signal is within the window, the high voltage at the transistor's base will turn Q_1 on, and the voltage at the collector V_C will be essentially zero. On the other hand, if the input is outside the window, the transistor is cut off, so the collector voltage is approximately +5 V. This action is graphed in Figure 8–18b. The collector resistor (R_C) is sized according to the transistor's maximum collector current rating, I_C(max). Using a 5-V supply, the minimum value for R_C is 5 V/I_C(max).

This type of circuit is frequently used in alarm circuits having high and low trip points. For example, a thermistor can be used to monitor the temperature

Figure 8–18
Window comparator.
(a) Circuit with transistor interface. (b) Transfer characteristic.

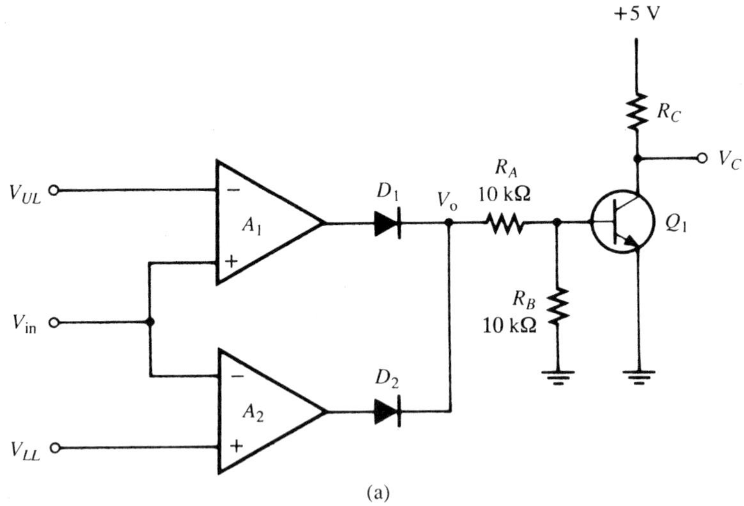

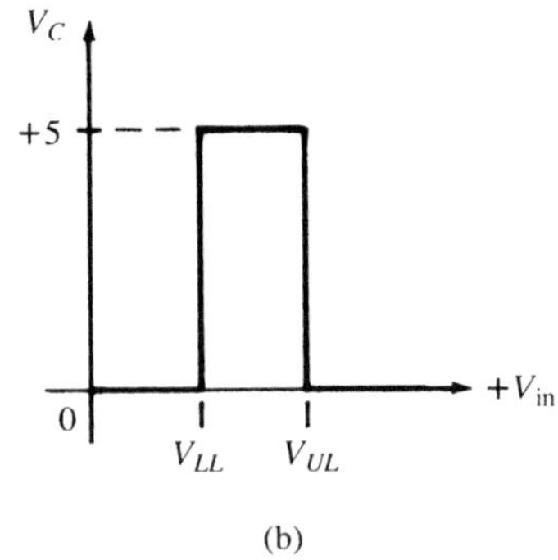

Figure 8–19
Window comparator circuit for Example 8–3.

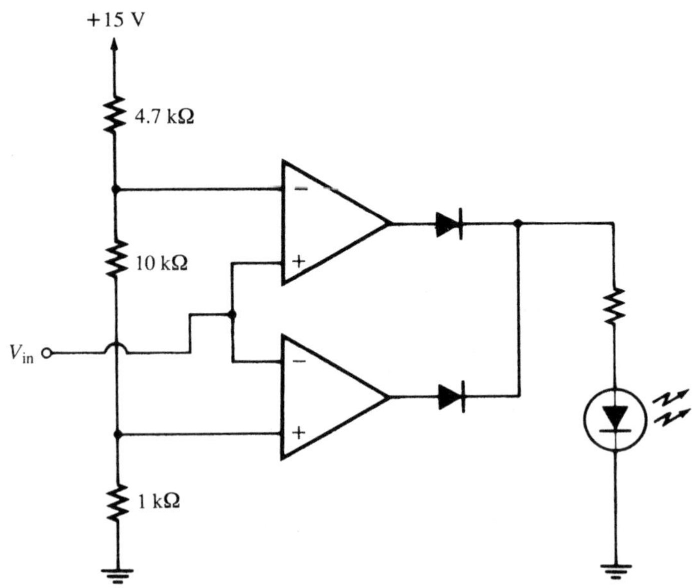

within an enclosure by producing an output voltage proportional to its resistance, which is dependent on the measured temperature. If the upper and lower reference voltage limits are chosen to correspond to voltage levels representing high and low temperature limits, then the window comparator can be used to monitor when the measured temperature is outside a given range. When outside the prescribed range, the collector voltage is then sufficient to drive a bell, siren, or some other type circuit.

Example 8–3

For the window comparator circuit of Figure 8–19, determine the range of input voltages for which the LED will be lighted.

Solution: The upper and lower threshold limits are set by the resistor divider string. The upper threshold is

$$V_{UL} = \frac{(10\ \text{k}\Omega + 1\ \text{k}\Omega)}{(4.7\ \text{k}\Omega + 10\ \text{k}\Omega + 1\ \text{k}\Omega)}(15\ \text{V})$$
$$= \mathbf{+10.51\ V}$$

whereas the lower threshold is

$$V_{UL} = \frac{1\ \text{k}\Omega}{(4.7\ \text{k}\Omega + 10\ \text{k}\Omega + 1\ \text{k}\Omega)}(15\ \text{V})$$
$$= \mathbf{+0.96\ V}$$

The LED will then be lighted whenever the input voltage is outside the window. For this circuit, the LED will be lighted for input levels less than +0.96 V as well as for those greater than +10.51 V.

Output Clipping

Clipping and **limiting** are interchangeable terms used to refer to those techniques that prevent an output signal from exceeding a certain level. Figure 8–20 shows a comparator circuit where the output is clipped, or limited, at −5 V but permitted to swing to $V_{+\text{SAT}}$.

If the input goes negative, the output swings to $V_{+\text{SAT}}$ and turns the diode off. If the input goes positive, the output swings negative until the diode is forward-biased. If we neglect the small diode drop, this will happen at $V_o = -5$ V. Once the diode turns on, very heavy negative feedback occurs and very little additional output swing can take place.

Figure 8–21 shows another comparator circuit with clipping, but this one will clip on both polarities. Here, the output swing in the positive direction is limited at about +5 V, whereas the swing in the negative direction is limited at one diode

Figure 8–20
Comparator with clipping at −5 V.

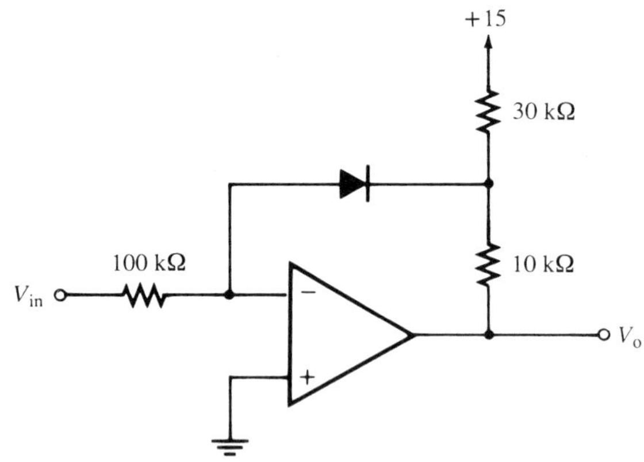

drop (about −0.7 V). The output of this clipped comparator circuit is directly compatible with 5-V TTL digital systems.

Norton Operational Amplifier Comparators

The Norton operational amplifier, or CDA, which is discussed in Chapter 5, is capable of being connected to perform several comparator operations. Unlike the conventional bipolar operational amplifier comparator, which compares the voltage levels of its two inputs, the CDA comparator compares the relationship between its two input currents such that

$$V_o = V_{+SAT}, \quad \text{for } I(+) > I(-) \qquad \textbf{(8–11a)}$$

$$= 0, \quad \text{for } I(+) < I(-) \qquad \textbf{(8–11b)}$$

The basic CDA comparator is shown in Figure 8–22, where the 1-MΩ input resistors define current to each input of the CDA. In terms of input voltage levels,

Figure 8–21
Comparator with clipping at −0.7 V and +5 V.

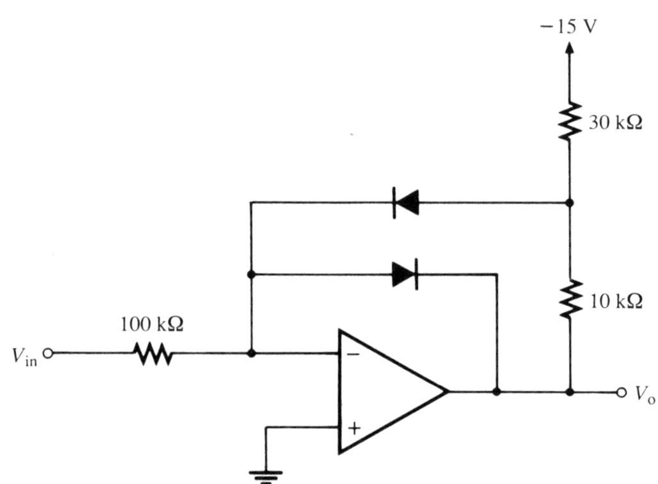

Figure 8–22
Norton amplifier noninverting comparator.

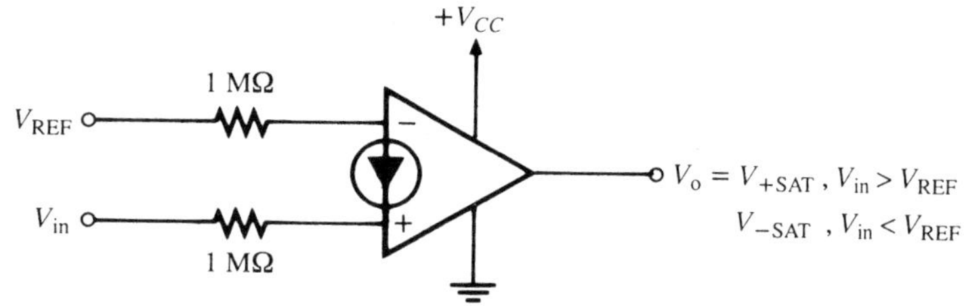

$$V_o = V_{+SAT}, \quad \text{for } V_{in} > V_{REF} \qquad (8\text{–}12a)$$
$$= 0, \quad \text{for } V_{in} < V_{REF} \qquad (8\text{–}12b)$$

Although the CDA is designed to be powered only by a positive supply voltage, it can nevertheless be made to compare negative voltages. In the circuit of Figure 8–23, resistors R_1 and R_2 establish a *common-mode bias current,* allowing V_{in} and/or V_{REF} to be negative voltage levels.

A CDA window comparator circuit is shown in Figure 8–24. Resistors R_1 and R_4 establish the required reference currents at the inverting input of A_1 and the noninverting input of A_2. For both currents to be equal, $R_1 = R_4$. The lower and upper threshold voltages are given by

$$V_{LT} = \frac{R_3}{R_4} V_{REF} \qquad (8\text{–}13)$$

$$V_{UT} = \frac{R_2}{R_1} V_{REF} \qquad (8\text{–}14)$$

Both outputs are tied to a third CDA comparator through R_4 and R_5. For proper biasing, $R_4 = R_5 = R_6/2$.

Figure 8–23
Norton amplifier noninverting comparator biased to compare negative voltages when the circuit is powered by a single positive supply voltage.

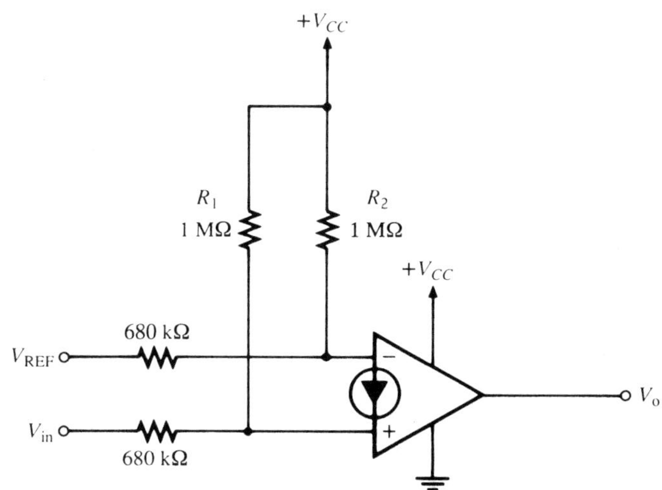

Figure 8–24
Norton amplifier window comparator.

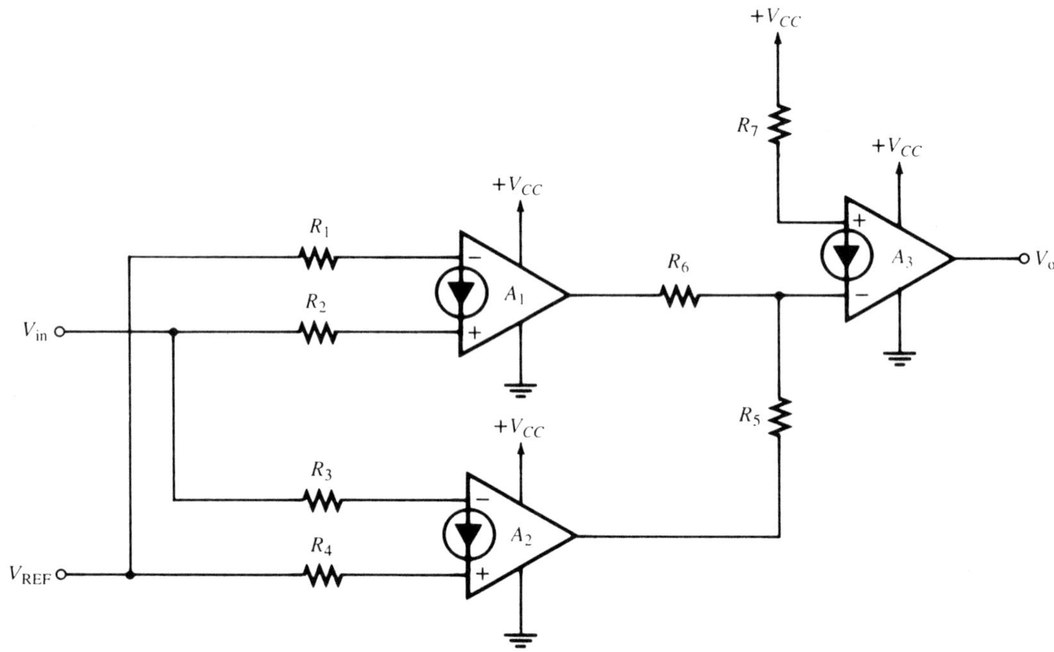

Example 8–4
For the CDA window comparator of Figure 8–25, determine the upper and lower threshold voltages.

Solution:
(a) Using Equation 8–13,

$$V_{LT} = \frac{220\ \text{k}\Omega}{680\ \text{k}\Omega}(15\ \text{V}) = \mathbf{4.85\ V}$$

(b) Using Equation 8–14,

$$V_{UT} = \frac{470\ \text{k}\Omega}{680\ \text{k}\Omega}(15\ \text{V}) = \mathbf{10.37\ V}$$

The output will be zero for input voltages between 4.85 V and 10.37 V. Otherwise, outside this range, $V_o = V_{+SAT}$.

Figure 8–25
Norton amplifier window comparator for Example 8–4.

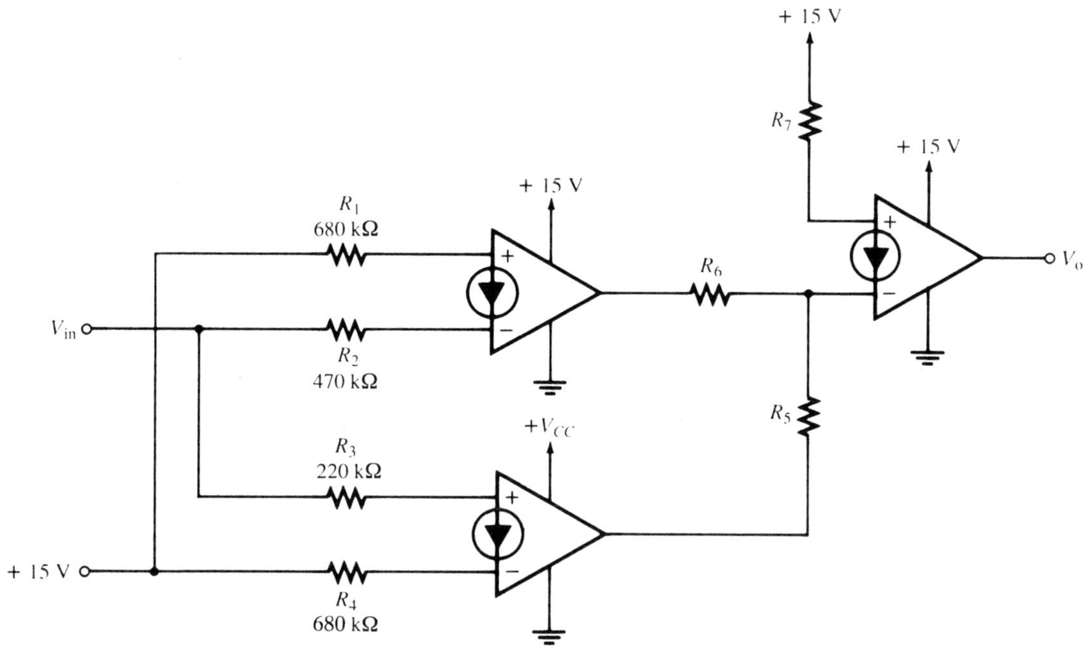

Limitations and Precautions

Several things must be kept in mind when selecting either an operational amplifier or CDA to be used as a comparator. First of all, the device must have a high slew rate so that the output will change as rapidly as possible. In addition, for high-speed operation, the unity-gain frequency should be as high as possible.

In order to obtain accurate switching levels, with or without hysteresis, the device that is chosen should also have low input offset voltage, low input bias current, and high CMR ratings. Despite these desirable factors, however, a general-purpose device such as the 741 is able to be used in low-speed, less critical applications.

Integrated-Circuit Comparators

Even though both conventional and Norton operational amplifiers can be wired to function in a number of comparator circuits, several manufacturers have designed precision comparators to overcome the limitations presented by many of the available operational amplifiers. In fact, the data sheet for such a comparator—for example, the 311—gives the impression that it might be an operational amplifier, as virtually all the parameters are common to both.

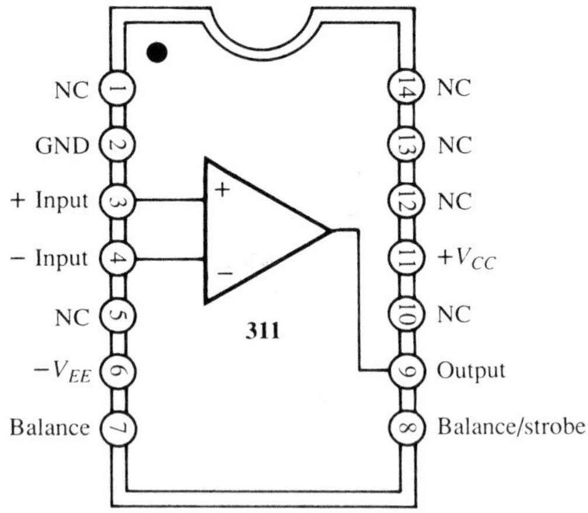

Figure 8–26
LM311 integrated-circuit comparator.

The 311, whose pinout is shown in Figure 8–26, is perhaps the most versatile for many applications. It can sink or source up to 50 mA and is capable of operating from supply voltages from 5 to 36 V. It is available in the following versions:

Commercial	311
Military	111
FET input	LF311
Low power	LP311

8–4 Clampers

A **clamper** circuit, sometimes called a *DC restorer,* adds a predetermined DC level to the output voltage. If the DC level is positive, the circuit is a *positive clamper.* On the other hand, if the DC level is negative, the circuit is called a *negative clamper.*

Figure 8–27a shows a circuit known as a *peak clamper,* since it clamps the output voltage at a level equal to V_{REF}. When V_{in} swings below V_{REF}, A_1 drives D_1 on and charges C_1 to the peak value of the input signal. After the negative peak has passed and D_1 is reversed-biased, the charge on C_1 is retained. The only possible discharge paths for C_1 are through the inputs of A_1 or A_2. Because of the high-impedance input of both operational amplifiers, the capacitor can reach the peak voltage quickly, whereas the discharge is very slow. As a result, the circuit has a fast-attack, slow-decay response to input level changes.

The clamping effect is to put a DC voltage (V_c) between the input signal and the noninverting input of A_2. This raises the average level of the output signal to a value equal to the peak value of V_{in}. As shown, the circuit of Figure 8–27a clamps the negative peaks of the output to V_{REF} (Figure 8–27b), but positive-peak clamping may be had by reversing D_1 (Figure 8–28).

Figure 8–27
Negative peak clamper. (a) Circuit. (b) Input and output waveforms.

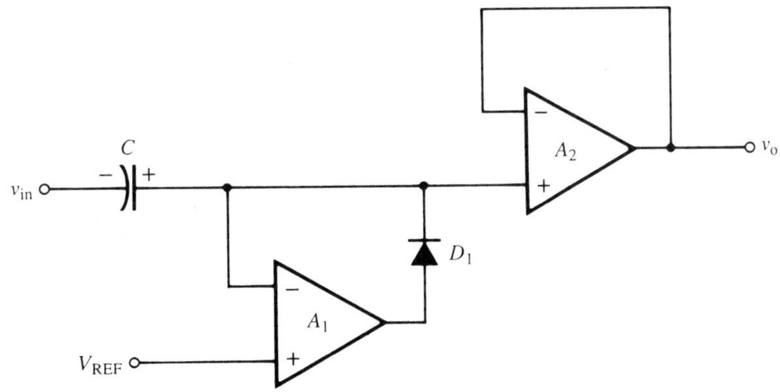

(a)

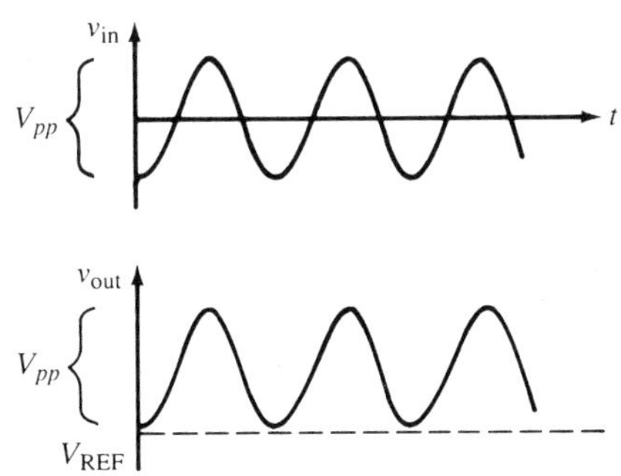

(b)

8–5 Sample-and-Hold Circuits

Sample-and-hold (S/H) amplifiers are used to (1) capture the value of a rapidly changing signal at a specific point in time and (2) hold it steady. This is important during analog-to-digital conversion operations (Chapter 12) as well as in the storage of the outputs of a multiplexer between updates in data-distribution systems. In this section, discrete S/H circuits using operational amplifiers will be discussed.

To understand the sample-and-hold operation, first consider a circuit network having (1) an input port, (2) an output port, and (3) a control line. When the control line is held at some high-level voltage, the output follows the input. When the control line is grounded or taken to a somewhat lower voltage, the output is held, or "frozen," at its current value. The output of the S/H circuit will change only when the control line is taken to the high-level voltage (Figure 8–29).

Figure 8–28
Positive peak clamper. (a) Circuit. (b) Input and output waveforms.

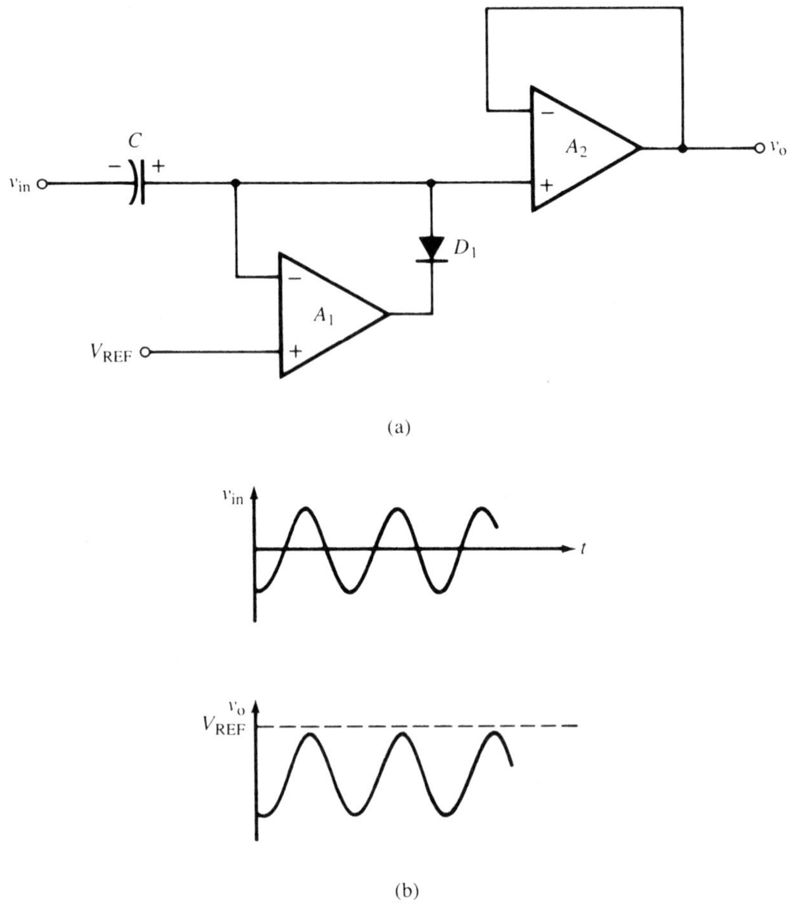

Figure 8–30 shows a simple sample-and-hold circuit. When the control signal is high, the enhancement MOSFET conducts, allowing the capacitor voltage to follow the input signal. When the control signal is low, the MOSFET is turned off, and the capacitor holds its voltage constant. The operational amplifier here merely acts as a high-input impedance buffer between the capacitor and the load.

A slightly more complex S/H circuit is shown in Figure 8–31. Here, the input signal is buffered from the capacitor by A_1, which provides a low-impedance source for charging and discharging the capacitor. As in Figure 8–29, A_2 serves as a high-impedance load for the capacitor to buffer it from the actual load. The overall accuracy and response time of the circuit is improved by using V_o as the negative feedback signal to A_1 instead of merely connecting the output of A_1 back to its inverting input. This arrangement also compensates for common-mode and offset errors of the output voltage follower.

Figure 8–29
Input and output waveforms illustrating the operation of a sample-and-hold amplifier.

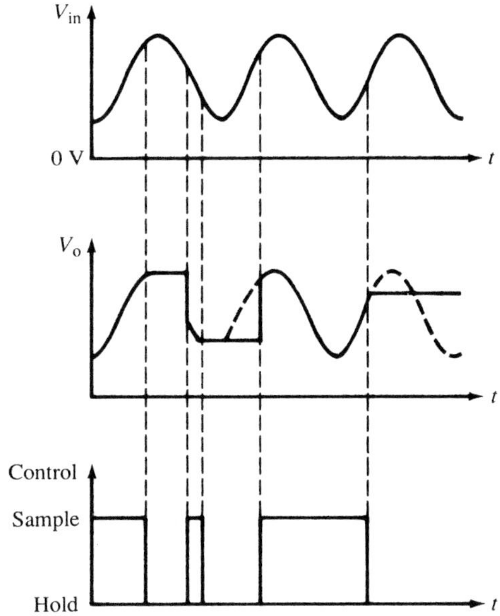

The principal errors associated with the use of discrete sample-and-hold circuits occur during the sample-to-hold and hold-to-sample intervals. The time that elapses between the signal to hold and the actual opening of the hold switch is called the **aperture time.** This delay results in an amplitude uncertainty that is dependent on the rate of change of the signal being sampled. Greater degrees of resolution for a given sampling rate require shorter aperture times.

Figure 8–30
Basic sample-and-hold amplifier circuit.

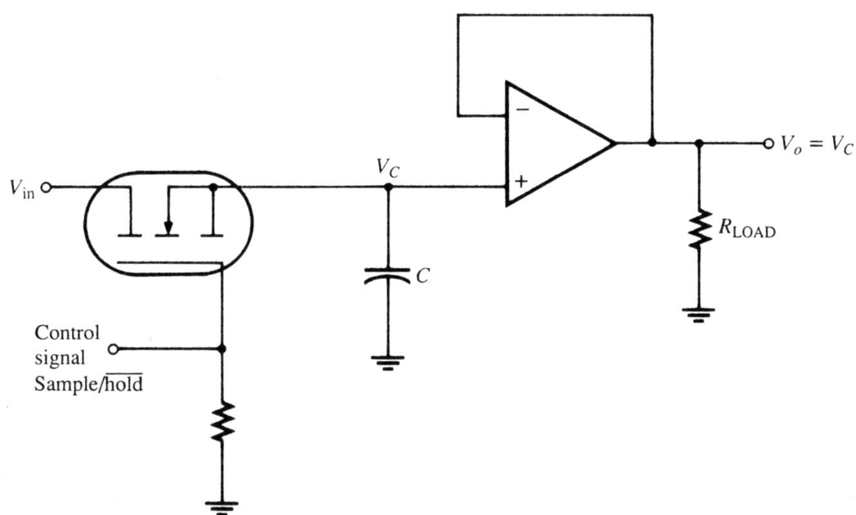

Figure 8–31
Improved sample-and-hold circuit using a buffer amplifier.

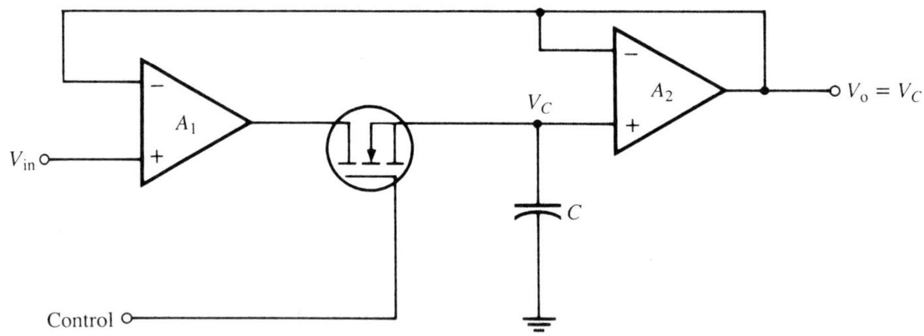

During the hold period, capacitor voltage droop and sample-hold switch feedthrough can contribute errors. Droop is caused by capacitor leakage and by the input bias requirement of the output amplifier, which is supplied by the capacitor during the hold interval. Droop is normally significant only for long hold intervals. **Feedthrough** is the fraction of the input that appears at the output during the hold period and is primarily due to capacitance across the switch. The hold-jump voltage is a constant voltage added to the output by the charge transferred to the capacitor during sample-to-hold switching.

One very important specification associated with sample-and-hold circuits is the **acquisition time.** This is the time that the input signal must be applied for the output to reach the desired accuracy and has a direct bearing on the maximum sampling rate. In general, sample-and-hold circuits with a short acquisition time cannot hold as long as sample-and-hold circuits with a longer acquisition time.

8–6 Operational Amplifier Logarithmic and Antilogarithmic Amplifiers

A **logarithmic amplifier,** or *log amp,* is a circuit that produces an output signal that is proportional to the logarithm of the input signal. Historically, logarithmic-type amplifiers were used in analog computers. Today, logarithmic amplifiers find many uses in circuits that include multipliers, dividers, RMS converters, and dynamic range compressors, just to name a few. Although it is possible to construct these circuits using discrete operational amplifier circuits, increased accuracy and flexibility can be obtained with those integrated circuit devices that are specifically designed for these purposes.

A semiconductor *pn* junction like those found in diodes and transistors is the key to all logarithmic amplifiers. Because of the increased dynamic range that can be obtained, however, transistors are generally preferred for discrete logarithmic amplifier circuits. The transistor base-emitter voltage (V_{BE}) is a logarithmic function of its collector current (I_C). For a silicon p-n junction,

$$V_{BE} = \frac{2.3\,kT}{q} \log_{10}\left(\frac{I_C}{I_{RS}}\right) \qquad (8\text{–}15)$$

where

k = Boltzmann's constant, 1.38×10^{-23} J/K
T = absolute temperature (K)
q = charge on an electron, 1.6×10^{-19} C
I_C = collector current (A)
I_{RS} = theoretical reverse saturation current (A)

The theoretical reverse saturation current is typically 0.1 pA at 25°C. Furthermore at 25°C (298 K), Equation 8–15 reduces to

$$V_{BE} = 0.0591 \log_{10}(I_C) + 0.768 \qquad (8\text{–}16)$$

Simply stated, this equation says that V_{BE} will increase about 60 mV for each tenfold increase in I_C. For low-leakage, high-gain transistors, this relationship holds over a wide range of voltages. Figure 8–32 shows the circuit of what is called a *transdiode logarithmic converter*. In this circuit, the feedback element is a grounded-base transistor. Here, a power transistor is often used to reduce the series feedback resistance.

The transistor's collector current equals the input current of the operational amplifier. Since the base of the transistor is grounded, the output of the operational amplifier is equal to V_{BE}. Consequently, the output voltage is proportional to the logarithm of the input current, which in turn is proportional to the input voltage. Because of the virtual ground at the inverting input, the collector and

Figure 8–32
Operational amplifier-transistor logarithmic amplifier.

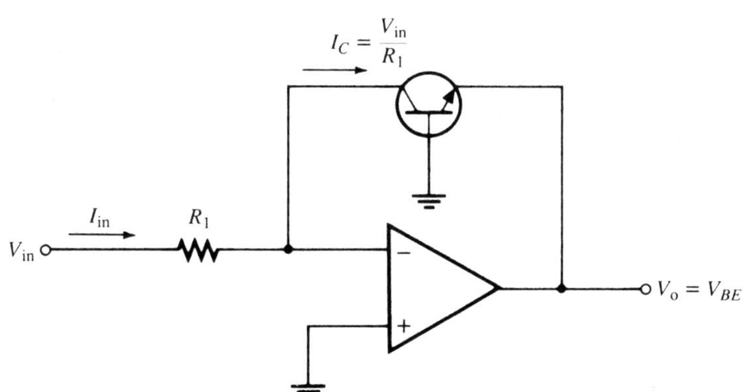

base voltages of the transistor are equal. At 25°C, Equation 8–15 can be rewritten as

$$V_{BE} = 0.0591 \log_{10}\left(\frac{V_{in}}{R_1}\right) + 0.768 \qquad (8\text{–}17)$$

If negative-level inputs are required, then a *pnp* transistor is used so that the output will be positive.

Even though the simple transdiode amplifier gives perhaps the truest output level that represents the logarithm of an input voltage, the circuit nevertheless has two major negative features. First, the logarithm relationship of the transistor is temperature-dependent. Secondly, the output change is only about 60 mV per decade, whereas 1 V/decade might be more desirable. These two affect primarily the scaling factor of 60 mV/decade. On the other hand, the current I_{RS}, which is temperature-sensitive, contributes a voltage offset in Equations 8–16 and 8–17 that is approximately 768 mV at 25°C.

All three of these problems can be compensated for with additional circuitry, such as the circuit of Figure 8–33, using a thermistor (R_2) to cancel changes due to temperature. Since the quantity kT/q in Equation 8–15 amounts to approximately a −0.3% change per degree Celsius, a thermistor should be chosen that has a temperature coefficient of +0.3%/°C to cancel the kT/q factor. In many situations, however, commercially available integrated-circuit logarithmic amplifier modules are designed to compensate automatically for these errors.

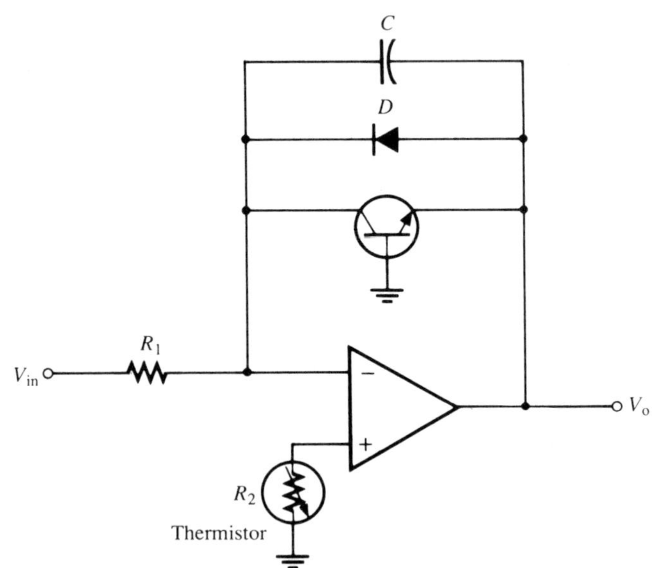

Figure 8–33
Temperature-compensated logarithmic amplifier.

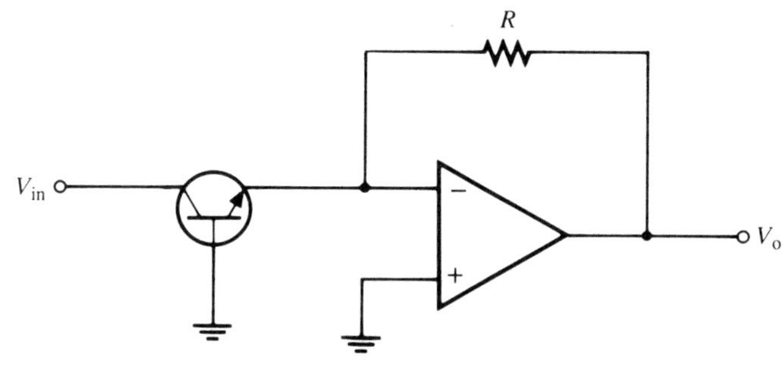

Figure 8–34
Antilogarithmic amplifier.

An amplifier circuit that has its output voltage proportional to the antilogarithm of the input voltage is called an **antilogarithmic amplifier.** Because of the identity

$$\ln^{-1}(X) = e^{\ln(X)} \quad \textbf{(8–18)}$$

the antilogarithmic amplifier is also often referred to as an **exponential amplifier.** Just as multiplication and division or integration and differentiation are inverse mathematical operations, so are computing the logarithm and antilogarithm inverse operations. For the case of the basic integrator and differentiator circuits, one circuit was changed to the other by merely transposing the input and feedback elements. The same is true for logarithmic and antilogarithmic amplifier circuits.

By interchanging the resistor and the transistor of the logarithmic amplifier circuit of Figure 8–33, the antilogarithmic amplifier of Figure 8–34 is formed. The output is proportional to the antilog, or inverse logarithm, of the input:

$$V_o = -RI_{RS}e^{(V_{in}q/kT)} \quad \textbf{(8–19)}$$

8–7 Multipliers and Dividers

Discrete Operational Amplifier Circuits

Using discrete logarithmic and antilogarithmic amplifiers and basic rules of logarithms, one is able to create analog circuits to give an output signal that is either the product or quotient of two input signals.

First, note that the logarithm of the product of two signals X and Y is the sum of the logarithms of the two numbers individually, or

$$\log(XY) = \log(X) + \log(Y) \quad \textbf{(8–20)}$$

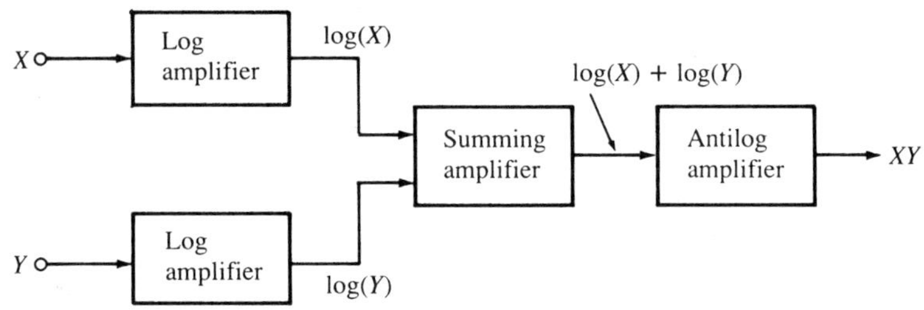

Figure 8–35
Block diagram showing how to multiply two signals.

As shown in block diagram form in Figure 8–35, if we sum the logarithms of two input signals, we get the logarithm of their product, log (XY). If we now pass the resulting signal through an antilogarithmic amplifier,

$$XY = \log^{-1}[\log(XY)] \qquad \textbf{(8–21)}$$

we obtain the product (XY) of the two signals.

In a similar manner, the quotient of two signals (X/Y) can be obtained. First, recall that the logarithm of the quotient of two signals X and Y is the difference of the logarithms of the two numbers individually, or

$$\log\left(\frac{X}{Y}\right) = \log(X) - \log(Y) \qquad \textbf{(8–22)}$$

As shown in block diagram form in Figure 8–36, if we take the difference of the logarithms of two input signals, we get the logarithm of their quotient, $\log(X/Y)$. If we now pass the resulting signal through an antilogarithmic amplifier,

$$\frac{X}{Y} = \log^{-1}\left[\log\left(\frac{X}{Y}\right)\right] \qquad \textbf{(8–23)}$$

we obtain the quotient (X/Y) of the two signals.

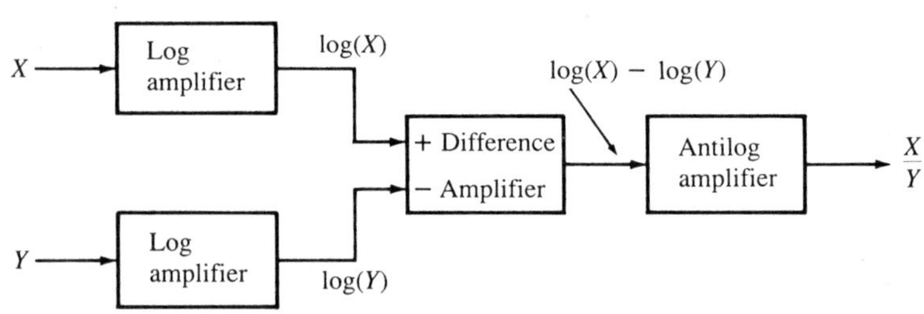

Figure 8–36
Block diagram showing how to divide two signals.

Figure 8–37
Schematic representation of an analog multiplier.

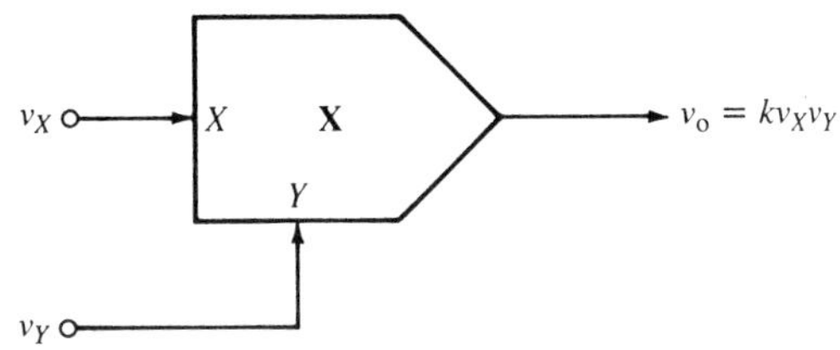

Integrated-Circuit Analog Multipliers

As mentioned previously, discrete operational amplifier logarithmic circuits are susceptible to several errors that affect their accuracy. Integrated-circuit **analog multipliers** are available that provide improved performance and versatility over their discrete counterpart.

The analog multiplier is somewhat different from the discrete logarithmic multiplier arrangement. Generally, logarithmic multiplication is restricted to only positive or negative DC levels, not both. On the other hand, the integrated-circuit analog multiplier is designed to process analog signals, regardless of polarity. For two input signals v_X and v_Y, the output of an analog multiplier is given as

$$v_o = kv_Xv_Y \tag{8–24}$$

where k is a scale factor. Figure 8–37 illustrates the multiplier schematic symbol showing both inputs and a single output. For most commercially available analog integrated-circuit multipliers that operate from a ±15-V supply, k is $\frac{1}{10}$. Since these devices, like the AD533, restrict input signal levels to the ±10-V range, the output then is limited to ±10 V.

Two- and Four-Quadrant Multipliers

Useful multipliers are generally available as either *two-quadrant multipliers* or *four-quadrant multipliers*. Figure 8–38a shows the input-output transfer characteristic of a typical two-quadrant multiplier to inputs X and Y. The upper right-hand 90° quadrant is usually referred to as the *first quadrant* and the remaining three quadrants are numbered in order in a counterclockwise fashion. For all possible combinations of polarities of the two input signals, the input-output characteristic line occupies only two quadrants (1 and 3). The X-input is capable of accepting positive or negative voltages, whereas the Y-input allows only positive voltages. In summary, the input output polarities for a two-quadrant multiplier are as follows:

Input Polarity		***Output***	
X	***Y***	**Polarity**	**Quadrant**
Positive	Positive	Positive	1
Negative	Positive	Negative	3

Figure 8–38
Analog multiplier transfer characteristic.
(a) Two-quadrant multiplier.
(b) Four-quadrant multiplier.

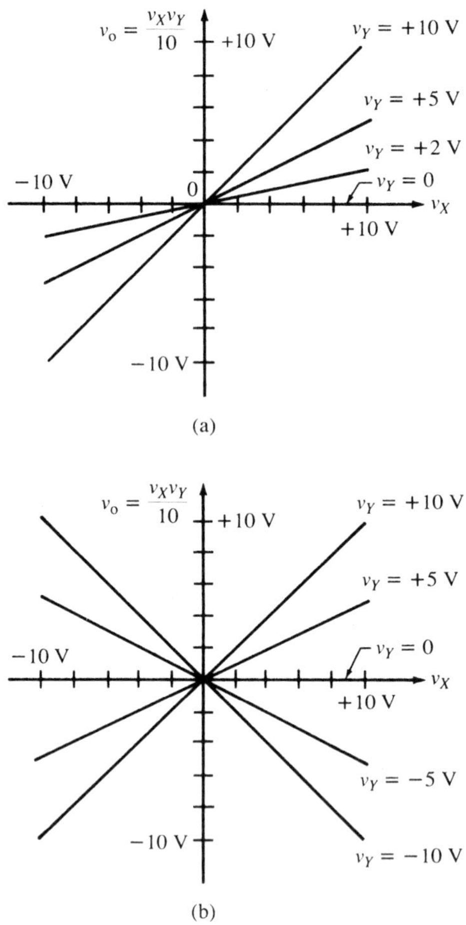

They follow established mathematical rules for the multiplication of signed numbers.

The four-quadrant multiplier is more versatile than the two-quadrant type. As shown in its input-output transfer characteristic (Figure 8–38b), there is now no restriction on the polarities of either input. In summary, the input-output polarities for a four-quadrant multiplier are as follows:

Input Polarity		***Output***	
X	***Y***	**Polarity**	**Quadrant**
Positive	Positive	Positive	1
Negative	Negative	Positive	2
Negative	Positive	Negative	3
Positive	Negative	Negative	4

Figure 8–39
Calibration adjustments for the XR2208 analog multiplier.

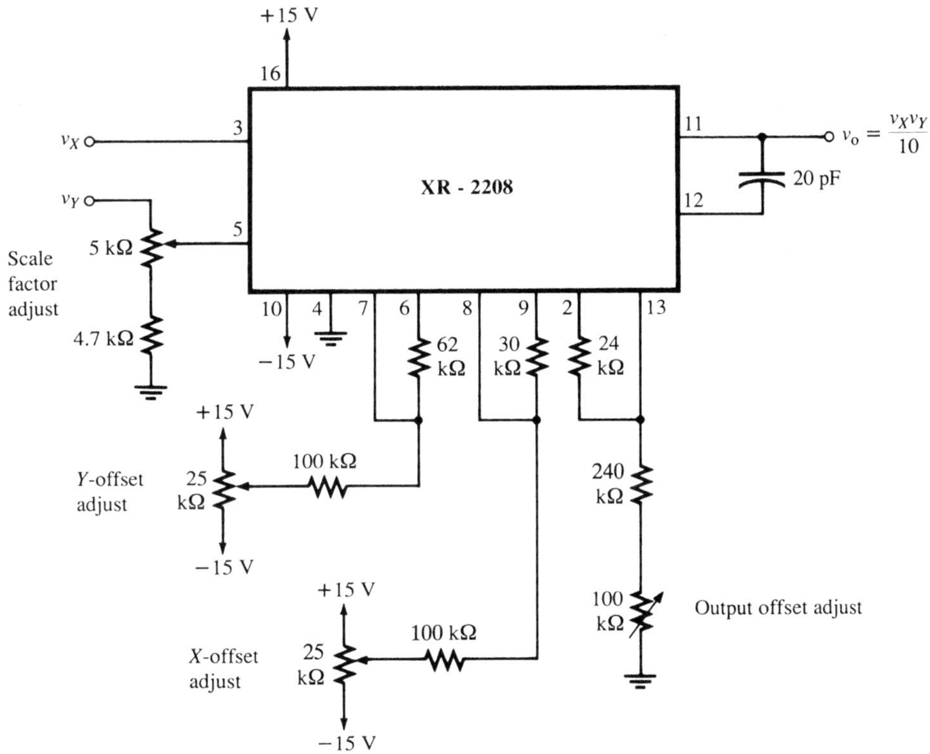

For example, if $v_X = -5$ V and $v_Y = 8$ V, the output voltage is then $(-5)(8)/10 = -4$ V (quadrant 3).

Low-cost multipliers, such as the AD533, XR2208, RC4200, and MC1495, require external adjustments for proper adjustment and calibration. Shown in Figure 8–39, four multiturn "trimpots" are set so that (1) the output is zero when both inputs are zero, (2) the output is zero when only the X-input is zero, (3) the output is zero when only the Y-input is zero, and (4) the scale factor is set such that the output is $\frac{1}{10}$ the product of both input voltages. There are other multipliers, such as the AD534, called *precision multipliers;* these cost more but require virtually no external settings.

Basic Arithmetic Functions

The multiplication process serves as the basis upon which many other operations are generated. The following subsections discuss how analog multipliers are wired to perform division, squaring and higher powers, and square roots.

Analog Division

Since division is the inverse, or reciprocal, of multiplication,

$$v_o = \frac{v_X}{v_Y} \qquad \textbf{(8–25a)}$$

can also be written as

$$v_o = v_X \left(\frac{1}{v_Y}\right) \qquad \textbf{(8–25b)}$$

The reciprocal of any mathematical operation can be obtained by placing it in the feedback loop of an inverting amplifier. Therefore, the division of two signals is achieved by placing an analog multiplier as the feedback element of an inverting amplifier, as shown in Figure 8–40. For two input signals v_X and v_Y, the output v'_o of the multiplier is

$$v'_o = \frac{v_X v_Y}{10} \qquad \textbf{(8–26a)}$$

But since v_X is the same as the output voltage v_o,

$$v'_o = \frac{v_o v_Y}{10} \qquad \textbf{(8–26b)}$$

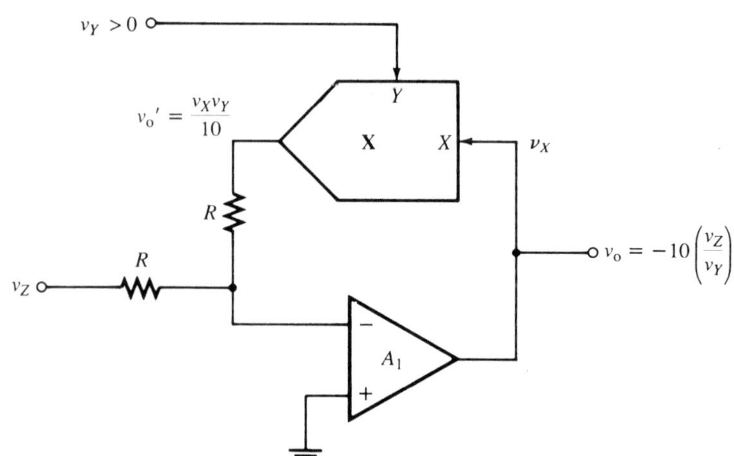

Figure 8–40
Analog division using an analog multiplier.

Since the multiplier output is one input of a two-input summing amplifier through R, then

$$v'_o = -v_Z \tag{8–27}$$

At the inverting input of the operational amplifier, both v_Z and v'_o must add up to zero. Equating Equations 8–26b and 8–27 and solving for v_o,

$$v'_o = \frac{-10v_Z}{V_Y} \tag{8–28}$$

However, for this analog divider circuit to function properly, voltage v_Y must always be a positive voltage; if it is negative, the operational amplifier will latch up due to the positive multiplier feedback.

Squaring and Higher Powers

As shown in Figure 8–41, a signal v_X can be squared,

$$v_o = \frac{(v_X)^2}{10} \tag{8–29}$$

by tying both multiplier inputs together. The generation of higher-ordered integer powers is accomplished by cascading additional multiplier sections.

Square Root

Again, recognizing the inverse relationships between certain mathematical functions, taking the square root is the inverse of squaring. Therefore, as shown by the circuit of Figure 8–42, the square root of an input signal is obtained by placing the squaring circuit of Figure 8–41 as the feedback element of an inverting amplifier circuit. The output of the multiplier is

$$-v'_o = \frac{(v_o)^2}{10} \tag{8–30}$$

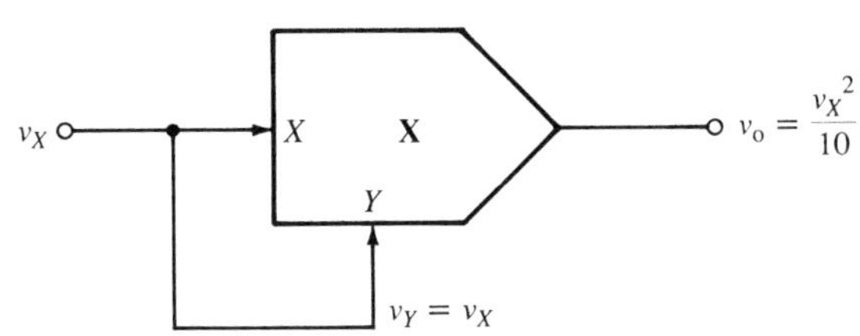

Figure 8–41
Squaring an analog signal.

Figure 8–42
Taking the square root of an analog signal.

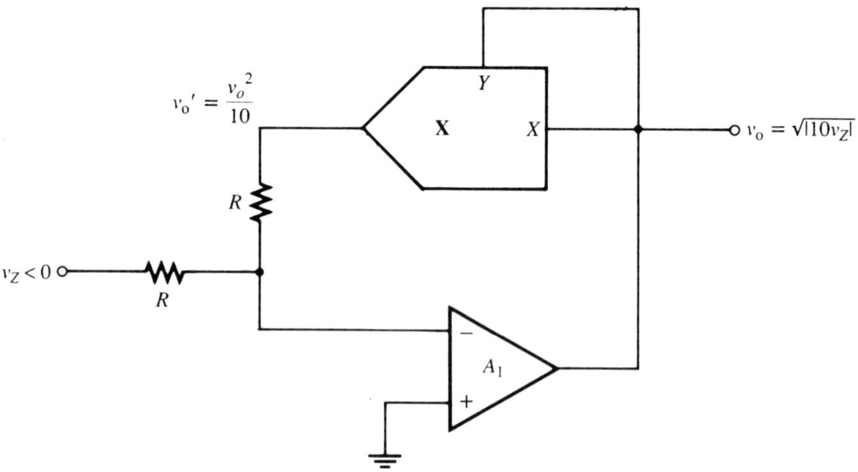

and if $v_Z < 0$,

$$v_o^2 = 10v_Z \qquad \textbf{(8–31)}$$

so that the inverse relation yields

$$v_o = \sqrt{(|10v_Z|)} \qquad \textbf{(8–32)}$$

An Application

Besides the traditional use for multipliers as computational elements for analog computers, the analog multiplier can be used to generate an amplitude modulated carrier. Amplitude modulation (AM) is, in fact, a multiplication process between a high-frequency sine-wave carrier signal and a much lower frequency complex audio signal. The carrier signal can be represented mathematically as

$$v_c(t) = V_c \sin(2\pi f_c t) \qquad \textbf{(8–33)}$$

For simplicity of analysis, the instantaneous modulating signal can be assumed to be a sine wave of frequency f_m, so that

$$v_m(t) = V_m \sin(2\pi f_m t) \qquad \textbf{(8–34)}$$

Figure 8–43
Amplitude modulator using an analog multiplier.
(a) Input carrier (v_X) and modulating (v_Y) signals. (b) Amplitude modulated output.

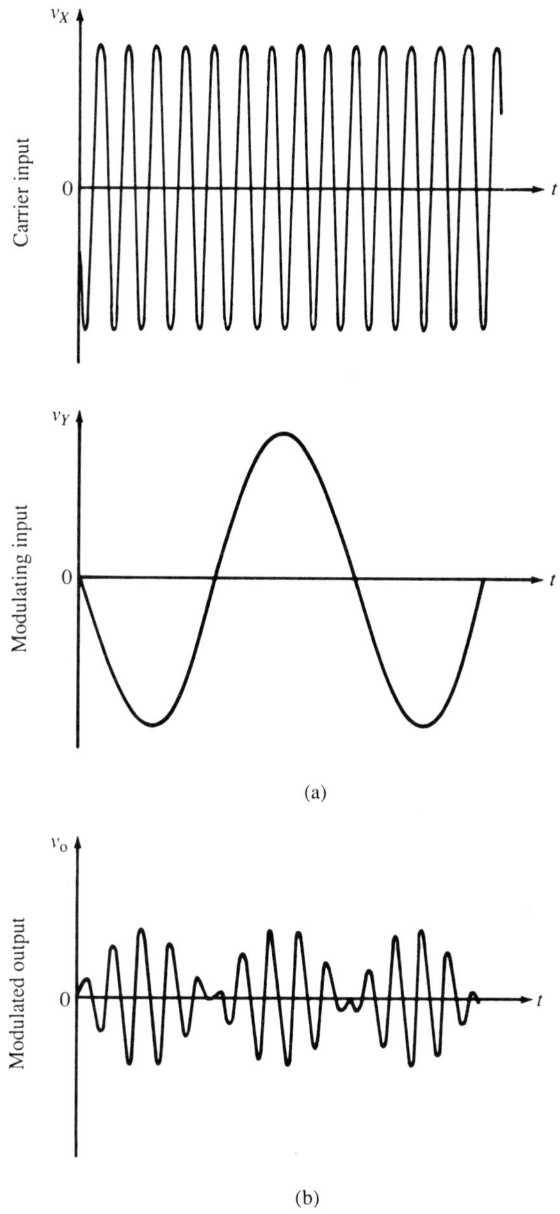

If the carrier and modulating signals are simultaneously applied to the *X*- and *Y*-inputs of four-quadrant analog multiplier (Figure 8–43a), the output is then

$$v_o = \frac{V_c V_m}{10} (\sin 2\pi f_c t)(\sin 2\pi f_m t) \qquad \textbf{(8–35)}$$

so the output is the product of two sine waves of unequal frequencies. Equation 8–35 can be simplified further by the trigonometric identity

$$(\sin X)(\sin Y) = \frac{1}{2}[\cos(X - Y) - \cos(X + Y)] \qquad \textbf{(8–36)}$$

Substituting $X = 2\pi f_c t$ and $Y = 2\pi f_m t$ and rewriting Equation 8–36,

$$v_o = \frac{V_c V_m}{20}\cos[2\pi(f_c - f_m)t] - \frac{V_c V_m}{20}\cos[2\pi(f_c + f_m)t] \qquad \textbf{(8–37)}$$

which is plotted in Figure 8–43b.

Equation 8–37 is composed of two cosine-wave components (sine waves that are 90° out of phase) having frequencies that are not equal to either the carrier or modulating frequencies. These two generated frequency components are referred to as the **sum frequency** ($f_c + f_m$) and the **difference frequency** ($f_c - f_m$). This is precisely the result for a *balanced* (i.e., no inserted carrier) modulator.

8–8 Integrated-Circuit Compandor

In any audio or communications system, the one factor that limits its performance is noise. Depending on the circumstances, noise effects may be significantly reduced, if not eliminated, in a variety of ways. With the advances made in integrated-circuit technology, the **compandor** is a system often used to achieve low-noise performance. The term *compandor* is the contraction of *com*pressor and ex*pandor,* the two basic functions involved.

A **compressor** is a circuit that is designed to compress, or reduce the signal levels between the smallest and largest input signals. This difference is often termed the **dynamic range.** Since this process is a nonlinear one, the analog input signal is deliberately distorted. High-level signals are attenuated more than low-level signals. On the other hand, an **expandor** has the opposite effect of the compressor. It *expands,* or increases, the signal's dynamic range and restores the proper levels of the original signal. Generally, these operations occur in pairs—the compressor might be located in the modulator section of a transmitter, whereas the expandor would be located after the detector in a radio receiver.

The effect of these two operations is illustrated in Figure 8–44. A complex input signal having minimum and maximum amplitudes over time of 5 mV and 10 V has a dynamic range of (10 V)/(5 mV), or 66 dB. When fed to a compressor, the dynamic range is reduced to, for example, 35 dB in the transmitter and then expanded back to the original 66 dB in the receiver. The **compression ratio** is then 66 dB − 35 dB, or 31 dB. When noise is present, the effect of companding increases the signal-to-noise ratio by the same number as the compression ratio.

Figure 8–44
Effect of compression and expansion.

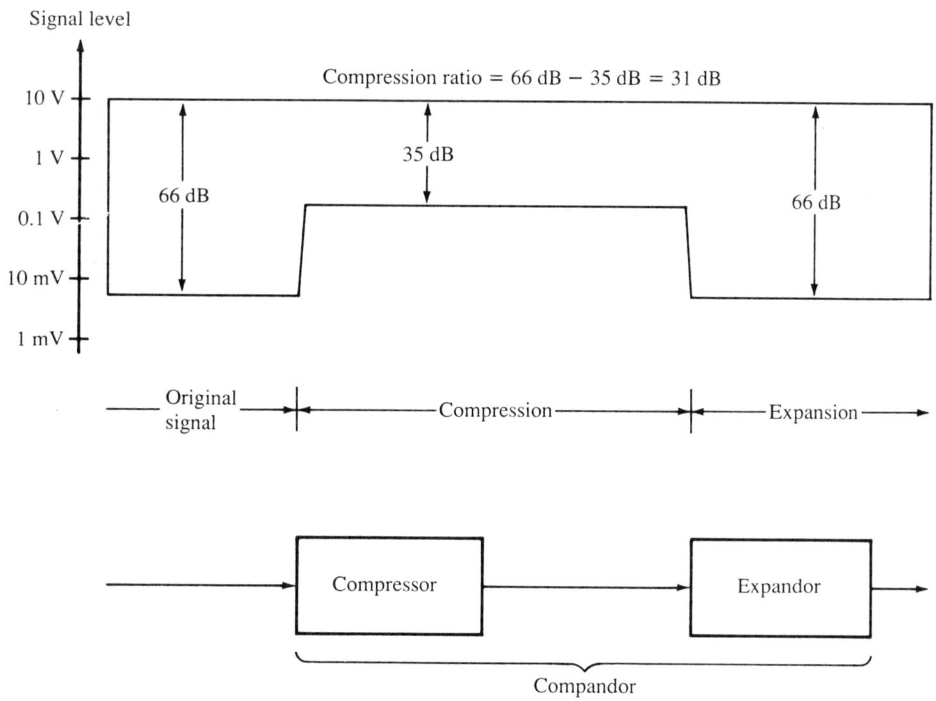

Because the compression and expansion transfer characteristics must be exactly opposite, this is difficult to achieve using operational amplifier circuits. However, the NE570/571 device (Figure 8–45a) overcomes many of the pitfalls with operational amplifier circuits. The NE570 operates with supply voltages between 6 and 24 V, whereas the NE571 is limited to those between 6 and 18 V. Figure 8–45b shows the NE570 wired as a compressor. The equations for component selection are those given by the manufacturer (Signetics), such that the output DC bias is given by

$$V_o(\mathrm{DC}) = (1.8\ \mathrm{V})\,\frac{2R}{30\ \mathrm{k}\Omega} \tag{8–38}$$

where the reactances of the capacitors, both bypass and coupling, are to be small at the lowest input frequency, $f_{\min}$. In most cases,

$$C = \frac{1}{2\pi f_{\min}(1\ \mathrm{k}\Omega)} \tag{8–39}$$

Figure 8–45
NE570/571 compandor. (a) Pin diagram. (b) NE570 wired as a compressor.

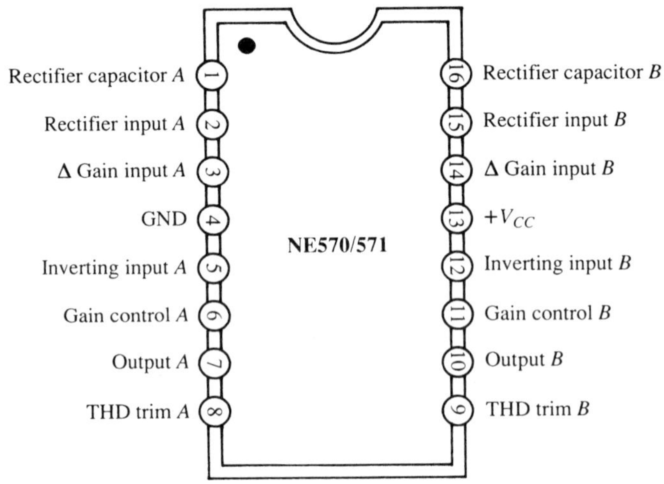

(a)

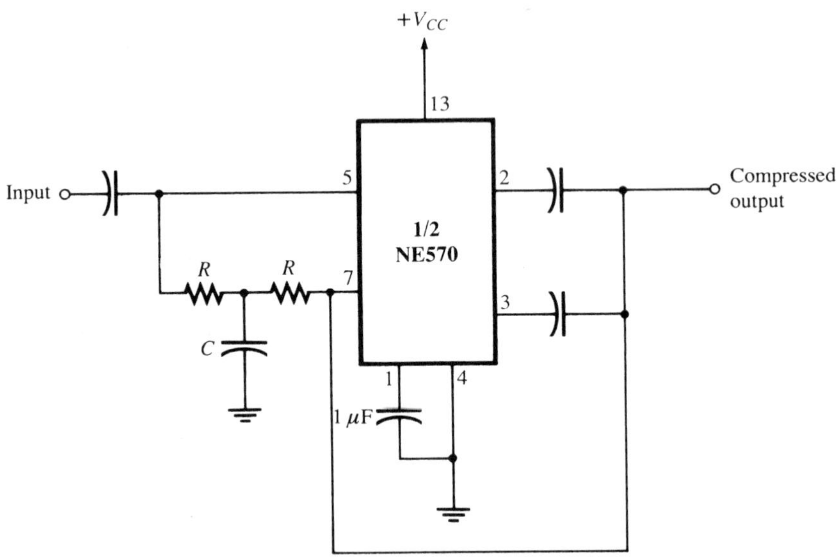

(b)

Figure 8–46
NE570 wired as an expandor.

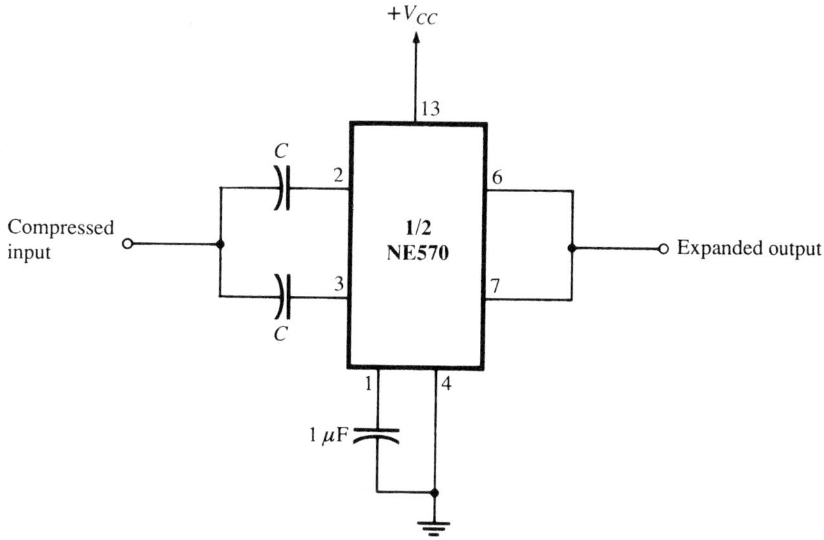

Figure 8–46 illustrates the associated expandor where the capacitors are determined using Equation 8–39.

Summary

This chapter discussed those circuits whose input signals must be processed in other ways besides linear amplification. These circuits comprise a broad category called signal processing circuits, and are often used to control another process, or transform signals to a form that allows for easy measurement. These circuits include precision half-wave and full-wave rectifiers, logarithmic type amplifiers, analog multipliers, compandors, comparators, peak detectors, sample-and-hold amplifiers, and clampers.

In many cases, these circuit functions were shown to be built using operational amplifiers. For others, specialized integrated-circuit modules offer improved circuit performance and versatility.

Glossary

acquisition time The time the input signal must be applied for the output in a sample-and-hold amplifier to equal the desired accuracy.

analog multiplier An analog circuit whose output is equal to the instantaneous product of the level of its two input signals.

antilogarithmic amplifier A nonlinear circuit whose output signal is proportional to the inverse, or antilogarithm, of its input signal. Also called an *exponential amplifier.*

aperture time The time that elapses between the signal to hold and the actual opening of the hold switch in a sample-and-hold amplifier.

clamper A circuit that adds a predetermined DC voltage to a signal. Also called a *DC restorer.*

clipping The acting that limits an output signal from exceeding a certain positive or negative level or both. Also called *limiting.*

compandor A contraction of the words compressor and expandor.

comparator A circuit that compares an input voltage to a predetermined DC reference voltage.

compression ratio The dB difference between the dynamic range of a compressor and an expandor.

compressor A nonlinear circuit that reduces the signal level between the smallest and largest input signal.

difference frequency The instantaneous frequency difference between a carrier frequency and the modulation frequency.

dynamic range Expressed in decibels, the difference between the largest and smallest input signal levels of a compressor.

expandor A nonlinear circuit that increases the dynamic range of an input signal.

exponential amplifier See *antilogarithmic amplifier.*

feedthrough The fraction of the input that appears at the output during the hold period in a sample-and-hold amplifier.

inverting comparator A circuit whose output goes negative when the input signal is more positive than the reference voltage.

limiting See *clipping.*

logarithmic amplifier A nonlinear circuit whose output signal is proportional to the logarithm of its input signal.

noninverting comparator A circuit whose output goes positive when the input signal is more positive than the reference voltage.

peak detector A diode circuit with an operational amplifier that captures the peak positive or negative excursion of an input signal for an infinite period of time until it is reset to zero.

polarity detector See *zero crossing detector.*

precision rectifier A rectifier circuit, either half-wave or full-wave, built around an operational amplifier.

sample-and-hold (S/H) amplifier A circuit that captures and holds the value of a rapidly changing input voltage at a specific point in time.

Schmitt trigger A comparator with hysteresis, triggered at predetermined setpoints.

sum frequency The instantaneous sum of a carrier frequency and the modulation frequency.

window comparator A circuit that detects the presence of a voltage between two specified voltage limits. Also called a *double-ended comparator.*

zero crossing detector A comparator circuit whose output changes when the input signal crosses zero volts. Also called a *polarity detector.*

Review Questions

1. Explain what is meant by a *precision* rectifier.
2. Draw the schematic diagram for a precision half-wave rectifier.
3. How is a precision rectifier made from a precision half-wave rectifier?
4. What is a peak detector?
5. Draw the schematic diagram of a peak detector using an operational amplifier.
6. What is a comparator?
7. Draw the schematic diagrams for and explain the operation of inverting and noninverting comparators.

8. Describe how a comparator can be made to function as a polarity detector.
9. Describe how a comparator can be made to function as a sine-wave-to-square-wave converter.
10. What is a Schmitt trigger? How does it differ from a comparator?
11. What is the *dead zone*?
12. Draw the schematic diagram for a Schmitt trigger.
13. Explain what a window comparator is.
14. What considerations should be kept in mind when selecting either operational amplifiers or Norton amplifiers for use as comparators?
15. Explain the difference between a clipper and a clamper.
16. What is a sample-and-hold amplifier?
17. What errors are generally associated with sample-and-hold amplifiers?
18. What is a logarithmic amplifier?
19. Describe how logarithmic and antilogarithmic amplifiers are used to create circuits to:
    (a) Multiply two voltages.
    (b) Take the quotient of two voltages.
20. Explain the differences between the operation of integrated-circuit two- and four-quadrant multipliers.
21. Explain how integrated-circuit analog multipliers can be wired to perform division, multiplication, squaring, higher powers, and square roots.
22. What is a compandor?
23. Explain what it means to compress or expand a signal.
24. What is the dynamic range of a compandor?

Exercises

1. Sketch the input and output waveforms of the precision half-wave rectifier circuit of Figure 8–2a for an input triangle wave having a peak amplitude of ±3 V. Assume that R_1 = 47 kΩ and R_2 = 33 kΩ and that silicon diodes are used (V_d = 0.7 V).
2. Repeat Exercise 1 if the polarities of both diodes are reversed.
3. Sketch the input and output waveforms for the precision full-wave rectifier circuit of Figure 8–3a for an input sine wave having a peak amplitude of 0.8 V.
4. Repeat Exercise 1 if the polarities of both diodes are reversed.
5. Determine the upper and lower threshold voltages for the Schmitt trigger circuit of Figure 8–14a if R_A = 56 kΩ and R_B = 8.2 kΩ if the operational amplifier's saturation voltage is ±10 V when using a ±12-V supply.
6. For the Schmitt trigger of Problem 5, above what input voltage level does the output equal the operational amplifier's positive saturation voltage? Below what input voltage level does the output equal the operational amplifier's negative saturation voltage? What is the hysteresis voltage?
7. Determine suitable component values for the Schmitt trigger of Figure 8–14a so that the upper and lower threshold levels are ±3.5 V if a saturation voltage of ±7 V is assumed when powered from a +9-V supply.
8. Determine the upper and lower threshold voltages and the hysteresis voltage for the inverting Schmitt trigger circuit of Figure 8–16a if R_A = 68 kΩ, R_B = 2.2 kΩ, and V_{REF} = 0.7 V if the operational amplifier's saturation voltage is ±13 V when using a ±15-V supply.
9. Repeat Exercise 8 for V_{REF} = −1.5 V.
10. For the Schmitt trigger of Exercise 9, above what input voltage level does the output equal the operational amplifier's positive sat-

Figure 8–47

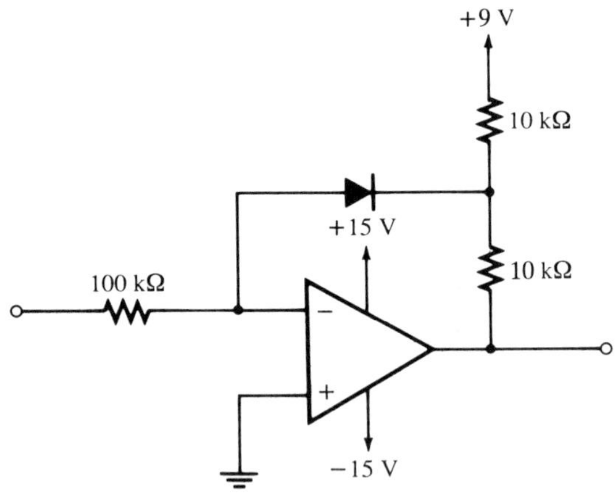

Figure 8–48

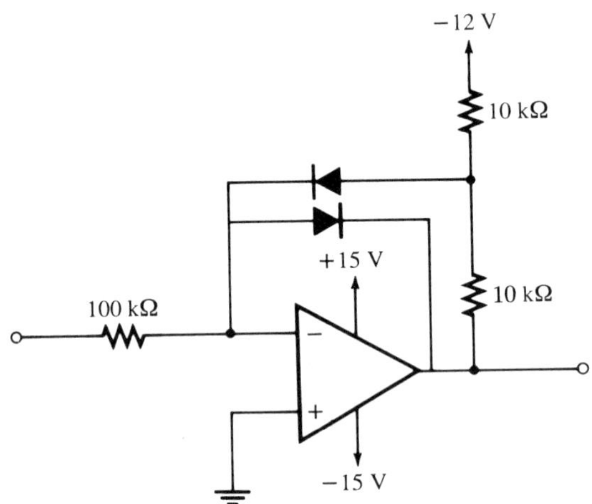

uration voltage? Below what input voltage level does the output equal the operational amplifier's negative saturation voltage?

11. Determine the upper and lower threshold voltages and the hysteresis voltage for the noninverting Schmitt trigger circuit of Figure 8–17a if $R_A = 47$ kΩ, $R_B = 10$ kΩ, and $V_{REF} = 2$ V, if the operational amplifier's saturation voltage is ±4 V when using a ±5-V supply.
12. The window comparator of Figure 8–18a has $V_{UL} = 4.5$ V and $V_{LL} = -2.6$ V. Between what input signal levels will the output of the operational amplifier equal zero? When will the output be equal to the operational amplifier's positive saturation voltage?
13. For the comparator circuit of Figure 8–47, determine the range of the output voltage, assuming that the saturation voltage of the operational amplifier is ±13 V when operated from a ±15-V supply.
14. For the comparator circuit of Figure 8–48, determine the range of the output voltage, assuming that the saturation voltage of the operational amplifier is ±13 V when operated from a ±15-V supply.
15. Determine the upper and lower threshold voltages for the Norton amplifier window comparator of Figure 8–24 if $R_1 = 1$ MΩ, $R_2 = 820$ kΩ, $R_3 = 330$ kΩ, $R_4 = 470$ kΩ, and $V_{REF} = +15$ V.
16. The NE570 is to be wired as a compressor using the circuit of Figure 8–45b. Determine suitable component values for R and C if the output DC bias is set at 25 mV and the lowest expected input frequency is 2 kHz.

9 Voltage and Current Regulators

Instructional Objectives

Upon completing this chapter, you will be able to

- Explain why voltage regulators are used.
- Draw the diagram and explain the operation of a series regulator.
- Draw the diagram and explain the operation of a shunt regulator.
- Explain how to implement constant-current limiting and fold-back current limiting as well as their advantages and disadvantages.
- Increase the output current of a regulator using a pass transistor.
- Explain how a switching regulator works and why it dissipates less power than linear regulators.
- Explain how to wire fixed and adjustable voltage three-terminal regulators.
- Describe the function of a crowbar circuit.
- Discuss the operation of a current regulator, or constant-current source.

Introduction

A **voltage regulator** is a circuit that provides a constant DC voltage level to its load. Ideally, its output voltage stays constant regardless of changes in source voltage or load current. A voltage regulator normally interfaces either between an unregulated or poorly regulated DC power source and a load that requires a constant voltage source.

The power supply and its need to produce a source of regulated voltage is perhaps the most important section of any active circuit design. Without a power supply, no active circuit could operate. Originally, voltage regulators were built around vacuum tube voltage regulators that were primarily of the high-voltage type (greater than 90 V). With the advent of solid-state technology, Zener diodes were created, which permitted the regulation of fixed voltages as low as 3 V to almost 100 V. However, the efficiency of such Zener diode **shunt regulators** varied with load requirements.

Further advances resulted in the introduction of the linear integrated circuit, a virtual "system on a chip," which has now greatly simplified the construction of power supplies and other circuits that require voltage and current regulation. As a result, it is now possible to build high-quality voltage or current regulators with a single integrated circuit and several additional components.

This chapter first discusses those positive and negative linear and switching-type regulators built around an operational amplifier. Depending on how the control element is connected, linear regulators are classified as either series or shunt. Besides regulating voltage, the operational amplifier can be also used to regulate current to a load. The latter part of this chapter concludes with a discussion of those three-terminal fixed and variable integrated-circuit regulators that deliver positive or negative regulated output voltages from 1.25 V to more than 30 V.

9–1 Series Regulators

The Basic Concept

Most linear voltage regulators are of the series type. As shown in the block diagram of Figure 9–1, a **series regulator** network is inserted in series with the unregulated voltage source and the load and drops a controlled portion of the DC source voltage across itself. It senses the DC voltage across its load and adjusts its own voltage drop to keep the DC load voltage at a predetermined level. The load voltage plus the voltage drop across the regulator must then always add up to the DC source voltage. If the DC source voltage should increase, the regulator will compensate by increasing its voltage drop by an equal amount. If the source voltage should decrease, the regulator will then compensate for this change by decreasing its voltage drop by an amount equal to the decrease.

Any ripple voltage that may be present on the DC source voltage is then greatly reduced by the voltage regulator action. Since ripple is a variation in the source voltage, the regulator will compensate for this variation with equal and opposite variations in the voltage drop across the regulator.

As shown in Figure 9–2, the series-pass voltage regulator has three basic sections:

- Control element
- Reference voltage
- Regulating circuitry

Figure 9–1
Block diagram showing a series-pass regulator in series with a load.

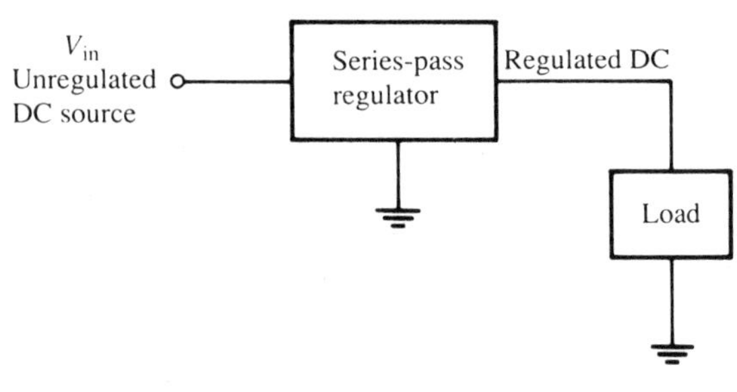

Figure 9–2
Elements of a series-pass regulator.

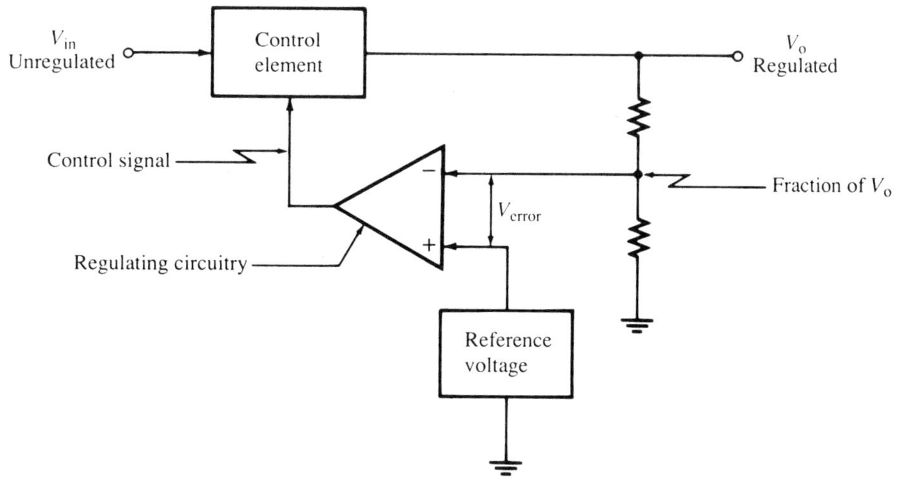

The control element is in series with the load and drops some of the source voltage. This can be either a single resistor or the collector-emitter terminals of a pass transistor. The reference voltage is the standard against which the output voltage is compared. It must therefore maintain a stable output voltage over time and with variations in temperature, for a voltage regulator can be only as good as its reference source. The regulating circuitry compares a fraction of the output voltage against the reference and produces a signal for the control element.

The series regulator is one of two classes of linear regulators because it smoothly and continuously varies its control element to maintain a constant voltage across the load and may be constructed using discrete transistors, operational amplifiers, or special-purpose integrated circuits. For low-current applications of 3 A or less, operational amplifiers and special-purpose integrated circuits are often used, since these devices provide stable regulation of circuits of simple design and low cost. For higher-current applications requiring load currents greater than 3 A, the series-pass element is usually a power transistor. Operational amplifiers and low-power transistors may be employed in the regulating circuitry that drives the series-pass control element.

Operational Amplifier Series Regulators

The circuit for a simple operational amplifier positive-voltage, low-current series regulator is shown in Figure 9–3. The operational amplifier here is wired as a voltage follower and supplies the load current directly. The voltage drop across a reverse-biased Zener diode is the reference voltage ($V_{REF} = V_Z$) and is applied to the noninverting input of the operational amplifier. The voltage to be regulated is fed to the operational amplifier's inverting input. Any voltage difference between

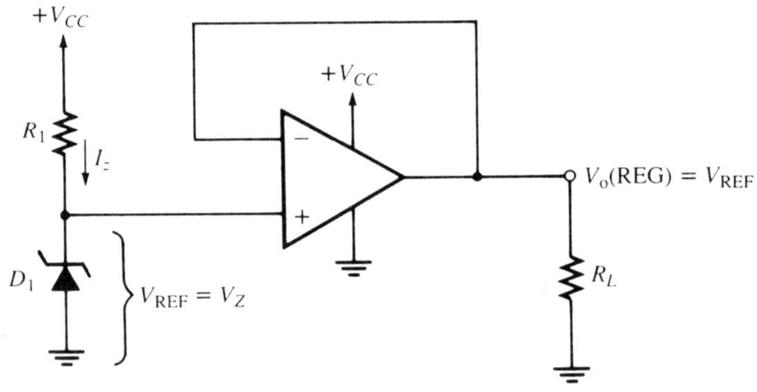

Figure 9–3
Positive voltage, low-current series regulator using a Zener diode.

these terminals is then amplified to bring the operational amplifier output to the same level as the Zener voltage. Here, the operational amplifier serves as both the control element and the regulating circuitry. Since a voltage follower is used, the regulated output voltage (V_o) will always be very close to the voltage applied to the noninverting input by the Zener reference. Normally, the output current capacity of this type of regulator is low and is limited by the output current and power dissipation ratings of the operational amplifier.

Note that the unregulated DC source voltage (V_{CC}) serves both to (1) reverse-bias the Zener diode and (2) furnish the necessary positive supply voltage for the operational amplifier. Resistor R_1 sets the proper current level through the Zener diode to give its rated Zener voltage, which is generally specified on the Zener's data sheet. Consequently,

$$R_1 = \frac{V_{CC} - V_{\text{REF}}}{I_Z} \qquad \textbf{(9–1)}$$

In addition, the negative supply pin is grounded. The value of the output voltage must be no greater than several volts below V_{CC}. This voltage limitation then prevents the operational amplifier output from saturating. Consequently, a sufficient amount of the supply voltage must be dropped across the regulator to ensure an adequate voltage regulation range without saturating the device.

This type of operational amplifier regulator is a big improvement over the regulated voltage that is obtained from a simple Zener diode–current-limiting resistor circuit. With the addition of the high input resistance of the operational amplifier, the current through the Zener diode is held essentially constant even though the load current may vary. Furthermore, the operational amplifier also provides a much lower source resistance for the load and a greater current capacity than that provided by the combination of a small Zener diode and a resistor standing alone.

It is possible to provide a higher output voltage by connecting the operational amplifier as a noninverting amplifier, shown in Figure 9–4. In this case, the unregulated input voltage supplies power to the operational amplifier as well as establishing the reference voltage across D_1. The operation of this circuit is very similar

Figure 9–4
Using a noninverting amplifier to generate a regulated output voltage greater than the Zener reference.

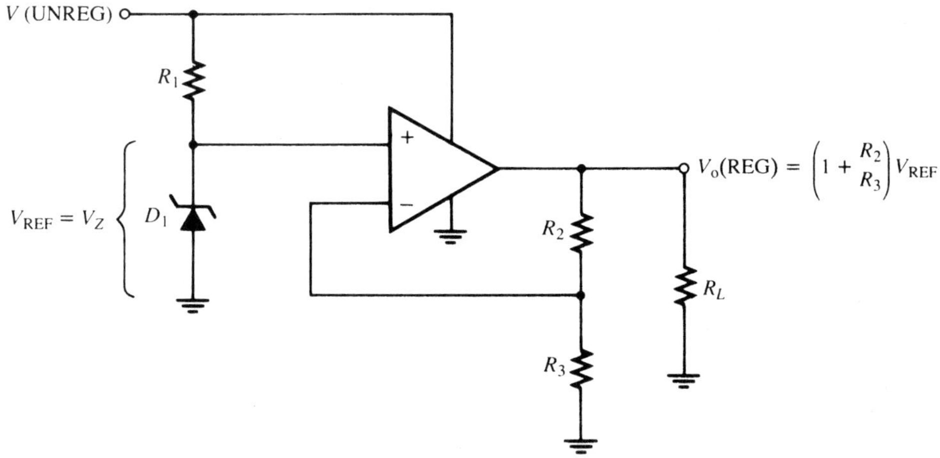

to the circuit in Figure 9–3 except that the reference voltage is compared with a fraction of the output voltage taken from a voltage divider. The voltage at the junction of the two *sampling,* or *sensing, resistors* R_2 and R_3 is kept equal to the reference voltage. The regulated output voltage is determined from:

$$V_o(\text{REG}) = \left(1 + \frac{R_2}{R_3}\right) V_{REF} \qquad \textbf{(9–2)}$$

Figure 9–5a illustrates the regulating action for increases in DC source voltage or load resistance. From Equation 9–2, the output voltage is determined by the two sampling resistors and the Zener voltage, and once either the input voltage or load resistance stabilizes at some new value, regulation will be maintained as long as the input voltage and load are within certain limits (Figure 9–5b). Figures 9–5c and 9–5d illustrate the opposite effect for decreases in DC source voltage or load resistance.

Example 9–1
Determine the regulated DC output voltage for the series voltage regulator of Figure 9–6.

Solution: Using Equation 9–2, the regulated output voltage is

$$V_o(\text{REG}) = \left[1 + \frac{4.7\ \text{k}\Omega}{10\ \text{k}\Omega}\right](6.2\ \text{V})$$
$$= \mathbf{9.11\ V}$$

Figure 9–5
Voltage regulator action. (a) Initial changes for increases in DC source voltage or load resistance. (b) Stabilized circuit. (c) Initial changes for decreases in DC source voltage or load resistance. (d) Stabilized circuit.

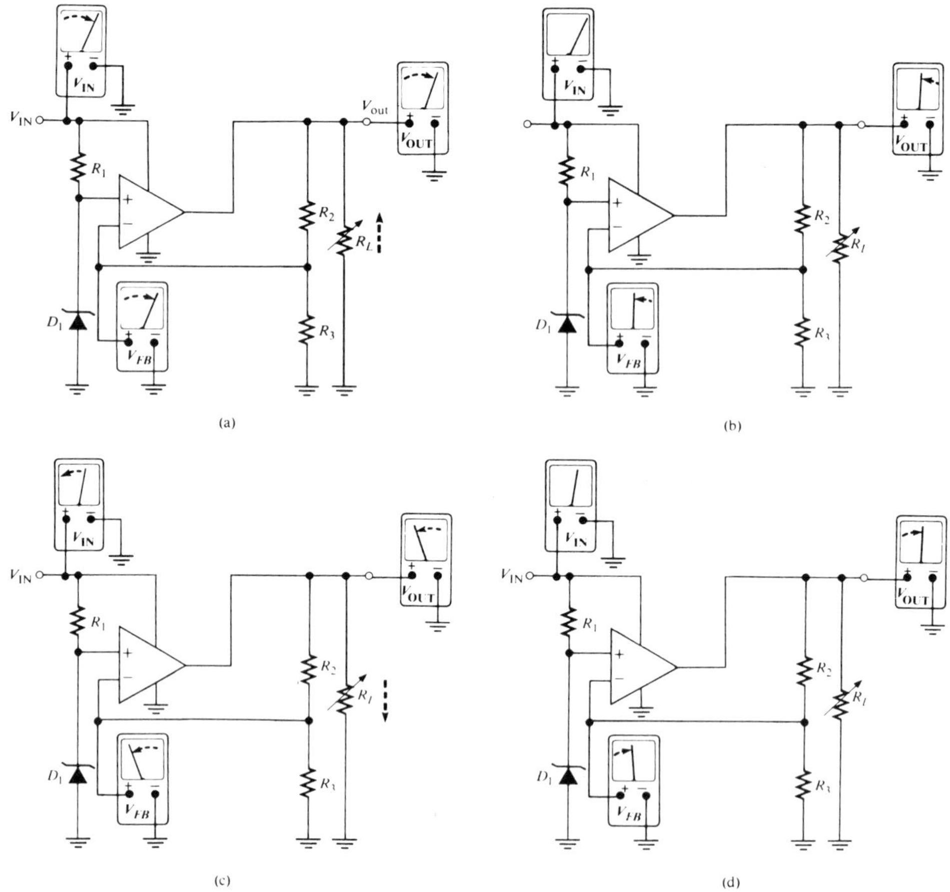

Figure 9–6
Series-pass regulator circuit for Example 9–1.

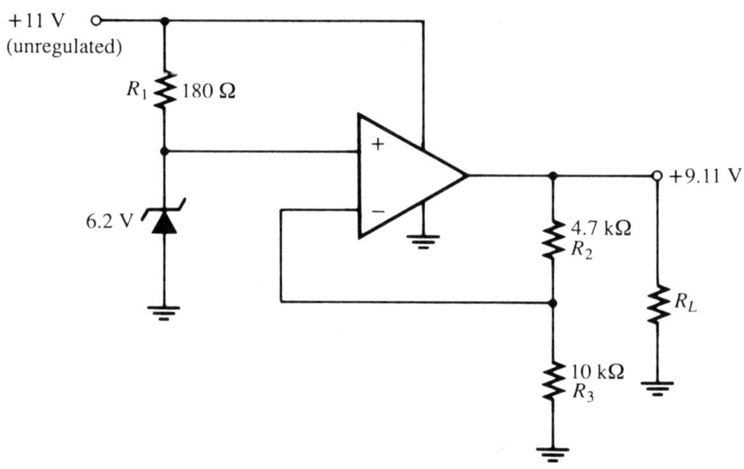

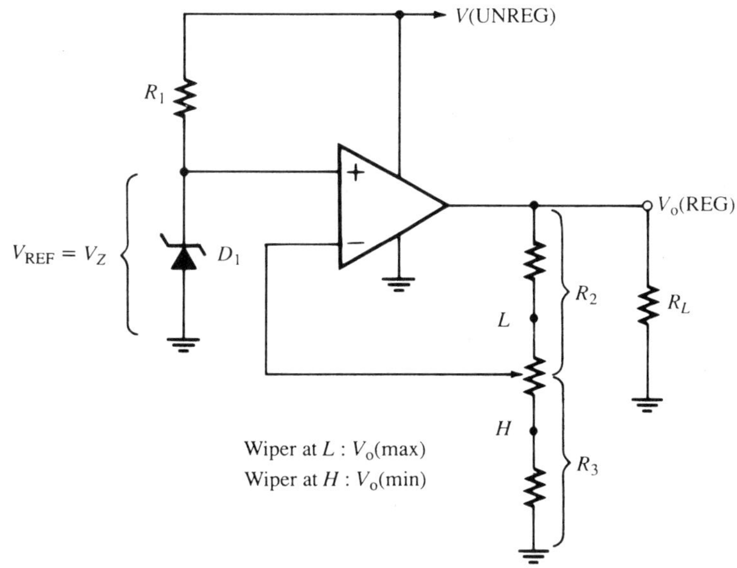

Figure 9–7
Series-pass voltage regulator with adjustable output voltage.

If, as shown in Figure 9–7, the fixed resistor voltage divider is replaced by a potentiometer, the regulated positive output voltage may then be varied over a limited range. With the potentiometer's wiper arm at the top at L, the output voltage will be at its minimum, whereas maximum output is obtained with the wiper arm at the bottom at H. To determine the voltage for a given setting of the potentiometer, the resistance above the arm is R_2 and the resistance below the arm is R_3 in Equation 9–2.

Since most operational amplifiers have rather limited current-output capabilities, the series circuits discussed so far are suitable only for applications that require small load currents. By having the operational amplifier drive an *npn* power transistor connected as an emitter follower and taking the feedback signal from the emitter, as shown in Figure 9–8a, the transistor behaves as though it is part of the output stage of the operational amplifier. This connection then gives the regulator a much greater output-current capacity, since the available output current is now limited mostly by the current and power ratings of the transistor and not by those of the operational amplifier. The control element has become, for all practical purposes, the transistor, and the operational amplifier now functions only as the regulating circuitry. The circuit is still a noninverting amplifier, but one with much greater output voltage and current. The only design calculations required are for the Zener diode reference (D_1 and R_1) and the specification of Q_1. High-power regulators of this type often require that the transistor be mounted on a heat sink in order to increase its power-handling capabilities (Chapter 7).

Figure 9–8b shows the same circuit, except that the regulated output voltage can be varied over a limited range with the potentiometer R_3. As illustrated in Example 9–2, the design calculations required are for the Zener diode reference (D_1 and R_1), the sensing resistors, and the specification of Q_1.

Figure 9–8
Using a transistor to increase the output current of a regulator.

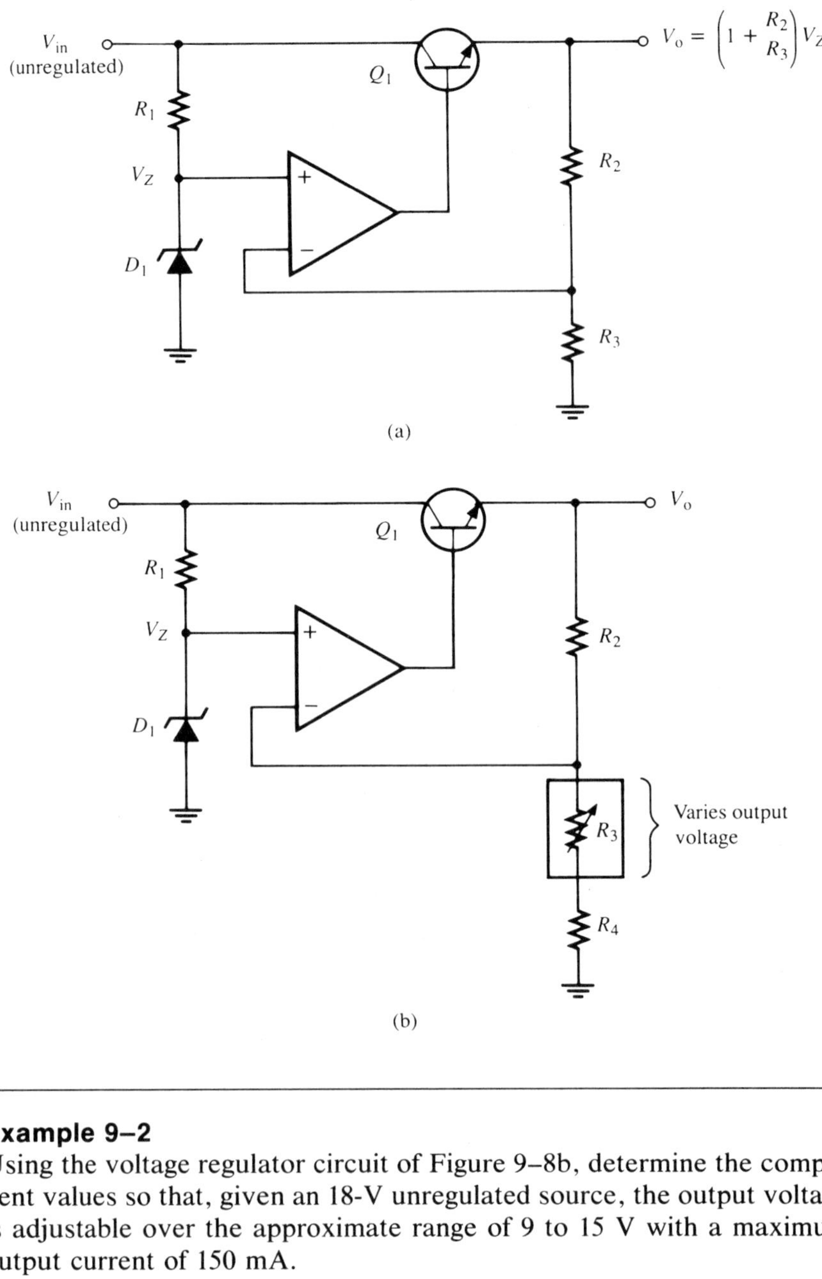

Example 9–2
Using the voltage regulator circuit of Figure 9–8b, determine the component values so that, given an 18-V unregulated source, the output voltage is adjustable over the approximate range of 9 to 15 V with a maximum output current of 150 mA.

Solution:
(a) As a general rule, the Zener diode voltage can be selected as approximately one-half the output voltage. Since the output is to vary from 9 V to 15 V, any Zener diode having a Zener voltage from 4.5 V to

7.5 will do. Here, assume that a 5.1-V Zener diode is chosen. From the diode's data sheet, this voltage typically requires that 20 mA flow through the Zener diode for 5.1 V. The value for R_1 is then determined on the basis of the source voltage (18 V) and the average output voltage (12 V),

$$R_1 = \frac{18\ \text{V} - 12\ \text{V}}{20\ \text{mA}} = \mathbf{300\ \Omega}$$

The power dissipated by R_1 is then

$$P = (20\ \text{mA})(6\ \text{V}) = 120\ \text{mW}$$

so that a **300-Ω, $\frac{1}{4}$-W** standard resistor can be used.

(b) The current flowing through the sensing resistors making up the voltage divider network must be greater than the input bias current flowing into the inverting input of the operational amplifier. In most cases then, the current flowing through the sensing network can be set to about 1 mA, since the input bias current is typically less than 100 nA.

When the output voltage is a minimum of 9 V, the wiper arm is at the top of the potentiometer R_3. Since the total voltage drop across R_3 and R_4 must be equal to the voltage reference of the Zener diode (5.1 V),

$$R_3 + R_4 = \frac{5.1\ \text{V}}{1\ \text{mA}} = 5100\ \Omega$$

On the other hand, the average voltage drop across R_2 is 12 V − 5.1 V, or 6.9 V. Since 1 mA flows through R_2,

$$R_2 = \frac{6.9\ \text{V}}{1\ \text{mA}} = 6900\ \Omega$$

for which a **6.8-kΩ** standard value can be used.

(c) When the output is at its maximum of 15 V, the wiper arm is now at the bottom of R_3. The maximum current flowing in the voltage divider string of R_2, R_3, and R_4 is now

$$I = \frac{15\ \text{V}}{(5.1\ \text{k}\Omega + 6.8\ \text{k}\Omega)} = 1.3\ \text{mA}$$

so that R_4 is found from

$$R_4 = \frac{5.1\ \text{V}}{1.3\ \text{mA}} = 3923\ \Omega$$

for which a **3.9-kΩ** standard value can be used. Consequently, R_3 is

$$R_3 = 5.1\ \text{k}\Omega - 3.9\ \text{k}\Omega = 1.2\ \text{k}\Omega$$

for which a **1-kΩ** potentiometer can be used without much loss of accuracy.

(d) When Q_1 is cut off, V_{CE} (max) = **18 V.** Using Kirchhoff's current law at the junction of Q_1's emitter, R_1, and R_2 yields

$$I_C(\text{max}) = 150 \text{ mA} + 1.3 \text{ mA}$$
$$= \mathbf{151.3 \text{ mA}}$$

so that the maximum power dissipation across the collector-emitter junction is

$$P_D(\text{max}) = (18 \text{ V} - 9 \text{ V})(151.3 \text{ mA})$$
$$= \mathbf{1.36 \text{ W}}$$

Consequently, any transistor having at least the minimum values above for V_{CE}(max), I_C(max), and P_D(max) will work. The final circuit is shown in Figure 9–9.

For those situations where a regulated negative voltage is required, the basic circuit of Figure 9–3 can be rearranged to operate from an unregulated negative supply voltage, as illustrated in Figure 9–10. Here, the polarity of the Zener diode is reversed; in addition, the positive supply pin of the operational amplifier is grounded and the DC source voltage is connected to the negative supply pin.

Constant-Current Limiting

In the event of a short circuit or overload on the output of a voltage regulator circuit, it is desirable to have some form of current limiting added to the series-pass regulator circuit of Figure 9–8a to prevent damage to the regulator compo-

Figure 9–9
Completed regulator circuit for Example 9–2.

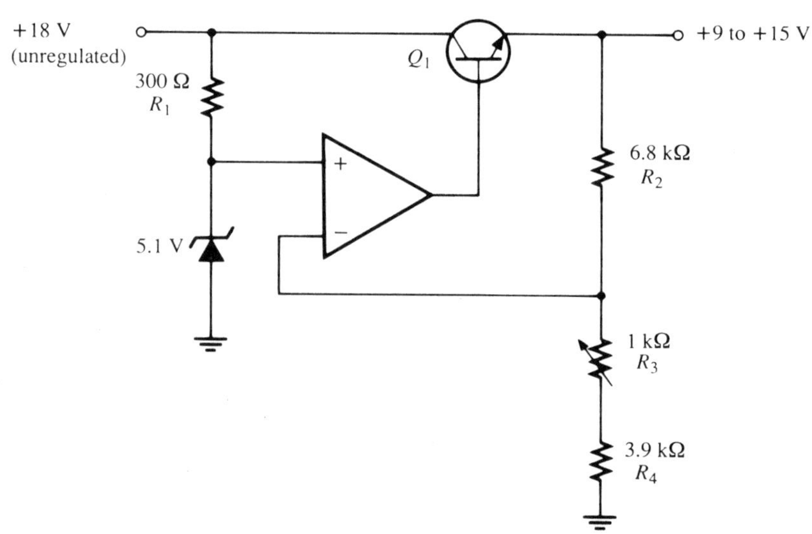

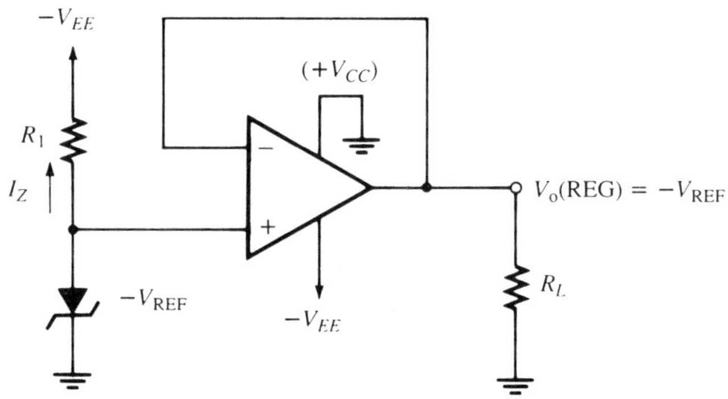

Figure 9–10
Series-pass voltage regulator with negative voltage output.

nents. Figure 9–11 shows a simple means of limiting output current during a short circuit or overload.

When the load current reaches a predetermined value, the voltage drop across the sense resistor R_3 reaches approximately 0.7 V. This is then sufficient to forward-bias transistor Q_2 and divert all the current away from the base of the series-pass transistor (Q_1) to the collector of Q_2. This action then puts an upper limit on the current that can be delivered by the regulator to slightly more than the value necessary to cause a 0.7-V drop across R_3,

$$I_L(\max) = \frac{0.7\text{ V}}{R_3} \tag{9–3}$$

At load currents below this value, the Q_2-R_3 current-limiting circuit has no effect on output voltage. Since R_3 is inside the feedback loop of the operational amplifier, any slight voltage drop caused by it is then compensated for by the voltage-regulating circuit. If the load current tries to increase above the limiting value, the

Figure 9–11
Series-pass voltage regulator with current limiting.

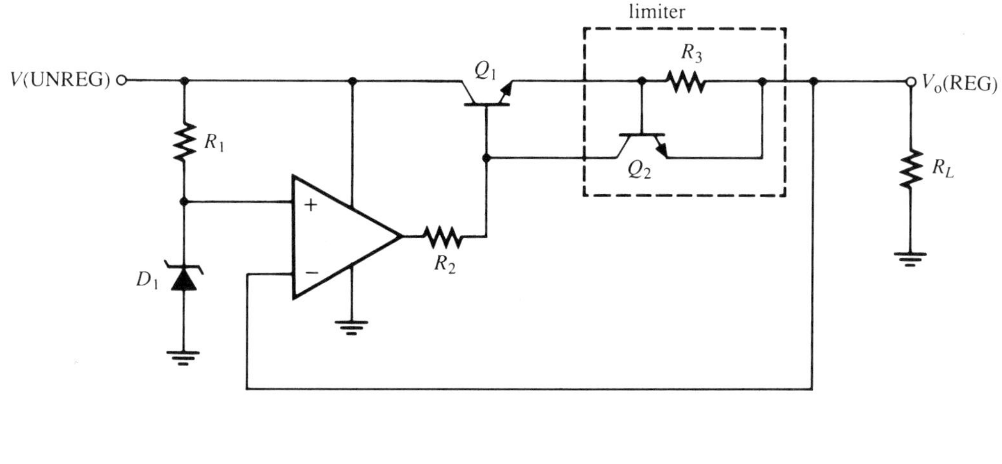

output voltage then decreases due to an increased collector-emitter voltage drop across Q_1, so that the output current will remain close to the limiting value despite further decreases in load resistance.

Foldback Current Limiting

The constant-current limiting technique just discussed nevertheless restricts the load current to a *maximum constant value*. Another type of current limiting, known as **foldback current limiting,** is used primarily in high-current regulators.

Figure 9–12
Series-pass regulator with foldback current limiting. (a) Circuit. (b) Foldback current characteristic.

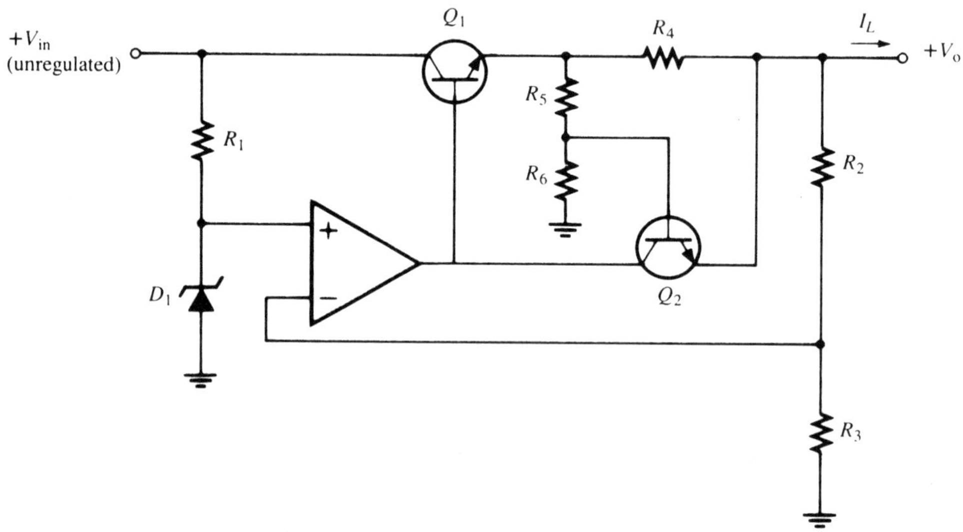

(a)

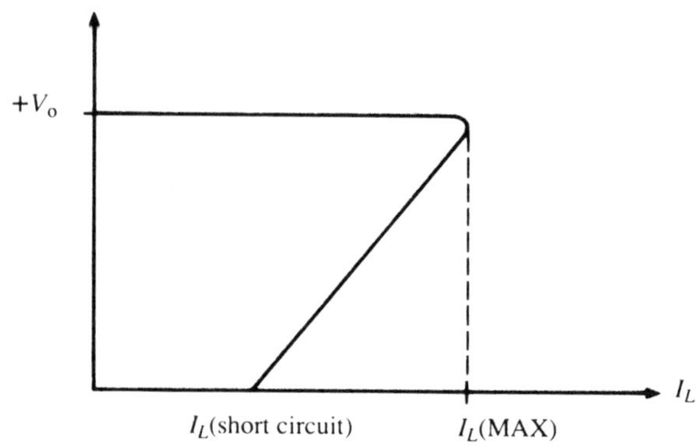

(b)

When subjected to a short circuit or overload, the output current of the regulator having foldback current limiting then decreases to a level that is below the peak load current capability. In this manner, the power dissipation of the circuit is not exceeded.

A series regulator circuit with provision for foldback current limiting is shown in Figure 9–12a. The voltage across R_4 due to the load current must be sufficient to (1) overcome the base-emitter voltage needed to forward bias Q_2 and (2) overcome the voltage across R_5. From Kirchhoff's voltage law,

$$V_{R_4} = V_{BE2} + V_{R_5} \qquad (9\text{–}4)$$

When the load current increases to a level that forces Q_2 to conduct, the load current is then limited at this point. The resulting decrease in the regulator's output voltage in turn forces a proportional decrease in voltage across R_5. Therefore, less current through R_4 is required to keep Q_1 turned on. The output characteristic curve of Figure 9–12b shows that the load current decreases as the output voltage decreases. The short-circuit load current then decreases to a safe level that prevents the device from overheating.

9–2 Shunt Regulators

A **shunt regulator** is the opposite of a series voltage regulator. Here, the output voltage is regulated by a parallel, or shunt, control element connected from the output to ground that, in turn, absorbs excess load current. The basic positive-voltage, low-power shunt regulator is shown in Figure 9–13. As with the series

Figure 9–13
Basic low-power shunt regulator.

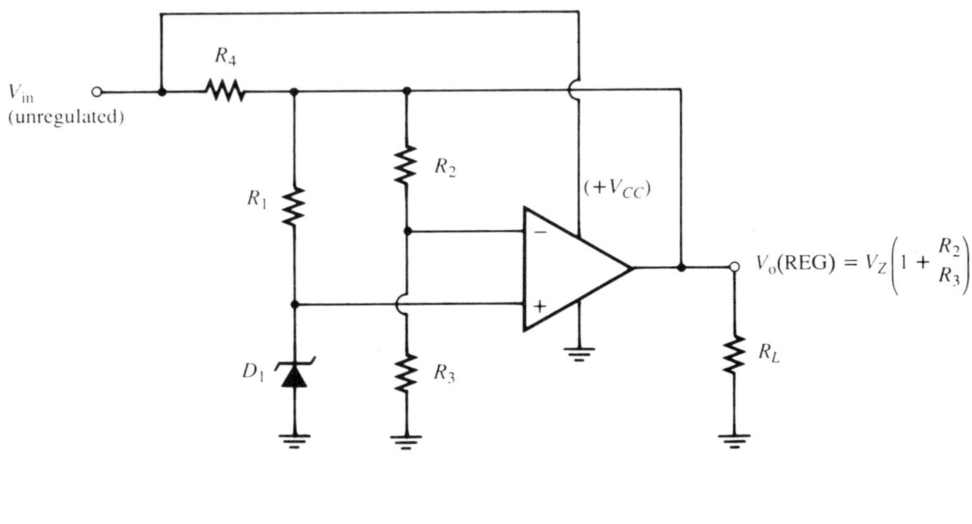

regulator, the output voltage of the operational amplifier is regulated at a voltage level equal to

$$V_o(\text{REG}) = \left(1 + \frac{R_2}{R_3}\right) V_{\text{REF}} \tag{9–5}$$

The current-limiting resistor R_1 for the Zener diode is found from

$$R_1 = \frac{V_o - V_{\text{REF}}}{I_Z} \tag{9–6}$$

If the output voltage tries to decrease in response to changes in either the input source voltage, load resistance, or temperature, sensing resistors R_2 and R_3 detect this level change and, in turn, feed it to the inverting input of the operational amplifier. Therefore, the voltage is maintained at a constant level. If, on the other hand, the output tries to increase, the opposite action occurs. Example 9–3 illustrates the procedure for determining the proper component values for the shunt regulator.

Example 9–3
Using the shunt regulator circuit of Figure 9–13, determine the component values necessary to give a regulated output of 12 V at a maximum load current of 10 mA, using a 20-V unregulated source voltage and a 6.2-V Zener diode biased at 5 mA.

Solution:
(a) For the current through D_1 to be 5 mA,

$$R_1 = \frac{(12 \text{ V} - 6.2 \text{ V})}{5 \text{ mA}} = \mathbf{1.2\ k\Omega}$$

(b) As with the series regulator of Example 9–2, the current through the voltage divider resistors R_2 and R_3 can be assumed to be set at approximately 1 mA in order not to be loaded down by the input of the operational amplifier. Since the voltage across R_3 is to be the same as the Zener voltage,

$$R_3 = \frac{6.2 \text{ V}}{1 \text{ mA}} = \mathbf{6.2\ k\Omega}$$

so that

$$R_2 = \frac{12 \text{ V} - 6.2 \text{ V}}{1 \text{ mA}} = 5.8 \text{ k}\Omega$$

Hence a **5.6-kΩ** standard value could be used.

(c) The current through R_4, using Kirchhoff's current law, is

$$I_4 = 10 \text{ mA} + 5 \text{ mA} + 1 \text{ mA}$$
$$= 16 \text{ mA}$$

so that R_4 is found from

$$R_4 = \frac{20 \text{ V} - 12 \text{ V}}{16 \text{ mA}}$$
$$= 500 \ \Omega$$

Here a **510-Ω** standard value can be used. The final circuit is shown in Figure 9–14.

9–3 Switching Voltage Regulator

The two major forms of linear regulators discussed have one serious drawback—the control element must dissipate a significant amount of power. An alternate solution is the *switching regulator*. Instead of the control element operating in the linear mode, where it constantly dissipates power, it is operated in the *switched mode*. The control element is then either switched (1) all the way on (saturated) or (2) all the way off (cutoff). When it is on, the current flowing through it is high, but the voltage drop is low. On the other hand, when it is off, the voltage drop is high, but the current is zero. Very little power is dissipated in either state.

By varying the ratio of on-time (t_{on}) to off-time (t_{off}), it is then possible to control the average (DC) value of output voltage. Mathematically, the average

Figure 9–14
Completed shunt regulator circuit for Example 9–3.

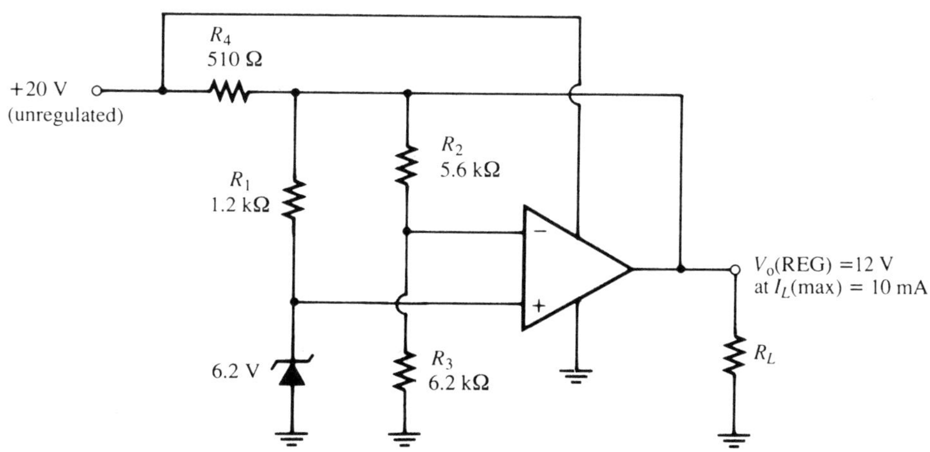

value of a switched waveform having a peak value V_p is given by

$$V_{DC} = \frac{t_{on}}{t_{on} + t_{off}} V_P \qquad (9\text{–}7)$$

The ratio $t_{on}/(t_{on} + t_{off})$ is called the *duty cycle* (D) and can vary from 0 to 1, although it is usually expressed as a percentage. Equation 9–7 can then be written as

$$V_{DC} = D \times V_P \qquad (9\text{–}8)$$

which states that the average output voltage is directly proportional to the duty cycle.

Figure 9–15 shows three switched waveforms together with their relative DC output levels. The waveforms illustrate duty cycles of less than 50%, exactly 50%, and greater than 50%, respectively. This action of varying the duty cycle is referred to as *pulse-width modulation* (PWM) and is used by the switching regulator to control the output voltage. By increasing the duty cycle, the average value of the switched waveform increases, which in turn increases the output voltage. By shortening the duty cycle, the output voltage decreases.

In order to eliminate most of the unwanted higher-frequency components that result as well as to supply the load with the DC component, the switched waveform is passed through an LC-filter network. The average value of the output voltage is given approximately by Equation 9–8.

Figure 9–16 shows the basic sections of a simple PWM-based switching regulator. Transistor Q_1 serves as the switched control element. The operational amplifier compares a sample of the output voltage taken from the junction of R_2 and R_3 with the reference voltage developed across Zener diode D_2. Any difference voltage generates an error signal at the output of the amplifier, which in turn

Figure 9–15
Effect of duty cycle on DC level. (a) 25% duty cycle. (b) 50% duty cycle. (c) 75% duty cycle.

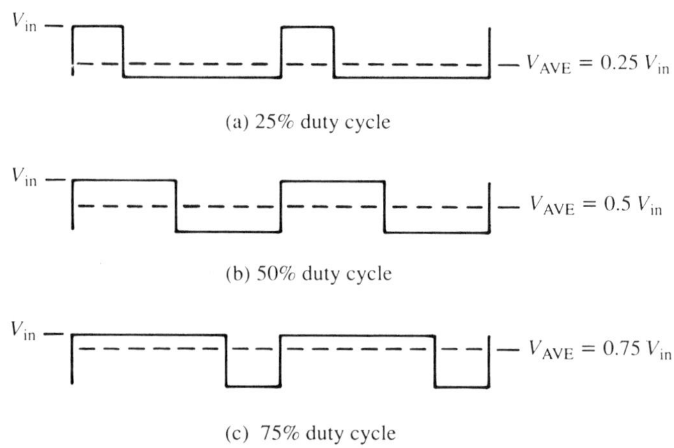

Figure 9–16
Pulse-width modulation based switching regulator.

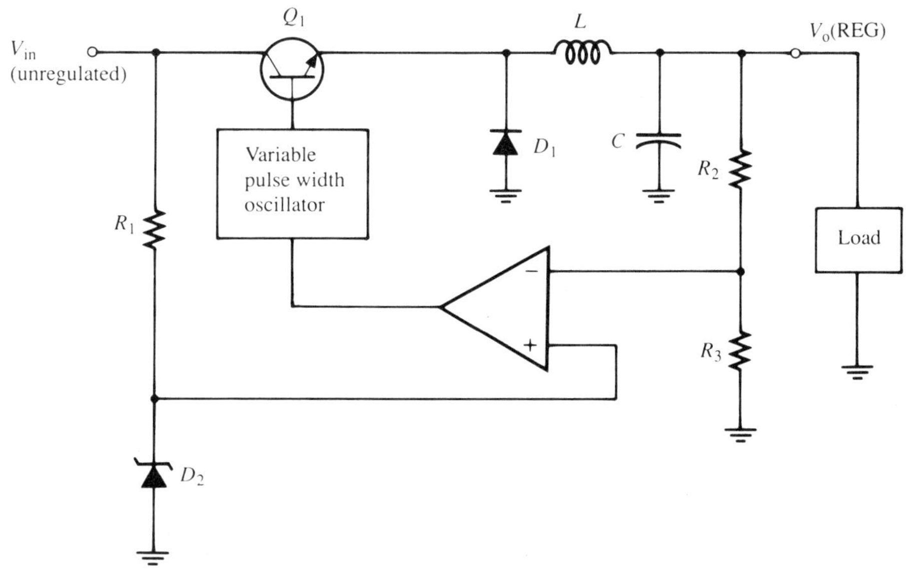

adjusts the variable pulse width oscillator. The variable pulse width oscillator in turn controls the switching duty cycle of Q_1.

The inductor's collapsing magnetic field tends to maintain a constant current during periods when Q_1 is cut off, whereas D_1 provides a current path to eliminate those high reverse voltages that would otherwise appear across Q_1 during this time. Filter capacitor C serves to dampen switching transients, and the LC-filter network serves to filter out all but the DC component of the switched waveform. Because of its reduced size and increased efficiency, this type of regulator is becoming more and more important for high-current, low-voltage computer power supplies. Its one drawback is the size, weight, and cost of the required inductor.

9–4 Integrated-Circuit Voltage Regulators

Instead of facing the trials and tribulations associated with constructing discrete voltage regulators, integrated-circuit voltage regulators now are the preferred approach to building regulated power supplies. These devices are simple to use and are available in wide ranges of fixed positive and negative voltages with output capacities up to 3 A. In addition, a variety of adjustable positive and negative voltage regulators are now available that serve as the basis of efficient, low-cost, variable DC power supplies.

In this section, the following three classes of regulators will be discussed:

1. Fixed-voltage regulators: positive and negative
2. Adjustable voltage regulators: positive and negative
3. Adjustable switching voltage regulators

Figure 9–17
Integrated-circuit, three-terminal voltage regulators. (a) Positive voltage. (b) Negative voltage. (c) Pin diagrams for TO-3 and TO-220 style packages.

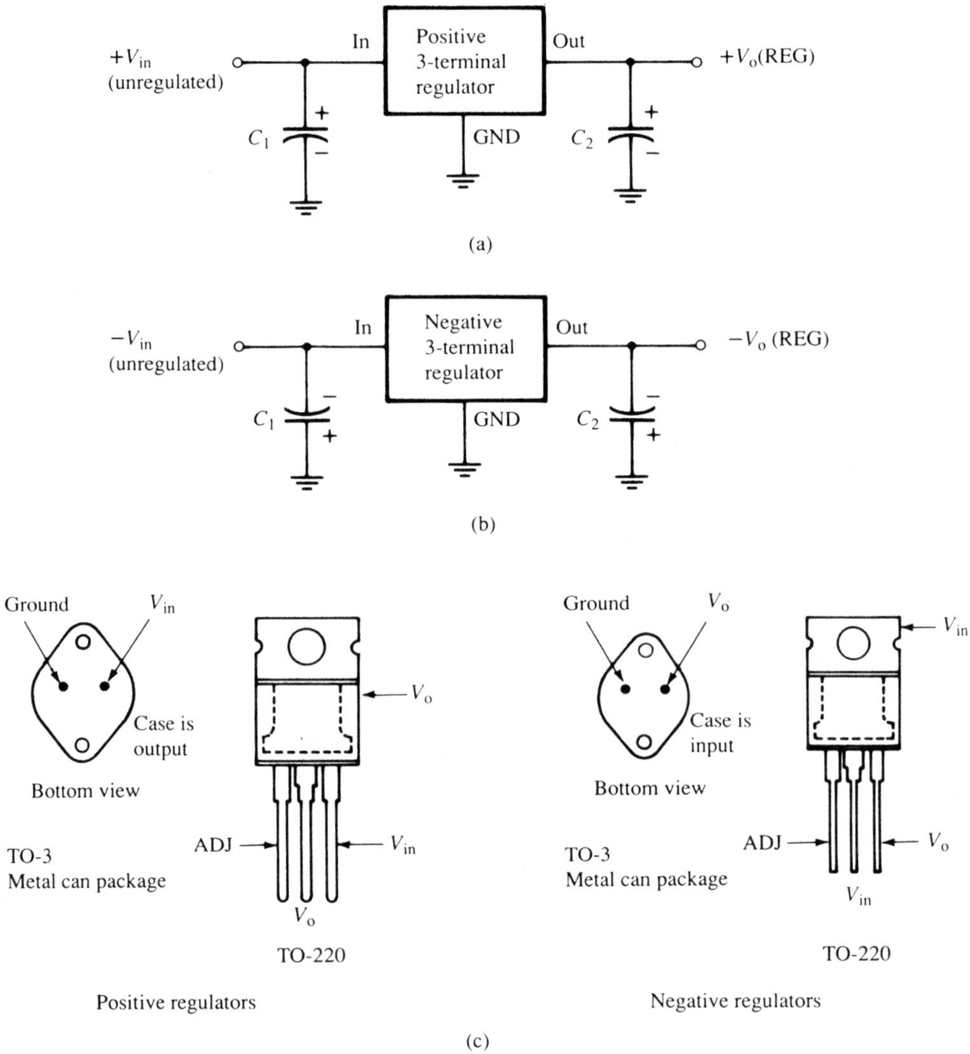

With the exception of the last class, these integrated-circuit voltage regulators are often referred to as *three-terminal regulators,* because, like a transistor, these devices have only three terminals.

Fixed-Voltage, Three-Terminal Regulators

Figure 9–17a shows the schematic representation for a typical three-terminal positive voltage regulator, and Figure 9–17b illustrates the corresponding negative-voltage device. In either case, the three terminals are labeled In, Out, and

Table 9–1
Popular three-terminal fixed positive regulators

Type	Output Voltage	Maximum Output Current
309	5 V	1.5 A
323	5 V	3 A
340-x series	See Note 1.	1.5 A
78xx series	See Note 2.	1.5 A

Notes:
1. Available in output voltages of 5, 12, and 15 V, whose voltage is the suffix for the model number.

2. Available in output voltages of 5, 6, 8, 10, 12, 15, 18, and 24 V, whose rated voltage is the last two digits of the model number (e.g., 7808 for 8-V regulator).

GND (ground). Three-terminal regulators are available primarily in TO-3 and TO-220 style packages, whose pin connections are shown in Figure 9–17c.

The unregulated input voltage is usually direct from the filtered output of either a half-wave or full-wave rectifier. Capacitor C_1 is required only if the regulator is located an appreciable distance (over 2 in.) from the power-supply filter capacitor. A second capacitor (C_2), although not always required, is often provided to improve transient response and give a lower AC output impedance. When used, these are to be connected as close as possible to the regulator terminals. The more popular fixed output voltages, which are summarized in Tables 9–1 and 9–2, vary from 5 to 24 V with output currents as high as 3 A with adequate heat sinking. In addition, these type regulators have internal thermal and short-circuit protection.

A number of performance parameters are normally specified for integrated-circuit voltage regulators by the manufacturers. Taking the popular LM309 5-V regulator as an example, these are as follows:

1. *Line, or input, regulation.* Line regulation is defined as the change in output voltage for a given change in input voltage. It is usually expressed in millivolts or as a percentage of output voltage. For the LM309, it is typically 5 mV when the input varies from 7 V to 25 V.
2. *Load regulation.* Load regulation is the change in output voltage over a given range of load currents. It is usually expressed in millivolts or as a percentage of output voltage. For the LM309, it is 20 mV for output currents from 5 to 500 mA.

Table 9–2
Popular three-terminal fixed negative regulators

Type	Output Voltage	Maximum Output Current
320 series	See Note 1.	1.5 A
345	−5 V	3 A
79xx series	See Note 2.	1.5 A

Notes:
1. Available in output voltages of −5, −12, and −15 V, whose voltage is the suffix for the model number.

2. Available in output voltages of −5, −6, −8, −10, 12, −15, −18, and −24 V, whose rated voltage is the last two digits of the model number (e.g., 7915 for 15-V regulator).

3. *Ripple rejection.* Ripple rejection is a measure of the regulator's ability to reject ripple voltage present at its input. It is defined as the ratio of the RMS input ripple voltage to the RMS output ripple voltage and is often expressed in decibels. For the LM309, it is typically 50 dB (i.e., approximately a 300 : 1 reduction).
4. *Dropout voltage.* The dropout voltage is the minimum voltage that must exist between the input and the output terminals for the regulator to operate properly. For most regulators, this is typically about 2 to 3 V.
5. *Maximum input voltage* (V_{in} max). This is the maximum input voltage that can be applied to the regulator without damage. For the LM309, it is 35 V.
6. *Maximum power dissipation* (P_D max). The maximum power dissipation rating is the maximum power that the regulator can safely dissipate. The actual power dissipated is approximately equal to the voltage drop between the input and the output terminals multiplied by the current through the regulator. For many regulators it is internally limited and varies whether or not a heat sink is used.
7. *Rated output current* (I_o). This is the maximum value of output current above which current limiting will occur and is usually dependent on the input-output voltage differential.
8. *Quiescent current* (I_Q). This is the current that must flow from the common (GND or Adjust) terminal for the regulator to operate. For the LM309, it is typically 5.2 mA.

For the most part, there is very little with which one has to be concerned in using these types of regulators. Other than the obvious need to stay within the maximum input voltage and maximum load current specifications, one parameter that should be kept in mind is the *dropout voltage*. This is the minimum voltage drop that must exist between the regulator's input and output terminals for the

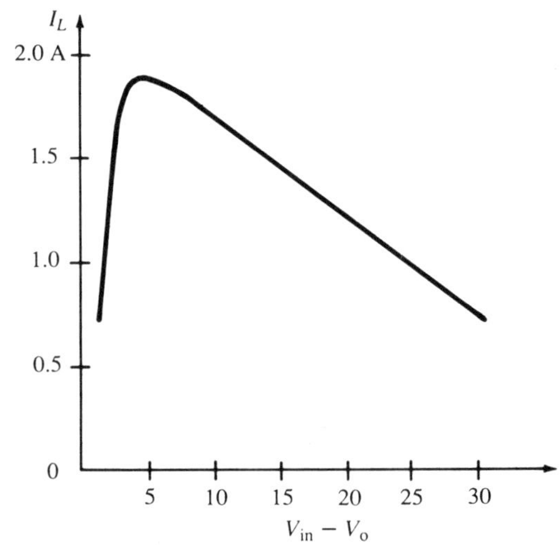

Figure 9–18
LM309 integrated-circuit voltage regulator dropout characteristic.

regulator to work properly and is approximately 2 to 3 V. For example, the regulated output voltage of a type 7812 regulator will be +12 V as long as the unregulated input voltage is at least 12 V + 3 V, or 15 V. For the corresponding 7912 negative regulator, the input voltage must be more negative than −15 V for the output to be properly regulated at −12 V. Generally, the magnitude of the difference between the input and output voltages affects the maximum output current that can be supplied. Shown in Figure 9–18 is the I_L versus $V_{in} - V_o$, or dropout, characteristic curve for the LM309 5-V regulator. For input voltages below approximately 3 V, the output current is zero, since the regulator is effectively disabled. The dropout voltage has not been exceeded. For higher input voltages, the output current peaks and then starts to drop off with increasing input voltage.

Figure 9–19a shows a line-operated regulated power supply that, with a proper heat sink, will provide a fixed output of +5 V at 1.5 A using the LM309 voltage regulator, and Figure 9–19b illustrates a circuit for a −12-V supply using an LM320-12.

Figure 9–19
Line-operated fixed voltage power supplies. (a) +5-V output using an LM309 regulator. (b) −12-V output using an LM320-12.

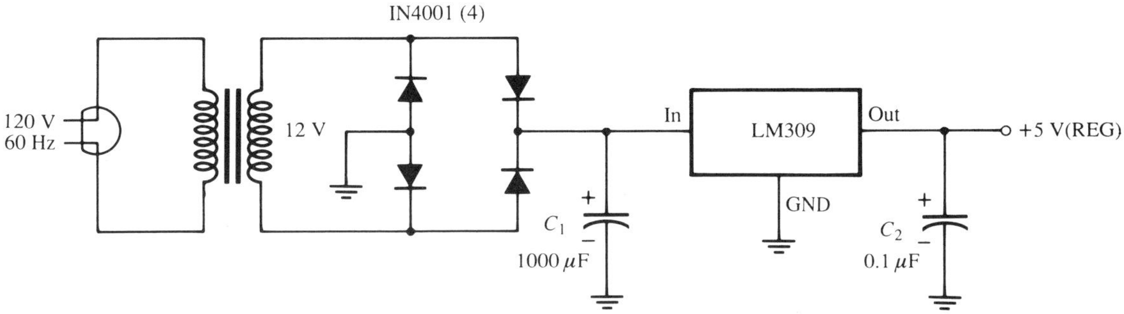

(a)

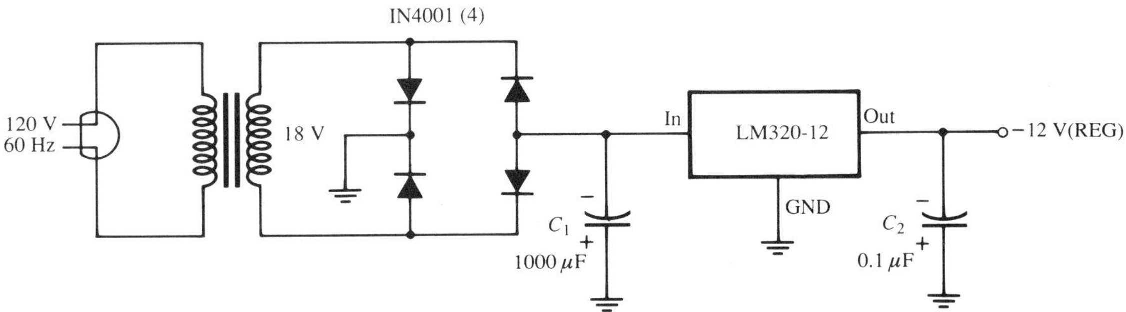

(b)

Figure 9–20
Obtaining a higher output voltage from a fixed-voltage, integrated-circuit regulator.

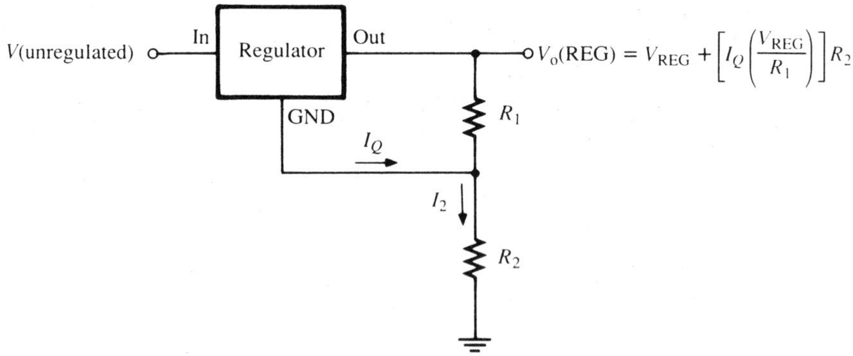

With these fixed-voltage regulators, it is a relatively simple procedure to increase the output voltage of a circuit that uses a fixed regulator above its nominal output with the addition of two resistors in the form of a voltage divider, as shown in Figure 9–20. Here, the regulator controls the voltage across R_1 and, therefore, the current through it, $V_o(\text{REG})/R_1$. The current through R_2 is equal to the current through R_1 plus a very small quiescent operating current (I_Q), which flows from the GND terminal of the integrated-circuit regulator:

$$I_2 = I_Q + \frac{V_{\text{REG}}}{R_1} \tag{9–9}$$

so that the voltage across R_2 is I_2R_2, or

$$V_2 = \left[I_Q + \left(\frac{V_{\text{REG}}}{R_1}\right)\right]R_2 \tag{9–10}$$

I_Q varies among different regulators. For the LM309, it is typically 5.2 mA.

The output voltage then is the sum of the voltages across the two resistors, so that

$$V_o(\text{REG}) = V_{\text{REG}} + \left[I_Q + \left(\frac{V_{\text{REG}}}{R_1}\right)\right]R_2 \tag{9–11}$$

For a given output voltage, an arbitrary value for R_1 is selected, and the resultant value for R_2 is determined from Equation 9–11.

If R_2 is replaced with a potentiometer, the output voltage can then be varied from as low as the rated output voltage of the regulator to a level approximately

3 V less than the input voltage. Again, the regulator's dropout voltage must be considered. For example, if a 7808 regulator is used to give a 14-V output using the circuit of Figure 9–20, then the input voltage must then be at least 14 V + 3 V, or +17 V, for the regulator-integrated circuit to operate properly. You cannot get more out of a regulator than the supplied voltage. In general, the higher the input voltage the better, as long as the regulator's maximum input voltage rating is not exceeded.

Example 9–4

Using an LM309 5-V regulator (I_Q = 5.2 mA) and the circuit of Figure 9–20, determine the values for R_1 and R_2 as well as the minimum input voltage required for an output voltage of 9 V.

Solution:

(a) Selecting R_1 = **1 kΩ** and using Equation 9–11, solving for R_2 yields

$$9\text{ V} = 5\text{ V} + \left[5.2\text{ mA} + \left(\frac{5\text{ V}}{1\text{ k}\Omega}\right)\right] R_2$$

or

$$R_2 = 392\ \Omega$$

for which a **390-Ω** standard value is used.

(b) For a 9-V output, the input voltage should be at least 12 V but no more than 35 V, as specified by its data sheet. The completed regulator circuit is shown in Figure 9–21.

Adjustable Three-Terminal Regulators

Besides fixed-voltage regulators, integrated-circuit voltage regulators are available that allow adjustment of output voltages from 1.25 V to as high as 57 V (Table 9–3). For the user, these types of regulators greatly reduce the number of different

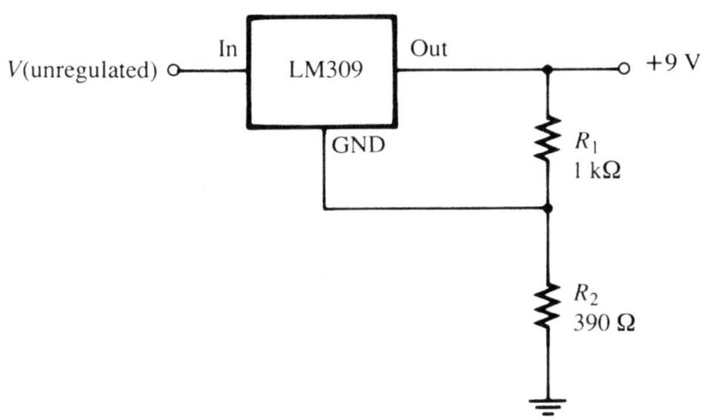

Figure 9–21
Completed integrated-circuit regulator network for Example 9–4.

Table 9–3
Three-terminal adjustable regulators

Type	Output Voltage	Maximum Output Current
317	+1.2 to 57 V	1.5 A
337	−1.2 to −47 V	1.5 A
338	+1.2 to 32 V	5 A
350	+1.2 to 32 V	3 A
396	+1.25 to 15 V	10 A

fixed regulators needed; they also obtain output voltages different from fixed regulators. Of these, the LM317 is the most popular for positive voltages, whereas the sole LM337 is used for negative-voltage applications.

Using the schematic representation of Figure 9–22a, both the 317 and 337 have terminals labeled In, Out, and ADJ (adjustment). Figure 9–22b shows the typical connections for the 317 for adjustable positive output voltage, and a similar circuit using the 337 for negative voltages is shown in Figure 9–22c. In both cases, these regulators require only two external resistors (R_1 and R_2) to set the output voltage.

Internally, the 317/337 develops a reference voltage of 1.25 V between the Out and ADJ terminals across R_1. Consequently, the current through R_1 is 1.2 V/R_1. Current I_{ADJ} flows out from the ADJ terminal and is similar to the quiescent current (I_Q) of the fixed-voltage regulators discussed previously. For the 317/337, I_{ADJ} is typically 100 μA. From Kirchhoff's current law, the current through R_2 is the sum of the current through R_1 and I_{ADJ}:

$$I_2 = 100\ \mu\text{A} + \frac{1.25\ \text{V}}{R_1} \qquad \textbf{(9–12)}$$

so that the voltage across R_2 is I_2R_2, or

$$V_2 = (100\ \mu\text{A})R_2 + \left(\frac{1.25\ \text{V}}{R_1}\right) R_2 \qquad \textbf{(9–13)}$$

Finally, the output voltage of the regulator is the sum of the voltages across R_1 and R_2:

$$V_o(\text{REG}) = 1.25\ \text{V} + \left[100\ \mu\text{A} + \left(\frac{1.25\ \text{V}}{R_1}\right)\right] R_2 \qquad \textbf{(9–14)}$$

As a consequence, the output voltage varies from 1.25 V to approximately V_{in} − 2 V. In terms of device specifications, the input voltage for the 317 can be as low

Figure 9–22
Integrated-circuit adjustable voltage regulators. (a) Schematic diagram for LM317/337. (b) Positive regulator using an LM317. (c) Negative regulator using an LM337.

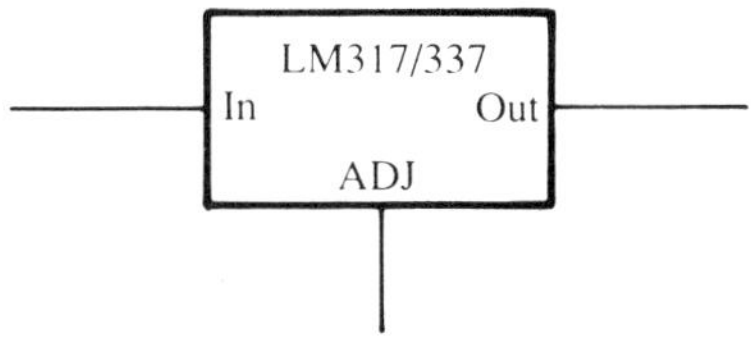

(a)

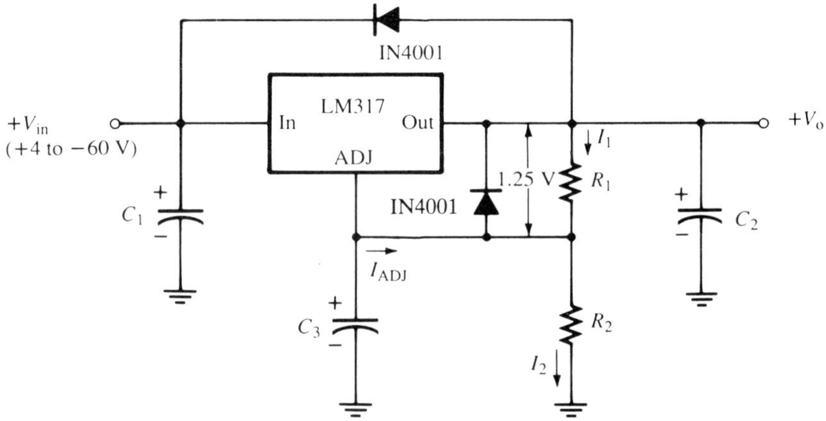

(b)

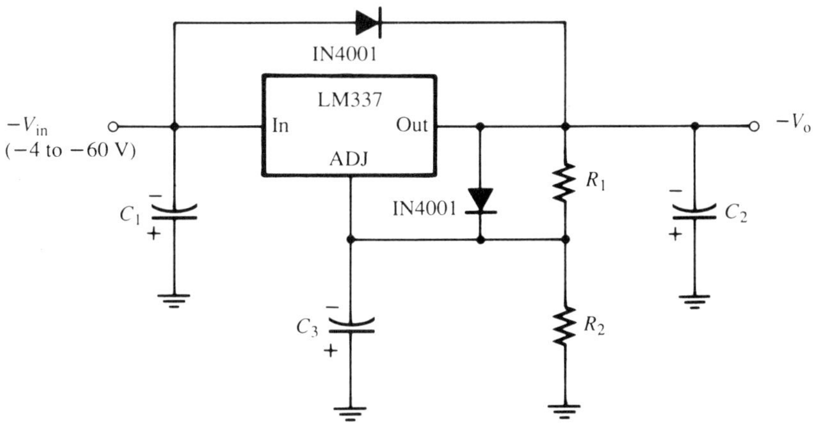

(c)

as 4 V to as much as 60 V to obtain the maximum range possible. For the 337, this range is −4 V to −50 V.

Apart from the rectifier filter capacitor, C_1 is used when the regulator integrated circuit is located more than several inches from the rectifier filter, whereas C_2 is used to improve the transient response. In addition, R_2 is bypassed with C_3 to achieve high ripple rejection. Typically, an 80-dB ripple rejection is obtained with $C_3 = 10\ \mu F$. When external capacitors C_1, C_2, and C_3 have values greater than 25 μF and regulated output voltages greater than 25 V, it is often necessary to include protection diodes D_1 and D_2 to prevent these capacitors from discharging through the regulator.

Example 9–5

Determine the output voltage range for the variable voltage power supply of Figure 9–23.

Solution: For a transformer secondary RMS voltage of 18 V feeding a full-wave bridge rectifier, the peak unregulated output voltage (accounting for two diode drops totaling 1.4 V for each half-cycle) across the 470-μF filter capacitor is

$$\begin{aligned} V_{\mathrm{UNREG}} &= (18\ \mathrm{VRMS})(1.414) - 1.4\ \mathrm{V} \\ &= 24.1\ \mathrm{V} \end{aligned}$$

Consequently, the maximum regulated output voltage possible is approximately 24.1 V − 2 V, or 22.1 V.

When the resistance of potentiometer R_2 is zero, the output voltage is a minimum, **1.25 V.** When R_2 is at its maximum 10-kΩ resistance, the output voltage using Equation 9–14 is

$$\begin{aligned} V_o(\mathrm{REG}) &= 1.25\ \mathrm{V} + \left[100\ \mu\mathrm{A} + \left(\frac{1.25\ \mathrm{V}}{820\ \Omega}\right)\right](10\ \mathrm{k}\Omega) \\ &= \mathbf{17.5\ V} \end{aligned}$$

The output voltage then varies from 1.25 V to 17.5 V. If, instead, R_1 was 470 Ω, then the maximum output voltage would try to be 28.9 V, which is more than the 22.1-V maximum possible. Therefore, in this case, the output voltage will vary from 1.25 V to 22.1 V.

Conversion to a Switching Regulator

As is discussed in Section 9–3, switching regulators dissipate less power compared with linear regulators. While it is possible to design switching regulators built around many of the popular three-terminal linear regulators listed in Tables 9–1, 9–2, and 9–3, more efficient operation is nevertheless achieved using one of several specialized multipin integrated-circuit regulators intended for this purpose. Of these, the most popular is the 723, whose pin configurations are shown in Figure 9–24 for both the 14-pin DIP and 10-pin, TO-100 metal can packages.

Figure 9–23
Positive regulator circuit for Example 9–5.

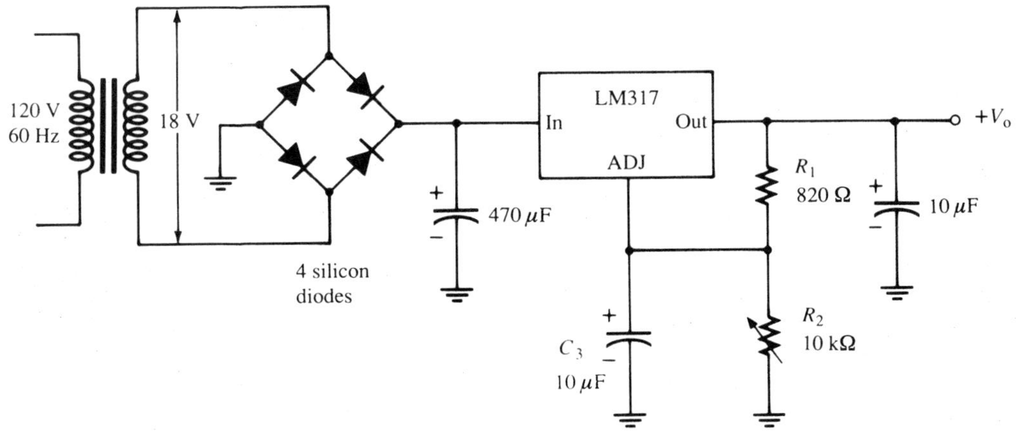

The 723 is quite versatile. Besides being able to be wired to function as a switching regulator having positive or negative outputs, it also can be wired to function simply as a linear regulator with low or high positive or negative output voltages. Although the maximum output current of the 723 is limited to 150 mA, it nevertheless can be increased upward to 10 A with suitable pass transistors.

Similar to circuits given in many data sheets of various manufacturers of the 723, Figure 9–25 shows a stock circuit using the 723 (14-pin DIP version) connected as a switching regulator to deliver 5 V at load currents up to 1 A from an unregulated 10-V source. The addition of the 2N4918 pass transistor (Q_1) as an external series-pass transistor allows the nominal 150-mA output limit to be in-

Figure 9–24
Pin diagrams for 723-type switching regulator.

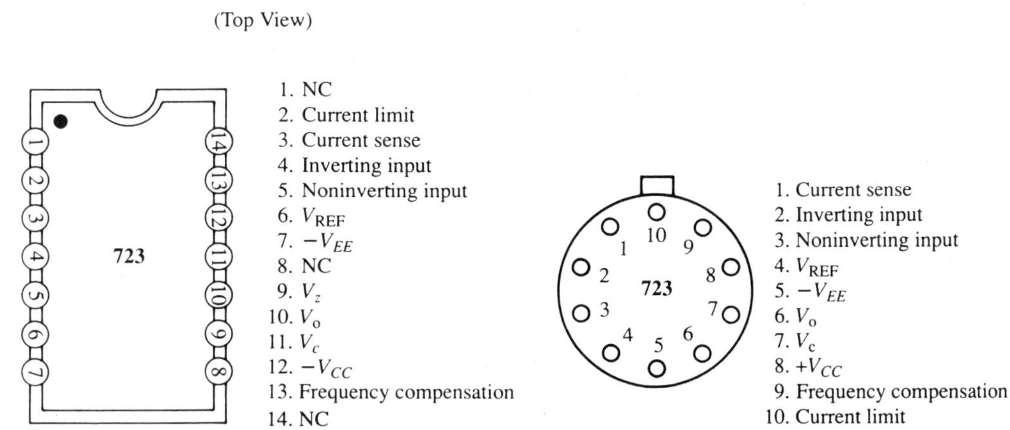

Figure 9–25
Switching regulator using the 723.

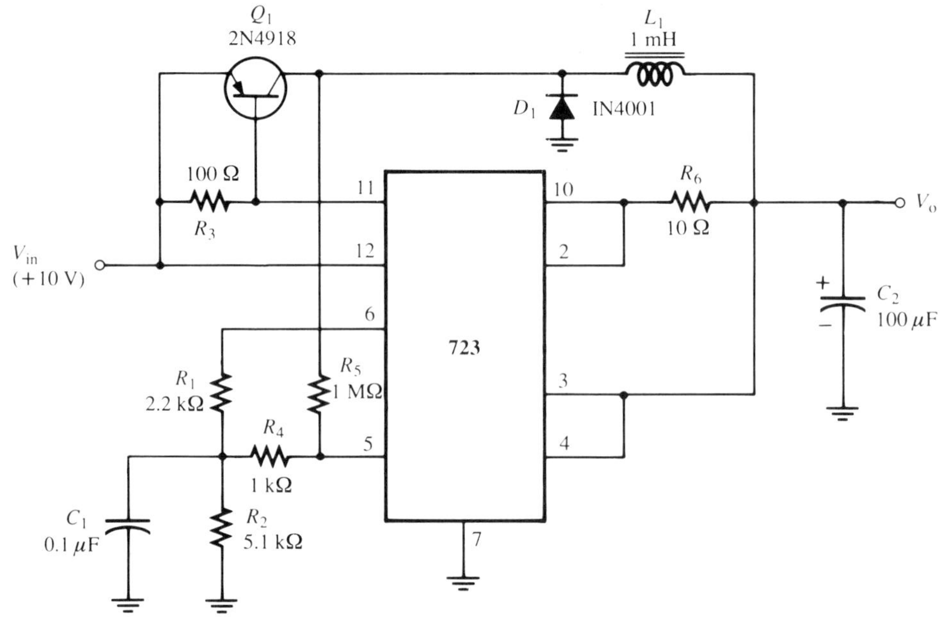

creased to 1 A. The 1-mH inductor (L_1) smooths out current pulses, and capacitor C_2 helps to maintain output voltage at a constant level. Resistor R_6 controls the output short-circuit current, whereas R_1 and R_2 control the output voltage based on the internal 7.15-V reference voltage at pin 6, so that

$$V_o(\text{REG}) = \left[\frac{R_2}{R_1 + R_2}\right](7.15 \text{ V}) \qquad \textbf{(9–15)}$$

Resistor R_3 controls how the load current is divided between the 723 and Q_1.

9–5 Crowbar Overvoltage Protection

If a voltage regulator circuit were to fail in such a way that the output voltage were to go to an abnormally high level, the load might be severely damaged. To prevent this possibility, a circuit called a **crowbar** is sometimes used. Its purpose is (1) to throw a dead short on the power supply if the output voltage exceeds a predetermined value and (2) to maintain that short circuit until a fuse or circuit breaker has time to operate and shut down the supply.

Figure 9–26 shows a circuit with a crowbar. When the voltage on the noninverting input of the operational amplifier exceeds 2.5 V, the output of the opera-

Figure 9–26
Crowbar protection circuit.

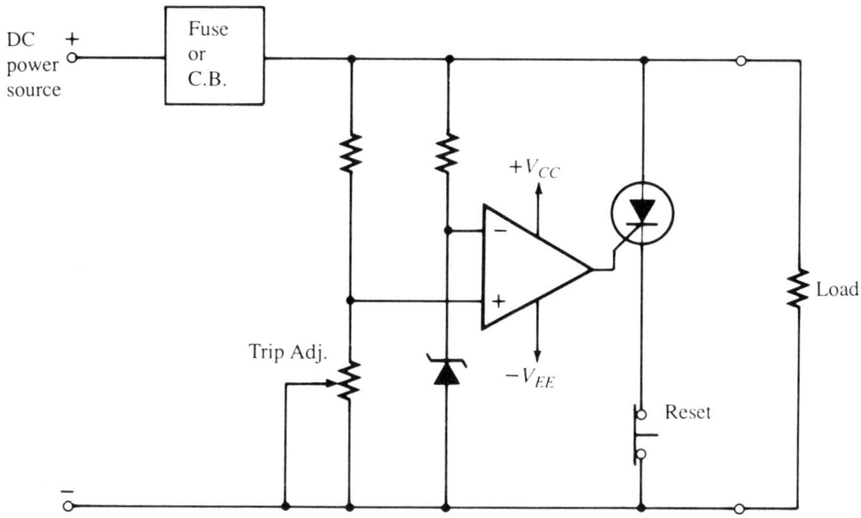

tional amplifier swings to its positive saturation voltage and triggers the silicon-controlled rectifier (SCR), putting a short across the output terminals. Once the SCR is turned on, it will continue to conduct until the current through it has been reduced below its hold-in value. By reversing the connections at the input of the operational amplifier, the crowbar circuit could then be used to protect against undervoltage conditions.

9–6 Current Regulators

Up to this point, this chapter has discussed circuits that maintain a constant output voltage independent of load and source variations. It is a simple matter to convert a voltage regulator to a **current regulator,** or a *constant-current source,* which delivers a fixed, constant current that is independent of load variations.

A simple constant-current source is shown in Figure 9–27. This is really the same circuit as the voltage regulator of Figure 9–4. Here the feedback resistor (R_2), of what is otherwise a noninverting amplifier, is the load. Since the input source voltage that forward-biases the Zener diode is larger than the regulated input voltage (V_Z), the current through either R_1 or R_2 is essentially constant. Therefore,

$$I_1 = I_2 = \frac{V_Z}{R_1} \qquad \textbf{(9–16)}$$

Figure 9–27
Operational amplifier current regulator.

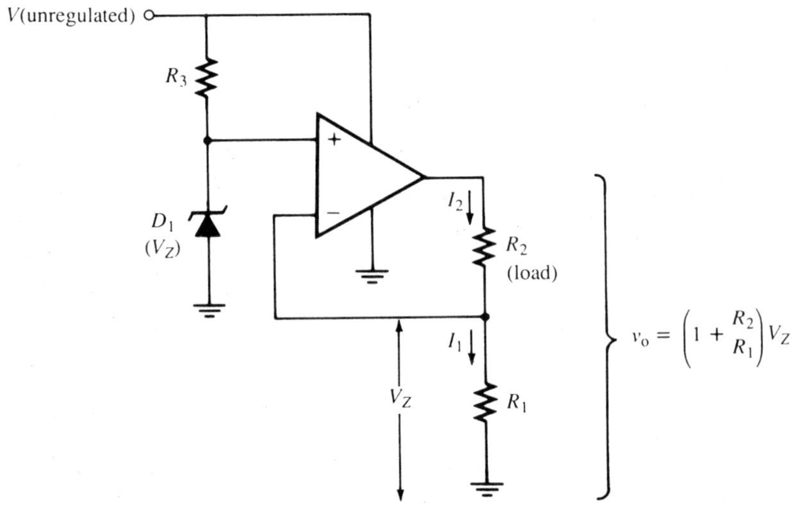

which shows that the current through R_2 is *independent* of R_2 even if R_2 changes, causing the closed-loop gain to change. Although the output voltage of the operational amplifier with respect to ground is always given as

$$V_o = \left(1 + \frac{R_2}{R_1}\right) V_Z \tag{9–17}$$

the voltage across R_1 is always equal to V_Z, and the resulting current is given by Equation 9–16. In effect, the circuit here is a form of a *voltage-to-current converter* circuit, which is further discussed in Chapter 12.

One problem with virtually all constant-current sources built around operational amplifiers is that the load is floating. That is, neither terminal of the load is connected to ground. A very simple way to achieve a constant-current source with one end of the load possibly connected to ground is to use a three-terminal voltage regulator as the heart of a constant-current source, as shown in Figure 9–28. Here, using a 7805 5-V regulator, the regulated current is set by R_1 such that

$$I_o(\text{REG}) = \frac{V_o}{R_1} \tag{9–18}$$

The voltage regulator will maintain a constant 5 V across the 10-Ω resistor (R_1), and this action results in a constant 0.5 A through R_1 as long as the DC source voltage is high enough to supply the sum of the necessary voltage drops. The load resistance R_L can even go to zero ohms, and the current will still remain at 0.5 A.

Figure 9–28
A 7805 (+5 V) regulator wired to function as a constant current source.

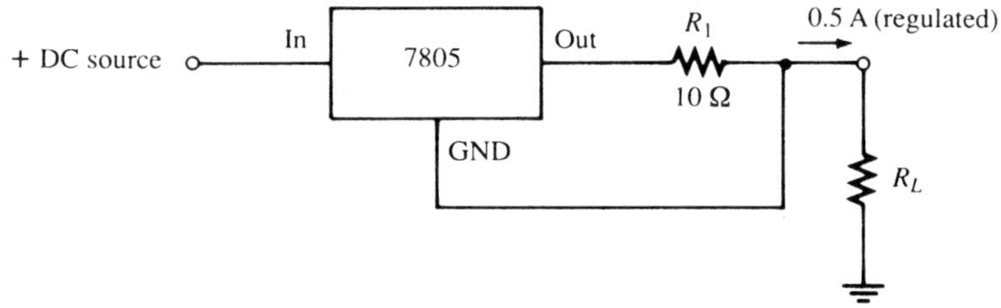

Summary

A voltage regulator is a circuit that provides a constant DC voltage level to its load from an unregulated voltage source. This chapter illustrated circuits for both series- and shunt-type regulators built around an operational amplifier. Although the primary emphasis was on various operational amplifier-transistor circuits, advances in solid-state technology have now greatly simplified the construction of power supplies and other circuits that require voltage and current regulation.

With the emphasis now on monolithic integrated-circuit regulators, the remainder of this chapter focused on those positive and negative linear and switching-type regulators built around a variety of fixed- and variable-voltage three-terminal regulators. These regulators have the capability of delivering positive or negative regulated output voltages from 1.25 V to more than 30 V and load currents up to 10 A.

Glossary

crowbar A low-resistance shunt quickly placed across the output terminals of a power supply as a protective measure.

current regulator A circuit that maintains its output current at a predetermined value independent of changes in the load current. Also called a *constant-current source*.

fold-back current limiting A regulated power supply circuit feature that protects against overloading and short circuits rather than maintaining a constant-current operation.

series regulator A voltage regulator circuit in which a transistor is placed in series with a power source to automatically vary its resistance, which in turn controls the output voltage or current.

shunt regulator A voltage regulator circuit in which a transistor is placed in parallel with a power source to vary its resistance automatically, which in turn controls the output voltage or current.

voltage regulator A circuit that maintains its output voltage at a predetermined value independent of changes in the load current or input voltage, as long as the input is greater than the predetermined output voltage.

Review Questions

1. Explain what a voltage regulator is and where it would normally be used.
2. Describe the function of the three basic components of a series regulator.
3. Draw the schematic diagram of a series regulator having a fixed positive output voltage using an operational amplifier.
4. Explain how the regulated output voltage can be varied in a series regulator.
5. How can one obtain a regulated negative output voltage from a series regulator?
6. Explain the difference between constant-current and foldback current limiting.
7. How does a shunt regulator differ from a series regulator?
8. What is a switching regulator and what advantage is there in using one?
9. What is pulse width modulation, and how does the duty cycle affect the DC value of a pulse train?
10. Name the three basic types of integrated-circuit voltage regulators.
11. What is the *dropout voltage*?
12. Draw the schematic diagram of a simple +5-V voltage regulator using an LM309 regulator.
13. Draw the schematic diagram of a simple variable-output voltage regulator using an LM317 regulator.
14. Explain what a *crowbar* is and why it is used.
15. Draw the schematic diagram of a constant-current regulator using an operational amplifier.
16. Draw the schematic diagram of a constant-current regulator using a three-terminal voltage regulator, such as an LM309.

Exercises

1. Determine suitable component values for the series regulator of Figure 9–4 to give a regulated output of +8.5 V from an unregulated +12-V source if a 5.1-V Zener diode is used and is biased at 20 mA.
2. Repeat Exercise 1 to give a regulated output voltage of −10 V from an unregulated −12-V source. Draw the schematic diagram of the completed circuit.
3. Determine suitable component values for the series regulator circuit of Figure 9–8a to give a regulated output of +12 V with a maximum current of 200 mA from a +16-V unregulated source. Assume that a 9.1-V Zener diode is used and is biased at 15 mA. Specify the minimum ratings for the transistor that is to be used for Q_1.
4. For the voltage regulator circuit of Figure 9–11, at what load current does current limiting begin if a 2-Ω resistor is used for R_3 and silicon-type *npn* transistors are used for Q_1 and Q_2?
5. The voltage regulator circuit of Figure 9–11 maintains an output voltage of 15 V. Determine the value of R_3 required to limit the short-circuit output current to 250 mA. Assume that silicon-type *npn* transistors are used for Q_1 and Q_2.
6. The voltage regulator circuit of Figure 9–12a uses foldback current limiting. Determine the value of R_4 required to limit the load current at 0.5 A if the regulated output voltage is 8 V, R_5 = 2 kΩ, and R_6 = 18 kΩ. Assume that silicon-type *npn* transistors are used for Q_1 and Q_2.
7. What is the regulated output voltage from the shunt regulator circuit of Figure 9–13 if R_1 =

1 kΩ, R_2 = 6.8 kΩ, R_3 = 8.2 kΩ, and a 7.5-V Zener diode is used?

8. Determine suitable component values for the shunt regulator circuit of Figure 9–13 to give a regulated output of 9 V at a maximum load current of 10 mA from an unregulated 13-V source. Assume that a 5.6-V Zener diode is used that is biased at 15 mA.

9. The switching regulator of Figure 9–16 is to maintain a constant output of 9 V even though the unregulated positive input voltage varies between 13 and 18 V. Determine the minimum and maximum duty cycles of the pulse-width modulator.

10. For the switching regulator of Figure 9–16, the output remains constant at 11.5 V. Determine the range of the unregulated input voltage if the duty cycle of the pulse-width modulator varies between 75% and 85%.

11. A 7805 three-terminal, 5-V regulator is used in the circuit of Figure 9–20 with R_1 = 470 kΩ and R_2 = 1.2 kΩ. What is the output voltage if the quiescent current for the 7805 is 8.5 mA?

12. A 7812 three-terminal, 12-V regulator is to be used in the circuit of Figure 9–20. Determine suitable values for R_1 and R_2 to give an output voltage of 15 V if the quiescent current for the 7812 is 8.5 mA.

13. If 0.1 A is drawn from the regulator circuit of Exercise 12, how much power is dissipated by the 7812 regulator if the unregulated DC input is 24 V? (Neglect I_Q).

14. What would be the minimum input voltage required for the regulator of (a) Exercise 11 and (b) Exercise 12?

15. An LM317 three-terminal adjustable regulator is used to make an adjustable voltage regulator. Using the basic circuit of Figure 9–23, determine suitable values for R_1 and R_2 so that the positive output voltage can be made to vary from 1.25 V to 24 V.

16. If a full-wave bridge rectifier is used in the regulator of Exercise 15, what is the minimum RMS voltage rating of the transformer's secondary that is necessary for proper operation?

17. Determine the load current for the constant-current regulator circuit of Figure 9–27 if R_1 = R_2 = 6.2 kΩ, R_3 = 680 Ω, and a 6.2-V Zener diode is used when the unregulated input is 13 V.

18. A 7812 three-terminal, 12-V regulator is used in the constant current regulator of Figure 9–28 to feed a 2-Ω load. If the DC input voltage to the regulator is +20 V and R_1 = 600 Ω, determine the constant current through the load.

19. For the regulator of Exercise 18, determine the power dissipated by the 7812 regulator.

20. Determine the required value of R_1 in the constant-current regulator circuit of Figure 9–28 using an LM340-8 if the load current is to be regulated at 0.5 A.

10 Oscillators and Function Generators

Instructional Objectives

Upon completing this chapter, you will be able to

- Explain the difference between an oscillator and a function generator.
- Explain how an amplifier becomes an oscillator.
- Design and determine the output frequency of both a phase-shift and a Wien-bridge oscillator using operational amplifiers, and describe how their outputs may be stabilized.
- Describe several methods for producing square waves, triangle waves, and sawtooth waveforms.
- Describe the operation of the 555 as a monostable and astable multivibrator.
- Describe how to program one or more XR-2240 timers as a very long time-delay monostable multivibrator.
- Create precision function generators capable of producing sine, triangular, and square waveforms of adjustable frequency and amplitude using either the XR-2206 or 8038 integrated circuit.
- Explain how a quartz crystal is used to control the stability of an oscillator's output frequency.

Introduction

When referring to signal sources, an **oscillator** generates sine waves only, whereas a **function generator** provides not only sine waves but also square waves, triangular waves, sawtooths, pulse trains, and any other type of periodic waveform. This chapter discusses operational amplifier circuits as well as specialized monolithic integrated-circuit devices that generate a variety of waveforms. Among these are the 555/556 and XR-2240 timers and the XR-2206 and 8038 precision waveform generators.

10–1 The Basic Oscillator

An oscillator is a sine-wave signal source of known frequency. It produces an output signal, yet it requires no external input signal. The only external input connections to an oscillator are for the DC power source. The sine-wave output signal normally has both its frequency and amplitude determined by the circuit type, power-supply voltage, and component values selected by the user. The most common form of sine wave oscillator consists of an amplifier with controlled *positive feedback*. As opposed to negative feedback in conventional amplifiers, which tends to reduce the amplifier gain while providing circuit stability, positive feedback creates overall circuit instability by increasing amplifier gain.

Positive feedback is applied to an amplifier by taking a fraction of the output signal and adding it to the incoming signal in such a way that the signal that is actually applied to the input of the amplifier is now the sum of the incoming signal and the signal being fed back from the output. By increasing the amount of feedback, it is possible to decrease the level of the incoming signal and still get the same output amplitude. If this is continued repeatedly, a point is eventually reached where an input signal is no longer needed to produce and sustain an output signal. Because the amplifier is now somewhat unstable, it oscillates in a sinusoidal fashion, supplying its own input signal.

Figure 10–1 shows the block diagram of an inverting amplifier stage with a positive feedback network path. The overall gain of the amplifier is given by

$$A_V = \frac{G}{1 - GB} \qquad \textbf{(10–1)}$$

where

G = amplifier voltage gain without feedback

B = feedback factor, the fraction of the output signal that is fed back to the input

If we start with a feedback factor of zero (i.e., no feedback), the output of the amplifier is 180° out of phase with its input. If the amount of feedback is slowly increased (i.e., making B larger), the denominator of Equation 10–1 becomes smaller. As a consequence, the circuit voltage gain (A_V) gets larger. When the

Figure 10–1
Block diagram showing an amplifier stage (G) with positive feedback (B).

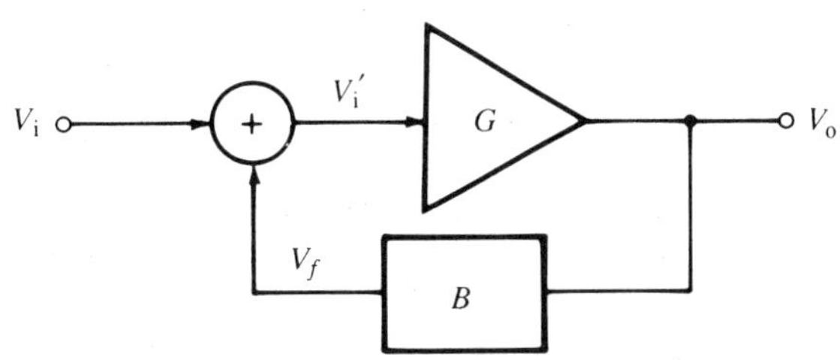

product of B and G, known as the **loop gain,** equals unity, the denominator of Equation 10–1 is then zero, and the circuit gain approaches infinity and becomes unstable. When this occurs, we have the condition necessary to support and sustain oscillation. No external input signal is then required to produce an output.

If the circuit is designed to make the product of the feedback factor and amplifier gain equal to unity at only a single frequency, the circuit will produce a sine wave. If the loop gain is less than 1, the circuit will not oscillate. Otherwise, if the loop gain is much greater than 1, the output will be somewhat distorted. The designer of a sine-wave oscillator then usually strives to produce a loop gain value just slightly greater than 1 at the frequency of oscillation but substantially less than 1 at all other frequencies.

The preceding conditions then require a feedback network that is frequency-selective. At oscillation, the phase shift of the feedback circuit is also 180°, so that the total phase shift around the feedback loop is 180° (from the amplifier) + 180° (from the feedback network) = 360°, or 0°. The loop gain and phase-shift conditions are collectively referred to as the **Barkhausen criteria** and must exist simultaneously for a circuit to function as an oscillator. In summary, the two conditions of the Barkhausen criteria are as follows:

1. The loop gain of an amplifier-feedback network must be 1.
2. The total phase shift around the feedback loop must be 0°.

The following section describes many types of circuits that fit these two conditions. They differ primarily in the configuration of the feedback circuit.

10–2 RC Oscillators

The **RC oscillator** is an oscillator that uses one or more resistor-capacitor networks as the frequency-selective feedback network. Because of the RC component values as well as the frequency response of the operational amplifier used, its use is primarily in fixed-frequency situations in the audio or low-RF range below approximately 30 kHz, where frequency stability is not much of a problem. The following are major types of RC oscillators:

- Phase-shift
- Twin-T
- Wien-bridge

Phase-Shift Oscillator

Figure 10–2a shows the circuit of the **phase-shift oscillator.** The feedback network consists of three identical RC networks that collectively produce a 180° phase shift at

$$f_o = \frac{1}{2\pi RC\sqrt{6}} \qquad \textbf{(10–2)}$$

Figure 10–2
Phase-shift oscillator. (a) Operational amplifier circuit. (b) Frequency response of the RC feedback network.

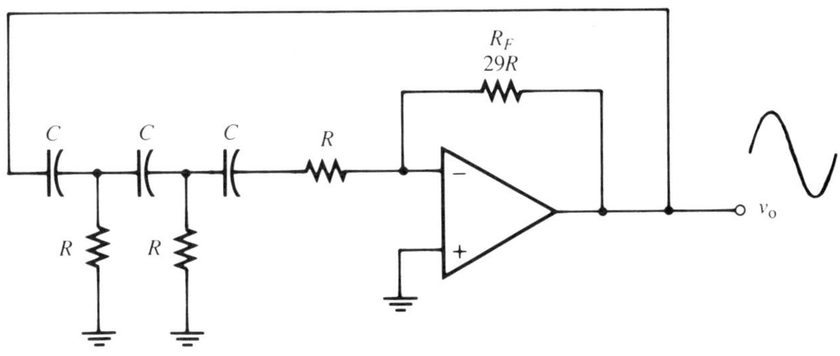

(a)

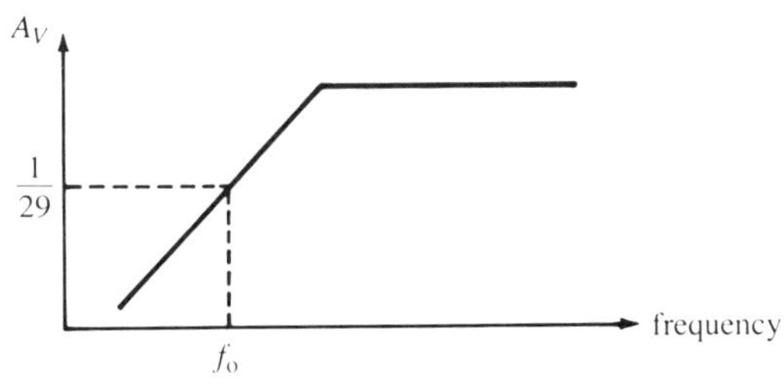

(b)

At this frequency the RC feedback network attenuates the amplifier's output by $\frac{1}{29}$. For the loop gain to be unity and the Barkhausen criteria to be satisfied for oscillation, the inverting amplifier's closed-loop voltage gain must then be 29. Consequently, the operational amplifier's feedback resistor (R_F) is then made equal to $29R$. In practice, however, the closed-loop gain is made slightly larger than 29 (i.e., 30 to 31) at the cost of some small amount of distortion in order to ensure that the circuit will immediately start to oscillate when power is applied to the circuit.

The maximum output voltage and oscillation frequency for the operational amplifier version depend on several factors. The peak-to-peak output voltage will be approximately 2 V less than the differential supply voltage ($V_{CC} + V_{EE}$). The maximum oscillation frequency is a function of the operational amplifier's gain-bandwidth product (GBP). As an example using a 741 operational amplifier (GBP = 1 MHz), the maximum oscillation frequency possible is then 1 MHz/29, or about 34.5 kHz.

As a general rule with most types of RC circuits, a standard capacitor value is first chosen and the corresponding value for the resistor is calculated, settling on the nearest standard value with a very slight loss in accuracy. This is done because the availability of 1%, 5%, or 10% resistor values is more extensive than available 10% capacitor values. Finally, because the phase-shift circuit requires that three equal components (either R or C) be simultaneously changed in a like manner to shift the oscillation frequency from one value to another, the phase-shift oscillator is primarily intended for fixed-frequency operation.

Example 10–1
For the operational amplifier phase-shift oscillator circuit of Figure 10–2a, determine the appropriate component values for a 650-Hz sine wave.

Solution: Assuming **0.1-μF** capacitors for C, then

$$R = \frac{1}{(2\pi)(650\ \text{Hz})(0.1\ \mu\text{F})(\sqrt{6})} = \mathbf{1000\ \Omega}$$

so that

$$R_F = (29)(1\ \text{k}\Omega) = \mathbf{29\ k\Omega} \qquad \text{(or use a 30-k}\Omega\text{ standard value)}$$

Twin-T Oscillator

Another low-frequency RC oscillator is the **twin-T,** or **parallel-T oscillator,** as shown in Figure 10–3a. The twin-T feedback network is borrowed from the classical passive notch filter circuit (Figure 10–3b), whereby its impedance at the resonant frequency is virtually infinite with a phase shift of 180°.

At resonance, the output frequency is given by

$$f_o = \frac{1}{2\pi RC} \qquad \textbf{(10–3)}$$

In this circuit, the twin-T network provides negative feedback, whereas the lamp (R_2) and resistor R_1 provide positive feedback. The lamp tends to regulate the circuit's output. If the output increases, the resistance of the lamp increases and, in turn, reduces the gain of the amplifier. Because of inherent component mismatches, the variable resistor is used to adjust the twin-T network for optimum performance. Since the bridge circuit requires that three components (either R or C) be changed simultaneously to shift the oscillation frequency from one value to another, the twin-T oscillator is primarily intended for fixed-frequency operation and is not easily tuned over a wide frequency range. Finally, the twin-T oscillator circuit is more stable than the phase-shift type but is also used primarily for fixed-frequency applications.

Figure 10–3
Twin-T oscillator.
(a) Operational amplifier circuit. (b) Passive RC notch filter network used as the feedback network.
(c) Frequency response of RC feedback network.

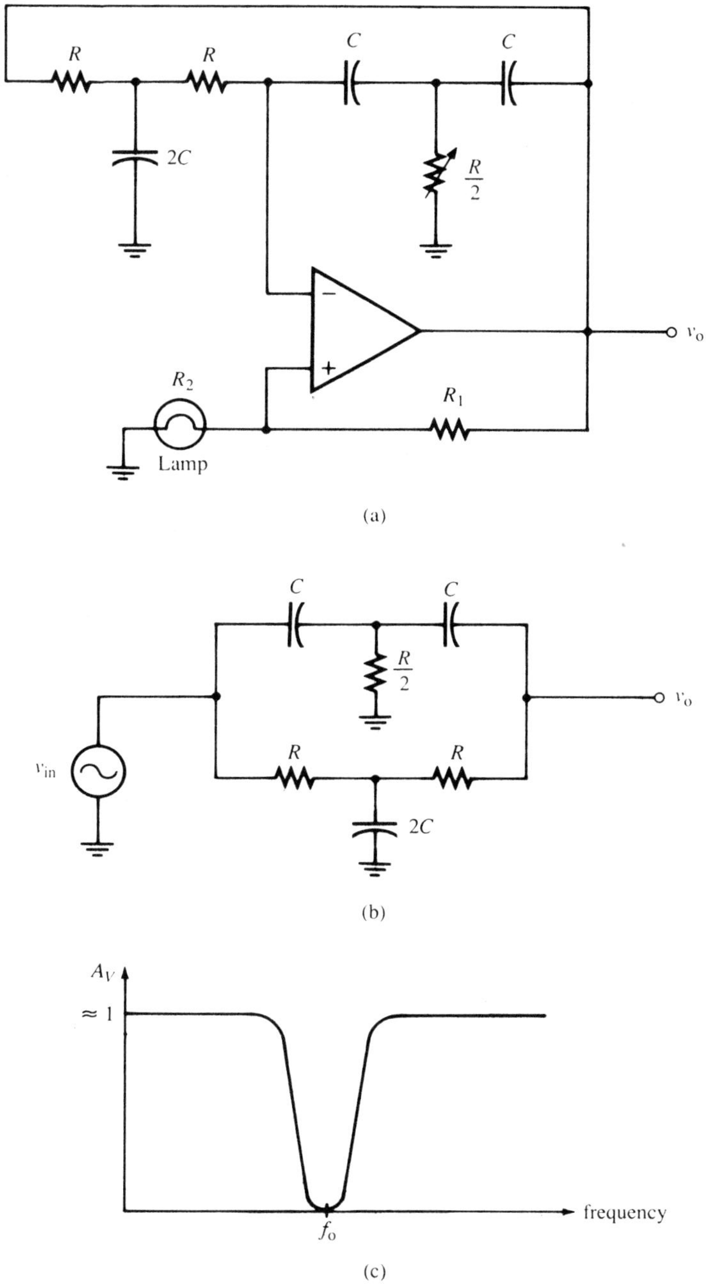

Wien-Bridge Oscillator

The final RC oscillator is the **Wien-bridge oscillator** of Figure 10–4a. Part of the Wien bridge is a dual RC network called a *lead-lag network*. Its name comes from the action of the network's output signal, which leads in phase (with the input signal) below a certain frequency and lags the input signal above that frequency.

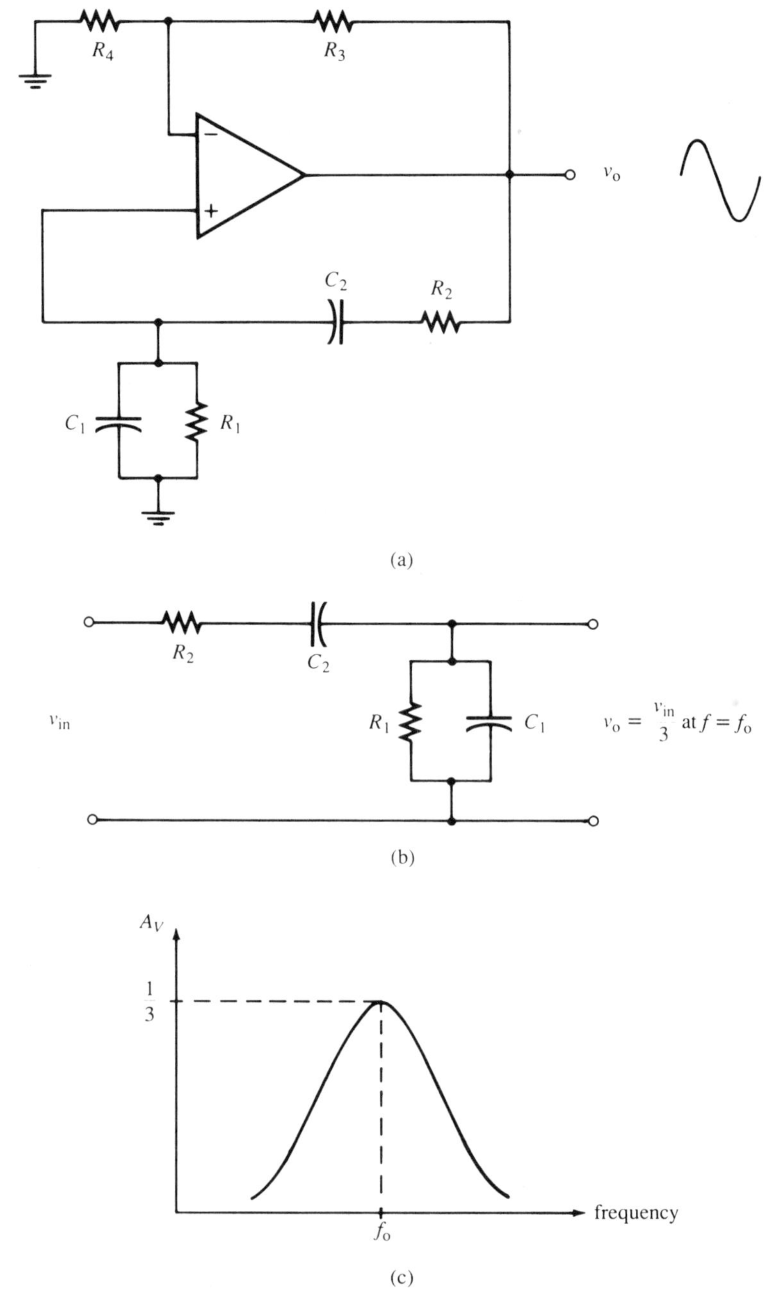

Figure 10–4
Wien-bridge oscillator. (a) Operational amplifier circuit. (b) RC lead-lag network used as the feedback network. (c) Frequency response of the lead-lag network.

Unlike the phase-shift and twin-T oscillator circuits, the operational amplifier basically functions as a noninverting amplifier, so that the output is always in phase with the input. The Wien bridge (Figure 10–4b), as the frequency-selective RC feedback network, has a phase shift that varies from +90° to −90° as the frequency increases. When the phase shift of the Wien-bridge network is zero at

f_o, the magnitude of the output reaches a maximum level of one-third of the input signal (Figure 10–4c). If the closed-loop gain of the amplifier stage is made equal to 3, the loop gain is $3 \times \frac{1}{3}$, or 1, and the circuit oscillates at a frequency given by

$$f_o = \frac{1}{2\pi\sqrt{R_1R_2C_1C_2}} \qquad \textbf{(10–4)}$$

To minimize the number of different component values possible in Equation 10–4, both the frequency-determining resistors (R_1 and R_2) are usually made equal, as are both frequency-determining capacitors. When $R_1 = R_2$ and $C_1 = C_2$, Equation 10–4 simplifies to:

$$f_o = \frac{1}{2\pi R_1C_1} \qquad \textbf{(10–5)}$$

By then using the lead-lag network as the positive feedback path for an amplifier with a gain of 3, the loop gain is 1, creating an oscillator. The lead-lag network forms one side of a Wien bridge, and the gain-setting resistors R_3 and R_4 form the other side. If the RC lead-lag network is omitted, what now remains is a noninverting amplifier with a closed-loop voltage gain of 3, so that

$$3 = 1 + \frac{R_3}{R_4} \qquad \textbf{(10–6a)}$$

or

$$R_3 = 2R_4 \qquad \textbf{(10–6b)}$$

At the oscillation frequency given by Equation 10–4 or 10–5, the Wien bridge is balanced and the amplifier's input voltage is in phase with the output voltage. At all other frequencies, the bridge is unbalanced, so the feedback signal does not have the amplitude and phase relationship of the Barkhausen criteria needed to sustain oscillation.

Resistors R_3 and R_4 are in the negative feedback path and set the amplifier gain at 3. As in Figure 10–4a, resistor R_4 is often a device with a *positive* voltage-versus-resistance characteristic, such as an incandescent lamp in the 3.4-kHz oscillator circuit of Figure 10–5. This type of characteristic then tends to regulate the output amplitude. If the amplitude of the output voltage (v_o) starts to increase, R_3 increases and reduces the gain of the amplifier. If v_o decreases, R_3 decreases and the gain of the amplifier is increased. This action serves to maintain the output amplitude at a relatively constant level.

Figure 10–6 shows another way of controlling the output amplitude of a Wien-bridge oscillator using back-to-back Zener diodes placed across the negative feed-

Figure 10–5
Using an incandescent lamp having a positive voltage versus resistance characteristic to regulate the output amplitude of a 3.4-kHz Wein-bridge oscillator.

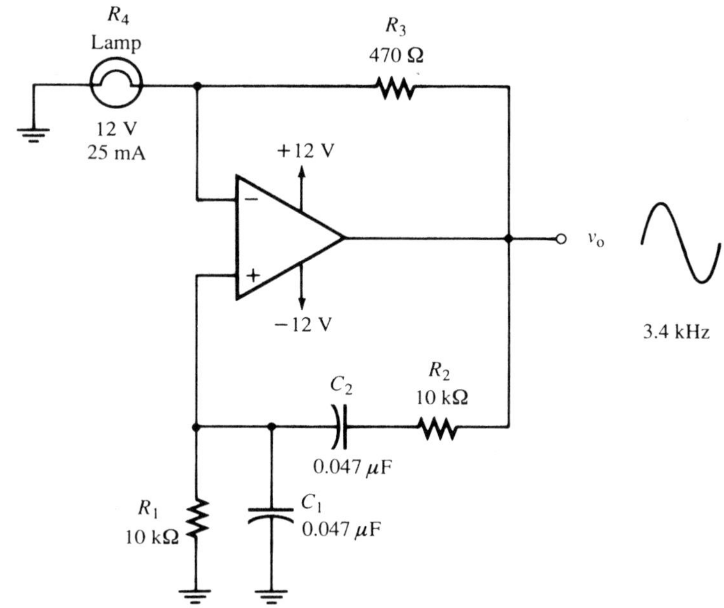

back resistor R_4. When the oscillator is first turned on and the signal is small, the Zener diodes do not conduct. Since potentiometer R_3 is initially adjusted to provide an amplifier gain of three, the signal increases in amplitude. When the signal peaks reach the Zener's breakdown voltage, their effective resistance is reduced, lowering the gain. This action then stabilizes the signal amplitude. The maximum output amplitude is controlled by the Zener voltage.

Figure 10–6
Wien-bridge oscillator with back-to-back Zener diodes used to limit output amplitude swings.

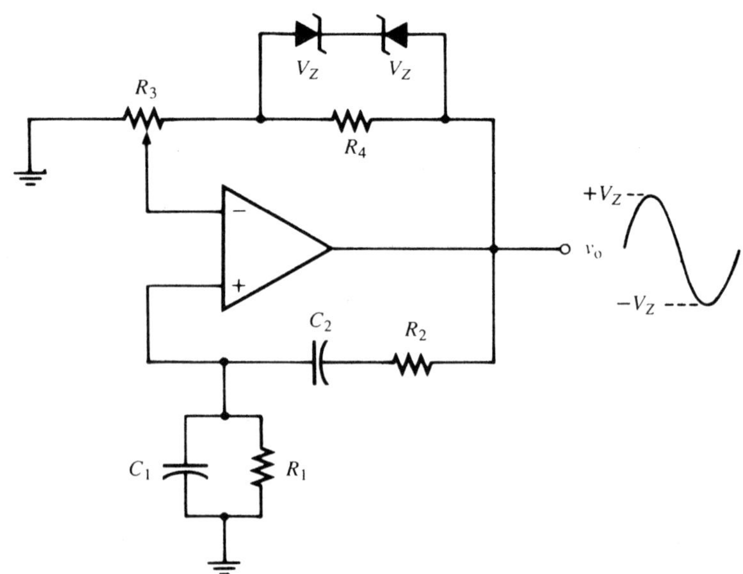

Figure 10–7
Wien-bridge oscillator with JFET amplitude control.

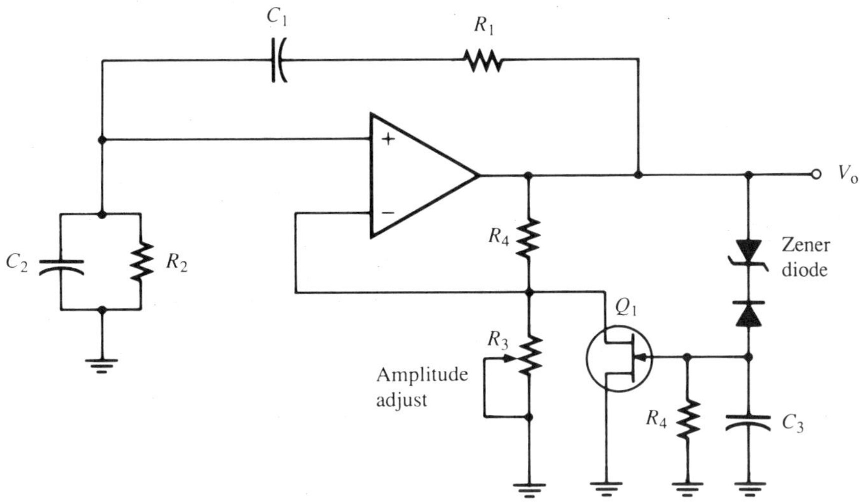

If R_3 is adjusted for a smaller output, oscillations will cease. On the other hand, adjusting R_3 for a larger output increases the signal amplitude but also increases the amount of distortion. Minimum distortion coincides with minimum amplitude.

An even better method of controlling the output level of a Wien-bridge oscillator is shown in Figure 10–7. Here, field-effect transistor Q_1 is operated in the ohmic region of that part of its characteristic curve where it looks like a resistor across the drain and source terminals. By varying the FET's gate voltage, its drain-source resistance is also varied. The output signal is passed through the Zener diode and the R_4-C_3 filter network to form the control voltage for Q_1. As the output voltage increases beyond the Zener voltage, the DC level on Q_1's gate goes more negative and increases the channel resistance. This in turn develops a larger negative feedback signal and automatically lowers the amplifier gain.

As a result, this circuit provides a low-distortion sine wave whose amplitude is determined by the Zener breakdown voltage and the diode drop. Finally, because the bridge circuit usually requires that two components (either R or C) be simultaneously changed in a like manner to shift the oscillation frequency from one value to another, the Wien-bridge oscillator is primarily intended for fixed-frequency operation and is not easily tuned over a wide frequency range.

Example 10–2

Find the output frequency of the Wien-bridge oscillator of Figure 10–4a with $C_1 = C_2 = 0.047\ \mu F$, $R_1 = R_2 = 10\ k\Omega$, $R_3 = 1.5\ k\Omega$, and $R_4 = 3\ k\Omega$.

Solution: From Equation 10–5,

$$f_o = \frac{1}{(2\pi)(10\ \text{k}\Omega)(0.047\ \mu\text{F})}$$
$$= \mathbf{339\ Hz}$$

Since the closed loop gain is 3, the loop gain equals 1 and the circuit oscillates.

10–3 Operational Amplifier Waveform Generators

Besides the sine-wave oscillator, operational amplifiers are capable of being wired to generate other types of waveforms such as square waves, pulse trains, and triangle waves. Some of these circuits used to generate nonsinusoidal waveforms are simple extensions of those already discussed.

Square-Wave Generators

Using any of the three oscillators discussed in Section 10–2, a simple scheme used to generate square waves of the same frequency is to add a comparator to the oscillator's output, as shown in the block diagram of Figure 10–8. The comparator stage can be built using either a high-slew-rate operational amplifier such as the LM318 (70 V/μs) or an integrated-circuit comparator such as an LM710. Unless diode limiting is used, the peak output voltage levels for both the sine-wave and square-wave outputs are approximately $-V_{EE}$ and $+V_{CC}$.

An **astable multivibrator,** or *free-running multivibrator,* is a circuit whose output has no stable state, or voltage level. Consequently, its output periodically "bounces" back and forth between two stable voltage levels. This then is precisely the action of a square wave, so that an astable multivibrator is in fact another name for a square-wave generator.

Figure 10–9a shows an operational amplifier–based square-wave generator. When the operational amplifier output is in positive saturation, the noninverting input is held at $V_{SAT}/2$ by voltage divider resistors R_1 and R_2 acting as a positive feedback network. Meanwhile, timing capacitor C_1 is charging through R_3. When the capacitor voltage reaches $V_{SAT}/2$, the inverting input becomes more positive

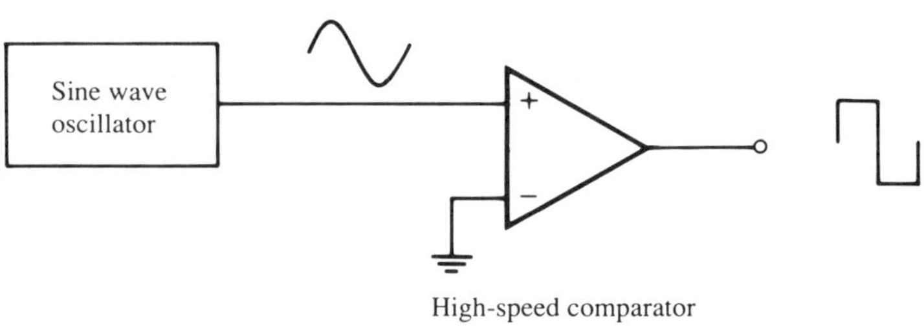

Figure 10–8
Using a comparator to convert a sine wave into a square wave.

Figure 10–9
Operational amplifier square-wave generator. (a) Basic circuit. (b) Output and capacitor voltage waveforms. (c) Circuit connections when using an electrolytic capacitor.

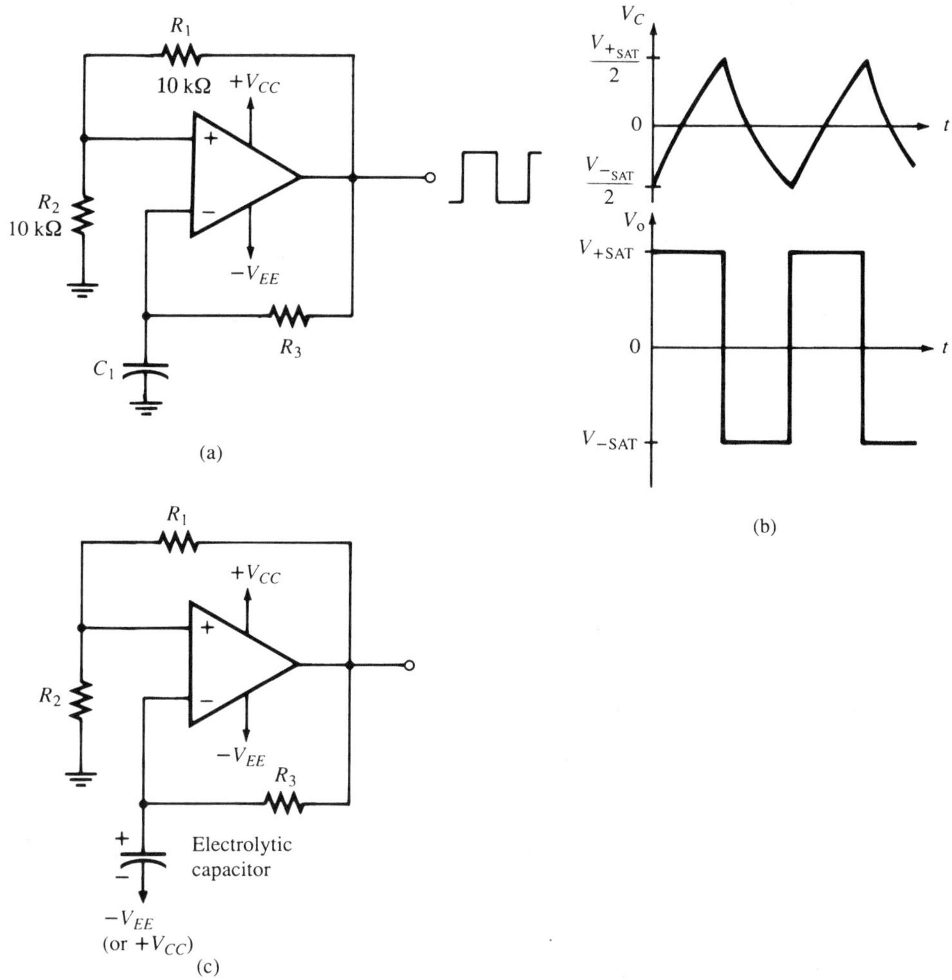

than the noninverting input, and the operational amplifier's output switches to V_{-SAT} (Figure 10–9b). This is similar to the action of an inverting comparator.

The noninverting input is now held at $V_{-SAT}/2$ until C_1 charges in the reverse direction to this level. When this occurs, the operational amplifier's output then switches back to positive saturation and the cycle repeats. This action produces a square wave that swings between the positive and negative saturation limits.

Besides the R_3-C_1 timing circuit, the output frequency, often called the *free-running frequency,* also depends on the positive feedback R_1-R_2 network, so that

$$f_o = \frac{1}{2R_3C_1 \ln[1 + (2R_2/R_1)]} \quad \textbf{(10–7)}$$

If $R_1 = R_2$, as shown in Figure 10–9a, then Equation 10–7 simplifies to

$$f_o = \frac{0.455}{R_3C_1} \quad \textbf{(10–8)}$$

For simplicity, many users prefer to adjust R_1 and R_2 so that Equation 10–7 simplifies to

$$f_o = \frac{1}{2R_3C_1} \quad \textbf{(10–9)}$$

which requires that $R_1 = 1.16R_2$. This can be verified by setting the natural logarithm term in the denominator of Equation 10–7 equal to 1 and solving for R_1 in terms of R_2, so that

$$1 = \ln\left[\left(1 + \frac{2R_2}{R_1}\right)\right]$$

Taking the inverse natural logarithm of both sides,

$$e = 1 + \frac{2R_2}{R_1}$$
$$= 2.718$$

or

$$R_1 = 1.16R_2$$

As a general rule, R_1 and R_2 are selected to be in the 1-kΩ to 100-kΩ range.

Several factors should be kept in mind for this type circuit to function properly. First, since the operational amplifier functions essentially as a comparator, it should have both a high slew rate and should not be frequency compensated. This is because compensation reduces the frequency response and is not necessary for stability. Finally, for low-frequency applications where C_1 is an electrolytic (polarized) type, the polarity of C_1 must be correct. As shown in Figure 10–9c, one end of C_1 must then be connected to either the negative or positive supply voltage, with the positive supply generally preferred.

Figure 10–10 shows a similar square-wave generator circuit but using a Norton-type operational amplifier. The free-running frequency is given by

$$f_o = \frac{0.707}{R_1C_1} \quad \textbf{(10–10)}$$

Figure 10–10
Norton amplifier square-wave generator.

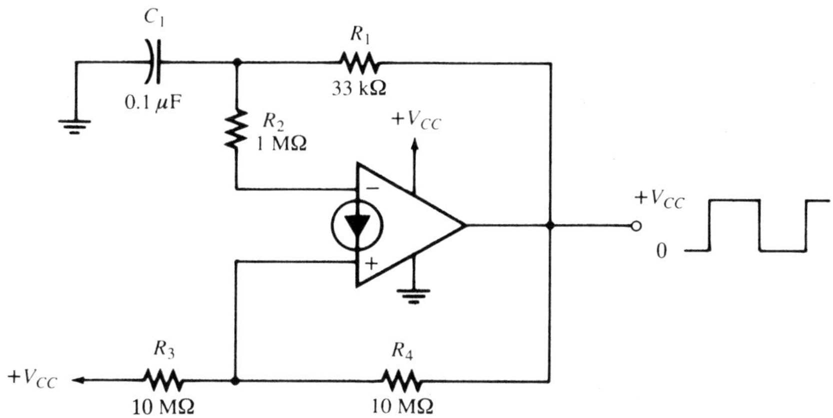

if $R_3 = R_4$ and R_2 is made much larger than R_1. With the components shown, the free-running frequency is 216 Hz.

In Figure 10–11, the basic square-wave circuit of Figure 10–9 is modified to produce an asymmetric pulse train by providing different charging rates for the timing capacitor during the positive and negative half-cycles. The addition of the parallel D_1-R_4 path across R_3 allows the capacitor to charge at a much higher rate when the output is positive then when it is negative. Thus the time that the output remains positive is shortened considerably.

Figure 10–11
Asymmetric pulse train generator.

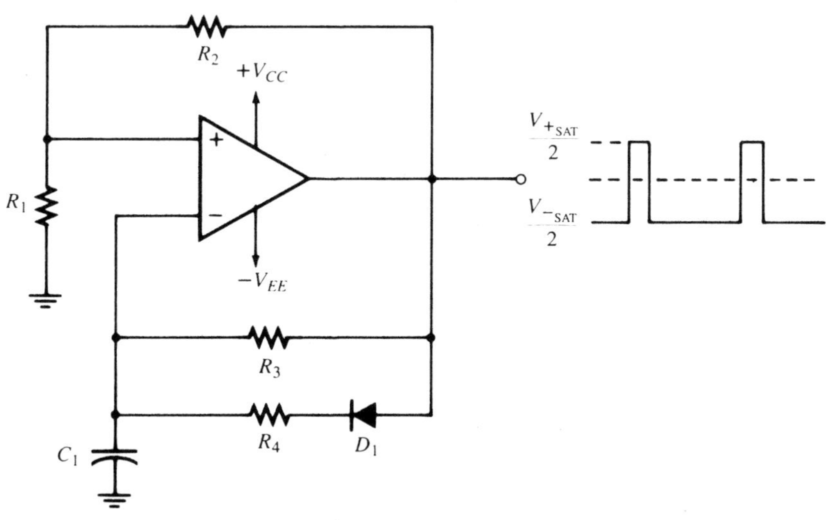

Figure 10–12
Triangular-wave generator. (a) Circuit. (b) Output waveforms at operational amplifiers A_1 and A_2.

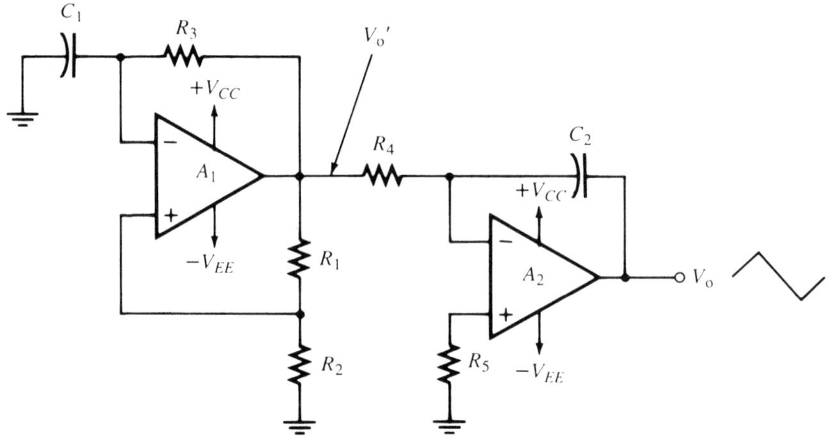

(a)

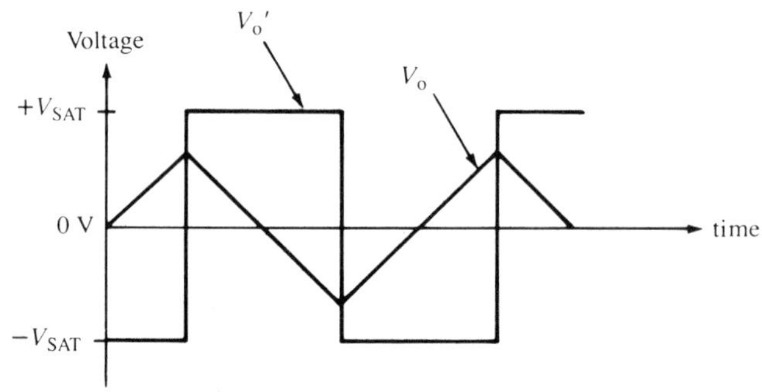

(b)

Figure 10–13
Triangular-wave generator using a comparator with hysteresis and an integrator.

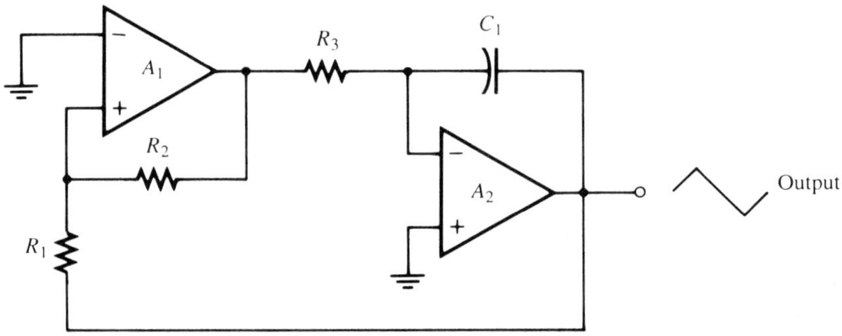

Triangular-Wave Generators

When a square wave is passed through an integrator, the output will be a triangular waveform having equal rise and fall times and a frequency equal to that of the astable's free-running frequency. Figure 10–12a shows the circuit for a triangular-wave generator using two operational amplifiers. The first operational amplifier (A_1) is connected as an astable multivibrator and produces a square wave. The second operational amplifier (A_2) is connected as a compensated integrator and converts the square wave to a triangular wave. Output errors due to A_2's input offset current are minimized by R_5, which is typically made equal to R_4.

Based on the level of the saturation voltage output of A_1, the peak-to-peak triangular output voltage of the integrator depends also on (1) the frequency of the free-running multivibrator and (2) the integrator's RC time constant, so that

$$V_o(\text{p-p}) = \frac{V_{\text{SAT}}}{R_4 C_2 f_o} \qquad \textbf{(10–11)}$$

The waveforms at the output of A_1 and A_2 are compared in Figure 10–12b.

A modification of the circuit of Figure 10–12a is shown in Figure 10–13, which is to have a comparator with hysteresis, or Schmitt trigger (A_1), in conjunction with an integrator (A_2). When power is first applied to the circuit, the output of A_1 will be either at its positive or negative saturation voltage. Assuming that the output is negative, the output of the integrator will be a positive-going ramp. When the output of the integrator reaches the upper threshold voltage of A_1, as given by

$$V_{UT} = V_{-\text{SAT}} \frac{R_1}{R_2} \qquad \textbf{(10–12)}$$

the output of the comparator will then switch to the positive saturation voltage, producing a negative-going ramp at the output of A_2. This continues until the output level of the integrator reaches A_1's lower threshold voltage, at which time the process repeats itself.

In terms of the circuit components, the output frequency of either the square or triangular waveforms is given by

$$f_o = \frac{R_2}{4R_1R_3C_1} \qquad \textbf{(10–13)}$$

Example 10–3

Using the circuit of Figure 10–13, determine the component values necessary for an output frequency of 10 kHz and a peak-to-peak output voltage of 12 V if a ±15-V supply is used. Assume that the positive and negative saturation voltages are ±13 V.

Solution: First, let's assume R_1 = **1 kΩ** and C_1 = **0.1 μF** as convenient starting points. From Equation 10–12, the lower (or upper) threshold level is one-half the desired peak-to-peak output, or ±6 V. Solving for R_2 yields

$$R_2 = \frac{13\text{ V}}{6\text{ V}}(1\text{ k}\Omega)$$
$$= \mathbf{2.17\text{ k}\Omega} \qquad \text{(use a 2.2-k}\Omega\text{ standard value)}$$

Finally, R_3 is found from Equation 10–13:

$$R_3 = \frac{2.2\text{ k}\Omega}{(4)(1\text{ k}\Omega)(0.1\ \mu\text{F})(10\text{ kHz})}$$
$$= \mathbf{550\ \Omega} \qquad \text{(use a 560-}\Omega\text{ standard value)}$$

The final circuit is shown in Figure 10–14.

Using Norton operational amplifiers, Figure 10–15 shows a similar approach of the circuit of Figure 10–13. Here, A_1 is an integrator that generates a linear ramp. Its output is fed to A_2, which acts as a comparator so that its peak-to-peak output is set approximately equal to $V_{CC}/3$; the output frequency is given by

$$f_o = \frac{1.5}{R_1C_1} \qquad \textbf{(10–14)}$$

For the best symmetry, $R_2 = R_1/2$, whereas $R_4 = R_5 = 3R_3$.

The sawtooth waveform is a general case of the triangular wave in which its rise and fall times are not equal. Figure 10–16 shows a circuit that is a modification of Figure 10–13 and produces adjustable rise and fall times. Operational amplifier

Figure 10–14
Completed triangular-waveform generator circuit for Example 10–3.

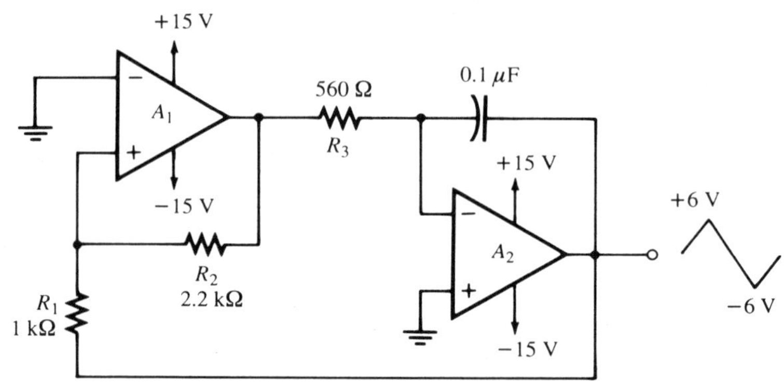

Figure 10–15
Norton amplifier triangular-wave generator.

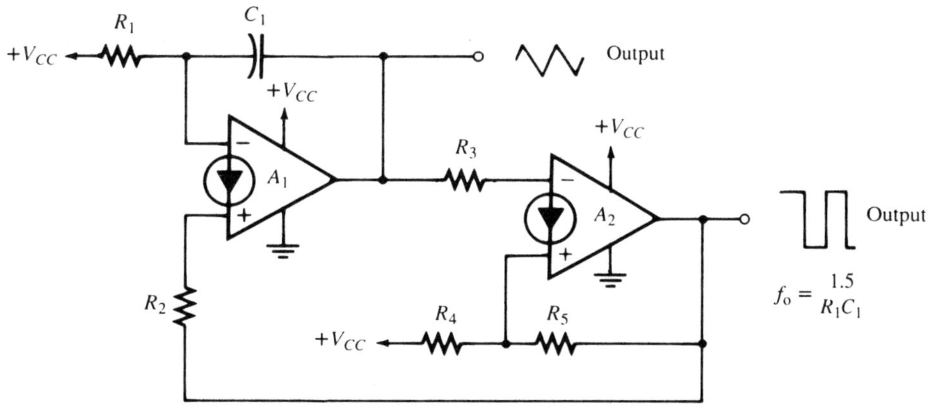

Figure 10–16
Sawtooth generator. (a) Circuit. (b) Output waveforms at operational amplifiers A_1 and A_2.

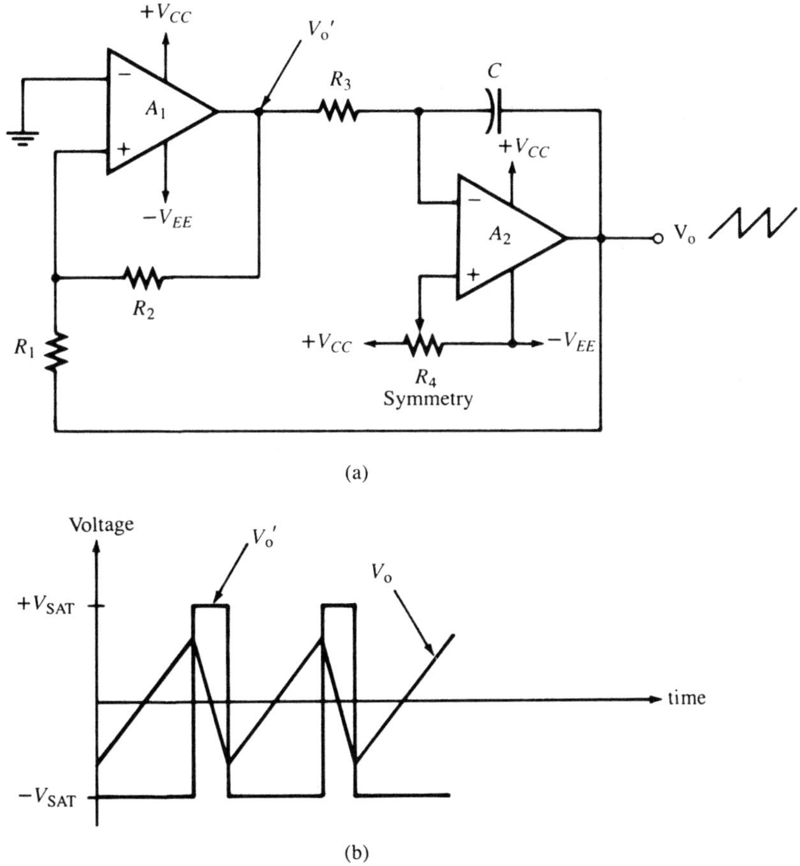

A_1 is connected as a noninverting comparator with hysteresis, whereas A_2 functions as an integrator.

To understand its operation, assume first that the output of A_1 is at positive saturation. This will cause the output of the integrator A_2 be a negative-going ramp riding on a DC level controlled by the setting of R_4. When A_2's output is sufficiently negative to pull the junction of R_1 and R_2 down through zero volts, the output of A_1 will switch to negative saturation. Now, the output of A_2 will start to ramp back up until the junction of R_1 and R_2 again crosses through zero. When this happens, the cycle starts all over again. If the DC level at the noninverting input of A_2 is negative, the rise time will be longer than the fall time. Conversely, if the DC level is positive, the fall time will be longer than the rise time. For the special case when the DC level is zero, the rise and fall times are equal.

10–4 Integrated-Circuit Timers

Since 1972, a number of specialized linear integrated-circuit devices have been developed to improve upon the design and performance of oscillators and waveform generators that could be built with operational amplifiers. These devices fall primarily into three broad categories. The first two—timers and precision function generators—are the most well known. The third class is the voltage-controlled oscillator; because of its relation to the phase-locked loop, it is discussed in Chapters 12 and 13.

The 555 Timer

One of the most well-known integrated circuits is the 555 timer, created in 1972 by H. R. Camenzind of Signetics. Although entire books have been written on the many uses to which this circuit has been put, this section concentrates simply on the basic 555 timer astable and monostable multivibrator configurations.

Although the 555 timer is also available in 14-pin DIP and 8-pin TO-99 metal can units, the most popular is the 8-pin mini-DIP unit, whose pin configuration is shown in Figure 10–17a. Although first produced by Signetics as the SE555/NE555, other manufacturers have since marketed their own pin-for-pin compatible versions of the 555 timer, which are summarized in Table 10–1. Furthermore, the 556, whose pin diagram is shown in Figure 10–17b, is nothing more than two independent 555-style timers in a single 14-pin DIP unit.

Figure 10–18 shows a block diagram of the 555 timer, which is the equivalent of more than 20 transistors, 15 resistors, and 2 diodes. Three equal resistors are connected in series as a voltage divider string between the supply voltage (V_{CC}) and ground. Connected to this divider string are two comparators, one whose reference is set at $\frac{2}{3}V_{CC}$ and one set at $\frac{1}{3}V_{CC}$. If the voltage on pin 6 exceeds $\frac{2}{3}V_{CC}$, the upper comparator (A_1) sets the output of the control flip-flop high (V_{CC}), which in turn drives transistor Q_1 on and the timer's output (pin 3) to ground. On the other hand, if the voltage on pin 2 drops below $\frac{1}{3}V_{CC}$, the lower comparator (A_2)

Figure 10–17
555 integrated-circuit timer. (a) Pin diagram. (b) Pin diagram of 556 dual timer.

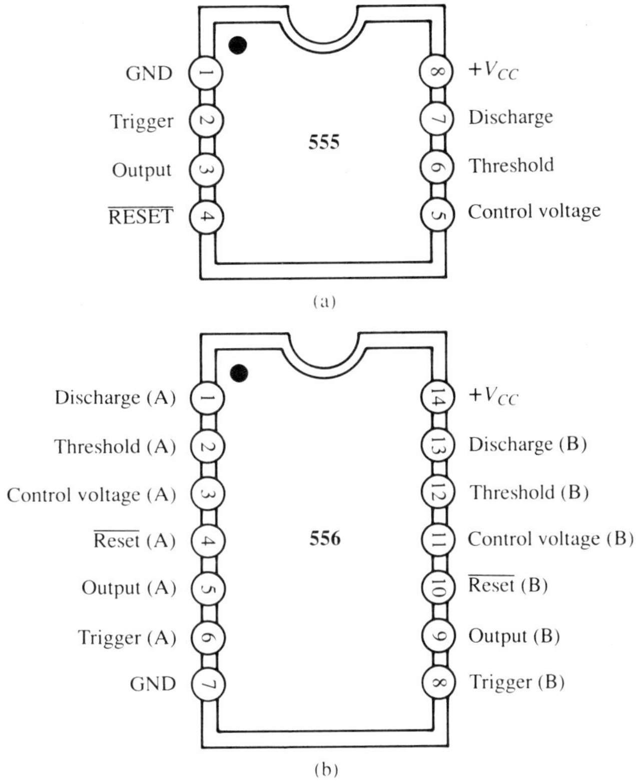

resets the flip-flop, and its output goes low (ground). As a result, Q_1 is turned off and the timer's output equals V_{CC}.

Pin 4 is the reset input. It is normally held at V_{CC}, but when it is grounded momentarily, the flip-flop is reset, causing output Q to go low. The control pin (pin 5), which is not shown in Figure 10–18, is used primarily for filtering in electrically noisy environments where a 0.01-μF capacitor is often connected between pin 5 and ground. The power-supply connections are pin 8 (V_{CC}) and pin 1 (ground), and the 555 timer can be powered from +4 V to +18 V.

Table 10–1
555 timer manufacturers

Manufacturer	Type Number
Exar	XR-555
Fairchild	NE555
Intersil	SE555/NE555
Lithic Systems	LC555
Motorola	MC1455/MC1555
National Semiconductor	LM555
Raytheon	RM555/RC555
RCA	CA555
Signetics	SE555/NE555
Texas Instruments	SN52555/SN72555

Figure 10–18
555 timer function block diagram.

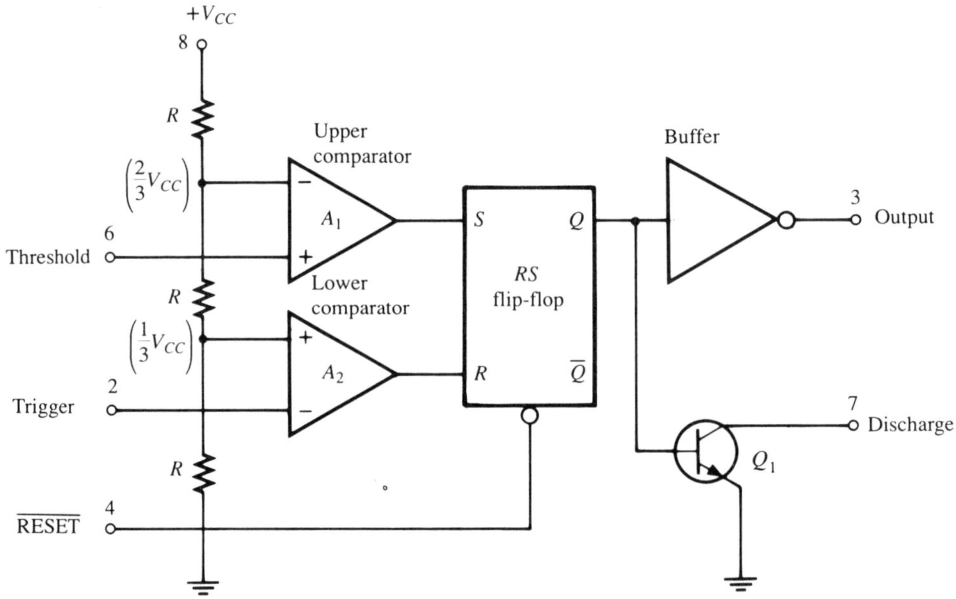

The 555 Timer as a Monostable Multivibrator

As was previously described in this chapter, an astable, or free-running, multivibrator basically generates a train of pulses without any input signal. On the other hand, a **monostable multivibrator,** or **one-shot,** has only one stable output state and needs an external signal, called a *trigger,* to cause the output to change states. When the one-shot is briefly triggered from an external pulse, its output changes to its unstable state for a given time period and then returns to the stable state. The time period during which the timer's output is at its unstable state is often referred to as the *time delay*. The stable state may be ground, whereas the unstable state usually equals the supply voltage.

The 555 timer is connected as a monostable multivibrator, as shown in Figure 10–19a. In its quiescent, or standby, state the internal control flip-flop of Figure 10–18 holds Q_1 on, holding the external timing capacitor (C) discharged so that the timer's output at pin 3 is at ground. When triggered only by a *negative-going* pulse, where pin 2 is taken below $\frac{1}{3}V_{CC}$, the lower comparator (A_2) resets the flip-flop, causing Q_1 to turn off, and lets the timing capacitor charge exponentially through R_A from ground to V_{CC} with a time constant equal to R_AC.

While the capacitor is charging, the timer's output at pin 3 remains at V_{CC}. When the capacitor voltage reaches $\frac{2}{3}V_{CC}$, the flip-flop changes state and again turns on Q_1, which discharges C and returns the circuit to its standby state so that the timer's output voltage is zero. Figure 10–19b illustrates the timing relationships between the input trigger, capacitor voltage, and the timer's output.

Figure 10–19
555 timer monostable multivibrator. (a) Circuit. (b) Waveforms.

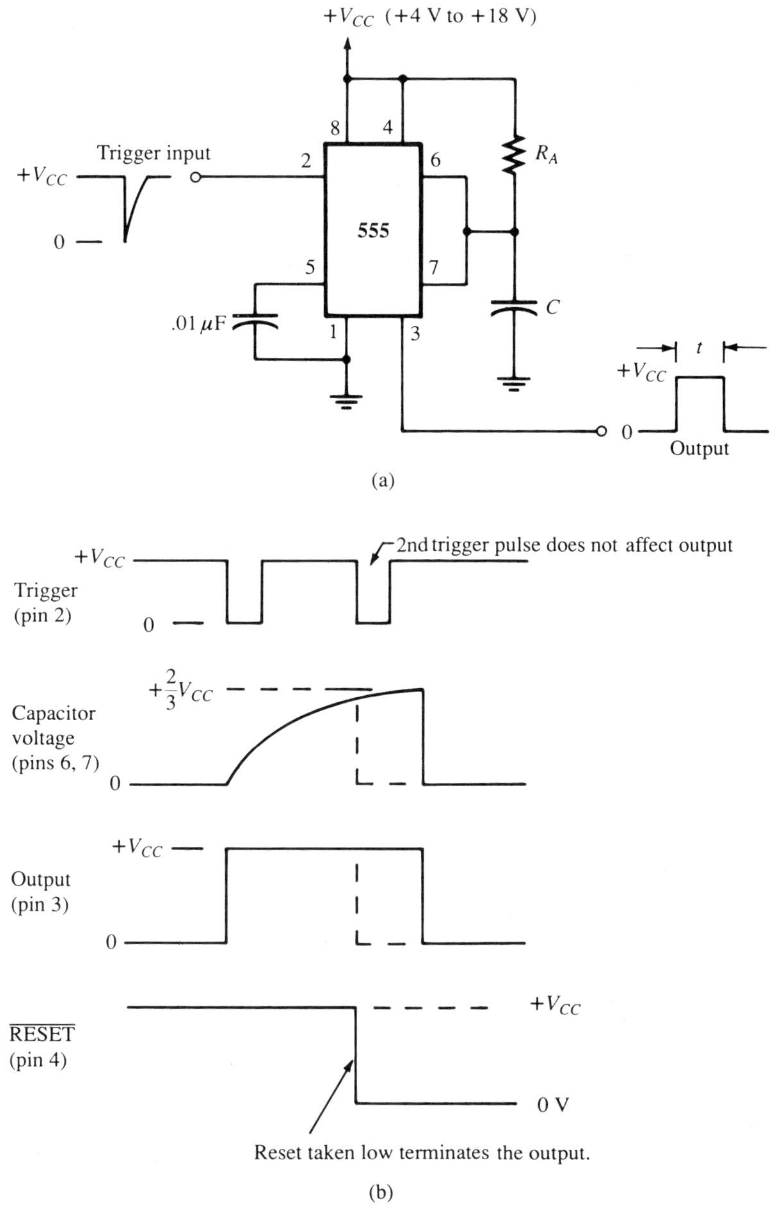

While the timing capacitor is charging, the instantaneous capacitor voltage at the junction of pins 6 and 7 is

$$V_C = V_{CC}(1 - e^{-t/R_AC}) \qquad \textbf{(10–15)}$$

The time that it takes for the capacitor to charge from ground to $+\frac{2}{3}V_{CC}$ is then found from Equation 10–15, so that

$$\frac{2}{3}V_{CC} = V_{CC}(1 - e^{-t/R_AC})$$

Solving for t gives

$$t = -R_AC \ln\left(\frac{1}{3}\right)$$

so that the *time delay,* or *output pulse width* (t), which is the time that the timer's output is high (i.e., the unstable state), is equal to

$$t = 1.1R_AC \qquad \textbf{(10–16)}$$

Figure 10–20 shows a graph of the various combinations of R_A and C necessary to produce a given time delay. As a general guideline, R_A is chosen to be between 1 kΩ and 10 MΩ, whereas C is made at least 0.001 μF. However, for long time delays requiring electrolytic capacitors greater than 1 μF, tantalum or low-leakage Mylar types should be used because capacitor leakage would severely

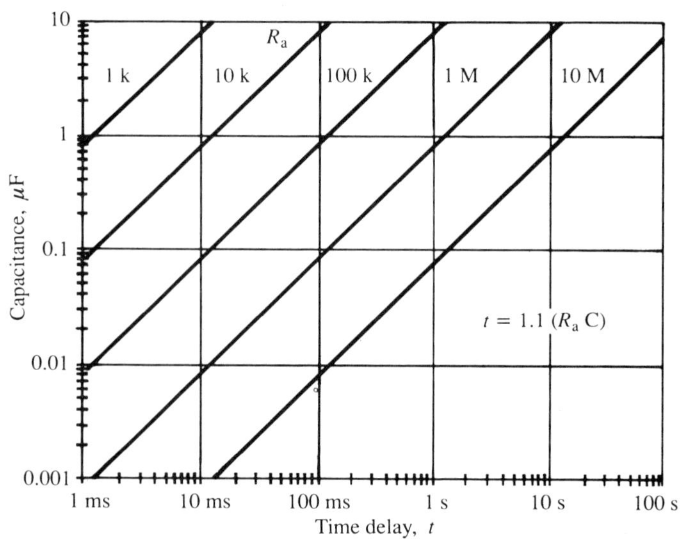

Figure 10–20 Graph of various combinations of timing components R_A and C necessary to produce a given time delay for the 555 timer.

affect the accuracy of the time delay given by Equation 10–16. Since the charging rate and comparator thresholds are directly proportional to the supply voltage, the time delay is independent of the supply voltage. Consequently, supply-voltage variations affect both in a manner that cancels any change in the time delay.

The input trigger pulse width, which can be as small as 30 μs, must always be kept short compared to the time delay; otherwise, the timer's output will remain high and the timing process will not start until the trigger pulse returns high. If the reset pin is grounded at any time during the timing cycle, the timing cycle is immediately ended and the output returns to zero. However, repeated triggering during the timing process has no effect on the output time delay. Very often a 0.01-μF capacitor is connected between pin 5 and ground to prevent any noise from affecting the calculated time delay.

The 555 Timer as an Astable Multivibrator

Figure 10–21a shows the 555 connected as an astable, or free-running, multivibrator. In this mode the output from pin 3 is a rectangular pulse train. As compared with the monostable multivibrator, the timing resistance is now split into two parts, R_A and R_B. When the power supply is turned on, the timing capacitor C

Figure 10–21
555 timer astable multivibrator. (a) Circuit. (b) Waveforms.

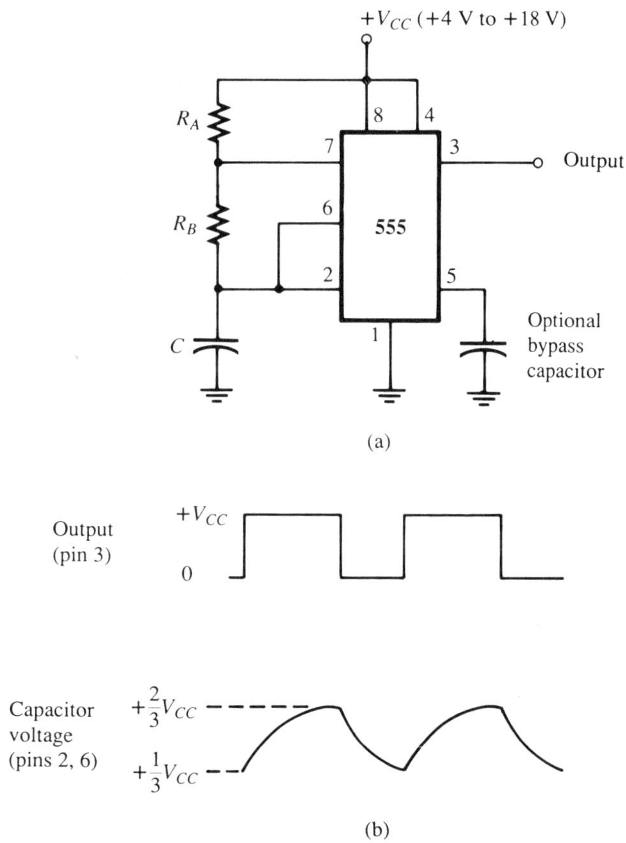

charges toward $\frac{2}{3}V_{CC}$ through both R_A and R_B. When the capacitor voltage reaches $\frac{2}{3}V_{CC}$, the upper comparator sets the flip-flop and the capacitor starts to discharge to ground through R_B. When the capacitor voltage discharges down to $\frac{1}{3}V_{CC}$, the timer's internal lower comparator then resets the flip-flop and a new cycle begins. Since the charge path is through $R_A + R_B$ and the discharge path is only through R_B, the output will be high for a longer period than it is low. The reset input is normally tied to the positive supply. Grounding the reset pin inhibits oscillation, so that it acts as an on-off control.

As shown in Figure 10–21b, the capacitor is then periodically charged and discharged between $\frac{2}{3}V_{CC}$ and $\frac{1}{3}V_{CC}$. Using Equation 10–15, the time that the timer's output waveform is high is given by

$$t_1 = (R_A + R_B)C \ln\left[\frac{V_{CC} - \frac{1}{3}V_{CC}}{V_{CC} - \frac{2}{3}V_{CC}}\right] \qquad \textbf{(10–17)}$$

or

$$t_1 = 0.693(R_A + R_B)C \qquad \textbf{(10–18)}$$

The time that the output waveform is low is given by

$$t_2 = 0.693R_BC \qquad \textbf{(10–19)}$$

so that the total period for one cycle is given by adding Equations 10–18 and 10–19, or

$$\begin{aligned} T &= t_1 + t_2 \\ &= 0.693(R_A + 2R_B)C \end{aligned} \qquad \textbf{(10–20)}$$

The free-running output frequency is $1/T$, or

$$f_o = \frac{1.443}{(R_A + 2R_B)C} \qquad \textbf{(10–21)}$$

Since the percent duty cycle of a rectangular pulse train is defined as the ratio of the time the output is high (t_1) to the time taken by the total cycle (T), then from Equations 10–18 and 10–20,

$$D = \left(\frac{R_A + R_B}{R_A + 2R_B}\right) \times 100\% \qquad \textbf{(10–22)}$$

By making R_B large compared to R_A, a percent duty cycle of nearly 50% can be obtained. On the other hand, if R_A is made much larger than R_B, then the percent duty cycle approaches 100%.

From Equations 10–21 and 10–22, it is then impossible to change the output frequency by varying R_A and R_B without also changing the duty cycle. If a specific duty cycle and free-running frequency are required, then Equations 10–21 and 10–22 are solved simultaneously by selecting a value for R_A and solving for R_B.

Figure 10–22a shows a graph of the required combinations of $(R_A + 2R_B)$ and C to produce a given output frequency. Figure 10–22b shows a similar graph for the duty cycle. Like the monostable configuration, the timing capacitor can be as low as 0.001 μF and limited to a maximum value by the capacitor's leakage resistance. The total resistance $R_A + 2R_B$ is selected to be in the 1-kΩ to 10-MΩ range. Although these component limits might suggest that a maximum free-running frequency of 1.443 MHz is possible, the 555 timer's maximum frequency is typically limited to about 250 kHz without serious waveform distortion.

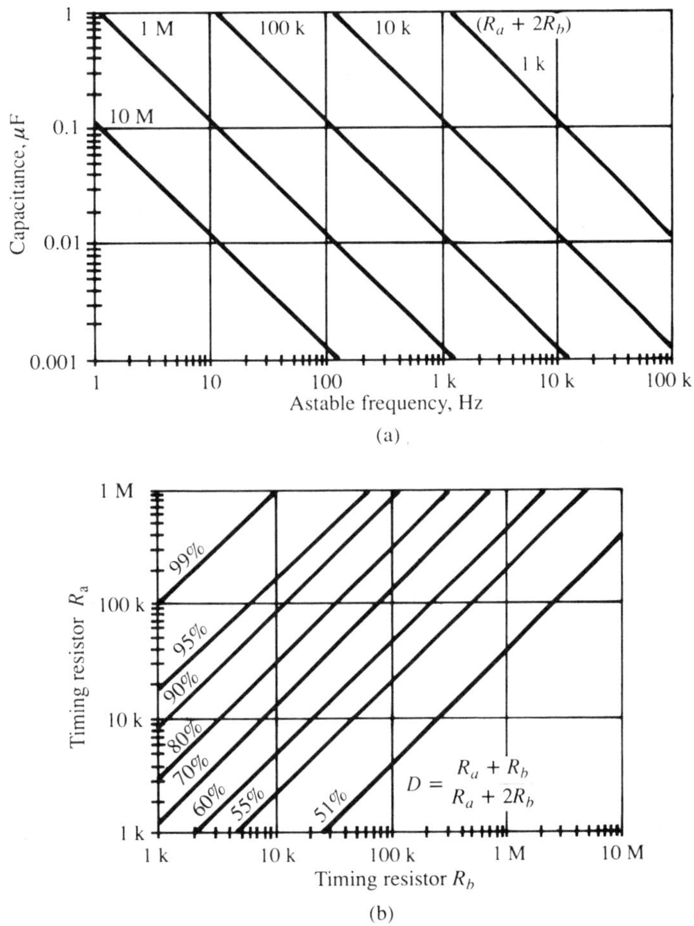

Figure 10–22
555 timer astable multivibrator design graphs.
(a) Combinations of timing components $R_A + 2R_B$ and C necessary to produce a given time delay. (b) Combinations of timing components R_A and R_B necessary to produce a given duty cycle.

Example 10–4

Determine the component values necessary for the 555 timer astable multivibrator circuit of Figure 10–21a having a free-running frequency of 900 Hz with a 65% duty cycle. Assume a timing capacitor of **0.22 μF.**

Solution: Using Equation 10–21,

$$900\text{ Hz} = \frac{1.443}{(R_A + 2R_B)(0.22\ \mu\text{F})}$$

so that

$$7288 = R_A + 2R_B$$

Using Equation 10–22,

$$65\% = \frac{R_A + R_B}{R_A + 2R_B} \times 100\%$$

so that

$$0 = -0.35R_A + 0.3R_B$$

Solving both equations simultaneously for R_B,

$$\begin{aligned} 7288 &= R_A + 2R_B \\ 0 &= -0.35R_A + 0.3R_B \end{aligned}$$

so that

$$R_B = \mathbf{2551\ \Omega} \qquad \text{(use a 2.4-k}\Omega\text{ standard value)}$$

and

$$R_A = \mathbf{2186\ \Omega} \qquad \text{(use a 2.2-k}\Omega\text{ standard value)}$$

Using these standard values and checking our results,

$$\begin{aligned} f_o &= \frac{1.443}{(2.2\text{ k}\Omega + 4.8\text{ k}\Omega)(0.22\ \mu\text{F})} \\ &= \mathbf{937\ Hz} \qquad (4.1\%\text{ error}) \end{aligned}$$

and

$$\begin{aligned} D &= \frac{2.2\text{ k}\Omega + 2.4\text{ k}\Omega}{2.2\text{ k}\Omega + 4.8\text{ k}\Omega} \times 100\% \\ &= \mathbf{65.7\%} \qquad (1.1\%\text{ error}) \end{aligned}$$

The final circuit is shown in Figure 10–23.

Figure 10–23
Completed 555-timer astable multivibrator circuit for Example 10–4.

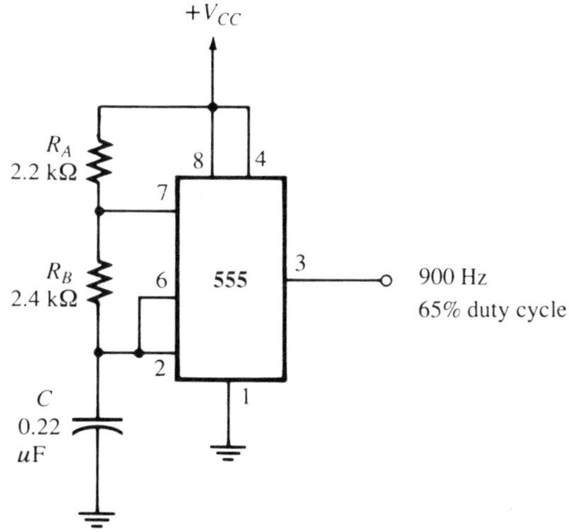

XR-2240 Programmable Timer

The XR-2240 programmable timer* made by Exar uses the basic 555 timer internal configuration discussed in the previous section to create a programmable timer whose time delay can be chosen to be any integral multiple from 1 to 255 times the RC time constant of its external timing components.

As shown in the logic diagram of Figure 10–24, the XR-2240 features a time-base oscillator, control logic, and an 8-bit binary counter. The time-base oscillator

Figure 10–24
XR-2240 programmable timer logic diagram.

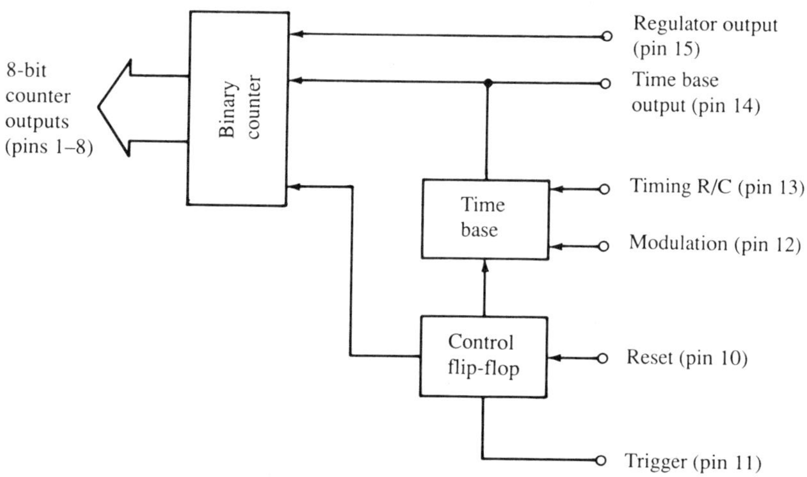

*The 8240 timer, made by Intersil, is identical in function to the XR-2240.

Figure 10–25
XR-2240 programmable timer. (a) Basic circuit used for monostable and astable operation controlled by switch S_1. (b) Connections required for a time delay of 32*RC*. (c) Connections required for a time delay of 83*RC*.

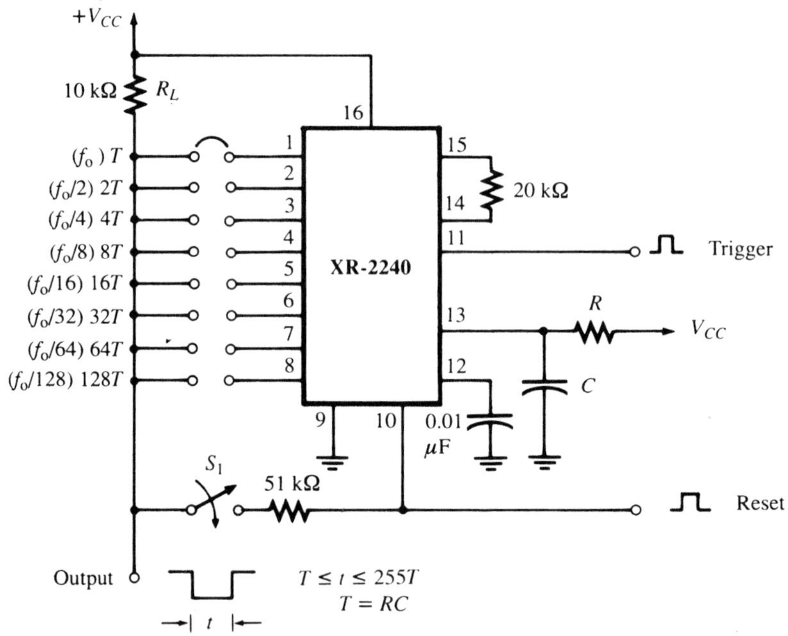

(a)

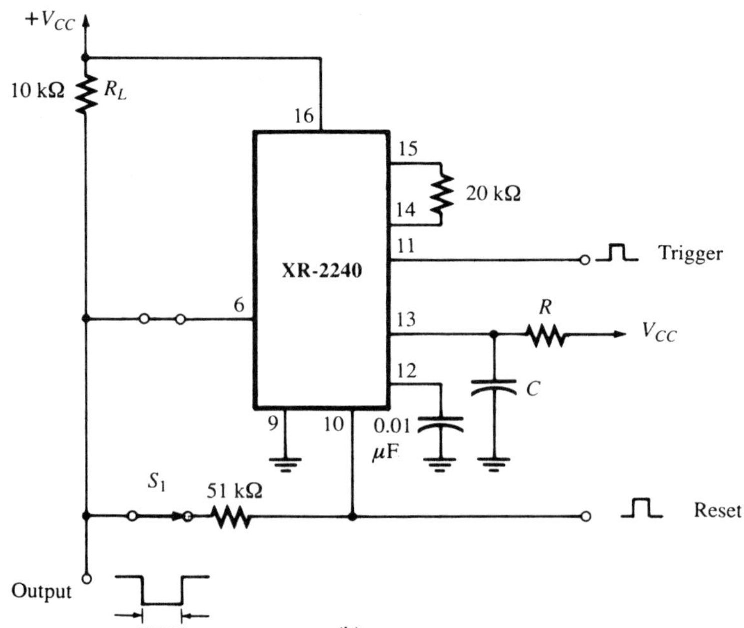

(b)

Figure 10–25 *continued*

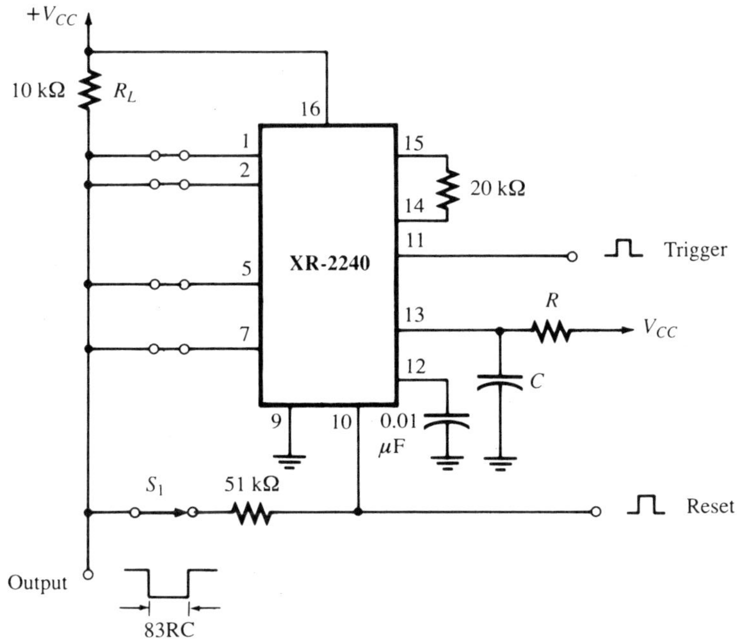

(c)

is similar to a 555-style astable multivibrator. The basic circuit is shown in Figure 10–25a and can be used for both monostable and astable operation. With switch S_1 closed, the circuit functions as a programmable monostable multivibrator. Unlike the 555 timer, a positive-going trigger is required to start the timing process. Timer programming is accomplished by simply connecting appropriate outputs to the common output bus tied to a pull-up resistor so that the output pulse width will be given by

$$t = nRC \tag{10–23}$$

where n is an integer from 1 to 255. Using an 8-bit binary counter, we can then have time delays ranging from $1RC$ to $255RC$. The reset input (pin 10) responds to positive-going pulses. When taken high, the time base is immediately disabled while all the internal counters are reset to their zero state.

As shown in Figure 10–25b, for example, if only pin 6 ($f_o/32$) is connected to the output bus, the total time for the output pulse width is $32RC$. Similarly, if pins 1 (f_o), 2 ($f_o/2$), 5 ($f_o/16$), and 7 ($f_o/64$) are connected (Figure 10–25c), the total time delay is then $t = (1 + 2 + 16 + 64)RC$, or $83RC$. In its standby state, the output is normally at V_{CC}, which can be set from 4 to 15 V. When the output is triggered by a positive-going pulse and the timing process starts, the output goes to ground for the duration of the timing cycle. The time-base oscillator will continue to run until the eight-bit counter reaches the programmed count, which in this case is a time period equal to $83RC$. When this count is reached, the output changes from ground to V_{CC}, which is also fed to the reset input to stop the time base and reset the counter stages. The XR-2240 timer is now back in its standby state waiting for the next trigger. In general, timing resistor R is chosen to be in the 1-kΩ to 10-MΩ range, whereas C is between 0.01 μF and 1000 μF. The maximum time constant allowed is approximately 10 s, so that the maximum time delay that can be programmed is 2550 s, or 42.5 min.

The connections for the astable operation mode for the XR-2240 are shown in Figure 10–26. Once started by a trigger pulse, the XR-2240 now continues to oscillate. With a single counter output connected to the pull-up resistor R_L, the timer will oscillate at a free-running frequency given by

$$f_o = \frac{1}{nRC} \tag{10–24}$$

where n is the count of the particular output selected (1, 2, 4, 8, 16, 32, 64, or 128). The maximum rated free-running frequency possible is approximately 130 kHz.

Unlike most astable multivibrators, this circuit will not oscillate until triggered at pin 11. If desired, the oscillations may be stopped by application of a reset pulse at pin 10, after which the output will go high. It will remain high until triggered again. Very often, this circuit can be made to be self-starting when power is

Figure 10–26
XR-2240 connected as an astable multivibrator.

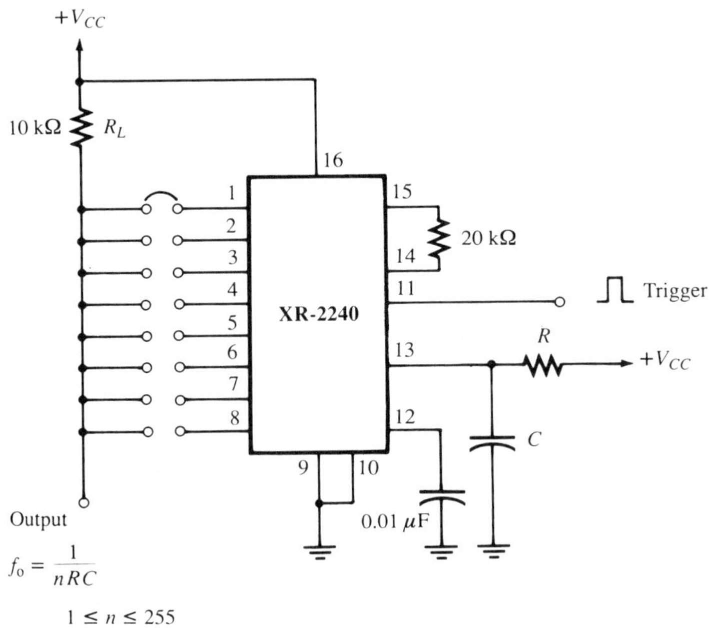

applied by connecting the trigger input (pin 11) to pin 15, which is the output of an internal voltage regulator.

At times, extra-long time delays—longer than those normally obtained using either the 555 or XR-2240 timers—are required. Simply increasing the timer's RC time constant may in theory produce long time delays, but it will not always give reliable and accurate time delays. With the XR-2240, it is possible to achieve time delays in excess of the 2550 s by cascading two 2240 timers, as shown in Figure 10–27. The first timer stage is used as the time base and has a frequency set by Equation 10–24. Here, the eighth bit of the internal counter section is used, so that $n = 128$. The output of the first timer stage serves as the timing pulses for the second timer stage. Here, all the outputs of the countersections of the second timer are wired to the output bus, giving a total time delay of 256 × 256, or 65,536*RC*. If, for example, the RC time constant of the first stage uses $R = 100$ kΩ and $C = 10$ μF, so that $t = 1$ s, then the overall time delay would then be 65,536 s, or 18 h 12 min!

Applications and Other Timers

There are virtually an unlimited number of applications possible for integrated-circuit timers such as the 555 and XR-2240. Besides the popular 555 and XR-2240 integrated-circuit time machines, there are several other lesser-known timers available. These include the XR-320, LM322 (or the equivalent LM3905), and

Figure 10–27
Cascading two XR-2240 timers to produce time delays up to 18 h.

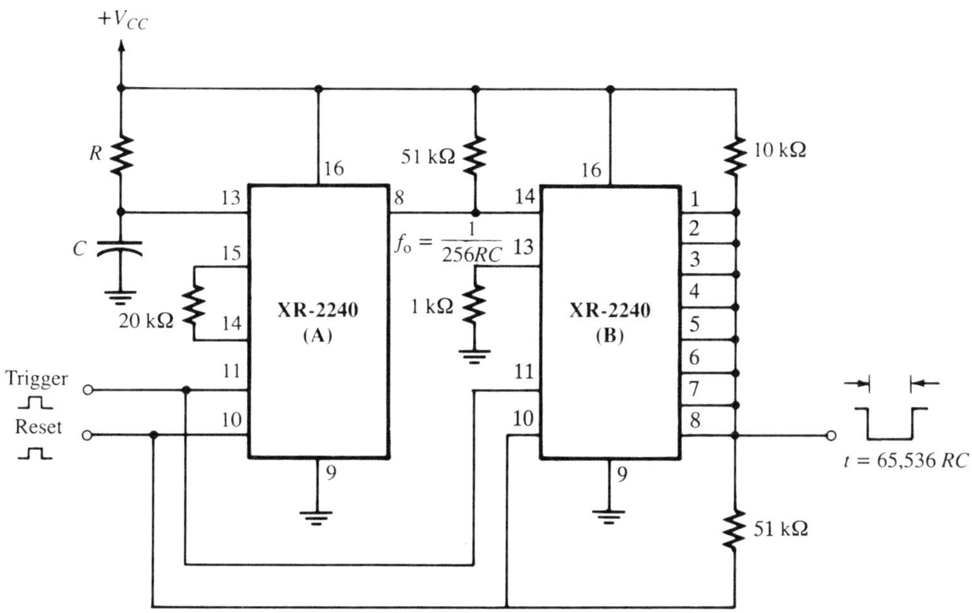

LM554, all of which exhibit differences in timing equations and circuit configurations from the 555 and XR-2240 devices.

There have been entire books as well as many articles in electronic hobby magazines written exclusively on the operation and applications of integrated-circuit timers. Some of these publications are as follows:

1. Howard M. Berlin, *IC Timer Review* 73 (January 1978):40–45.
2. Howard M. Berlin, *The 555 Timer Applications Sourcebook, with Experiments,* Indianapolis, Ind.: Howard W. Sams & Co., Inc., 1978.
3. Joseph J. Carr, *IC Timer Handbook,* Blue Ridge Summit, Penn.: TAB Books, Inc., 1981.
4. Walter J. Jung, *IC Timer Cookbook,* 2d ed., Indianapolis, Ind.: Howard W. Sams & Co., 1983.

Using the 555 timer, the following four applications demonstrate only a very small fraction of the various circuits possible.

Missing-Pulse Detector

Figure 10–28a shows a circuit called a *missing-pulse detector*. As long as the input trigger pulse train at pin 2 keeps arriving at a certain rate, the timer stays triggered and the output remains at V_{CC}. Any change in the triggering frequency, or the omission of a pulse, then allows completion of the time cycle and the output returns to ground, as shown in the timing diagram of Figure 10–28b. As a general rule, the timer's time delay is set at approximately 33% to 50% longer than the expected time between triggering pulses.

Audible Continuity Tester

The *audible continuity tester* circuit shown in Figure 10–29 is a simple adaptation of the astable multivibrator for which free-running frequency is set to operate in the 500-Hz to 2-kHz range. The output is capacitively coupled to an inexpensive permanent-magnet 8-Ω speaker, which can be salvaged from an inexpensive transistor radio. When there is continuity between the test probes, power is applied to the circuit, causing it to oscillate so that the tone is heard in the speaker.

Two-Trace Oscilloscope Display

When two or more analog signals referenced to a common baseline are simultaneously displayed on an oscilloscope, it is often difficult to identify each waveform. This can be especially true for signals that are nearly in phase with each other.

The 555-timer astable multivibrator circuit shown in Figure 10–30a* adds dotted- or dashed-line markers to one of the analog signals (Figure 10–30b and c). The oscilloscope's instantaneous vertical trace position is determined solely by the sum of the voltages across R_4 and R_5. When the timer's output pulse train

---

*This circuit is covered by U.S. Patent 4,132,298 (January 2, 1979), "Device for Identifying Waveforms in a Multiple Oscilloscope Display," by Howard M. Berlin.

Figure 10–28
555-timer missing pulse detector. (a) Circuit. (b) Waveforms.

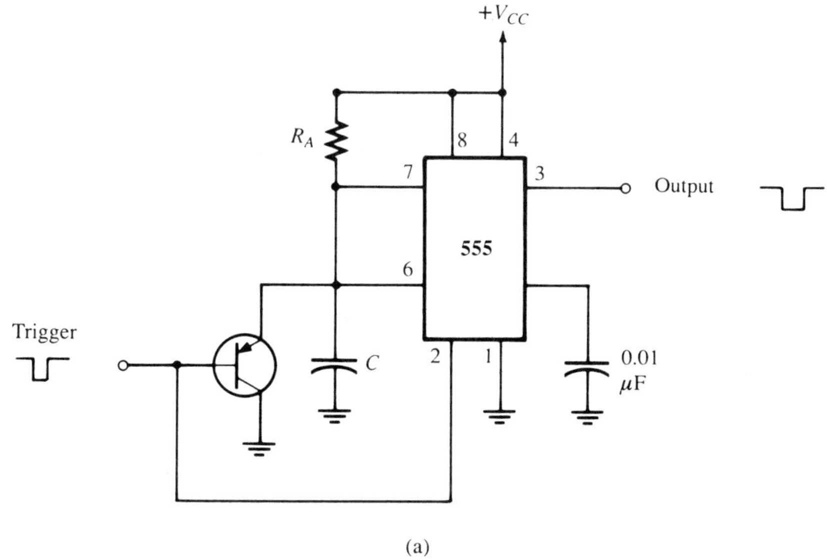

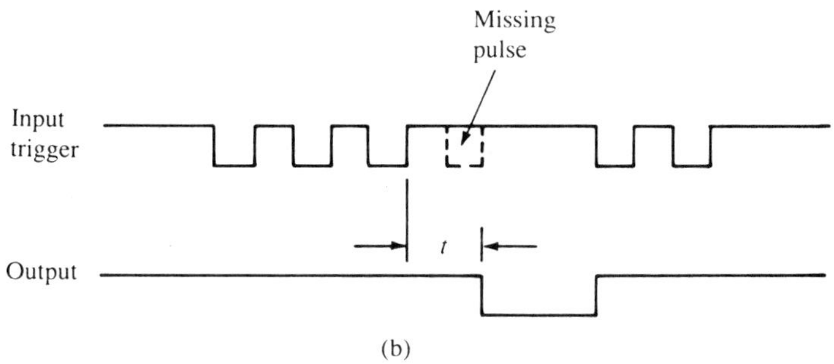

Figure 10–29
Audible continuity tester.

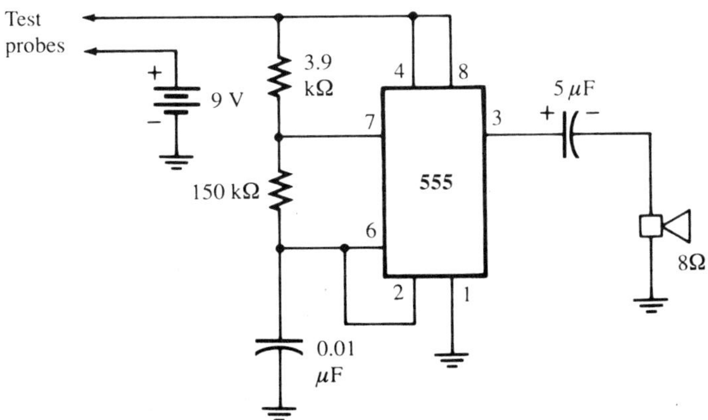

Figure 10–30
Two-channel oscilloscope display marker. (a) Circuit. (b) Waveforms without circuit. (c) Waveforms with circuit.

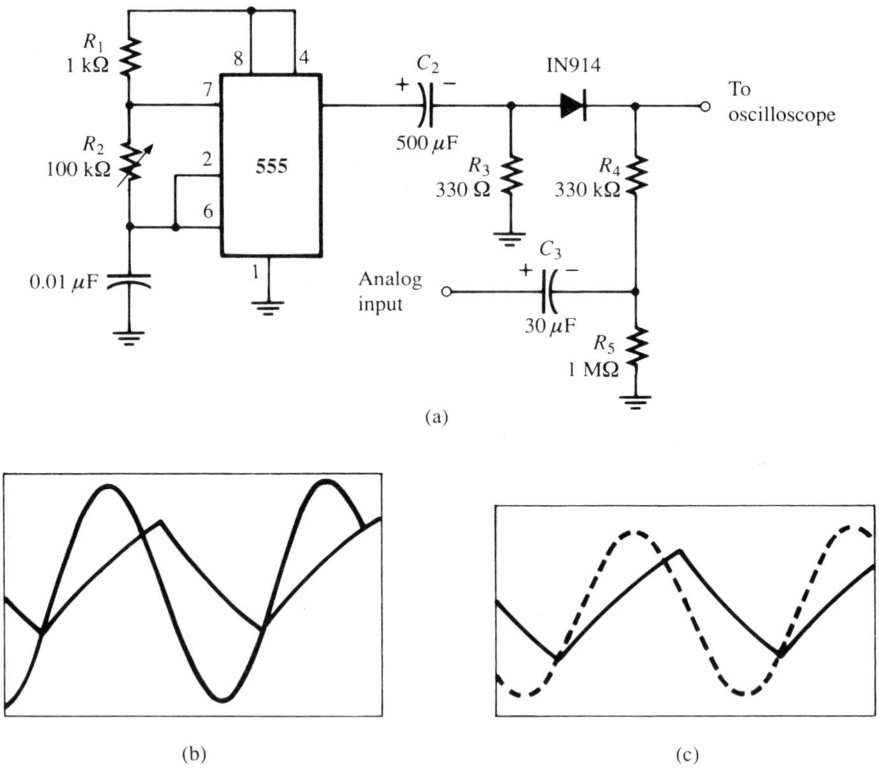

across R_4 is zero, the vertical position is determined solely by the analog input signal. When the timer's output signal is not zero, it then drives the trace off the screen. This off-screen trace transition cannot be seen at normal oscilloscope intensities. The width of line markers can be adjusted from dashed to dotted by adjusting the duty cycle of the 555 timer's output, whose frequency should be at least ten times greater than the analog input signal for proper operation.

Two-Note Generator

Using the circuit of Figure 10–31, two 555 timers can be made to generate alternately two frequencies. The first timer stage (U_1), wired as an astable with a free-running frequency of 7 Hz, is then connected through a 100-kΩ resistor to another 555-connected astable stage (U_2). Each time the output of the first timer goes high, the 100-kΩ resistor is effectively placed in parallel with the 330-kΩ timing resistor of the second astable section. With the components shown, this then gives a free-running frequency of about 1.5 kHz. When the output of the timer U_1 is low, the 100-kΩ resistor, because of its large resistance compared with the 10-kΩ timing

Figure 10–31
555-timer two-note generator.

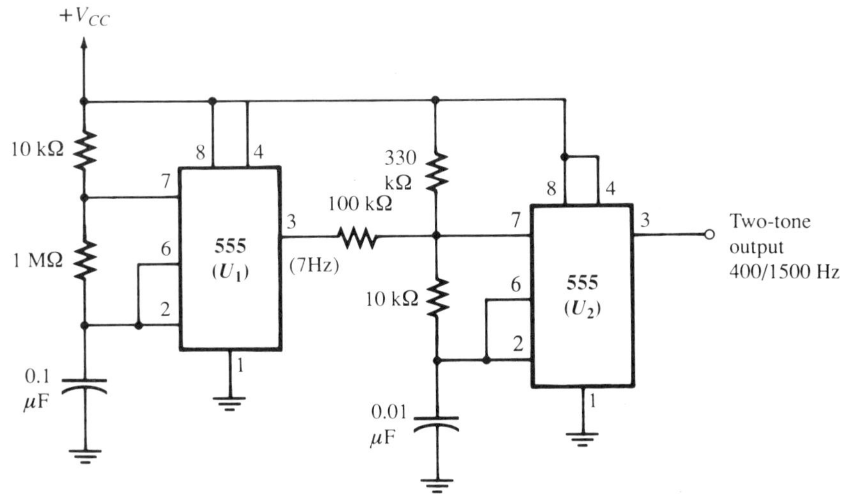

resistance of U_2, is essentially removed from the circuit and the second timer stage oscillates approximately at 400 Hz. Consequently, an audible tone at the output of U_2 shifts back and forth between 400 Hz and 1.5 kHz approximately seven times each second.

10–5 Precision Waveform Generators

The development of the 555 timer ultimately paved the way for the creation of more sophisticated linear integrated circuits, the so-called systems on a chip. A natural progression from operational amplifiers and timers was the **precision waveform generator.** This device is nothing more than a single-chip function generator capable of producing sine, square, sawtooth, and pulse waveforms with variable frequency and amplitude using a minimum number of external components and adjustments. The two most popular devices available are the XR-2206 and the 8038. Like most integrated-circuit devices, the interest and emphasis should not be on what is actually inside the device (which we cannot change) and what makes it work, but rather how to wire the device. In general, the manufacturer's data sheets gives all the information necessary for proper operation.

The 8038 waveform generator is made by Intersil and Exar. Its operating frequency can range over nine decades from 0.001 Hz to 1 MHz by the choice of external RC timing components. The function generator circuit of Figure 10–32 has provisions for variable frequency, output amplitude, and sweep-rate controls; external modulation provisions are also available without affecting the quality of the output waveform. The three basic waveforms (sine, triangular, and square waves) are available simultaneously from independent output terminals.

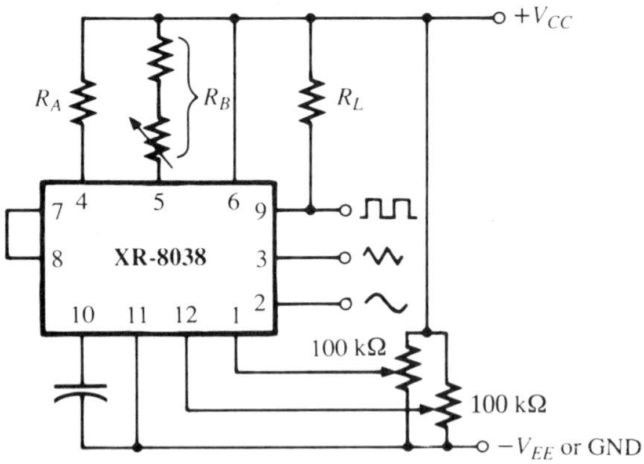

Figure 10–32
8038 precision function generator producing square, triangular, and sine-wave outputs.

A more sophisticated function-generator circuit using the XR-2206 is shown in Figure 10–33. The circuit is a function-generator system capable of producing AM or FM modulated sine, square, triangular, ramp, and pulse waveforms of high stability and accuracy. The circuit shown is designed to operate with a single 12-V supply. However it can be powered by a ±6-V split supply, which is preferred for most applications, since it produces an output with almost no DC component. As shown, this circuit produces four overlapping frequency ranges of from 1 Hz to 100 kHz. Throughout each range selected by S_1, the frequency may be varied over a 100 : 1 tuning range by R_{13}.

The sine or triangular output, having an output resistance of 600 Ω, can be varied from 0 to 6 V peak-to-peak with R_{12}. A square-wave output signal is available at the "sync" output pin for synchronizing an oscilloscope or for driving logic circuits.

The symmetry of the sine-wave output signal is adjusted with R_{11}, whereas R_{10} is adjusted to minimize sine-wave distortion.

10–6 The Crystal-Controlled Oscillator

Quartz crystals are often used when it is necessary to hold the frequency of an oscillator to very tight tolerances. They are commonly found in radio transmitters, instrument time bases, and computer clock sources.

Quartz crystals may be operated in either the *series resonant mode* or the *parallel resonant mode*. In the series resonant mode, the crystal behaves as a very high-Q, stable series resonant circuit. In the parallel resonant mode, the crystal behaves as a very stable parallel resonant circuit. Figure 10–34 shows one section of an LM339 quad comparator connected as a 100-kHz oscillator. Here, the crystal is operated in the series resonant mode and is in the positive feedback loop path. This arrangement permits sufficient positive feedback to occur only at the crystal frequency and, unlike most crystal-controlled oscillators, produces a square-wave output, which would be suitable for clocking synchronous logic circuits.

Figure 10–33
XR-2206 precision function generator.

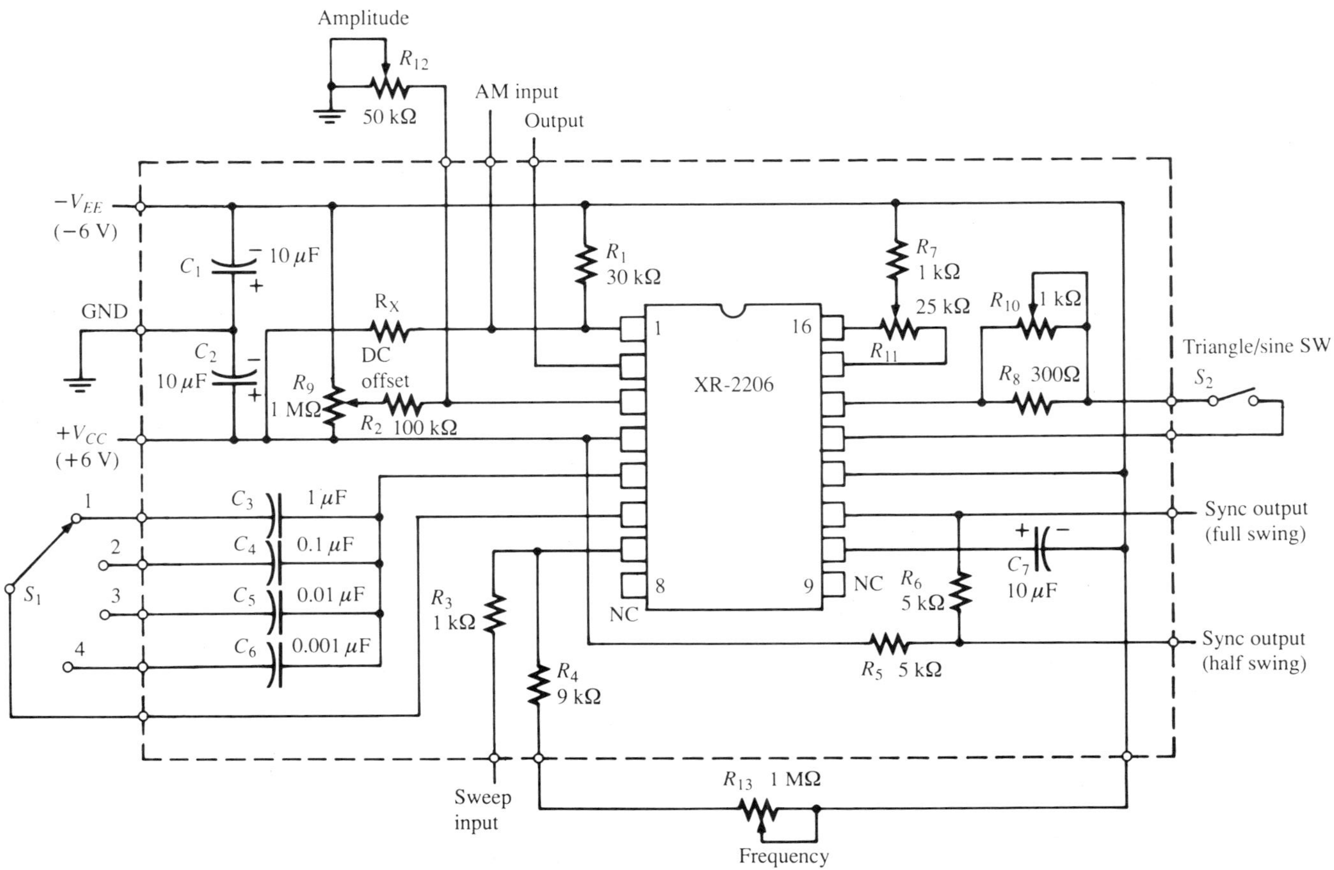

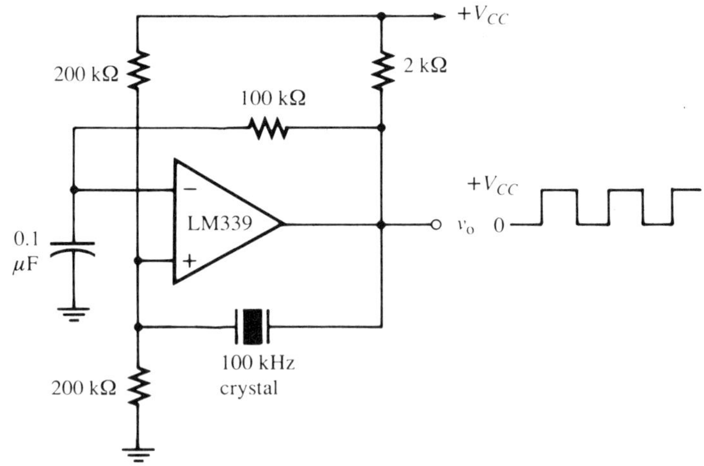

Figure 10–34
Crystal-controlled oscillator using one section of an LM339 quad comparator.

Summary

This chapter discussed a variety of circuits that are capable of generating sine, square, triangular, pulse, and sawtooth waveforms. When referring to signal sources that require no input, the oscillator is one that generates sine waves only, whereas a function generator is a more versatile circuit that provides any type of periodic waveform.

Using operational amplifiers, oscillation is achieved by controlling the amount and phase of positive feedback. Here, a class of low-frequency RC oscillators included the phase-shift, twin-T, and Wien-bridge networks. In addition, this chapter described those networks using several specialized monolithic integrated-circuit devices that generate a variety of waveforms. Among these are the 555/556 and XR-2240 timers and the XR-2206 and 8038 precision waveform generators.

Glossary

astable multivibrator A free-running circuit whose output has no stable output state, such that it generates a train of pulses of known frequency.

Barkhausen criteria The two conditions required for oscillation whereby the loop gain is unity and the total phase shift around the feedback loop is zero.

function generator A signal source capable of producing waveforms of shapes other than sine waves.

loop gain The product of an amplifier's voltage without feedback and the fraction of the amplifier output signal that is fed back to its input.

monostable multivibrator A circuit whose output has only one stable output state and requires an external trigger to cause the its output to change states.

one-shot See *monostable multivibrator*.

oscillator A signal source that generates only sine waves.

parallel-T oscillator See *twin-T oscillator*.

phase-shift oscillator A low-frequency oscillator circuit that uses three identical RC networks as the feedback element to produce a 180° phase shift at the desired frequency of oscillation.

precision waveform generator A single integrated-circuit function generator capable of producing sine, square, and triangular waveforms of variable frequency and amplitude using a minimum of external components and adjustments.

RC oscillator A class of oscillators whose output frequency is controlled by one or more RC networks.

twin-T oscillator A low-frequency oscillator circuit that uses a pair of RC networks, each arranged in a T shape and connected in parallel as the feedback element, so that its impedance is zero and phase shift is 180° at the desired frequency of oscillation. Also called a parallel-T oscillator.

Wien-bridge oscillator An RC oscillator circuit that uses an RC lead-lag network as the feedback elements to achieve oscillation.

Review Questions

1. Define the term *oscillator*.
2. How does an oscillator differ from a function generator?
3. How is positive feedback used to produce oscillation?
4. What are the Barkhausen criteria for oscillation?
5. Draw the schematic diagram and write the equation for the output frequency of a phase-shift oscillator.
6. What must the closed-loop gain be for a phase-shift oscillator to oscillate?
7. Draw the schematic diagram and write the equation for the output frequency of a twin-T oscillator.
8. Draw the schematic diagram and write the equation for the output frequency of a Wien-bridge oscillator.
9. What must the closed-loop gain be for a Wien-bridge oscillator to oscillate?
10. What is the function of using a nonlinear resistance, such as a lamp, in a Wien-bridge oscillator?
11. Identify the positive and negative feedback networks of a Wien-bridge oscillator.
12. What is an astable multivibrator?
13. Draw the schematic diagram and write the equation for the output frequency of an astable multivibrator using an operational amplifier.
14. Draw the schematic diagram and write the equation for the output frequency of an operational amplifier circuit that generates triangular waveforms.
15. Draw the schematic diagram and write the equation for the output pulse width of a 555 timer wired as a monostable multivibrator.
16. Draw the schematic diagram and write the equations for the output frequency and duty cycle of a 555 timer wired as an astable multivibrator.

Exercises

1. Determine the frequency of oscillation for the phase-shift oscillator of Figure 10–2a if $R = 1.2\ \text{k}\Omega$ and $C = 0.022\ \mu\text{F}$.
2. For the phase-shift oscillator of Figure 10–2a, determine the value of R required if

$C = 0.0033\ \mu F$ and the frequency of oscillation is 12.5 kHz.

3. Determine the frequency of oscillation for the twin-T oscillator of Figure 10–3a if $R = 5.6$ kΩ and $C = 0.047\ \mu F$.
4. For the twin-T oscillator of Figure 10–3a, determine the value of R required if $C = 0.022$ μF and the frequency of oscillation is 1500 Hz.
5. For the Wien-bridge oscillator of Figure 10–4a, $R_1 = R_2 = 2.2$ kΩ and $C_1 = C_2 = 0.01$ μF. Determine the frequency of oscillation and the required value for R_3 if $R_4 = 1$ kΩ.
6. For the Wien-ridge oscillator of Figure 10–4a, where $R_1 = R_2$ and $C_1 = C_2$, determine the value of R_1 required if $C_1 = 0.082$ μF and the frequency of oscillation is 500 Hz.
7. Determine suitable component values for the astable multivibrator circuit of Figure 10–9a with a free-running output frequency of 850 Hz. Assume a capacitor value of 0.022 μF.
8. Repeat Exercise 7, but use the Norton amplifier circuit of Figure 10–10.
9. Determine suitable component values for the triangular wave generator circuit of Figure 10–13 with a free-running frequency of 2.5 kHz and a peak-to-peak output voltage of 9 V if a ±15-V supply is used. Assume $C = 0.1$ μF, $R_1 = 1$ kΩ, and saturation voltages of ±12 V for the operational amplifier.
10. Assuming a capacitor value of 0.82 μF, determine the value for R_A in a monostable multivibrator using a 555 timer (Figure 10–19a) to produce an output pulse width of 135 ms when triggered by a negative-going input pulse.
11. Determine the free-running frequency and output duty cycle for the 555 timer astable multivibrator circuit of Figure 10–21a if $R_A =$ 2.7 kΩ, $R_B = 12$ kΩ, and $C = 0.033\ \mu F$.
12. Repeat Exercise 11 if the values of R_A and R_B are reversed.
13. For the 555 timer astable multivibrator circuit of Figure 10–21a, determine the component values for R_A and R_B for a free-running frequency of 43 kHz and a duty cycle of 71%. Assume a capacitor value of 0.0047 μF.

11 Active Filters

Instructional Objectives

Upon completing this chapter, you will be able to

- Describe the passband and stopband characteristics of the basic low-pass, high-pass, bandpass, and notch filters.
- Compare the frequency response characteristics of Butterworth, Chebyshev, elliptic, and Bessel filters.
- Determine the order required of either a Butterworth or Chebyshev filter.
- Explain the relationship between Q, bandwidth, and center frequency in bandpass and notch filters.
- Calculate the decibel frequency response of bandpass and notch filters.
- Draw and calculate the required component values for a variety of second-order RC active filters using Sallen-Key, multiple-feedback, twin-T, and state-variable network designs.

Introduction

A **filter** network is a device that passes electric signals at certain frequencies or frequency ranges while preventing the passage of others. Specifically, an **active filter** is a network that is built around an active device such as an operational amplifier. On the other hand, a passive filter is a filter network composed solely of passive devices—resistors, capacitors, or inductors.

When compared with passive filters, active filters offer the following advantages:

- *No insertion loss:* Since the operational amplifier is capable of providing gain, the input signal will not be immediately attenuated as the filter passes those frequencies of interest.
- *Lower cost:* On the average, active filters will cost less than passive filters because inductors are expensive and are not always readily available.

- *Easier tuning:* The frequency response of active filters may be easily tuned, or adjusted, over a wide frequency ranges without significantly changing the desired frequency response.
- *Isolation:* When using operational amplifiers, active filters will have a high input impedance and a low output impedance, virtually guaranteeing almost no interaction between the filter and either its signal source or load.

On the other hand, there are several disadvantages or limitations when using active filters:

- *Frequency response:* The upper frequency response is limited by the frequency response of the transistor or operational amplifier. In general, the frequency response of transistors is greater than that of operational amplifiers. Active filters are generally used for audio frequencies, whereas most passive filters are suited for radio frequency (RF) applications.
- *RF applications:* Active filters are not suited for RF applications, such as a low-pass filter used in series with a transmission line connected between a transmitter and antenna.
- *Power supply:* Unlike passive filters, active filters require some form of power supply for their operation.

In this chapter, filters are discussed on the basis of their type, frequency response, and shape, as well as certain quantitative parameters such as cutoff frequency, bandwidth, and quality factor. Filters may be used to intentionally remove unwanted noise or other frequency components that interfere with a desired signal, or they may be used to alter the shape of waveforms, as in synthesizing musical sounds.

11–1 Filter Basics

There are four basic filter frequency responses:

- Low-pass
- High-pass
- Bandpass
- Notch

In reality, it is necessary to learn only two of these functions because the remaining two are opposite functions. A high-pass filter has the opposite function of a low-pass filter, and a bandpass filter is the opposite of a notch filter. In the following subsections, these opposite relationships are discussed.

Low-Pass Filter

A **low-pass filter** allows incoming signals to be passed through with little or no attenuation up to a given frequency. Above this frequency, the filter rejects or greatly attenuates the level of the input signal. Now consider the ideal low-pass

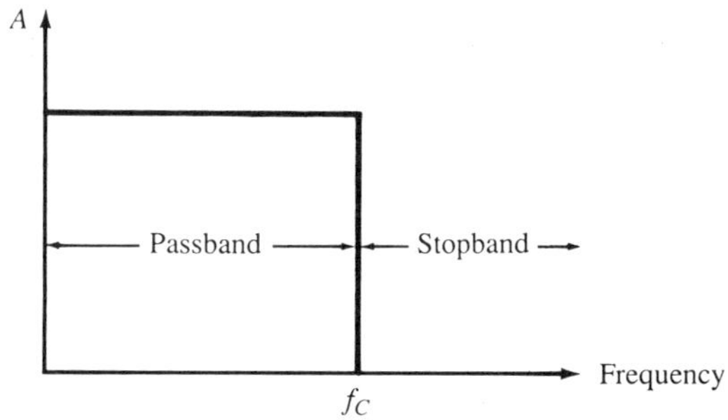

Figure 11–1
Frequency response of an ideal low-pass filter.

characteristic response curve shown in Figure 11–1. With filters, we are primarily interested in the relationship between the level, or amplitude, of the filter's output voltage compared with its input voltage at various frequencies. This type of response curve relating the filter's output-to-input voltage ratio (v_{out}/v_{in}) as a function of signal frequency is called the filter's *amplitude,* or *frequency response*. For convenience, as is shown later, the voltage ratio is expressed in terms of decibels (dB), so that the dB-amplitude response is

$$\text{Amplitude response (dB)} = 20 \log_{10}\left(\frac{v_{out}}{v_{in}}\right) \qquad \textbf{(11–1)}$$

When the output level is greater than the input's, we have gain, and the dB-amplitude response is a positive number. On the other hand, if the output level is less than the input's, there is a loss, or *attenuation,* and the dB-amplitude response will be a negative number. Since the filter's response may extend over a very wide frequency range—for example, from 20 Hz to 20 kHz—filter response curves are usually plotted on *semilog graph paper*. The vertical axis, dB-amplitude response is linear, whereas the horizontal axis, frequency, is logarithmic.

For input frequencies less than f_C, called the **cutoff frequency,*** the frequency range of the relatively constant amplitude response is called the **passband.** In the passband, input signal levels are passed through the filter virtually unchanged, except for possible gain. The frequency range above the cutoff frequency, at which the filter's amplitude response decreases with frequency, is called the **stopband.** Here the filter network stops, rejects, or attenuates the input level as an unwanted signal. The higher the input frequency is above the cutoff frequency, the better the rejection. Simply stated, a low-pass filter passes low frequencies and rejects high frequencies.

In reality, no low-pass filter approaches the ideal response, such as that of Figure 11–1. As will be explained shortly, the change from the filter's passband to the stopband will be more gradual and will not involve the sudden, abrupt change of Figure 11–1. This transition will apply also to high-pass, bandpass, and notch filters.

---

*The cutoff frequency is also referred to as either the *corner frequency, critical frequency, break frequency,* or 3-*dB frequency*.

High-Pass Filter

As shown in Figure 11–2, the frequency response characteristic of a high-pass filter is the opposite of a low-pass filter. A **high-pass filter** allows incoming signals to be passed through with little or no attenuation above the filter's cutoff frequency. Below this frequency, the filter rejects or greatly attenuates the level of the input signal. In reference to the cutoff frequency, the positions of the passband and stopband are interchanged. The frequency range below the cutoff frequency, where the filter's amplitude response decreases with decreasing frequency, is the stopband, whereas frequencies above the cutoff frequency are in the passband.

Simply stated, a high-pass filter passes *high* frequencies and rejects low frequencies. Although a high-pass filter's passband in theory extends up to infinity, the upper frequency limit of the passband of active filters is limited by the frequency response of the active device used. Table 11–1 summarizes the opposite relationships between low-pass and high-pass filters.

Bandpass Filter

A **bandpass filter** is a combination high-pass filter in series with a low-pass filter (Figure 11–3a). As shown in Figure 11–3b, the bandpass filter allows incoming signals to be passed through with little or no attenuation within a given median frequency range. Below and above this median range, the filter rejects or greatly attenuates the level of the input signal. Although there is one passband, there are *two* stopbands.

The passband extends from a given lower cutoff frequency (f_L) to a higher cutoff frequency (f_H). The range below the lower cutoff frequency to the passband is similar in function to a high-pass filter, whereas the remaining frequency response resembles the response of a low-pass filter. The cutoff frequency (f_L) associated with the "high-pass" response is lower than the cutoff frequency (f_H) associated with the "low-pass" response.

The frequency difference between the lower and upper cutoff frequencies is called the **bandwidth** (BW):

$$BW = f_H - f_L \qquad (11\text{–}2)$$

Figure 11–2
Frequency response of an ideal high-pass filter.

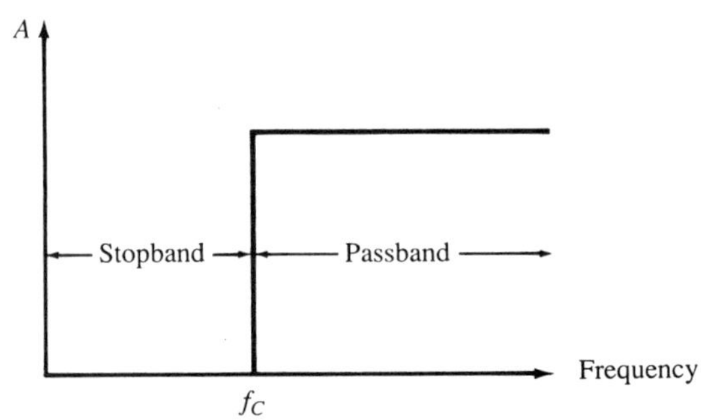

Table 11–1
Comparison of low-pass and high-pass filter characteristics

Parameter	**Low-Pass**	**High-Pass**
Passband	$f_{in} < f_C$	$f_{in} > f_C$
Stopband	$f_{in} > f_C$	$f_{in} < f_C$
Rolloff	negative	positive

BW can be as small as several hertz or relatively wide when compared to the center frequency.

Because the filter's frequency response of Figure 11–3a is plotted on a logarithmic scale, the frequency response appears to be symmetrical about a median frequency (f_o) called the **center frequency.** However, the logarithmic axis distorts the actual relationship between the two cutoff frequencies and the center frequency. If we were not aware of the logarithmic axis, we might then be strongly tempted to conclude by observation that the bandpass filter's center frequency is the *algebraic* average of the lower and upper cutoff frequencies,

$$f_o = \frac{f_L + f_H}{2} \qquad \textbf{(11–3)}$$

Figure 11–3
Bandpass filter. (a) Ideal frequency response. (b) Frequency response composed of low-pass and high-pass filter responses.

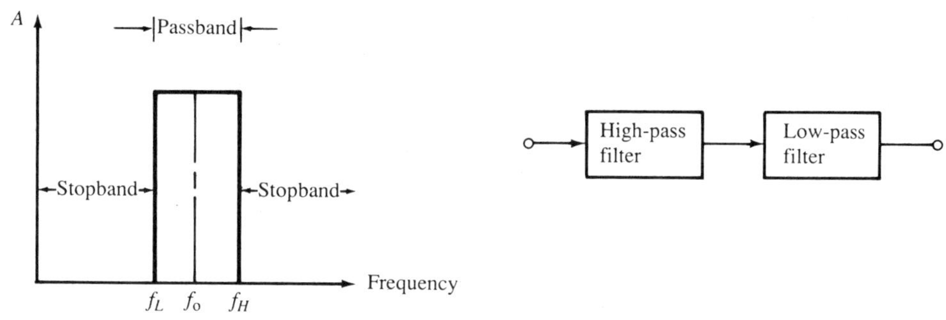

(a)

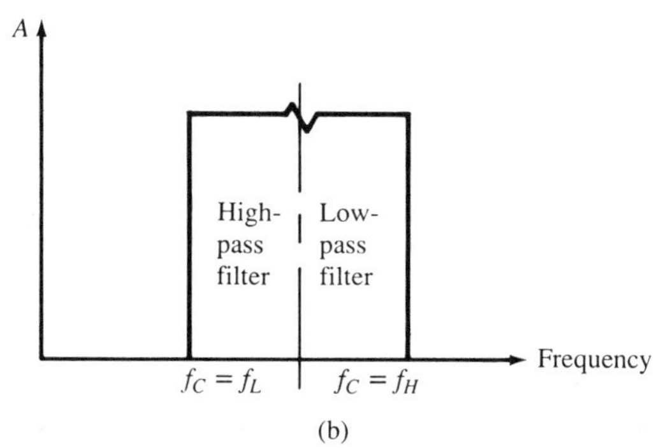

(b)

as it appears to be exactly midway between the two. However, the center frequency is actually found by taking the *geometric average:*

$$f_o = \sqrt{f_L \times f_H} \quad \textbf{(11–4)}$$

The following example illustrates this difference.

Example 11–1

Determine the center frequency of a bandpass filter when the lower and upper cutoff frequencies are 600 Hz and 1200 Hz.

Solution: Using the geometric average formula of Equation 11–4, the correct center frequency is

$$f_o = \sqrt{(600\text{ Hz})(1200\text{ Hz})}$$
$$= \mathbf{848.5\ Hz}$$

However, if we (incorrectly) compute the algebraic average, we obtain

$$f_o = \frac{600\text{ Hz} + 1200\text{ Hz}}{2}$$
$$= 900\text{ Hz}$$

which is a difference of approximately 6% from the true value.

If the filter's bandwidth and center frequency are known, then both 3-dB cutoff frequencies can be determined from

$$f_L = \frac{-\text{BW} + \sqrt{\text{BW}^2 + 4f_o^2}}{2} \quad \textbf{(11–5)}$$

and

$$f_H = f_L + \text{BW} \quad \textbf{(11–6)}$$

Example 11–2

Determine the upper and lower cutoff frequencies of a bandpass filter having a center frequency of 1.2 kHz and a bandwidth of 200 Hz.

Solution: Using Equations 11–5 and 11–6, we have

$$f_L = \frac{-200\text{ Hz} + \sqrt{(200\text{ Hz})^2 + (4)(1.2\text{ kHz})^2}}{2}$$
$$= \mathbf{1104\ Hz}$$

and

$$f_H = 1104\text{ Hz} + 200\text{ Hz}$$
$$= \mathbf{1304\ Hz}$$

The bandpass filter's center frequency and bandwidth are related to each other by a figure of merit called the *quality factor Q,* so that

$$Q = \frac{f_o}{\text{BW}} \qquad \textbf{(11–7a)}$$

In terms of f_L and f_H,

$$Q = \frac{f_o}{f_H - f_L} \qquad \textbf{(11–7b)}$$
$$= \frac{\sqrt{f_L f_H}}{f_H - f_L} \qquad \textbf{(11–7c)}$$

Another useful measure is the reciprocal of the quality factor, $1/Q$, called the **fractional bandwidth.** Expressed as a decimal, the fractional bandwidth is the bandwidth expressed as a percentage of the center frequency. From Example 11–2, the value of Q for the bandpass filter is 1.2 kHz/200 Hz, or 6, and the fractional bandwidth is $\frac{1}{6}$, or 0.167. The fractional bandwidth tells us that the filter's 200-Hz bandwidth is 0.167, or 16.7%, of the 1.2-kHz center frequency.

As was briefly mentioned earlier, no filter has a frequency response like that of the ideal response. Most practical bandpass filters have a response similar to that shown in Figure 11–4, whereby the response peaks with a center frequency gain at the center frequency and decreases on either side of the center frequency in a symmetrical fashion. In terms of the filter's Q, center frequency, and center

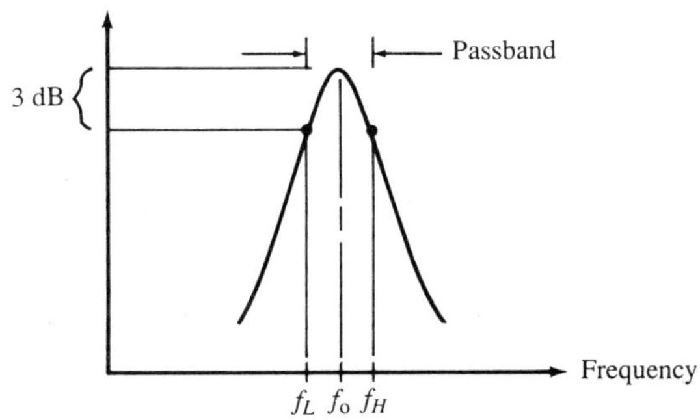

Figure 11–4
Realistic bandpass filter frequency response.

frequency gain (A_o), the dB-amplitude response of such a bandpass filter is given by

$$A_{dB} = A_o(\text{dB}) - 10 \log_{10} \left\{1 + Q^2 \left[\frac{(f_{in}/f_o)^2 - 1}{(f_{in}/f_o)}\right]^2\right\} \qquad \textbf{(11–8)}$$

Since the response curve is symmetrical about the center frequency, the dB-amplitude response will be the same for the ratio f_{in}/f_o as it is for its reciprocal, f_o/f_{in}. For example, a bandpass filter with a center frequency of 1000 Hz will have the same response at 800 Hz ($f_{in}/f_o = 0.8$) as where the input frequency equals $(1/0.8) \times 1000$ Hz, or 1250 Hz. That is, the response is the same at 0.8 times the center frequency as it is at 1.25 times the center frequency.

Example 11–3

If a bandpass filter has a Q of 7, passband gain of zero decibels, and center frequency of 700 Hz, determine the dB-amplitude response if the input frequency is 900 Hz. At what other frequency will the dB-amplitude response equal that at 900 Hz?

Solution:

(a) First, calculate f_{in}/f_o:

$$\frac{f_{in}}{f_o} = \frac{900 \text{ Hz}}{700 \text{ Hz}} = 1.286$$

Substituting into Equation 11–8 gives

$$\begin{aligned} A_{dB} &= 0 \text{ dB} - 10 \log_{10} \left\{1 + (7)^2 \left[\frac{(1.286)^2 - 1}{1.286}\right]^2\right\} \\ &= -10 \log_{10} \left\{1 + (49) \left[\frac{(0.654)}{1.286}\right]^2\right\} \\ &= -10 \log_{10}(13.7) \\ &= \mathbf{-11.4 \text{ dB}} \end{aligned}$$

(b) The same response of -11.4 dB will also occur at (700 Hz/1.286), or **544 Hz.**

Notch Filter

The frequency response of a **notch filter** is shown in Figure 11–5a and is the opposite of that of a bandpass filter. The notch filter is also referred to as either a *band-reject, band-stop,* or *band-elimination filter*. Input signals are stopped, rejected, or attenuated within a given median frequency range. Below and above

Figure 11–5
Notch filter. (a) Ideal frequency response. (b) Frequency response composed of summing the outputs of low-pass and high-pass filters having the same cutoff frequency.

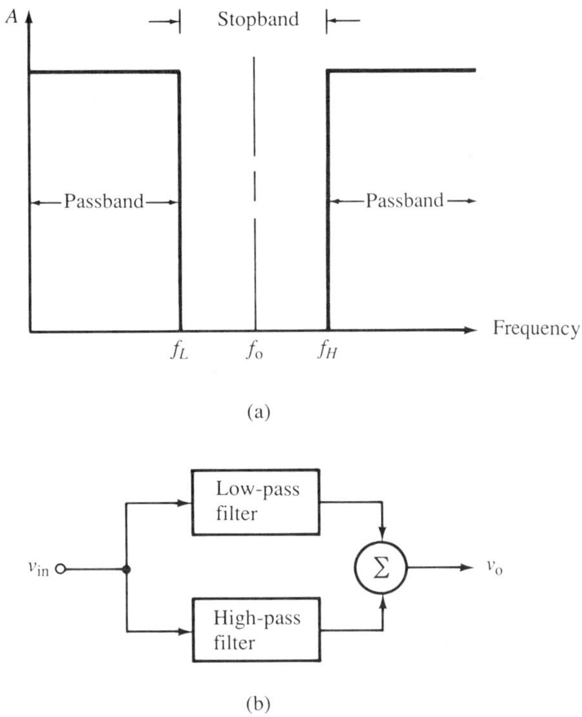

this median range, the filter passes the level of the input signal with no attenuation, although there may be some gain. Here, there are *two passbands* and only *one stopband*. The notch filter response is basically created by summing the outputs of both a low-pass and a high-pass filter having the same cutoff frequency (Figure 11–5b).

Unlike the bandpass filter, the stopband extends from a given lower cutoff frequency (f_L) to the higher cutoff frequency (f_H). The two passbands are on either side of the stopband. Furthermore, the notch filter is primarily used to remove a signal at a single frequency, such as 60-Hz noise from an audio amplifier. Like bandpass filters, however, the frequency difference between the lower and upper cutoff frequencies is the bandwidth. Table 11–2 summarizes the opposite relationships that exist between bandpass and notch filters.

Unlike the ideal notch response of Figure 11–5a, practical notch filters have a response like that of Figure 11–6. Here, the response bottoms out at the center frequency (f_o), which is also often referred to as the *notch*, or *null frequency*. The value of the decibel response at the notch frequency is sometimes called the *notch depth*, or *depth of null*. On either side of the notch frequency, the response

Table 11–2
Comparison of bandpass and notch filter characteristics

Parameter	Bandpass	Notch
Passband	$f_L < f_{in} < f_H$	$f_L < f_{in}, f_{in} > f_H$
Stopband	$f_L < f_{in}, f_{in} > f_H$	$f_L < f_{in} < f_H$

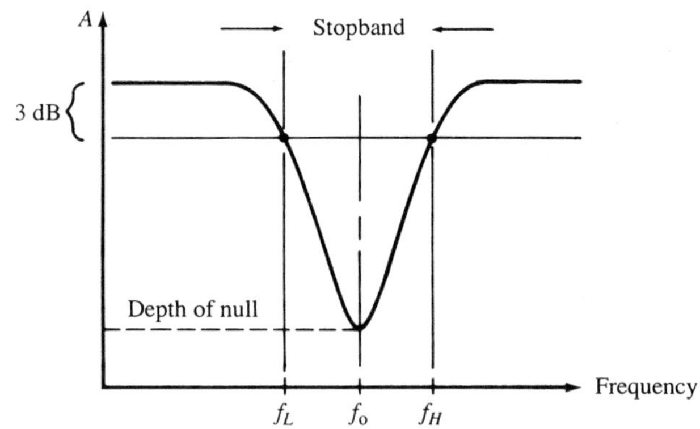

Figure 11–6
Realistic notch filter frequency response.

increases in a symmetrical fashion. In terms of the filter's Q, center frequency, and passband gain (A_o), the dB-amplitude response of such a notch filter is given by

$$A_{dB} = A_o(\text{dB}) + 20 \log_{10} \left\{ \frac{1 - (f_{in}/f_o)^2}{[1 - (f_{in}/f_o)^2]^2 + (f_{in}/Qf_o)^2} \right\} \qquad \textbf{(11–9)}$$

Since the response curve is symmetrical about the center frequency, the dB-amplitude response, as with the bandpass filter, will be the same for the ratio f_{in}/f_o as for its reciprocal, f_o/f_{in}.

11–2 Filter Response and Selectivity

How well a given filter rejects undesired input signals is called *selectivity*. The ability of an active filter to be more selective in passing the desired input signals depends on several factors. The frequency responses for the four basic filter types shown earlier in the chapter are very idealized. In the following subsections, we discuss the comparisons among various filter response shapes, which involve tradeoffs between passband and stopband response characteristics, as well as a discussion of how the filter's selectivity can be improved by increasing the roll-off. The various major response shapes are

1. Butterworth
2. Chebyshev
3. Elliptic
4. Bessel

Of these four, the first two are the most often used and are discussed in more detail than the remaining two types.

Butterworth Filters

The most common class of filters exhibits a frequency response shape known as the **Butterworth,** or **maximally flat, response.** As illustrated in Figure 11–7a and b, the Butterworth low-pass and high-pass responses are both characterized by a very flat passband. For low-pass filters, the passband is flat up to the vicinity of

Figure 11–7
Butterworth, or maximally flat frequency responses. (a) Low-pass filter. (b) High-pass filter.

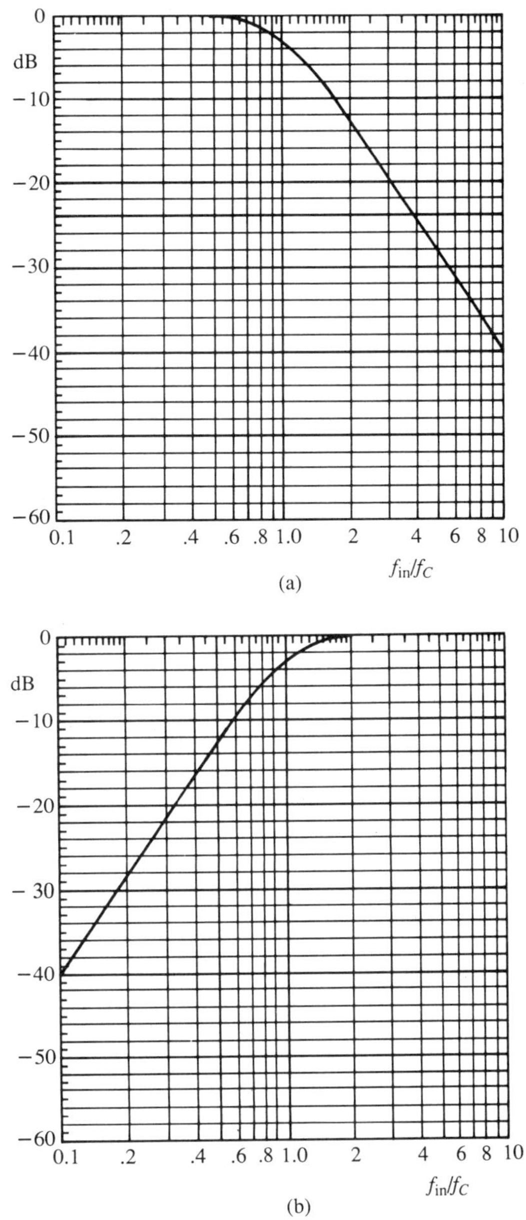

the filter's cutoff frequency and then decreases *linearly* with increasing frequency. For a Butterworth high-pass filter, the passband is flat down to the vicinity of the filter's cutoff frequency and decreases linearly with decreasing frequency. The passband voltage gain is usually 1 or greater. Furthermore, it is always assumed that the axis on which frequency is graphed is logarithmic.

The cutoff frequency is the dividing line between the filter's passband and stopband. In the stopband for low-pass filters, the amplitude response decreases at a virtual constant rate (in high-pass filters, it *increases* linearly). This increase or decrease is called the **roll-off** and is calculated as the slope of the line. In terms of octaves or decades, the filter's roll-off is standardized in terms of a given number of decibels per octave or decibels per decade. Since the decibel represents the unit of the vertical axis (amplitude response, A) and the octave or decade represents the measure of the horizontal axis (frequency, f), the roll-off in units of decibel per octave or decibel per decade then represents the slope of the line, $\Delta A/\Delta f$. The roll-off, in turn, depends on several factors. As is discussed in this and later subsections of this chapter, the filter's roll-off depends on (1) the *order* of the filter and (2) the type of filter response.

The general form of the frequency response of a Butterworth low-pass filter is given by

$$A = \frac{A_{PB}}{\sqrt{1 + (f_{\text{in}}/f_C)^{2n}}} \tag{11–10}$$

or, in terms of decibels,

$$A_{\text{dB}} = A_{PB}(\text{dB}) - 10 \log_{10}[1 + (f_{\text{in}}/f_C)^{2n}] \tag{11–11}$$

where

$$\begin{aligned} f_{\text{in}} &= \text{input frequency} \\ f_C &= \text{filter cutoff frequency} \\ A_{PB} &= \text{filter passband voltage gain} \\ n &= \text{order of the filter} \end{aligned}$$

The **order** of the filter controls the slope of the stopband. In effect, it is a measure of how selective the filter is. The higher the order, the more selective the filter is. The roll-off is therefore controlled by the filter's order. For Butterworth low-pass and high-pass filters, the roll-off is an integer multiple of $\pm 6n$ dB/octave or $\pm 20n$ dB/decade. A first-order low-pass Butterworth filter ($n = 1$) has a roll-off of -6 dB/octave or -20 dB/decade. A second-order ($n = 2$) filter has a roll-off that is twice as fast as first-order filter, or -12 dB/octave (-40 dB/decade), and so on. An infinite-order low-pass filter has a roll-off that is perfectly vertical, giving the ideal low-pass response of Figure 11–1.

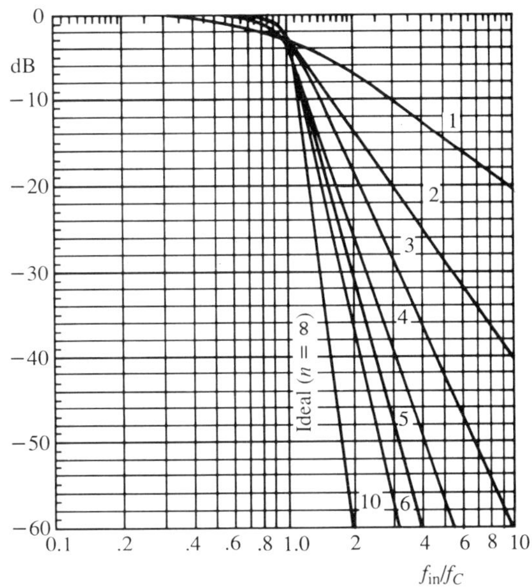

Figure 11–8
The effect of the filter's order on the selectivity of a Butterworth low-pass filter.

The effect of the order on the frequency response of a Butterworth low-pass filter is shown in Figure 11–8. In each case, the response is flat for most of the passband and starts to drop off near the cutoff frequency. At the cutoff frequency, where $f_{in} = f_C$, Equation 11–10 reduces to a value that is 0.707 times the passband voltage gain and is equivalent to 3 dB less than the decibel passband gain (Equation 11–11). Regardless of the filter's order, all responses intersect at the 3-dB point, where $f_{in} = f_C$. Although the higher-order filters are more effective in minimizing signals above the cutoff frequency, they are somewhat limited by their cost and the tolerance of the component values needed to *realize,* or make such a filter.

Example 11–4
A given third-order Butterworth low-pass filter is to have a passband voltage gain of 1. If the input is a 1.5-V RMS, 1800-Hz sine wave, what is the cutoff frequency needed to have an output voltage of 0.32 V RMS?

Solution: At 1800 Hz, the filter response must then be 0.32 V/1.5 V, or 0.213, or −13.43 dB. From Equation 11–10,

$$0.213 = \frac{1}{\sqrt{1 + (1800/f_C)^6}}$$

Taking the reciprocal and squaring both sides,

$$22.041 = 1 + \left(\frac{1800}{f_C}\right)^6$$

or

$$21.041 = \left(\frac{1800}{f_C}\right)^6$$

Taking the log of both sides,

$$1.323 = 6 \log\left(\frac{1800}{f_C}\right)$$
$$0.221 = \log(1800) - \log(f_C)$$
$$0.221 = 3.255 - \log(f_C)$$
$$3.034 = \log(f_C)$$

Finally, solving for f_C by taking the antilog (inverse log) of both sides, we obtain (after rounding)

$$f_C = \mathbf{1081\ Hz}$$

The frequency response of a Butterworth high-pass filter is calculated by reversing the f_{in}/f_C term in the low-pass filter equation (Equation 11–10), so we now have

$$A = \frac{A_{PB}}{\sqrt{1 + (f_C/f_{in})^{2n}}} \qquad \textbf{(11–12)}$$

or, in terms of decibels:

$$A_{dB} = A_{PB}(dB) - 10 \log_{10}[1 + (f_C/f_{in})^{2n}] \qquad \textbf{(11–13)}$$

Example 11–5

A Butterworth high-pass filter has a passband gain of +4.5 dB and a cutoff frequency of 800 Hz. What would be the order of the filter required so that the response would be 26 dB less than the passband gain at 500 Hz?

Solution: A response that is 26 dB less than a 4.5-dB passband is 4.5 dB − 26 dB, or −21.5 dB. Using Equation 11–13,

$$-21.5\ dB = 4.5\ dB - 10 \log_{10}\left[1 + \left(\frac{800\ Hz}{500\ Hz}\right)^{2n}\right]$$

$$-26\ dB = -10 \log_{10}\left[1 + \left(\frac{800\ Hz}{500\ Hz}\right)^{2n}\right]$$

Dividing by 10 and rearranging yields

$$2.6\text{ dB} = \log_{10}\left[1 + \left(\frac{800\text{ Hz}}{500\text{ Hz}}\right)^{2n}\right]$$

$$398.1 = 1 + \left(\frac{800\text{ Hz}}{500\text{ Hz}}\right)^{2n}$$

$$\log_{10}(397.1) = 2n \log_{10}(1.6)$$

$$2.599 = 2n(0.204)$$

Solving for n yields

$$n = \frac{2.599}{(2)(0.204)} = 6.4$$

Since n must be an integer, n must then be rounded up to $n = 7$ to give a minimum response that is 26 dB less than the passband gain of 4.5 dB. In fact, for a seventh-order filter at 500 Hz, the response will be -24.1 dB, which is $-24.1 - 4.5$, or -28.6 dB, down from the passband. A sixth-order filter would not have the required minimum response.

Compared with other types of filter responses, the Butterworth response is classified as having the best overall characteristics. It has the flattest passband characteristic as well as a moderately fast rolloff rate.

Chebyshev Filters

Another class of low-pass and high-pass filters exhibits a frequency response shape known as the **Chebyshev,*** or **equal-ripple response.** As illustrated in Figure 11–9a and b, the Chebyshev low-pass and high-pass responses are both characterized by a passband having ripples of equal amplitude and a faster initial rolloff rate than the Butterworth response.

The general form of the frequency response of a Chebyshev low-pass filter is given by

$$A = \frac{A_{PB}}{\sqrt{1 + (\varepsilon C_n)^2}} \qquad \textbf{(11–14)}$$

*The basis of the equal-ripple response is from the work of P. L. Chebyshev, who developed what are now known as *Chebyshev polynomials*. Because of the inexact transliteration from the Russian language, Chebyshev is also spelled as Chebychev, Chebishev, and Tchebysheff in other books and magazine articles.

Figure 11–9
Chebyshev, or equal-ripple frequency responses. (a) Low-pass filter. (b) High-pass filter.

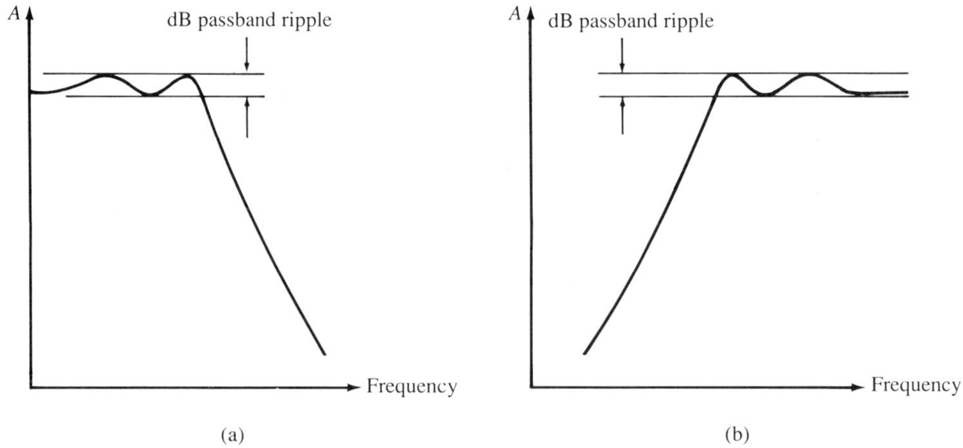

or, in terms of decibels,

$$A_{\text{dB}} = A_{PB}(\text{dB}) - 10\log_{10}[1 + (\varepsilon C_n)^2] \qquad \textbf{(11–15)}$$

where C_n is a Chebyshev polynomial* of order n and ε is a constant that relates the decibel amplitude of the passband ripple (A_r):

$$\varepsilon = \sqrt{10^{(0.1)(A_r)} - 1} \qquad \textbf{(11–16)}$$

The passband ripples are measured from their minimum to maximum values and are of constant amplitude. They are selected to vary from as low as 0.1 dB to

---

*Letting $\alpha = f_{\text{in}}/f_C$ for a low-pass filter and $\alpha = f_C/f_{\text{in}}$ for a high-pass filter, the n ordered Chebyshev polynomials C_n are defined as:

$$C_0 = 1$$
$$C_1 = \alpha$$
$$C_2 = 2\alpha^2 - 1$$
$$C_3 = 4\alpha^3 - 3\alpha$$
$$C_4 = 8\alpha^4 - 8\alpha^2 + 1$$
$$C_5 = 16\alpha^5 - 20\alpha^3 + 5\alpha$$
$$C_6 = 32\alpha^6 - 48\alpha^4 + 18\alpha^2 - 1$$
$$\vdots$$
$$C_{n+1} = 2\alpha C_n - C_{n-1}$$

Figure 11–10
Determining the order of a Chebyshev filter by counting the number of peaks and valleys in the passband. (a) Third-order filter. (b) Fourth-order filter.

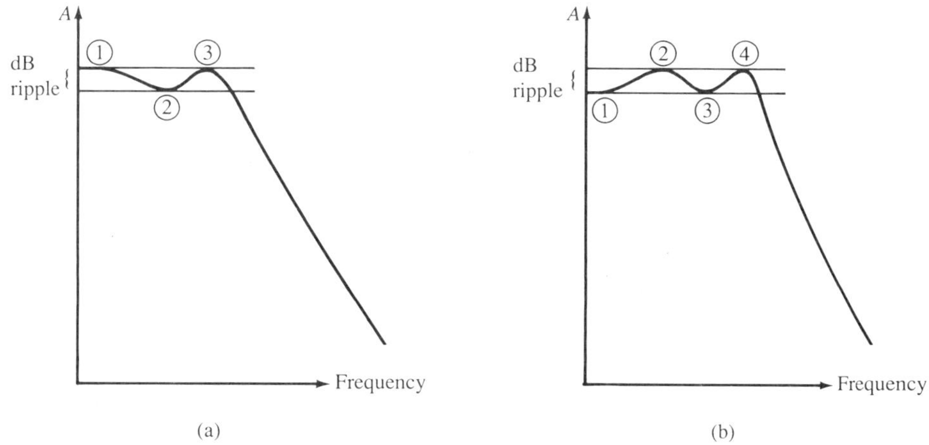

as much as 3 dB in amplitude. As shown in Figure 11–10a and b, the number of ripples in the passband equals the order of the filter. For odd-ordered Chebyshev filters, the passband response (i.e., the first ripple at DC) starts at its maximum value. As the frequency approaches the cutoff frequency, the ripple distance between the ripples increases. For even-ordered filters, the first ripple starts at the minimum value of the ripple's amplitude.

Although the mathematics concerning Chebyshev filters is beyond the scope of this text, some of the important results can nevertheless be summarized here without derivation. First, for large values of f_{in}/f_C in the stopband, Equation 11–13 can be simplified to the approximate relation

$$A_{dB} \simeq -20 \log_{10}(\varepsilon) - 20n \log_{10}\left(\frac{f_{in}}{f_C}\right) - 6(n - 1) \qquad \textbf{(11–17)}$$

Clearly, this approximation is dependent on values of ε (from Equation 11–16) and n.

Example 11–6
A fourth-order Chebyshev low-pass filter has a passband ripple of 1.0 dB and a cutoff frequency of 500 Hz. Determine the dB-amplitude response at 1.5 kHz.

Solution: First, using Equation 11–16,

$$\varepsilon = \sqrt{10^{(0.1)(1\ \text{dB})} - 1}$$
$$= 0.509$$

Then substituting ε into Equation 11–17,

$$\begin{aligned} A_{dB} &\simeq -20\log_{10}(0.509) - (20)(4)\log_{10}\left(\frac{1500}{500}\right) - (6)(4-1) \\ &\simeq 5.87\text{ dB} - 38.17\text{ dB} - 18\text{ dB} \\ &\simeq \mathbf{-50.3\ dB} \end{aligned}$$

The formula used to calculate the order n of a filter necessary for realizing a Chebyshev low-pass response is given as

$$n = \frac{\cosh^{-1}[(10^{0.1A_r} - 1)/(10^{0.1A_{dB}} - 1)]^{1/2}}{\cosh^{-1}(f_{in}/f_C)} \qquad \textbf{(11–18)}$$

where A_{dB} is the dB-stopband response at f_{in}.

When the Chebyshev low-pass response is compared to the Butterworth response having the same order, cutoff frequency, and input frequency, we find that the attenuation of the Chebyshev filter will be greater than the Butterworth filter. Figure 11–11a and b illustrate the comparisons of Chebyshev and Butterworth responses for third- and fourth-order low-pass filters. After an initial stopband attenuation of $20\log_{10}(\varepsilon) + 6(n-1)$ dB, the Chebyshev response also rolls off at $-6n$ dB/octave ($20n$ dB/decade).

As both the order and passband ripple of a Chebyshev filter are increased, the filter has a faster initial roll-off rate but a poorer transient response. In comparison with Butterworth filters, a lower-order Chebyshev filter will usually give the same stopband attenuation. However, the flat Butterworth characteristic passband is traded off against the faster roll-off and a small amount of passband ripple.

Elliptic Filters

As discussed previously, the Chebyshev filter yields a greater stopband response than the Butterworth filter. However, for both, the stopband attenuation continues at a rate of $6n$ dB/octave or $20n$ dB/decade beyond the cutoff frequency. The **elliptic filter,*** on the other hand, has equal-amplitude ripples in the passband, as does the Chebyshev, but also has ripples in the stopband, as shown in Figure 11–12.

The stopband roll-off for the elliptic filter is much steeper than either the Butterworth or Chebyshev. However, the trade-off is the response in the stopband is allowed to bounce back up and then continue either at a constant attenuation with frequency or gradually fall off with frequency. As in the case of the Chebyshev filter, the mathematics behind the elliptic filter is well beyond the scope of this text.

*The elliptic filter is often referred to as a **Cauer filter.** Because it is based on the ratio of two polynomials called *Jacobian elliptical functions*, this type filter is sometimes called a *Jacobian filter*.

Figure 11–11
Comparison of Chebyshev low-pass filter frequency response with a Butterworth filter of the same order and cutoff frequency. (a) Third order. (b) Fourth order.

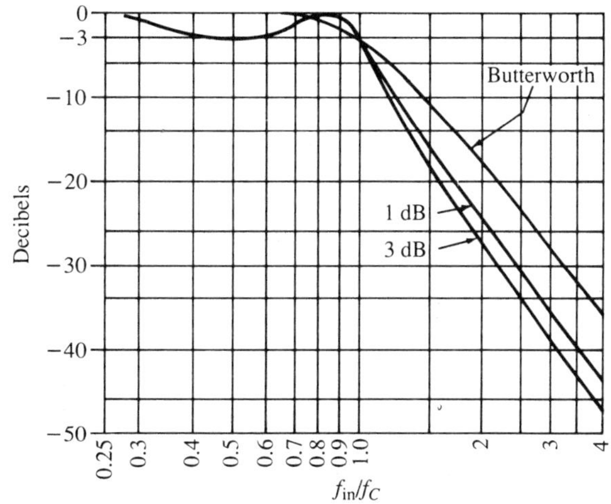

(a)

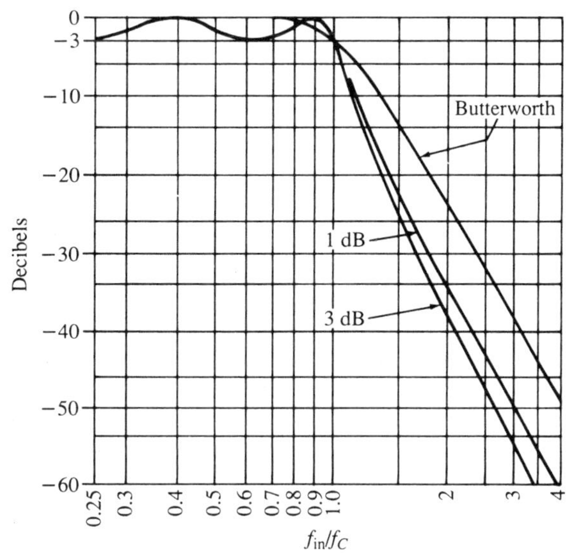

(b)

Bessel Filters

The last of the major filter types is the **Bessel filter,** which is also referred to as the *best time-delay filter*. Unlike the Butterworth, Chebyshev, and Cauer filters, the Bessel filter is not intended for sharp stopband responses; instead, it is used to provide a relatively linear time delay, or phase-frequency response. Here, the phase response with frequency is of primary importance rather than the amplitude. Figure 11–13 compares the amplitude responses of fourth-order Butterworth, Chebyshev (1- and 3-dB ripple), and Bessel low-pass filters.

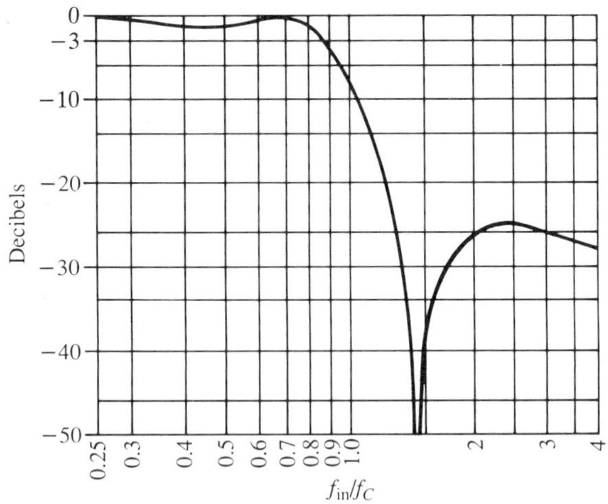

Figure 11–12
Frequency response of an elliptic, or Cauer, low-pass filter.

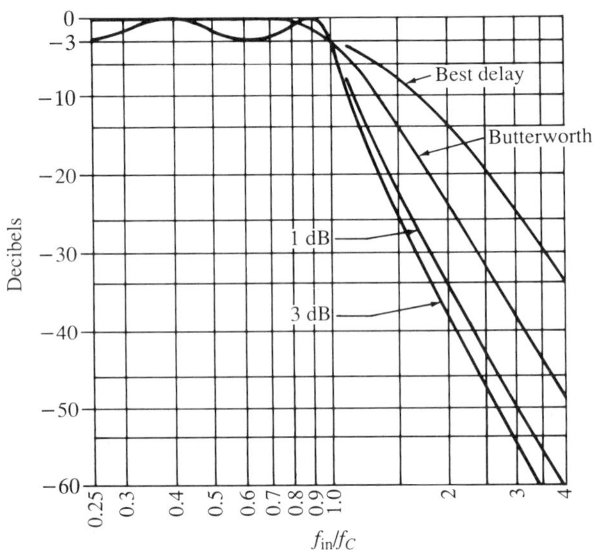

Figure 11–13
Comparison of the frequency responses of 4th-order Butterworth, Chebyshev (1 and 3-dB ripple), and Bessel low-pass filters.

11–3 Active Filter Circuits

The basic active filter networks with which it is possible to realize various responses are

- Sallen-Key: low-pass and high-pass
- Multiple-feedback: low-pass, high-pass, and bandpass
- Twin-T: bandpass and notch

These are all basic 2nd-order RC networks used as building blocks in the formation of higher-order filters. In addition, there are other specialized filter circuits, such as the state-variable filter, that are available. All active filter circuits that are discussed here use the operational amplifier powered by a dual-supply voltage.

Sallen-Key Networks

The basic **Sallen-Key filter** configuration, also called a *voltage-controlled voltage source* (VCVS), is shown in Figure 11–14a. Depending on the location of the specific frequency-determining resistors and capacitors for the four impedances (Z_1 through Z_4), it is possible to have a low-pass filter section (Figure 11–14b) and, by interchanging the position of the RC components, create the equivalent high-pass network (Figure 11–14c).

Figure 11–14
2nd-order Sallen-Kay filter networks. (a) General network. (b) Low-pass filter. (c) High-pass.

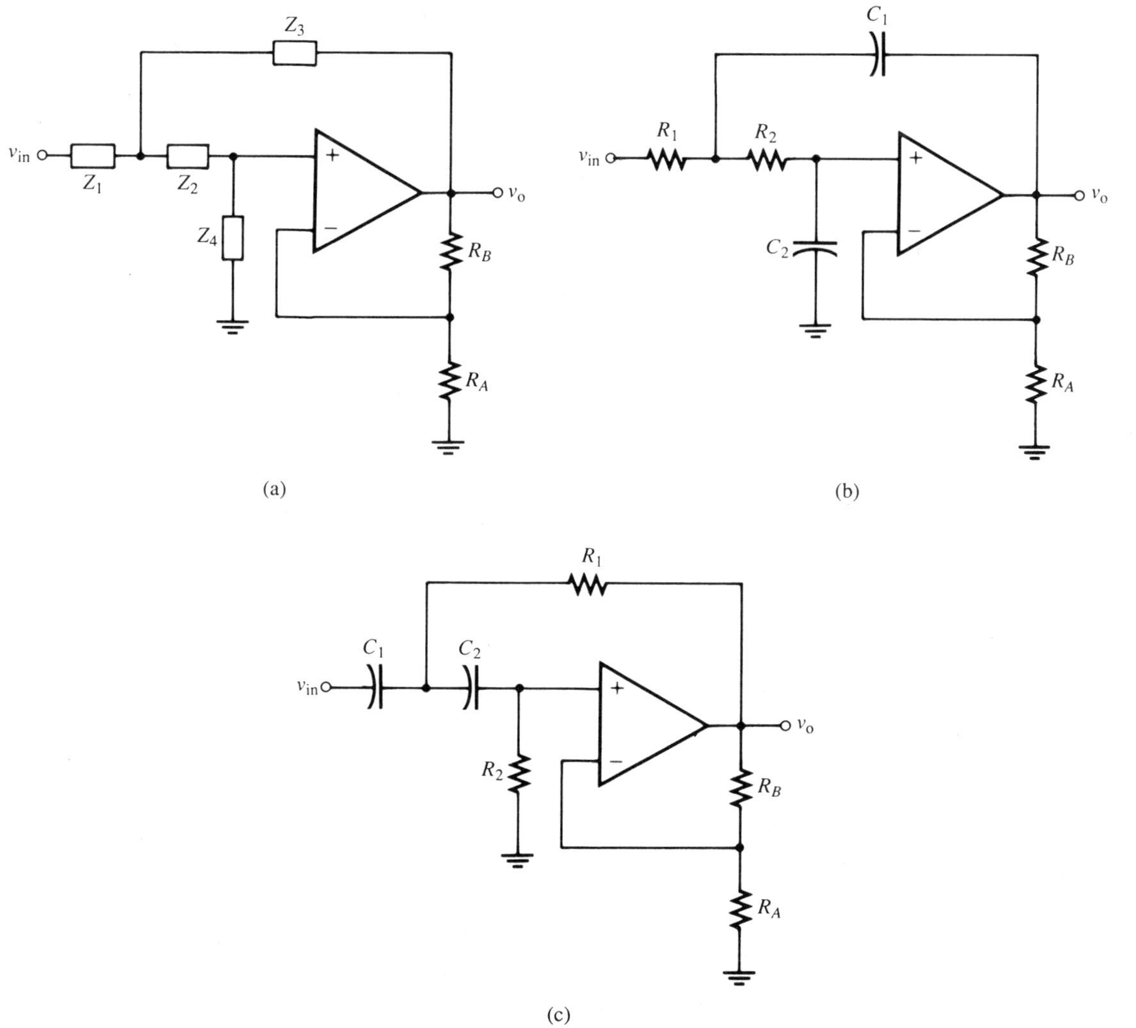

Based on the two resistors and capacitors, the cutoff frequency for both the low-pass and high-pass filters is given by

$$f_C = \frac{1}{2\pi\sqrt{(R_1 R_2 C_1 C_2)}} \tag{11–19}$$

As a general rule for convenience, R_1 is made equal to R_2 and C_1 is made equal to C_2. This then results in what is called an *equal-component-value VCVS filter*. Equation 11–19 then reduces to

$$f_C = \frac{1}{2\pi R_1 C_1} \tag{11–20}$$

A standard value for C_1 is chosen, and the resulting value for R_1 is calculated for a given cutoff frequency from Equation 11–20. The *gain-setting resistors* R_A and R_B are chosen to give either a Bessel, Butterworth, or Chebyshev response. Usually these two resistors are determined from the ratio R_B/R_A. For second-order low-pass and high-pass filters, these ratios are summarized in Table 11–3. The required cutoff frequency for the filter section, as a factor of the desired cutoff frequency, is also listed in Table 11–3.

Although Table 11–3 gives the necessary parameters for second-order Bessel and Chebyshev filters, these second-order responses are rarely used by themselves alone. The effectiveness of Bessel and Chebyshev filters is more pronounced in higher-order low-pass and high-pass filters.

Example 11–7
Using the Sallen-Key circuit of Figure 11–14b, determine the component values for a second-order 1-dB Chebyshev low-pass filter having a cutoff frequency of 1500 Hz.

Solution:

(a) For the low-pass filter, the cutoff frequency of the filter network must be 0.863 × 1500 Hz, or 1294.5 Hz. Assuming a standard value for C_1, such as **0.022 μF,** then from Equation 11–20,

$$R_1 = \frac{1}{(2\pi)(0.022\ \mu\text{F})(1294.5\ \text{Hz})}$$
$$= \mathbf{5591\ \Omega} \quad \text{(use a 5600-}\Omega\text{ standard value)}$$

(b) Assuming a standard value for R_A, such as **12 kΩ,** then R_B must be 0.955 × 12 kΩ, or **11.5 kΩ.** The final circuit is shown in Figure 11–15.

Table 11–3
Second-order VCVS filter parameters

Response	R_B/R_A	Low-pass	High-pass	Passband Gain
Bessel	0.268	$1.274f_C$	$0.785f_C$	1.27 (2.1 dB)
Butterworth	0.586	$1.000f_C$	$1.000f_C$	1.59 (4.0 dB)
1-dB Chebyshev	0.955	$0.863f_C$	$1.159f_C$	1.96 (5.8 dB)
2-dB Chebyshev	1.105	$0.852f_C$	$1.174f_C$	2.11 (6.5 dB)
3-dB Chebyshev	1.233	$0.841f_C$	$1.189f_C$	2.23 (7.0 dB)

Higher-Order Filters

To achieve rolloff rates higher than that afforded by the basic 2nd-order VCVS filter, higher-order filters are formed by cascading, or connecting in series, 1st- and 2nd-order filter sections, or stages.

The basic first-order low-pass and high-pass filter sections are shown in Figure 11–16. In both cases, the active filter section is nothing more than a passive RC filter network ahead of a voltage follower. Here, the passband voltage gain is unity and the roll-off is 6 dB/octave (or 20 dB/decade). In both cases, the cutoff frequency is the same as Equation 11–20.

Figure 11–17 illustrates the scheme used to create higher-order low-pass and high-pass filters using first- and second-order sections. Odd-order filters are achieved by cascading one first-order section (Figure 11–16) with one or more second-order sections (Figure 11–14). For example, a fifth-order high-pass filter is

Figure 11–15
Completed 2nd-order low-pass filter for Example 11–7.

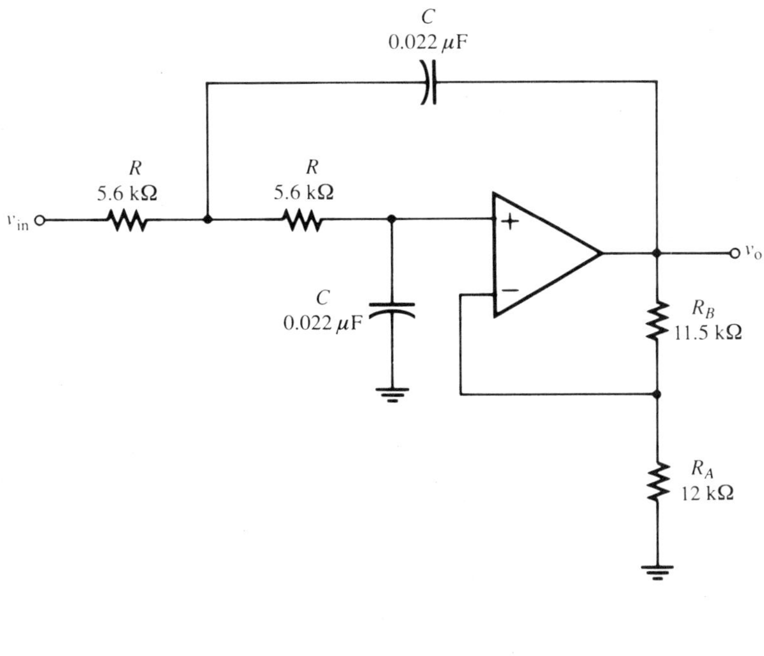

Figure 11–16
1st-order filter networks. (a) Low-pass. (b) High-pass.

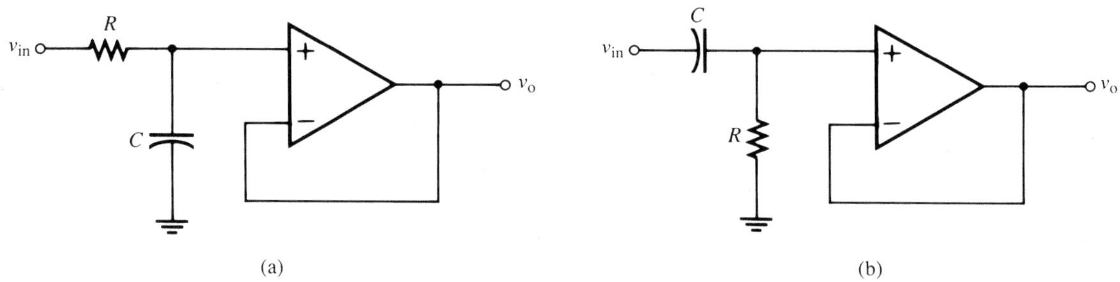

built by cascading one first-order high-pass filter section with two second-order high-pass filter sections. Here the order of the resultant filter ($n = 5$) is equal to the sum of the orders of the individual sections, so that $1 + 2 + 2 = 5$.

For even-ordered filters, only second-order filter sections are used. Therefore, for an nth-order low-pass (or high-pass filter), only $n/2$ second-order sections are required. For example, a sixth-order high-pass filter requires 6/2, or 3, second-order high-pass sections.

The design of higher-order active filters is easily done by using tables that summarize the required design factors, as shown in Table 11–4. Here, the required cutoff frequencies for each filter stage are given for low-pass filters as a

Figure 11–17
Creation of higher-order filters by cascading first- and second-order filter sections.

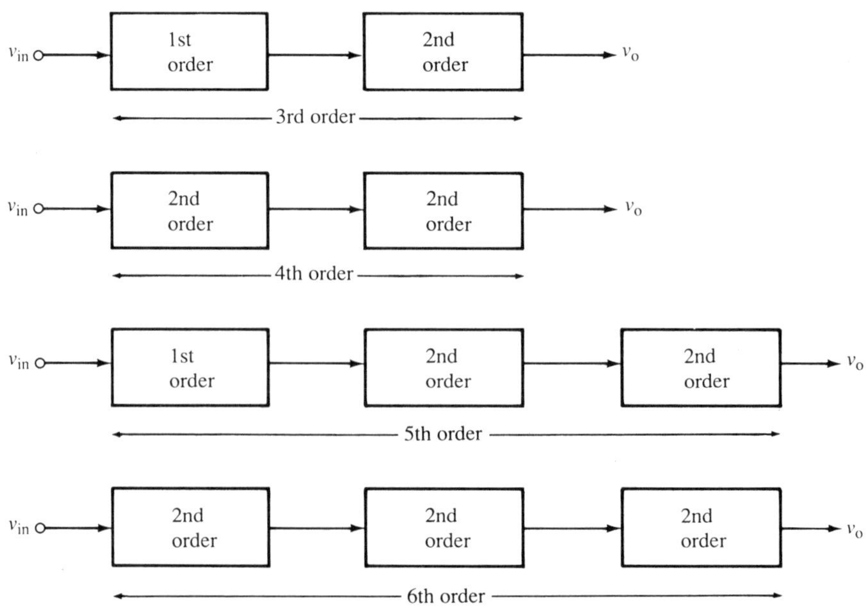

factor of the overall filter's cutoff frequency. From the table, the following characteristics should be noted:

1. To obtain the required frequencies for the corresponding high-pass filter, the reciprocal of the cutoff frequency factor is taken.
2. The first stage for all odd-order filters is always a first-order stage. The rest are second-order stages.
3. For first-order filter stages, there is no value for R_B/R_A given because the operational amplifier is connected as a voltage follower.
4. For Butterworth filters, the cutoff frequency for all stages is identical and is the same as the cutoff frequency of the overall higher-order filter.

The following three examples illustrate the use of Table 11–4.

Example 11–8

Determine the circuit and component values for a third-order, 1-dB Chebyshev low-pass filter having a cutoff frequency of 847 Hz.

Solution:

(a) From Table 11–4, the cutoff frequency for the first-order first stage (Figure 11–16a) is

$$f_{C'} = (0.452)(847 \text{ Hz}) = 382.8 \text{ Hz}$$

Assuming a value for C_1, such as $C_1 = \mathbf{0.033\ \mu F}$, then

$$R_1 = \frac{1}{(2\pi)(382.8 \text{ Hz})(0.033\ \mu\text{F})} = \mathbf{12.6\ k\Omega} \quad \text{(use a 12-k}\Omega\text{ standard value)}$$

(b) For the second-order VCVS section, the cutoff frequency for the second stage is

$$f_{C'} = (0.911)(847 \text{ Hz}) = 771.6 \text{ Hz}$$

Assuming a value for C_1, such as $C_1 = \mathbf{0.01 \mu F}$, then

$$R_1 = \frac{1}{(2\pi)(771.6 \text{ Hz})(0.1\ \mu\text{F})} = \mathbf{2.06\ k\Omega} \quad \text{(use a 2-k}\Omega\text{ standard value)}$$

(c) If a value for R_A is chosen, such as $R_A = \mathbf{10\ k\Omega}$, then

$$R_B = (1.504)(10\ \text{k}\Omega) = \mathbf{15.04\ k\Omega} \quad \text{(use a 15-k}\Omega\text{ standard value)}$$

The completed third-order filter circuit composed of the first- and second-order stages cascaded together is shown in Figure 11–18.

Table 11–4
Higher-order low-pass filter parameters using cascaded filter sections

Bessel Response

	First Stage		Second Stage		Third Stage		
Order	R_B/R_A	$f_{C'}$*	R_B/R_A	$f_{C'}$*	R_B/R_A	$f_{C'}$*	**Overall Passband Decibel Gain**
3	—	1.328	0.553	1.454			3.8
4	0.084	1.436	0.759	1.610			5.6
5	—	1.557	0.225	1.613	0.909	1.819	7.4
6	0.041	0.298	0.364	0.722	1.023	0.975	9.2

Butterworth Response

	First Stage		Second Stage		Third Stage		
Order	R_B/R_A	$f_{C'}$†	R_B/R_A	$f_{C'}$†	R_B/R_A	$f_{C'}$†	**Overall Passband Decibel Gain**
3	—	1	1.000	1			6.0
4	0.152	1	1.235	1			8.2
5	—	1	0.382	1	1.382	1	10.3
6	0.068	1	0.586	1	1.482	1	12.5

1-dB Chebyshev Response

	First Stage		Second Stage		Third Stage		
Order	R_B/R_A	$f_{C'}$‡	R_B/R_A	$f_{C'}$‡	R_B/R_A	$f_{C'}$‡	**Overall Passband Decibel Gain**
3	—	0.452	1.504	0.911			8.0
4	0.725	0.502	1.719	0.943			13.4
5	—	0.280	1.286	0.714	1.820	0.961	16.2
6	0.686	0.347	1.545	0.733	1.875	0.977	21.8

2-dB Chebyshev Response

	First Stage		Second Stage		Third Stage		
Order	R_B/R_A	$f_{C'}$‡	R_B/R_A	$f_{C'}$‡	R_B/R_A	$f_{C'}$‡	**Overall Passband Decibel Gain**
3	—	0.322	1.608	0.913			8.3
4	0.924	0.466	1.782	0.946			14.6
5	—	0.223	1.437	0.624	1.862	0.964	16.9
6	0.879	0.321	1.637	0.727	1.901	0.976	23.2

*The required factor for corresponding high-pass filter sections is obtained by taking the reciprocal.

†The required factor for corresponding high-pass filter sections is the same for all low-pass sections.

‡The required factor for corresponding high-pass filter sections is obtained by taking the reciprocal.

Table 11–4
continued

3-dB Chebyshev Response

	First Stage		*Second Stage*		*Third Stage*		
Order	R_B/R_A	$f_{C'}$‡	R_B/R_A	$f_{C'}$‡	R_B/R_A	$f_{C'}$‡	**Overall Passband Decibel Gain**
3	—	0.299	0.553	0.916			8.5
4	0.084	0.443	0.759	0.950			15.3
5	—	0.178	0.225	0.614	0.909	0.967	17.3
6	0.041	0.298	0.364	0.722	1.023	0.975	24.2

Figure 11–18
Completed third-order low-pass filter for Example 11–8.

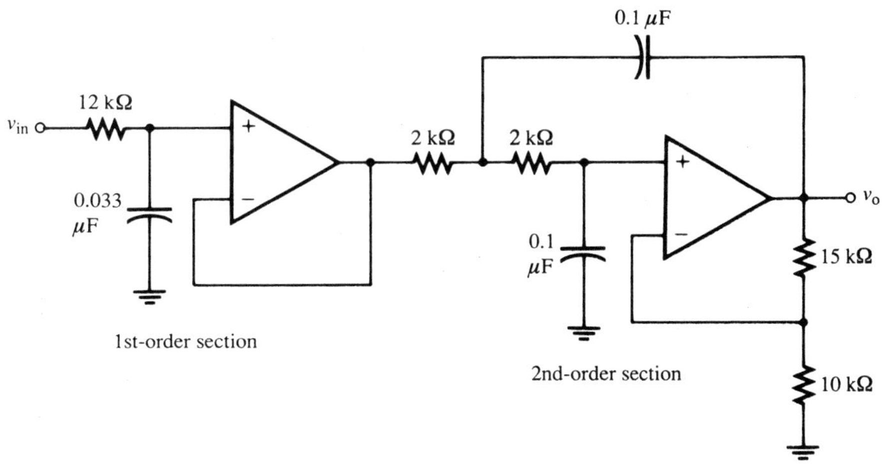

Example 11–9
Determine the circuit and component values for a fourth-order Butterworth low-pass filter having a cutoff frequency of 1200 Hz.

Solution:
(a) For both second-order sections, the cutoff frequency is the same as that for the overall fourth-order response, or 1200 Hz. Then assuming a value for C_1, such as C_1 = **0.01 μF** for both sections,

$$R_1 = \frac{1}{(2\pi)(1200\text{ Hz})(0.01\ \mu\text{F})}$$
$$= \mathbf{13.3\ k\Omega} \qquad \text{(use a 13-k}\Omega\text{ standard value)}$$

(b) For the first section, $R_B/R_A = 0.152$. If a value for R_A is chosen so that R_A = **10 kΩ,** then

$$R_B = (0.152)(10\ \text{k}\Omega)$$
$$= \mathbf{1.52\ k\Omega} \qquad \text{(use a 1.5-k}\Omega\text{ standard value)}$$

(c) For the second section, $R_B/R_A = 1.235$. If a value for R_A is chosen, such as R_A = **12 kΩ,** then

$$R_B = (1.235)(12\ \text{k}\Omega)$$
$$= \mathbf{14.8\ k\Omega} \qquad \text{(use a 15-k}\Omega\text{ standard value)}$$

The completed fourth-order filter circuit composed of two second-order stages cascaded together is shown in Figure 11–19.

Figure 11–19
Completed fourth-order low-pass filter for Example 11–9.

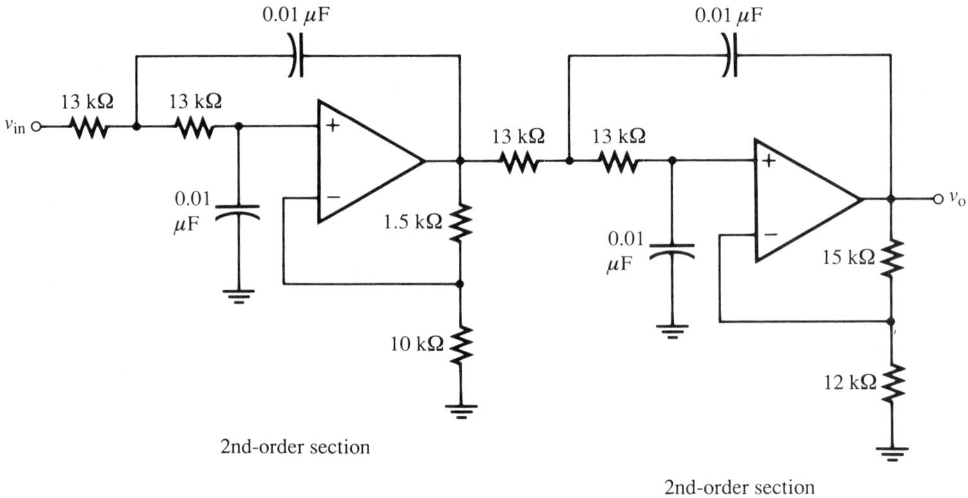

Example 11–10

Determine the circuit and component values for a fifth-order, 1-dB Chebyshev high-pass filter having a cutoff frequency of 350 Hz.

Solution:

(a) From Table 11–4, the cutoff frequency for the first stage is

$$f_{C'} = \left(\frac{1}{0.280}\right)(350\ \text{Hz})$$
$$= 1250\ \text{Hz}$$

Assuming a value for C_1, such as $C_1 = \mathbf{0.01\ \mu F}$, then

$$R_1 = \frac{1}{(2\pi)(1250\ \text{Hz})(0.01\ \mu\text{F})}$$
$$= \mathbf{12.7\ k\Omega} \qquad \text{(use a 13-k}\Omega\text{ standard value)}$$

(b) For the second stage, the cutoff frequency is

$$f_{C'} = \left(\frac{1}{0.714}\right)(350\ \text{Hz})$$
$$= 490.2\ \text{Hz}$$

Assuming a value for C_1, such as $C_1 = \mathbf{0.033\ \mu F}$, then

$$R_1 = \frac{1}{(2\pi)(490.2\ \text{Hz})(0.033\ \mu\text{F})}$$
$$= \mathbf{9.84\ k\Omega} \qquad \text{(use a 10-k}\Omega\text{ standard value)}$$

If a value for R_A is chosen so that $R_A = \mathbf{2.4\ k\Omega}$, then

$$R_B = (1.285)(2.4\ \text{k}\Omega)$$
$$= \mathbf{3.08\ k\Omega} \qquad \text{(use a 3-k}\Omega\text{ standard value)}$$

(c) For the last stage, the cutoff frequency is

$$f_{C'} = \left(\frac{1}{0.961}\right)(350\ \text{Hz})$$
$$= 364.2\ \text{Hz}$$

Assuming a value for C_1, such as $C_1 = \mathbf{0.022\ \mu F}$, then

$$R_1 = \frac{1}{(2\pi)(364.2\ \text{Hz})(0.022\ \mu\text{F})}$$
$$= \mathbf{19.9\ k\Omega} \qquad \text{(use a 20-k}\Omega\text{ standard value)}$$

If a value for R_A is chosen so that $R_A = \mathbf{1.2\ k\Omega}$, then

$$R_B = (1.820)(1.2\ \text{k}\Omega)$$
$$= \mathbf{2.18\ k\Omega} \qquad \text{(use a 2.2-k}\Omega\text{ standard value)}$$

The completed fifth-order filter circuit composed of first- and second-order stages cascaded together is shown in Figure 11–20.

Multiple-Feedback Networks

Besides the Sallen-Key arrangement, another basic circuit that can be used for low-pass, high-pass, and bandpass filters is the **multiple-feedback filter** network shown in Figure 11–21a. There are two obvious differences between this and the Sallen-Key circuit:

1. There is an additional feedback path—hence the name multiple-feedback filter.
2. The operational amplifier is connected in the inverting mode.

Figure 11–20
Completed fifth-order high-pass filter for Example 11–10.

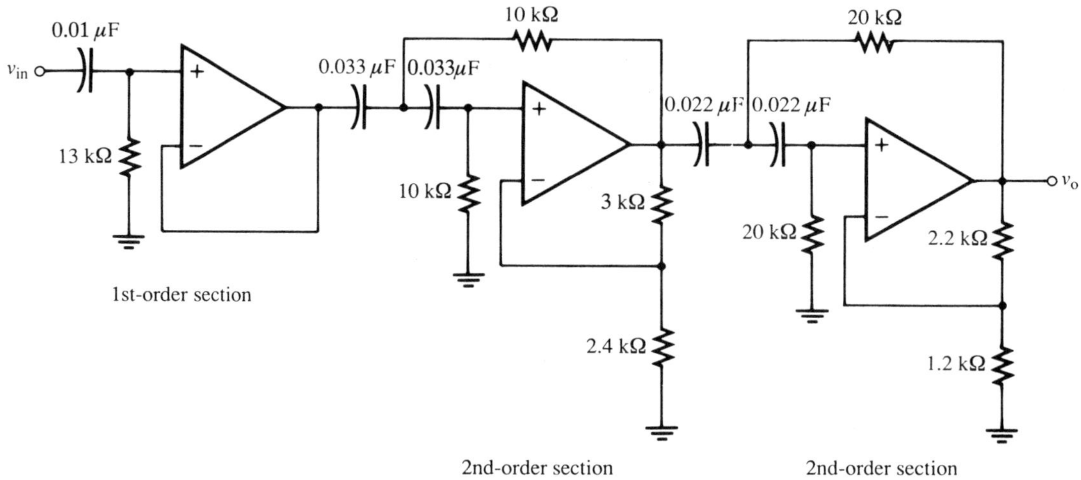

Figure 11–21
Second-order multiple-feedback filter networks. (a) General network. (b) Low-pass filter. (c) High-pass. (d) Bandpass filter.

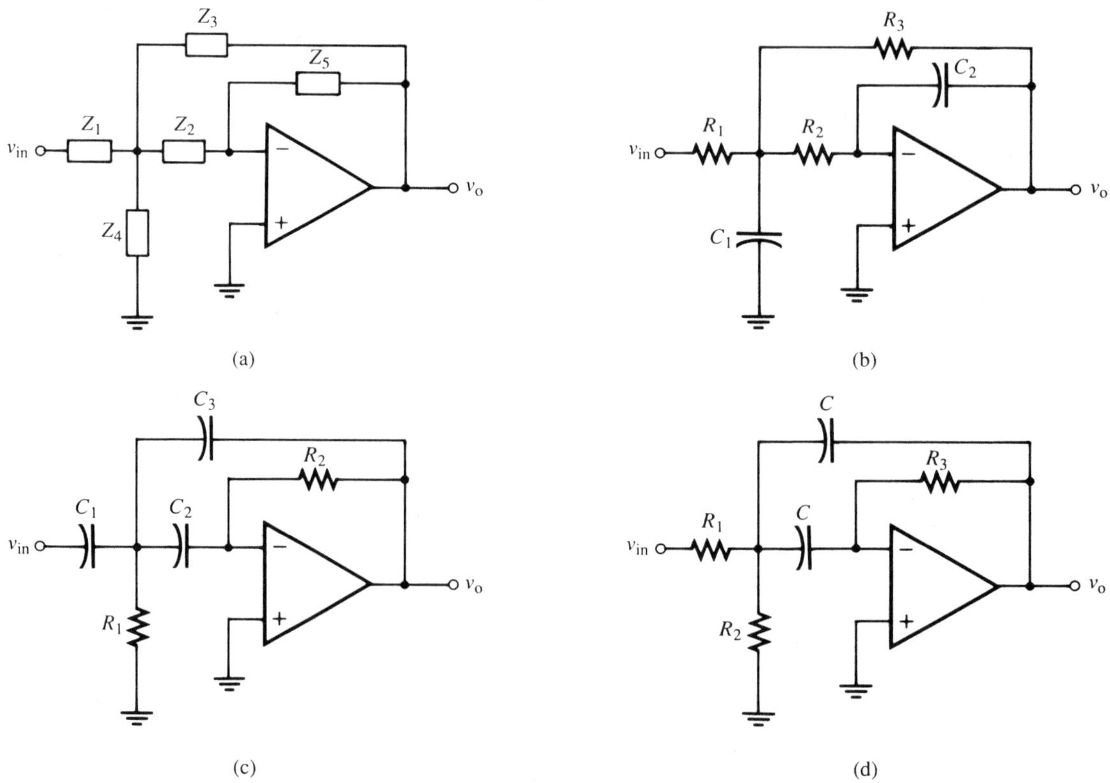

The cutoff frequency is determined from

$$f_C = \frac{1}{2\pi\sqrt{Z_2Z_3Z_4Z_5}} \tag{11–21}$$

However, because there are five component values that must be chosen to satisfy simultaneously requirements of passband gain, cutoff frequency, and the choice of either a Bessel, Butterworth, or Chebyshev response, there are no easy simplifications that can be made in the relationships among the five components. Whenever a simplification such as equal capacitor values is tried, some of the remaining requirements are then hard to achieve.

For these reasons, the multiple-feedback filter network is best suited for bandpass filters (Figure 11–21d) having values of Q up to approximately 15. Here, both capacitors are chosen to have the same value. The center frequency is given by

$$f_o = \frac{1}{2\pi C}\left(\frac{R_1 + R_2}{R_1R_2R_3}\right)^{1/2} \tag{11–22}$$

For given values of capacitance (C), Q, center frequency (f_o), and passband voltage gain A_o, the values of the three resistors are found from

$$R_1 = \frac{Q}{2\pi f_o C A_o} \tag{11–23a}$$

$$R_2 = \frac{Q}{2\pi f_o C(2Q^2 - A_o)} \tag{11–23b}$$

$$R_3 = \frac{Q}{\pi f_o C} \tag{11–23c}$$

Because of the interaction between Q and the center frequency gain in Equation 11–23b, the following must hold:

$$Q > \sqrt{\frac{A_o}{2}} \tag{11–24}$$

The center frequency gain is based on

$$A_o = \frac{R_3}{2R_1} \tag{11–25}$$

Example 11–11

Determine the required component values for the multiple-feedback bandpass filter of Figure 11–21d having a center frequency of 750 Hz, Q of 4.2, and a center frequency of 1.32.

Solution:

(a) We must first check using Equation 11–24 if we can have a Q of 4.2 with a passband gain of 1.32:

$$Q > \sqrt{\frac{1.32}{2}}$$

$$4.2 > 0.81$$

which is sufficient.

(b) Assuming a standard capacitor value of **0.01 μF,** the three resistors are determined using Equations 11–23a, 11–23b, and 11–23c (or Equation 11–25):

$$R_1 = \frac{4.2}{(2\pi)(750\ \text{Hz})(0.01\ \mu\text{F})(1.32)}$$

$$= \mathbf{67.6\ k\Omega} \qquad \text{(use a 68-k}\Omega\text{ standard value)}$$

$$R_2 = \frac{4.2}{(2\pi)(750\ \text{Hz})(0.01\ \mu\text{F})[(2)(4.2)^2 - 1.32)]}$$

$$= \mathbf{2.6\ k\Omega} \qquad \text{(use a 2.7-k}\Omega\text{ standard value)}$$

$$R_3 = (2)(67.6\ \text{k}\Omega)(1.32)$$

$$= \mathbf{178\ k\Omega} \qquad \text{(use a 180-k}\Omega\text{ standard value)}$$

The final circuit is shown in Figure 11–22.

From classical filter theory, one way of forming a notch filter is to subtract the output signal of a bandpass filter from its input signal. The basic multiple-feedback bandpass filter of Figure 11–21d can be combined with a two-input summing

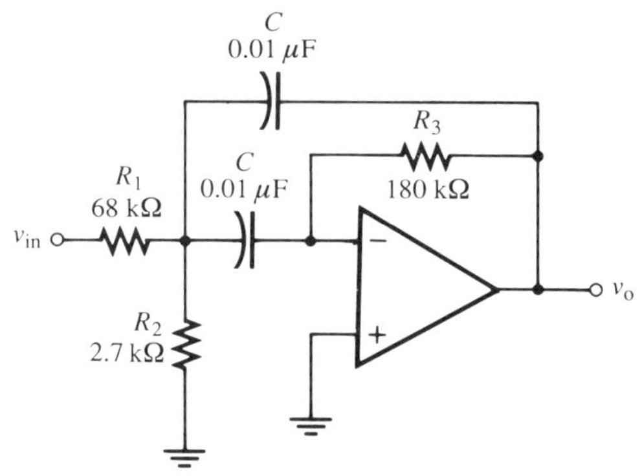

Figure 11–22
Completed multiple-feedback bandpass filter for Example 11–11.

Figure 11–23
Notch filter formed by summing the output of a bandpass filter network with the filter's input signal.

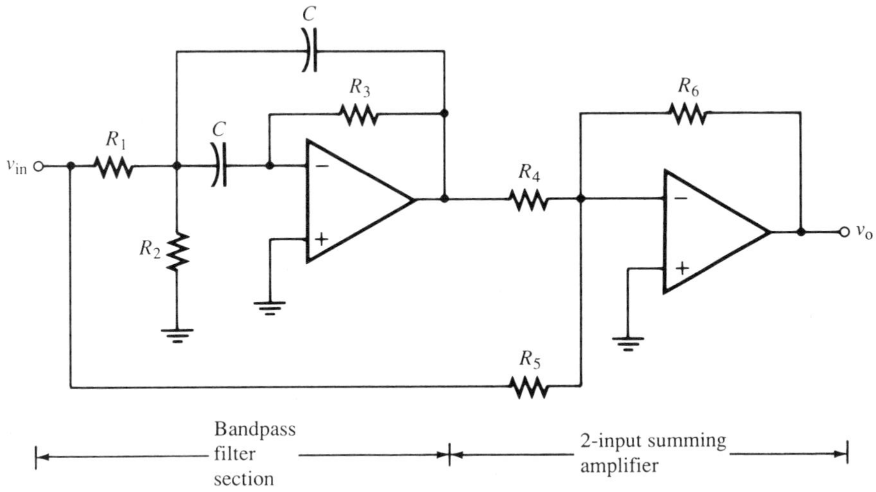

amplifier to form a notch filter circuit, as shown in Figure 11–23. Here, the design of the circuit is identical to the bandpass circuit for Q and f_o. At the bandpass filter section's center frequency, the output signal is 180° out of phase with the input. The summing amplifier, in effect, then subtracts the inverted bandpass output signal from the input signal of the filter. To produce the deepest null possible at the filter's center frequency, the following must hold:

$$\frac{R_5}{R_4} = \frac{R_3}{2R_1} \qquad (11\text{–}26)$$

and

$$R_5 = R_6 \qquad (11\text{–}27)$$

However, the right-hand term of Equation 11–26 is simply the same as the center frequency passband gain of the bandpass filter section (Equation 11–25). The overall passband gain of the resulting notch filter circuit is then equal to unity.

Twin-T Filters

The **twin-T,** or **parallel-T, network** is a popular one that can be used in the formation of both bandpass (Figure 11–24a) and notch filters (Figure 11–24b). For these circuits to work properly, the two capacitors in the "upper T" should each be exactly twice the value of the capacitor in the "lower T" network. For resistors, the opposite is true. Based on R and C, the center frequency is found from

Figure 11–24
Twin-T filters. (a) Bandpass. (b) Notch.

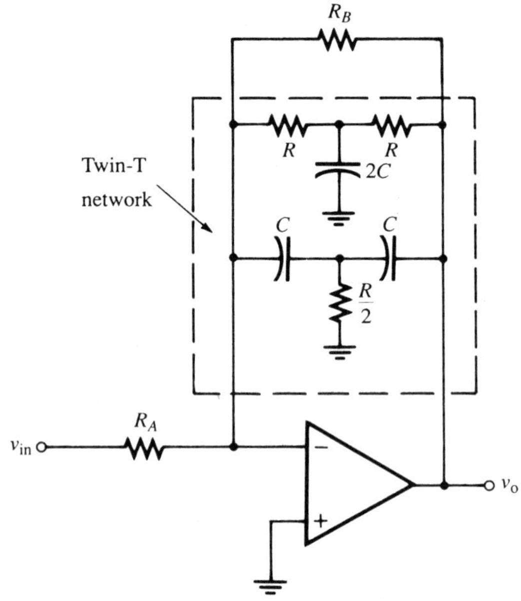

(a)

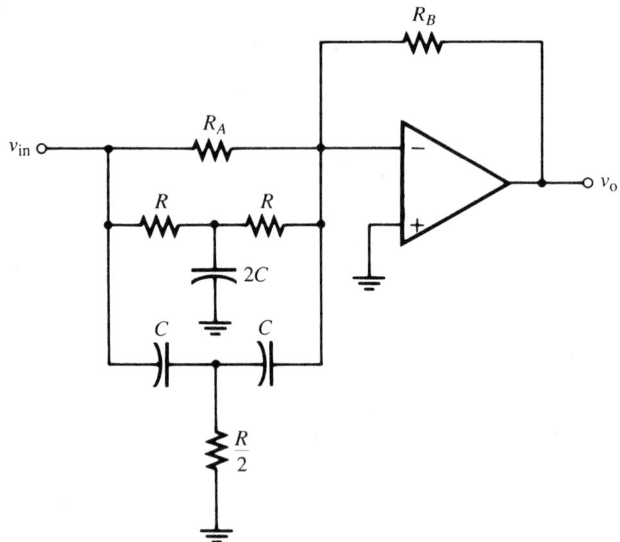

(b)

$$f_C = \frac{1}{2\pi RC} \qquad \textbf{(11–28)}$$

At the center frequency of the twin-T bandpass filter of Figure 11–24a, the current of the upper T section (*R-R-2C*) is 180° out of phase, with the current in the lower T section (*R*/2-*C-C*) at the operational amplifier's inverting input. Consequently, the twin-T network looks like an open circuit, and no current flows in the operational amplifier's feedback loop. At frequencies other than f_o, current flows through the feedback loop. In effect, the closed-loop voltage gain will approach infinity at the center frequency. Here, the gain is limited at f_o by the ratio R_B/R_A, which is the closed-loop inverting gain of the operational amplifier.

The twin-T notch filter of Figure 11–24b is similar to the bandpass version except that the twin-T network is placed in the input path instead of the feedback path. At the center frequency, the closed-loop voltage gain is zero, giving infinite rejection. At frequencies other than f_o, the closed-loop voltage gain is limited by the ratio R_B/R_A.

11–4 The State-Variable Filter

By properly connecting three operational amplifiers, we are able simultaneously to provide second-order low-pass, high-pass, and high-*Q* bandpass filter outputs. Such a filter, whose circuit is shown in Figure 11–25, is called a **state-variable,** or **universal, filter.** It is basically composed of a difference amplifier (A_1) and two identical integrators (A_2 and A_3).

For such a circuit, the center and cutoff frequencies are the same, as given by

$$f_C = \frac{1}{2\pi RC} \qquad \textbf{(11–29)}$$

The filter's *Q* is set solely by resistors R_A and R_B, so that

$$R_A = (3Q - 1)R_B \qquad \textbf{(11–30)}$$

With such a filter, values of *Q* up to several hundred can be achieved with stability. For the low-pass and high-pass outputs, the passband gain is unity. However, for the bandpass response, the center frequency gain equals *Q*, so that high *Q* results in high gain; care must then be taken to ensure that the operational amplifier's gain-bandwidth product specification is observed.

Because this type of filter gives a second-order response for both the low-pass and high-pass outputs, it is not possible to obtain optimum performance with all three outputs *simultaneously*. For a Butterworth low-pass or high-pass response,

Figure 11–25
State-variable filter having simultaneous low-pass, high-pass, and bandpass outputs.

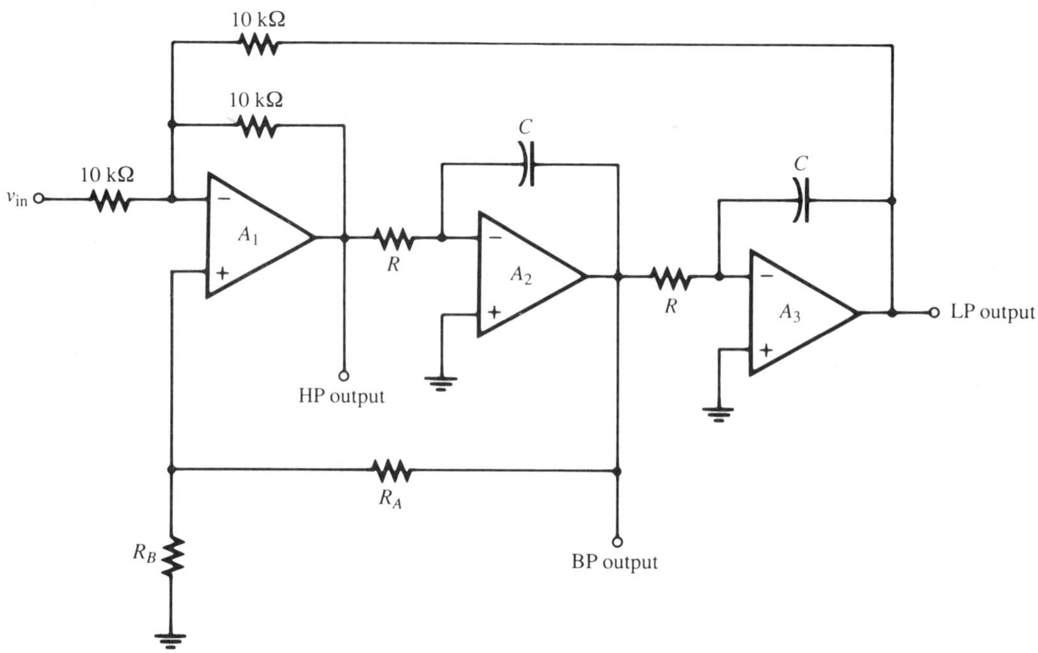

$Q = 0.707$. As a result, one designs either for a (1) Butterworth low-pass/high-pass response ($Q = 0.707$) or a (2) high-Q bandpass response.

Example 11–12
Determine the component values for the state-variable filter circuit of Figure 11–25 for a bandpass response having a center frequency of 600 Hz and a Q of 50. Assume that type-741 operational amplifiers (GBP = 1 MHz) are to be used.

Solution:

(a) A center frequency gain of 50 at 600 Hz results in a gain-bandwidth product of

$$\begin{aligned} \text{GBP} &= 50 \times 600 \text{ Hz} \\ &= 30 \text{ kHz} \end{aligned}$$

which is less than the 1 MHz specified for the 741. Consequently, the 741's closed-loop response will be sufficient to provide a closed-loop gain of 50 at 600 Hz.

(b) From Equation 11–29, assuming a **0.022-μF** capacitor is selected for C, then

$$R = \frac{1}{(2\pi)(0.022\ \mu\text{F})(600\ \text{Hz})}$$
$$= \mathbf{12.1\ k\Omega} \qquad \text{(use a 12-k}\Omega\text{ standard value)}$$

(c) From Equation 11–30,

$$R_A = [(3)(50) - 1]R_B$$
$$= 149R_B$$

If $R_B = 1$ kΩ, then $R_A =$ **149 kΩ,** for which a 150-kΩ standard value is used.

One very nice feature of the basic state-variable filter of Figure 11–25 is that we can simultaneously add the low-pass and high-pass outputs equally, as shown in Figure 11–26, which results in a state-variable notch filter. This requires only an additional operational amplifier and three equal-valued resistors, which are typically 10 kΩ for convenience.

The passband is unity, the notch frequency is the same as the filter's cutoff frequency (or center frequency), and the filter Q is the same as for the bandpass filter response.

Summary

This chapter discussed various active filters as frequency-selective networks and the relative merits of each. Filters can be classified as either low-pass, high-pass, bandpass, or notch. Furthermore, the low-pass and high-pass responses were then divided into various response shapes that are the result of trade-offs between passband and stopband responses. The Butterworth, or maximally flat, response was shown to have the best overall response when compared with either the Chebyshev, elliptic (Cauer), and Bessel responses having the same order filter and cutoff frequency.

Practical active filter networks discussed included the multiple-feedback, Sallen-Key, twin-T, and state-variable networks.

Figure 11–26
State-variable notch filter formed by summing low-pass and high-pass outputs.

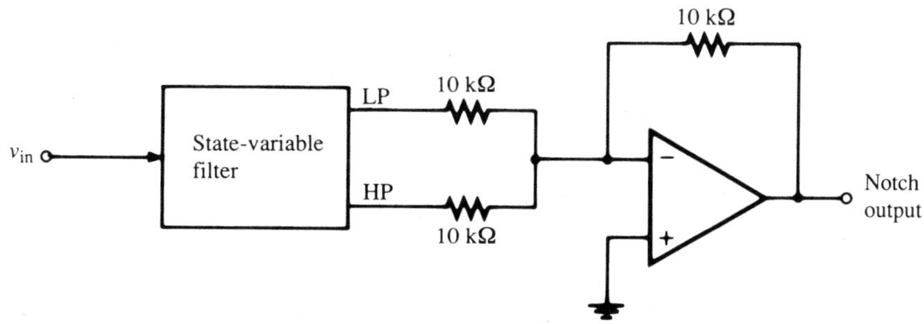

Glossary

active filter Any filter network built around an active device such as a transistor or operational amplifier to produce gain in the filter's passband.

bandpass filter A filter network that allows signals between two frequencies to pass through unaffected but that rejects signal frequencies above and below this range.

bandwidth The frequency range between frequencies where the response is 3 dB less than the passband gain.

Bessel filter A filter response whose low-pass or high-pass phase-shift characteristic is linear with frequency.

Butterworth response A filter response characterized by a flat passband. Also called a maximally flat response.

Cauer filter See *elliptic filter*.

center frequency The median frequency of bandpass and notch filters.

Chebyshev response A filter response characterized by equal ripples of constant amplitude in the passband. Also called an equal-ripple response.

cutoff frequency The frequency of low-pass and high-pass filters where the response is 3 dB less than the passband gain.

elliptic filter A filter network having a response characterized by equal ripples of constant amplitude in the passband and stopband. Also called a Cauer filter.

equal-ripple response See *Chebyshev response*.

filter Any frequency selective passive or active network that passes certain frequencies while rejecting all others.

fractional bandwidth Expressed as a percentage, the ratio of the filter's bandwidth to its center frequency. Equal to the reciprocal of the filter's quality factor.

high-pass filter A filter network that allows signals above a certain frequency to pass through unaffected but rejects signal frequencies below this frequency.

low-pass filter A filter network that allows signals below a certain frequency to pass through unaffected but rejects signal frequencies above this frequency.

maximally flat response See *Butterworth response*.

multiple feedback filter A type of operational amplifier active filter having multiple feedback paths connected between the output and the operational amplifier's inverting input such that the passband output signal is 180° out of phase with the input.

notch filter A filter network that rejects signals between two frequencies but allows signals at frequencies above and below this range to pass through unaffected.

order An integer value that relates the roll-off, or selectivity, of a low-pass or high-pass filter.

parallel-T network See *twin-T network*.

passband The filter's frequency range through which the input signal passes with no attenuation.

rolloff The rate at which the stopband response of a filter network changes with frequency, expressed in terms of decibels per octave or decibels per decade.

Sallen-Key filter A type of low-pass or high-pass active filter network whose passband output is in phase with the input signal.

state-variable filter A type of active filter using three operational amplifiers to achieve simultaneous low-pass, high-pass, and bandpass responses.

stopband The filter's frequency range in which the input signal is attenuated.

twin-T network An RC network that consists of two parallel T-networks, with the capacitor and resistor portions interchanged in one of them. By itself, it functions as a passive notch filter. Also called a parallel-T network.

universal filter See *state-variable filter*.

Review Questions

1. What is a filter?
2. What is a passive filter? An active filter? Compare the advantages and disadvantages of both.
3. List the four basic filter frequency responses, and sketch their corresponding amplitude response curves.
4. For each of the four filter frequency responses, identify the cutoff frequency, passband, and stopband.
5. How are the center frequency and bandwidth related in a bandpass filter?
6. Explain how a bandpass response can be achieved using low-pass and high-pass filter sections.
7. What is meant by the term *fractional bandwidth,* and how is it related to the quality factor of either a bandpass filter or notch filter?
8. What is meant by an *octave?* A *decade?*
9. List four major filter response shapes and describe their characteristics.
10. What is another name for a Butterworth filter response?
11. Explain the effect of the order of a filter on its selectivity.
12. What is another name for a Chebyshev filter response, and how does its response compare with a Butterworth filter of the same order?
13. What are other names for an elliptic filter?
14. How does the response of an elliptic filter compare with those of Butterworth and Chebyshev filters?
15. How does the response of a Bessel filter compare with other filter responses?

Exercises

1. A fourth-order Butterworth high-pass filter is to have a passband voltage gain of 5 dB. The cutoff frequency is 1570 Hz.
   (a) Determine the decibel response at 632 Hz.
   (b) If the input is 4.6 V peak-to-peak, what is the output voltage?
2. A Butterworth low-pass filter has a passband gain of unity and a cutoff frequency of 942 Hz. Determine the order of the filter required so that the response is 35 dB down from the passband gain at 2000 Hz.
3. What cutoff frequency of a second-order Butterworth low-pass filter is required for a passband gain of +5.2 dB if the decibel response is −18.5 dB at 1500 Hz?
4. What would be the input frequency to a fifth-order Butterworth high-pass filter having a passband gain of +8 dB and a cutoff frequency of 1.2 kHz for a decibel response of −62 dB?
5. Determine the decibel response at 1345 Hz of a third-order Chebyshev low-pass filter that has a passband ripple of 2 dB and a cutoff frequency of 650 Hz.
6. Repeat Exercise 5 for a passband ripple of 0.5 dB.
7. The 3-dB frequencies of a bandpass filter are 647 Hz and 832 Hz. Determine each value.
   (a) The center frequency
   (b) The 3-dB bandwidth
   (c) The filter's Q
8. A bandpass filter has a center frequency of 2.6 kHz and a Q of 8.3. Determine the upper and lower 3-dB frequencies.

9. The filter of Exercise 8 has a passband gain of 3 dB. What is the dB response at each value?
   (a) 1.8 kHz
   (b) 4 kHz
10. If a bandpass filter has a Q of 10, passband gain of zero decibels, and center frequency of 1100 Hz, determine the dB-amplitude response if the input frequency is 800 Hz. At what other frequency will the dB-amplitude response equal that at 800 Hz?
11. Determine the center frequency and Q of a bandpass filter if the lower and upper cutoff frequencies are 1205 Hz and 1762 Hz.
12. What is the frequency in each case?
   (a) Three decades above 400 Hz
   (b) Two decades below 3.5 kHz
   (c) Four octaves above 831 Hz
   (d) Three octaves below 2.2 kHz
13. Determine the cutoff frequency of the second-order Sallen-Key low-pass filter of Figure 11–14b for $R_1 = R_2 = 18\ \text{k}\Omega$ and $C_1 = C_2 = 0.022\ \mu\text{F}$.
14. Determine the required value for $R_1 = R_2$ for the second-order Sallen-Key high-pass filter of Figure 11–14c if the cutoff frequency is to be 554 Hz with $C_1 = C_2 = 0.47\ \mu\text{F}$.
15. Using the Sallen-Key circuit of Figure 11–14b, determine the component values for a second-order Butterworth low-pass filter having a cutoff frequency of 1200 Hz. Assume 0.01-μF capacitors.
16. Determine the circuit and component values for a fourth-order Butterworth high-pass filter having a cutoff frequency of 650 Hz. Assume 0.01-μF capacitors.
17. Determine the circuit and component values for a fifth-order Butterworth low-pass filter having a cutoff frequency of 1300 Hz. Assume 0.01-μF capacitors.
18. Determine the circuit and component values for a fourth-order, 1-dB Chebyshev high-pass filter having a cutoff frequency of 700 Hz. Assume 0.01-μF capacitors.
19. If, for the multiple bandpass filter of Figure 11–21d, $R_1 = 47\ \text{k}\Omega$, $R_2 = 3.3\ \text{k}\Omega$, $R_3 = 150\ \text{k}\Omega$, and $C = 0.015\ \mu\text{F}$, determine each of the following.
   (a) The center frequency of the filter
   (b) The passband gain in dB
   (c) Q
   (d) The 3-dB bandwidth
20. Determine the decibel response at 850 Hz for the bandpass filter of Exercise 19.
21. Determine the required component values for the multiple-feedback bandpass filter of Figure 11–21d having a center frequency of 1100 Hz, Q of 5, and a center frequency gain of 2. Assume 0.01-μF capacitors.
22. Determine the notch frequency and passband gain for the twin-T notch filter of Figure 11–24b if $R = 5.6\ \text{k}\Omega$, $R_A = 100\ \text{k}\Omega$, $R_B = 220\ \text{k}\Omega$, and $C = 0.47\ \mu\text{F}$.
23. Determine the component values for the state-variable filter circuit of Figure 11–25 for a bandpass response having a center frequency of 950 Hz and a Q of 40. Assume 0.01-μF capacitors.
24. Determine the component values for the state-variable filter circuit of Figure 11–25 such that the second-order low-pass response has a cutoff frequency of 700 Hz. Assume 0.01-μF capacitors.

12 Operational Amplifier and Integrated-Circuit Converters

Instructional Objectives

Upon completing this chapter, you will be able to

- Define the term *conversion gain*.
- Design operational amplifier circuits that will convert current to voltage, voltage to current with floating load, and voltage to current with grounded load.
- Design a V/F converter using either a 566 or a 4151.
- Design an F/V converter using either a 4151 or LM331.
- Explain the operation of digital-to-analog converters built around an operational amplifier using either a weighted summing amplifier or an *R*-2*R* ladder network.
- Determine the resolution of a given digital-to-analog converter.
- Draw the block diagram and explain the basic operation of the ramp-type (single slope), staircase ramp, dual-slope, and successive approximation analog-to-digital converters.
- Calculate the conversion time of a successive approximation type analog-to-digital converter.

Introduction

Operational amplifiers and specific analog linear integrated circuits are used extensively for the purposes of converting the following parameters:

- Voltage to voltage
- Voltage to current
- Current to voltage
- Current to current
- Voltage to frequency
- Frequency to voltage

- Analog voltage to digital
- Digital to analog voltage

These conversion circuits are, in effect, transducers, which take a signal in one form—either voltage, current, frequency, or digital data—and convert it to a linearly proportional signal of the other.

Often when analog signals must be transmitted over a considerable distance by hard-wired cable circuits, a voltage signal is first converted to a proportional current signal at the transmitting end and then converted back to the corresponding voltage at the receiving end. This technique then eliminates problems due to voltage drops in a long circuit path. Since the current at all points in a series circuit must be the same, the signal level is then independent of circuit length.

As with the voltage-to-current conversion, voltage-to-frequency conversion is also often used where the voltage to be measured is physically located some distance from the actual source, as it is easier to transmit the pulse train with more reliability than it is with a low-level DC signal. In addition, it is well suited for use in tone generators, frequency-shift keying (FSK), FM modulators, clock generators, signal generators, function generators, and phase-locked loops (Chapter 13).

Whenever an analog signal must be read and processed further by a computer, it must first be converted to an equivalent digital signal by a circuit called an *analog-to-digital converter*. Conversely, when a computer must control an external device that requires an analog input, the digital output from the computer must be passed through another circuit called a *digital-to-analog converter*.

As a general rule, the simpler circuit functions, such as the current-to-voltage and voltage-to-current converters and current amplifiers, are easily made using operational amplifiers. The more complex functions, such as voltage-to-frequency, frequency-to-voltage, analog-to-digital, and digital-to-analog converters, are best achieved using one of the specialized integrated-circuit devices designed for these purposes.

12–1 Conversion Gain

The output parameter of a linear converter is related to its input parameter by a proportionality constant called the **conversion gain** (K), sometimes called the *transfer function*. The conversion gain is simply equal to the output divided by the input. For example, if the output of a voltage-to-current converter is 2 mA when 0.5 V is applied to the input, the conversion gain is 2 mA/0.5 V, or 4 mA/V. In effect, this is similar to the slope of a straight line, so that here an output current of 4 mA is converted from each 1 V of input. To find the value of the output for a given input, the input parameter is then multiplied by the conversion gain.

Example 12–1

A voltage-to-frequency converter has a conversion gain of 0.052 kHz/V. Determine the output frequency if the input voltage is 3.2 V.

Solution: The output frequency of the voltage-to-frequency converter is then

$$f_o = (0.052 \text{ kHz/V}) \times (3.2 \text{ V}) = \mathbf{166 \text{ Hz}}$$

12–2 Current-to-Voltage Converter

The basic **current-to-voltage converter** circuit of Figure 12–1 is nothing more than an inverting amplifier without the input resistor. Since the inverting input is at a virtual ground, the magnitude of the input current (I_{in}) and current (I_F) through the feedback resistor R_F are equal. Using Ohm's law, the output voltage is the voltage across R_F, so that

$$V_o = -I_{\text{in}}R_F \qquad \textbf{(12–1)}$$

In reality, the operational amplifier's input bias current must be added to the applied current input. Consequently, the operational amplifier chosen must have a small input bias current compared to the applied current input. For this circuit, the conversion gain is V_o/I_{in}, or $-R_F$. Because the inverting input appears as a virtual ground, the input to this circuit looks like a short circuit to the input current source, whereas the output looks like a voltage source to the load.

Figure 12–2 shows a current-to-voltage converter that functions as an electronic milliammeter. For the values shown, an input current of 10 μA would produce a output voltage of 1 V. Therefore, the conversion gain is 1 V/10 μA, or 0.1 V/μA. By using this circuit, a conventional voltmeter is then easily adapted for the measurement of small currents. An added feature is that the series resistance inserted into the current path is quite small compared to that of a conventional moving-coil ammeter suitable for this current range.

Other common uses for the current-to-voltage converter are with photodetectors. These are light-sensitive devices, such as a photoconductive cell whose

Figure 12–1
Current-to-voltage converter.

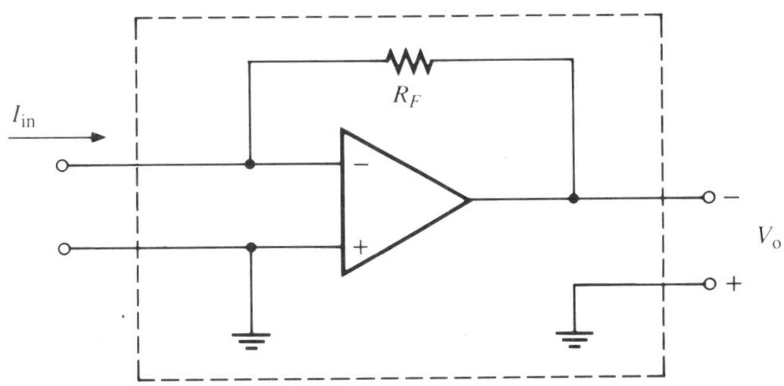

Figure 12–2
Current-to-voltage converter with a voltmeter to form an electronic ammeter.

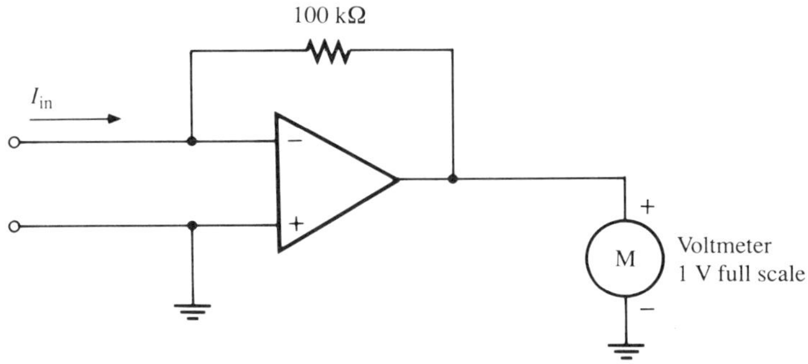

Figure 12–3
Using a current-to-voltage converter with a photoresistor to measure changes in light intensity. (a) Circuit.
(b) Norton equivalent circuit.

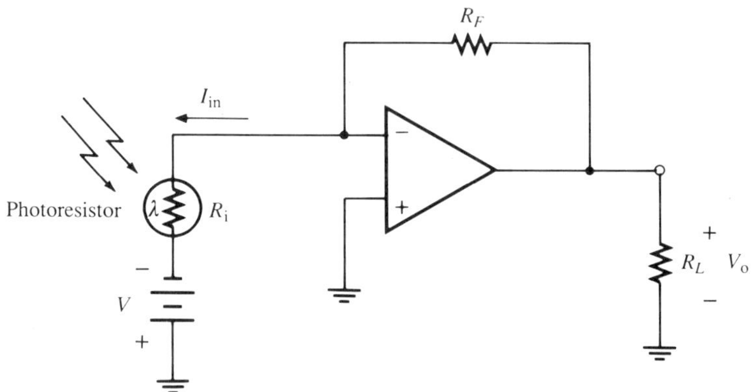

(a)

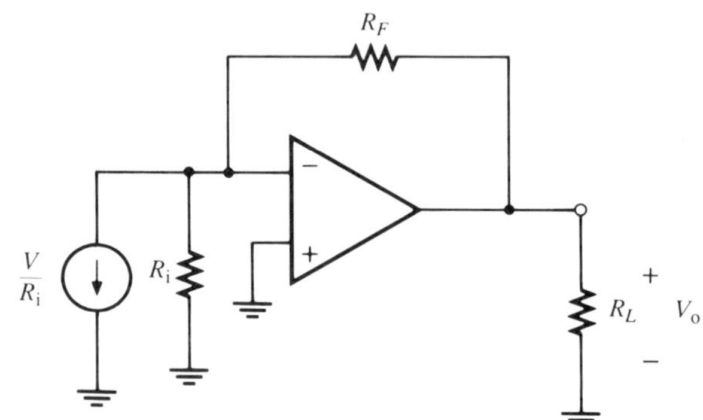

(b)

internal parameter (usually a resistance) varies with the light intensity that shines on it. In the case of a photoresistor, its resistance is inversely proportional to the light intensity.* Figure 12–3a shows the circuit of a current-to-voltage converter using a photoconductive cell connected in series with a negative voltage with

*In general, the internal resistance of most photoconductive cells does not decrease linearly with increasing light intensity.

respect to ground. Because of the virtual ground of the operational amplifier's inverting input, the current flowing through the photoresistor for a given light intensity is, in effect, the short-circuit current flowing through it:

$$I_{\text{in}} = I_{SC} = -\frac{V_o}{R_i} \tag{12–2}$$

where R_i is the internal resistance of the photoresistor at a given light intensity. Since the conversion gain (V_o/I_{in}) of the circuit is $-R_F$, the negative input current produces a positive output voltage. Figure 12–3b shows the equivalent Norton circuit of the photoresistor-voltage source as a current source.

By measuring the output voltage, it is then possible to determine the photoresistor's internal resistance or current flowing through it for any given lighted condition. A calibration curve relating light intensity and resistance is sometimes provided by the manufacturer so that it is then possible to determine indirectly the light intensity.

Example 12–2

For the circuit of Figure 12–3a with $V = 1.5$ V and $R_F = 10$ kΩ, determine the internal resistance of the photoresistor in each case.

(a) Under fully lighted conditions, the output voltage is +7.88 V.

(b) The photoresistor is completely covered from all light so that the output voltage is +152 mV.

Solution:

(a) Under fully lighted conditions, the current flowing through the photoresistor is

$$I_{\text{in}} = \frac{7.88 \text{ V}}{10 \text{ k}\Omega} = 788\ \mu\text{A}$$

The internal resistance of the photoresistor is a minimum, so that

$$R_i = \frac{1.5 \text{ V}}{788\ \mu\text{A}} = \mathbf{1904\ \Omega}$$

(b) When the photoresistor is completely covered from light, the internal resistance of the photoresistor (its *dark resistance*) is a maximum, so the current flowing through it is then

$$I_{\text{in}} = \frac{152 \text{ mV}}{10 \text{ k}\Omega} = \mathbf{15.2\ \mu A}$$

The internal resistance of the photoresistor is then

$$R_i = \frac{1.5\ \text{V}}{15.2\ \mu\text{A}} = \mathbf{98.7\ k\Omega}$$

12–3 Voltage-to-Current Converters

The basic inverting amplifier configuration can also be used to create a **voltage-to-current converter** if the input resistor is retained and a current-driven load is substituted for the feedback resistor, as shown in Figure 12–4. Here, the input voltage is converted to an output current that flows through the load, R_L.

Ideally, the current through the load must equal

$$I_L = \frac{V_{in}}{R_1} \qquad (12\text{–}3)$$

so that the load current is independent of the load resistance. However, the one major disadvantage of this circuit is that neither end of the load is grounded, and this may present a problem in some applications. The conversion gain is I_L/V_{in}, or $1/R_1$, and is typically expressed in units of milliamperes per volt (mA/V). Since this represents the basic units for conductance, the voltage-to-current converter is often referred to as a *transconductance-amplifier*.

Another form of the voltage-to-current converter uses the noninverting amplifier configuration, as shown in Figure 12–5. Here, the load current equals V_{in}/R, since the voltage across R is approximately equal to the input voltage V_{in}. The conversion gain, like that for the circuit of Figure 12–4, is $1/R_L$. With this circuit, we could make a high-impedance voltmeter by using a suitable D'Arsonval meter movement as the load.

Figure 12–4
Voltage-to-current converter, or transconductance amplifier.

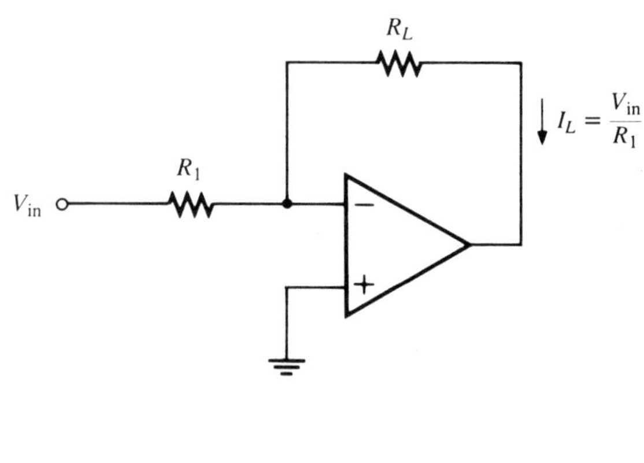

Figure 12–5
Noninverting voltage-to-current converter with floating load.

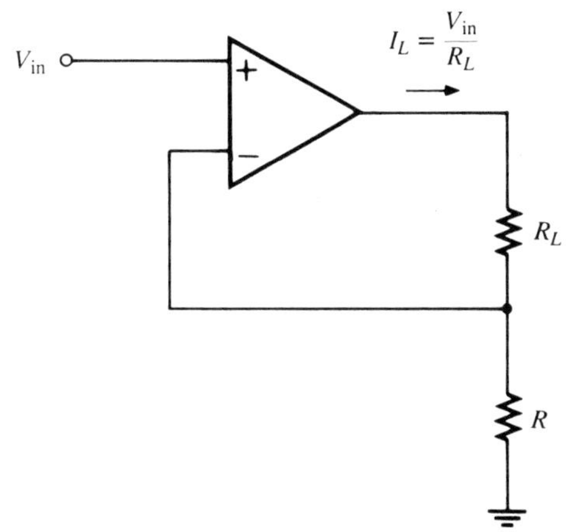

Unlike the two previous circuits, the voltage-to-current converter of Figure 12–6 has the advantage that one side of the load is grounded, as is customarily the practice. By making all four resistors equal in value, the load current is equal to V_{in}/R. Using Kirchhoff's current law at the operational amplifier's noninverting input,

$$I_L = I_1 + I_2 \tag{12-4}$$

Since

$$I_1 = \frac{V_{in} - V_1}{R} \tag{12-5}$$

and

$$I_2 = \frac{V_o - V_1}{R} \tag{12-6}$$

Substitution of Equations 12–6 and 12–5 into 12–4 yields

$$I_L R = V_{in} + V_o - 2V_1 \tag{12-7}$$

As a noninverting amplifier based on the two resistors tied to the operational amplifier's inverting input, the closed-loop voltage gain is

$$\frac{V_o}{V_1} = 1 + \frac{R}{R} \tag{12-8}$$

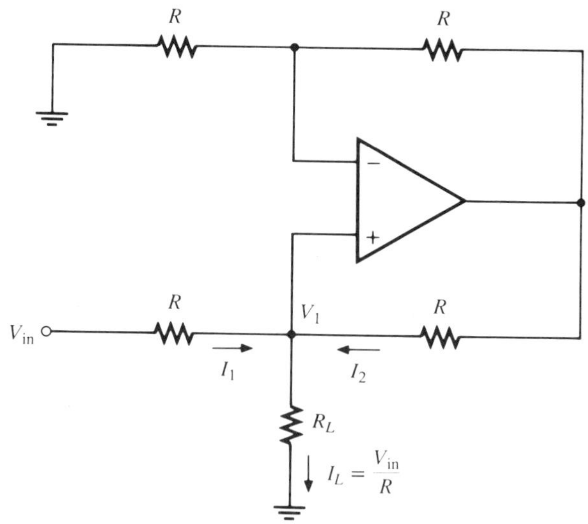

Figure 12–6
Voltage-to-current converter with grounded load.

so that

$$V_o = 2V_1 \tag{12–9}$$

Substituting Equation 12–9 into Equation 12–7 and solving for I_L yields

$$I_L = \frac{V_{in}}{R} \tag{12–10}$$

Again, the load current is controlled primarily by the input voltage and is independent of the load. In addition, the conversion gain is $1/R$. If the input voltage is highly regulated, this circuit also makes an excellent constant-current source.

12–4 Current Amplifier

Throughout this text, the term *amplifier* has been assumed to mean *voltage* amplification. However, the operational amplifier can also be used to achieve *current amplification*. One case in point is the necessity to increase the level of current produced by a signal source such as a solar cell. As an example, Figure 12–7 shows a current amplifier circuit that uses an ammeter to measure the output current. Using Kirchhoff's current law, the current flowing through the ammeter (I_M) is the sum of the shunt current (I_S) flowing through R_1 and the input current (I_{in}) flowing through the feedback resistor R_2:

$$I_M = I_S + I_{in} \tag{12–11}$$

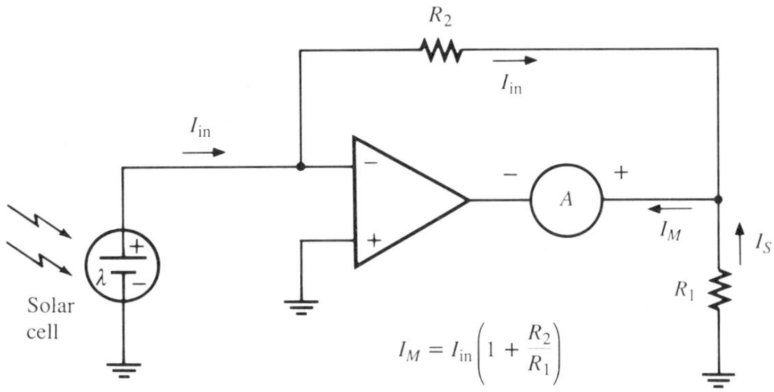

Figure 12–7
Current amplifier with ammeter to measure the current generated by a solar cell.

Since resistors R_1 and R_2 form a current divider, the meter current can then be expressed in terms of I_{in} and the two resistors as

$$I_M = \left(1 + \frac{R_2}{R_1}\right) I_{in} \qquad (12\text{–}12)$$

12–5 Voltage-to-Frequency Converters

The **voltage-to-frequency (V/F) converter** changes a DC input voltage into a pulsed waveform whose output frequency, or repetition rate, is proportional to the magnitude of an input control voltage. It is also referred to as a *voltage-controlled oscillator*, or VCO. Although the term *oscillator* as used in Chapter 10 is specifically defined as a sine-wave generator, the VCO has been generally taken to mean that its output can be of any waveshape. The conversion gain relates output frequency to input voltage and is expressed in units of hertz per volt. Since random noise tends to have a zero average value, the V/F converter usually has an excellent immunity to noise and is routinely used in tone generators, frequency-shift keying (FSK) circuits, FM modulators, waveform generators, and phase-locked loops. The basic concepts are discussed in this chapter; specific applications involving its role with the phase-locked loop are discussed in Chapter 13.

The Basic Concept

Figure 12–8 shows the elements of a basic V/F conversion system. The input voltage (V_{in}) generates a linear ramp with slope proportional to the amplitude of V_{in}. Each time the amplitude of the ramp exceeds the absolute value of the comparator's reference voltage, V_{REF}, the comparator output goes high and triggers the monostable multivibrator, which in turn discharges the capacitor and produces an output pulse to the counter. This cycle repeats itself at a frequency that is a linear function of the amplitude of V_{in}. For example, an input voltage of

Figure 12–8
Block diagram of a voltage-to-frequency converter.

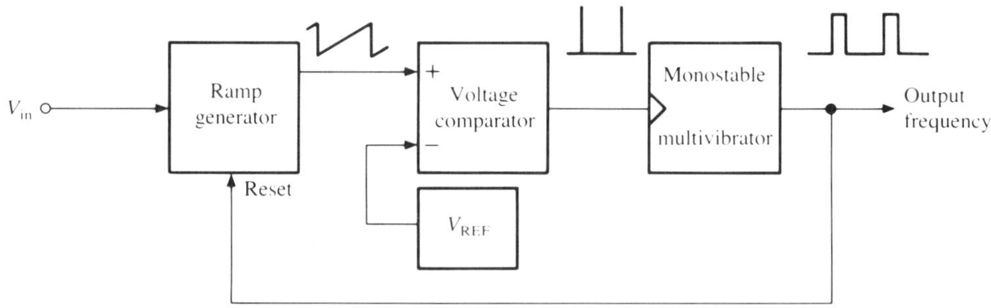

2 V causes the ramp to reach the switching level of the comparator in half the time required by a 1-V level. Thus, a DC input of 2 V produces pulses at twice the frequency of that produced by 1 V.

However, the V/F converter is not practical when it is necessary to measure very small voltage levels, as the time between pulses would become excessive. To overcome this problem, a V/F converter often has a minimum output frequency, or frequency offset, specified even at zero volts, which increases linearly with increasing values of V_{in}. In terms of an equation of a straight line, the output frequency of a V/F converter is given by

$$f_{out} = f_{min} + K_{VCO} V_{in} \qquad \textbf{(12–13)}$$

where

f_{min} = output frequency when $V_{in} = 0$
K_{VCO} = F/V converter (VCO) conversion gain, Hz/V.

The V/F converter is often used where the voltage to be measured is located some distance from the display, as it is easier to transmit a pulse train or sine wave with more reliability than a low-level DC signal.

Integrated-Circuit V/F Converters

The construction of practical V/F converters is best achieved using one of many integrated-circuit devices specifically designed for this purpose. This section discusses the 566 and 4151 as typical examples.

The 566 Function Generator*

The 566 is a basic building block used in the 560-series of phase-locked loop integrated-circuit devices made by Signetics as well as several other manufactur-

*Very often, manufacturer's data sheets refer to the 566 as a *function generator* rather than a VCO or V/F converter, since it is capable of generating more than one type of waveform. However, its primary use is that of a voltage-controlled oscillator, or voltage-to-frequency converter.

Table 12–1
566 VCO parameters

Supply Voltage	10 to 24 V
VCO	
Maximum frequency	1 MHz
Drift	100 ppm/°C
Maximum sweep rate	1 MHz
Sweep range	10 to 1
Control input impedance	1MΩ
Square Wave	
Output impedance	50 Ω
Output voltage	5.4 V peak-to-peak
Duty cycle	45 to 55%
Rise time	20 ns
Fall time	50 ns
Triangle Wave	
Output impedance	50 Ω
Output voltage	2.4 V peak-to-peak
Linearity	0.2%

ers (Chapter 13). It is illustrated here because (1) it has exceptional stability and linearity, and (2) it includes both buffered square-wave and triangular-wave outputs. The free-running frequency of oscillation is controlled by an external timing resistor and capacitor to a maximum of 1 MHz. In addition, the VCO's frequency can be programmed over a 10 : 1 frequency range by the adjustment of the external timing resistor, or it can be modulated (shifted) over a 10 : 1 range by the external control voltage. Table 12–1 summarizes several of the typical characteristics of the 566.

Figure 12–9a shows the pin configuration of the 8-pin mini-DIP version, whereas Figure 12–9b shows the block diagram of its internal equivalent circuit. The basic output level of the current source is set by R_1 and can be further varied by the control, or modulating voltage, at pin 5. This current is then used to charge and discharge the external timing capacitor (C_1) at pin 7, which is under control of the Schmitt trigger. The constant charge and discharge current generates a triangular voltage waveform across the external timing capacitor. The output of the Schmitt trigger, which is at pin 3, is a square wave, which in turn is buffered. This is the primary output signal of the V/F converter, although the buffered triangle wave output at pin 4 can be used in some instances.

Figure 12–9c shows a typical connection diagram, which can be used to generate simultaneous square-wave and triangular-wave outputs up to frequencies of about 1 MHz. The control voltage terminal (pin 5) must be externally biased with a control voltage (V_C). In this case, the bias voltage is set by the voltage divider formed by R_2 and R_3. The modulating signal is then AC-coupled through capacitor C_2. On the other hand, the modulating signal could also be directly coupled if the proper DC bias voltage is present in the modulating signal. The VCO output frequency is given approximately by

$$f_o = \frac{2(V_{CC} - V_C)}{R_1 C_1 V_{CC}} \qquad \textbf{(12–14)}$$

Figure 12–9
566 integrated-circuit function generator. (a) Pin diagram. (b) Internal logic diagram. (c) General operational circuit. (d) Waveforms.

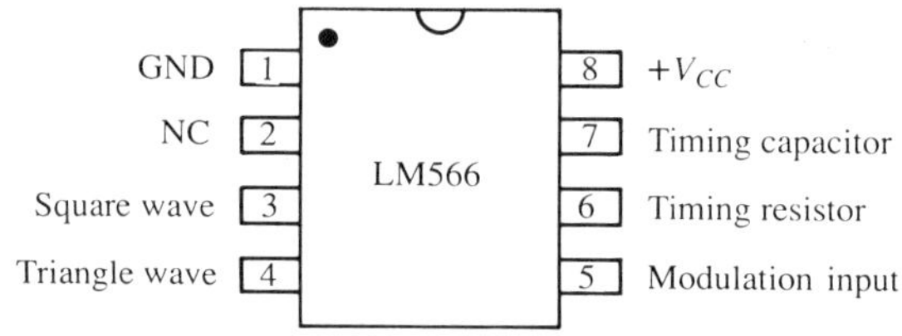

(a)

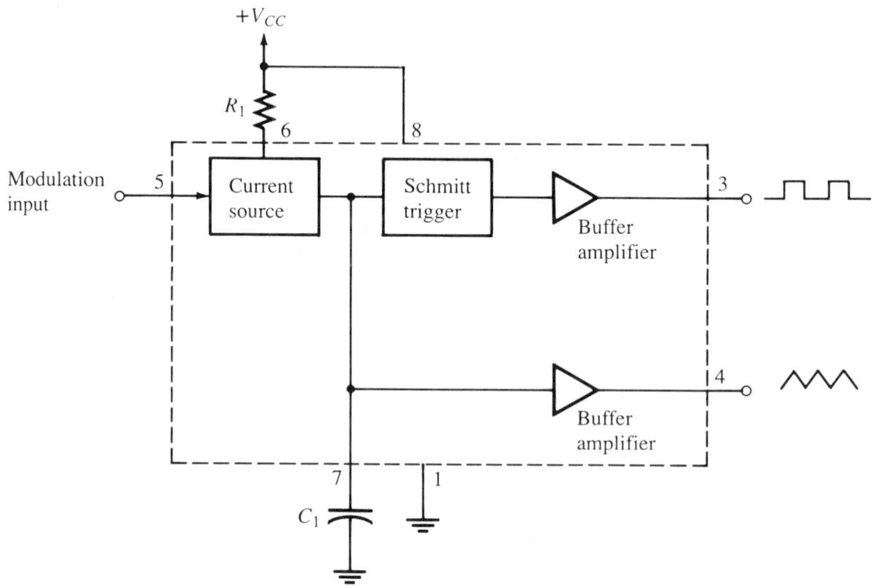

(b)

For proper operation, the following guidelines (as recommended by the manufacturer) must be followed:

- External timing resistor (R_1): 2 kΩ to 20 kΩ
- Supply voltage (V_{CC}): 10 V to 24 V
- Control voltage (V_C): $0.75V_{CC}$ to V_{CC}

In addition, a small capacitor (typically 0.001 μF) is usually connected between pins 5 and 6 to eliminate possible oscillations in the internal control current source. By operating the VCO from a ±5-V dual-power supply, as shown in Figure 12–10, the square-wave output has the proper DC levels to drive TTL digital integrated circuits.

Figure 12–9
continued

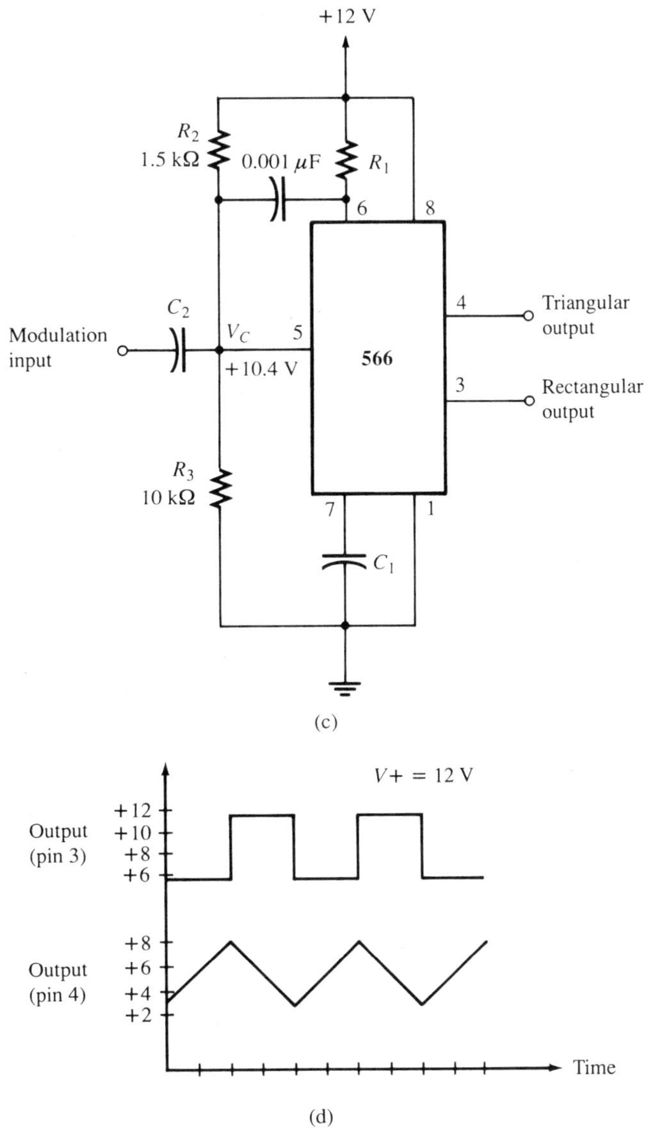

Example 12–3

For the 566 VCO circuit of Figure 12–9c with $R_1 = 8.2\ \text{k}\Omega$ and $C_1 = 220$ pF, determine the output frequency range if the control voltage is varied from 12 V to 14 V and a 15-V supply voltage is used.

Solution:

(a) Using a control voltage of 12 V in Equation 12–14, the maximum output frequency will be

$$f_o = \frac{(2)(15\text{ V} - 12\text{ V})}{(8.2\text{ k}\Omega)(220\text{ pF})(15\text{ V})}$$
$$= \mathbf{221.7\ kHz}$$

(b) At a control voltage of 14 V, the minimum output frequency will be

$$f_o = \frac{(2)(15\text{ V} - 14\text{ V})}{(8.2\text{ k}\Omega)(220\text{ pF})(15\text{ V})}$$
$$= \mathbf{73.9\ kHz}$$

Figure 12–10
566 circuit powered from a ±5-V dual supply so that the square-wave output at pin 3 has the proper DC levels to drive TTL digital integrated circuits.

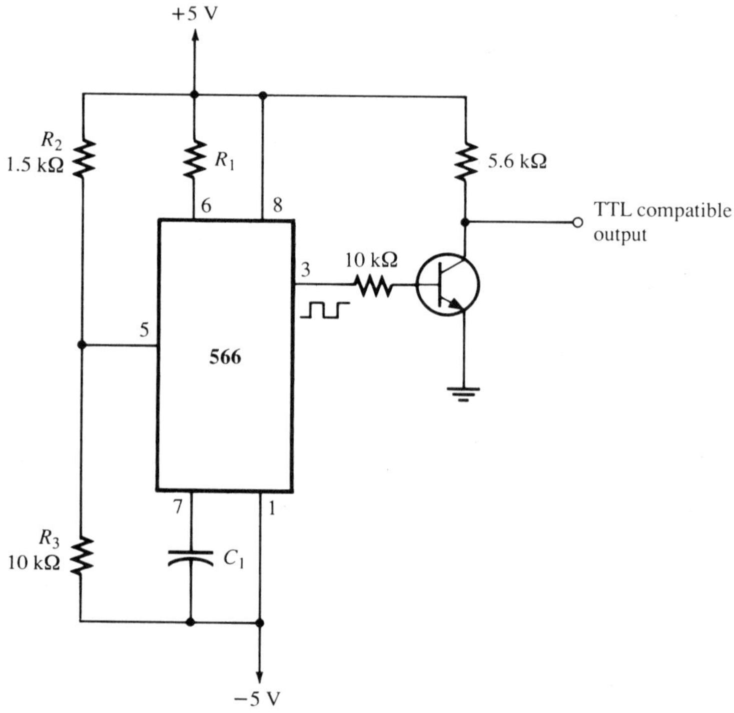

Figure 12–11
4151 integrated-circuit V/F converter pin diagram.

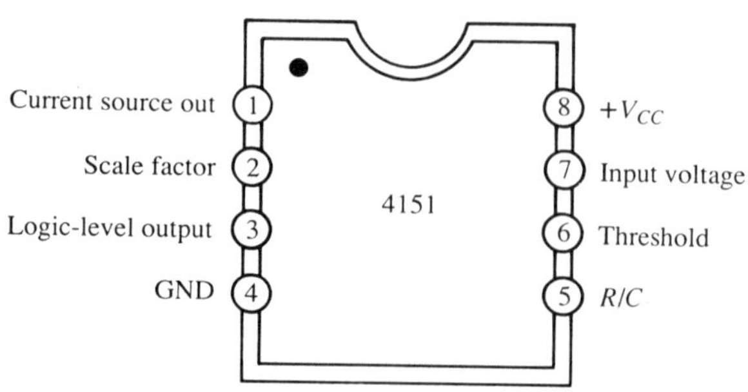

Table 12–2
4151 V/F converter parameters

Supply voltage	8 to 22 V
Conversion scale factor	1 kHz/V
Drift with temperature	±100 ppm/°C
Nonlinearity	<1%

4151 V/F Converter

Besides the 566, another popular V/F converter is the 4151, whose pin configuration is shown in Figure 12–11. It was originally made by Raytheon but is also now manufactured by Exar as the XR-4151. Along with some of the major parameters listed in Table 12–2, the 4151, using a single-supply voltage between +8 V and +22 V, has a typical linearity of 1% with a temperature stability of ±100 ppm/°C. Unlike virtually all other V/F converters, the 4151 also has the capability of performing frequency-to-voltage conversion (see Section 12–6).

Although more accurate V/F converter circuits using the 4151 are possible, the circuit shown in Figure 12–12 is the simplest type, having an output frequency that varies from 0 to 10 kHz when the input control voltage is varied from 0 to +10 V. Consequently, the conversion gain is typically 1 kHz/V when properly adjusted.

Figure 12–12
V/F converter circuit having a conversion gain of 1 kHz/V and a maximum output frequency of 10 kHz.

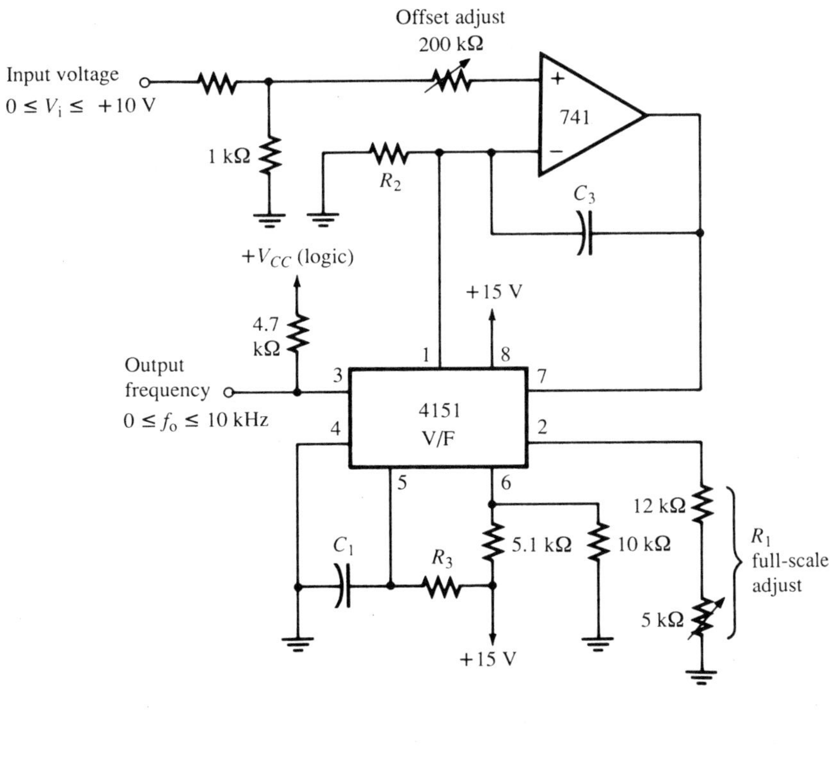

The full-scale frequency adjustment is done by applying +10 V and adjusting the 5-kΩ potentiometer in series with the 12-kΩ resistor (R_1) at pin 2 for a 10-kHz output frequency at pin 3. Typically, R_1 is approximately 14 kΩ to provide operation over the 10-V/10-kHz input/output range. The output frequency is given by

$$f_o = 0.486\ V_1 \left(\frac{R_1}{R_2 R_3 C_1}\right) \tag{12–15}$$

so that the output pulse train will have a pulse width (T) equal to

$$T = 1.1 R_3 C_1 \tag{12–16}$$

Based on the manufacturer's data sheet, the required design equations are as follows:

$$T = 1.1 R_3 C_1 = \frac{0.75}{f_o} \tag{12–17}$$

such that R_3 is in the range of 6.8 kΩ to 680 kΩ and C_1 is in the range of 0.001 μF to 1 μF.

$$C_2 = \frac{0.01}{f_o} \tag{12–18}$$

$$C_3 = \frac{5 \times 10^{-5}}{f_o} \tag{12–19}$$

The use of these equations is illustrated by the following example.

Example 12–4

Determine the required component values for the 4151 circuit of Figure 12–12 for a 10-kHz output frequency when +10 V is applied as the input control voltage.

Solution:

(a) Using Equation 12–17,

$$T = \frac{0.75}{10\text{ kHz}} = 75\ \mu\text{S}$$

From Equation 12–16, select a value for R_3 and solve for C_1. Here, assume that R_3 = **6.8 kΩ,** so that

$$C_1 = \frac{75\ \mu S}{(1.1)(6.8\ k\Omega)} = \mathbf{0.01\ \mu F}$$

(b) From Equation 12–18,

$$C_2 = \frac{0.01}{10\ kHz} = \mathbf{1\ \mu F}$$

(c) Finally, C_3 is determined from Equation 12–19:

$$C_3 = \frac{5 \times 10^{-5}}{10\ kHz} = \mathbf{0.005\ \mu F} \text{ (use a 0.0047-}\mu\text{F standard value)}$$

Such a circuit can then be used to measure DC voltages using a digital frequency counter.

12–6 Integrated-Circuit Frequency-to-Voltage Converters

The **frequency-to-voltage (F/V) converter** is the opposite of the V/F converter, although not as common. Here, the output voltage is linearly proportional to an input frequency so that the conversion gain is in units of volts per hertz.

As with V/F converters, practical F/V converters are available as integrated circuits specifically designed for this purpose. This section discusses the 4151 and LM331 as typical examples.

4151 F/V Converter

The 4151, besides being able to function as a V/F converter, is also capable of being wired as an F/V converter using a single-supply voltage, as shown in Figure 12–13. The output voltage, as a function of the input frequency, is obtained using Equation 12–15, so that

$$V_o = 2.058\, f_i \left(\frac{R_2 R_3 C_1}{R_1}\right) \tag{12–20}$$

The component values for R_1, R_2, R_3, and C_1 are determined in a like manner for the V/F configuration using Equations 12–16 through 12–19. Such a circuit could allow a digital voltmeter to read frequencies up to 10 kHz directly. Because of the typical conversion gain of 1 V/kHz when properly adjusted, an input frequency of 3270 Hz produces an output of 3.27 V.

Figure 12–13
4151 connected as a frequency-to-voltage converter.

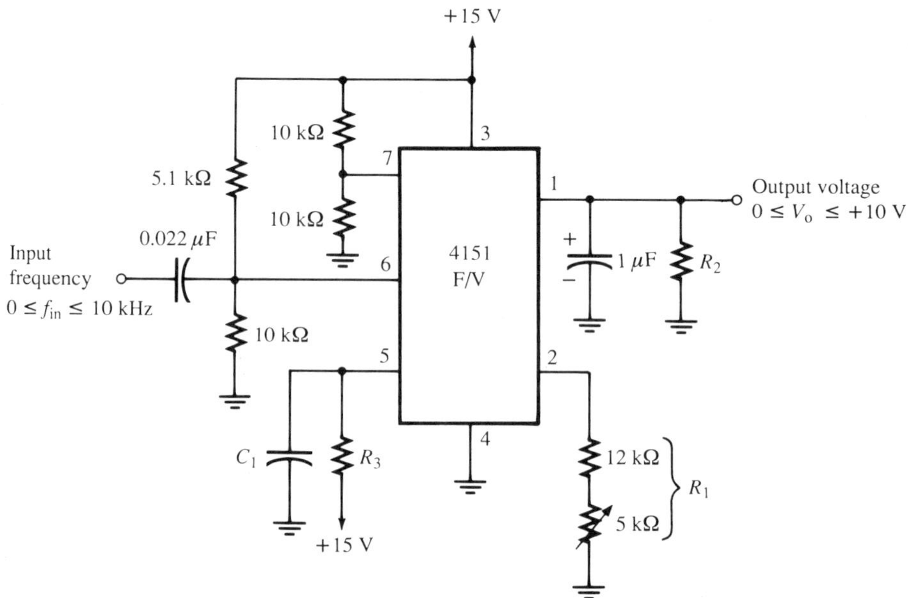

The circuit of Figure 12–13 responds only to square-wave or pulse-type inputs. If a slower-changing input waveform such as a sine wave is used, then a high-speed comparator circuit should be used ahead of the F/V converter to first convert the input to a square-wave-like input signal.

LM331 F/V Converter

As shown in Figure 12–14, the LM331 (made by National Semiconductor) is wired as an F/V converter. As outlined by the manufacturer's data sheet, the selection of the component values is given by the following stepwise process.

1. A standard value for C_T is first chosen and R_T is computed from

$$R_T = \frac{0.727}{C_T f_{in}} \qquad (12\text{–}21)$$

where f_{in} is the maximum input frequency to the LM331.

2. For the input frequency having a given duty cycle (D),

$$R_D \leq \frac{0.1D}{5C_D f_{in}} \qquad (12\text{–}22)$$

Figure 12–14
LM331 F/V converter circuit.

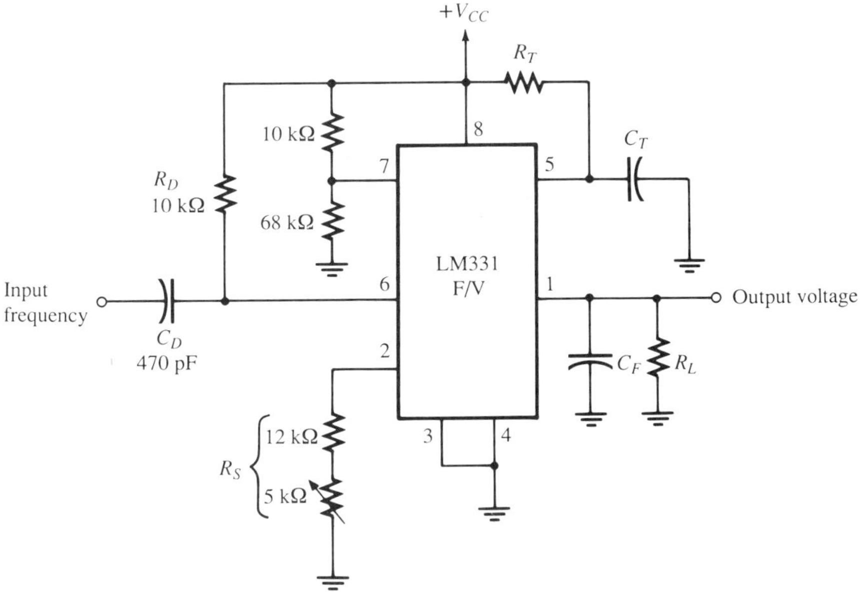

3.

$$R_S = \frac{2.2 R_T C_T R_L f_{\text{in}}}{V_{\max}} \tag{12–23}$$

where $V_{\max}$ is the maximum output voltage required at $f_{\max}$ for the LM331.

4. For a given degree of ripple filtering,

$$C_F \geq -\frac{10\text{ ms} - (0.8/f_{\text{in}})}{\ln(1 - \text{ripple})R_L} \tag{12–24}$$

Example 12–5

Determine the required component values for the LM331 F/V converter circuit of Figure 12–14 for the following specifications:

Input frequency range: 100–1 kHz (50% duty cycle)
Maximum output voltage (at 1 kHz): 10 V
Output ripple: less than 10% at 100 Hz
Load resistance: 100 kΩ

Solution:

(a) Selecting a standard value of **0.33 μF** for C_T, then R_L is determined from Equation 12–21:

$$R_T = \frac{0.727}{(0.33\ \mu\text{F})(1\ \text{kHz})}$$
$$= \mathbf{2.2\ k\Omega}$$

(b) From Equation 12–22 with **0.001 μF** selected for C_D,

$$R_D \leq \frac{(0.1)(0.5)}{(5)(0.001\ \mu\text{F})(1\ \text{kHz})}$$
$$\leq \mathbf{10\ k\Omega}$$

(c) From Equation 12–23,

$$R_S = \frac{(2.2)(2.2\ \text{k}\Omega)(0.33\ \mu\text{F})(100\ \text{k}\Omega)(1\ \text{kHz})}{10\ \text{V}}$$
$$= \mathbf{16\ k\Omega}$$

(d) For 10% ripple using Equation 12–24,

$$C_F \geq -\frac{10\ \text{ms} - (0.8/1\ \text{kHz})}{\ln(1 - 0.1)(100\ \text{k}\Omega)}$$
$$\geq -\mathbf{0.87\ \mu F} \qquad \text{(use a 1-}\mu\text{F standard value)}$$

The final circuit is shown in Figure 12–15.

Figure 12–15
Completed LM331 F/V circuit for Example 12–5.

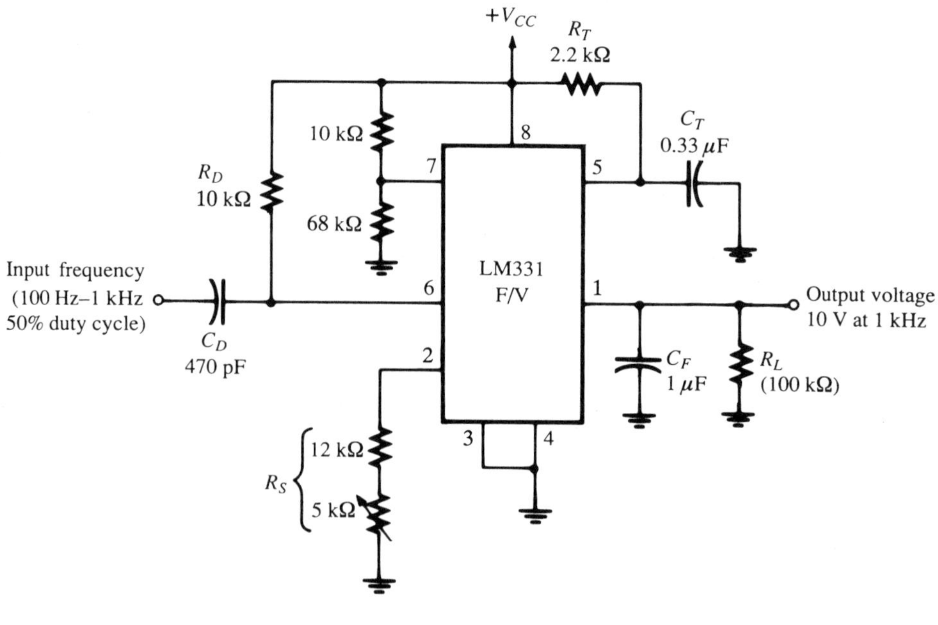

12–7 Digital-to-Analog Converters

A **digital-to-analog (D/A) converter,** or **DAC,** is a special linear circuit designed to produce an analog voltage that is proportional to a weighted parallel binary input. The output voltage varies in steps, with the smallest step being controlled by the least significant bit (LSB) of the binary input. For example, an 8-bit DAC would have 2^8, or 256, possible output levels. However, in order to use DACs properly, it is first necessary to understand the two main conversion methods that are used. These are (1) the weighted summing amplifier and (2) the *R*-2*R* ladder network.

Weighted Summing Amplifier D/A Converter

The first method, as shown in Figure 12–16 is simply a summing amplifier with weighted inputs. Here, the circuit converts a 4-bit binary number to an equivalent analog output with 16 discrete voltage levels (i.e., a 4-bit DAC). The binary inputs are simply four switches connected to a DC source of +1 V. Each switch is labeled with its corresponding binary weight. By using superposition, it is easy to determine the analog output for a given combination of switch closures.

For example, if the switch labeled 2^2 is closed, it will produce a component of the output voltage (V_o) equal to $-8R/2R$, or -4 V. If on the other hand, the switch labeled 2^1 is closed, it will produce an output component equal to $-8R/4R$, or -2 V. If both of these switches are closed at the same time, the output will then be the sum of the two components, or -6 V. Table 12–3 summarizes the analog outputs for all combinations of 16 possible binary inputs. If positive analog levels are desired, the +1-V reference is replaced with a -1-V reference, or a unity-gain inverting amplifier could be used to invert the output.

In an actual circuit, the mechanical switches are replaced with solid-state switches, each controlled by a binary input line. The +1-V reference would be derived from a regulated voltage source. When any of the binary input lines goes high, the corresponding solid-state switch closes. Conversely, when an input line returns low (ground), the switch opens. Using this scheme, then, any number of

Figure 12–16
Weighted-summing amplifier type digital-to-analog converter.

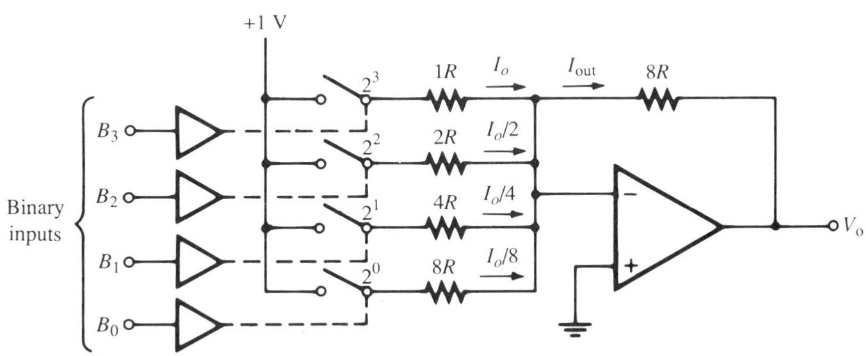

Table 12–3

Inputs				
2^3	2^2	2^1	2^0	**Output Voltage**
0	0	0	0	0
0	0	0	1	−1
0	0	1	0	−2
0	0	1	1	−3
0	1	0	0	−4
0	1	0	1	−5
0	1	1	0	−6
0	1	1	1	−7
1	0	0	0	−8
1	0	0	1	−9
1	0	1	0	−10
1	0	1	1	−11
1	1	0	0	−12
1	1	0	1	−13
1	1	1	0	−14
1	1	1	1	−15

inputs could be added to handle any required number of binary bits. However, as the number of inputs increase, it becomes more difficult to maintain the proper resistor ratios. As a result, this type of DAC is normally limited to a relatively small number of input lines.

R-2R Ladder Network D/A Converter

A circuit that is much more practical for converting binary numbers with a larger number of bits is shown in Figure 12–17a. This converter uses a circuit called a *ladder network* to give the proper weight to the binary input lines. The big advantage here is that only two different resistor sizes are required regardless of how many binary input lines are used.

To simplify the diagram, mechanical switches are shown here, but like the summing amplifier-type converter, an actual circuit would utilize solid-state switches controlled by the binary input lines. The switches have two positions. A binary 1 throws the corresponding switch to the reference voltage, whereas a binary 0 grounds the switch. An analysis of this network scheme would show that the switches have a direct binary weighting with the most significant bit (MSB) controlling the switch closest to the amplifier. The LSB, of course, would then control the switch furthest from the amplifier.

By driving this DAC with the output of a 4-bit binary counter that continually sequences the outputs from 0 to 15, a waveform like that of Figure 12–17c is generated. This is called a *staircase generator;* it has a number of applications, including use in some analog-to-digital converter circuits.

Figure 12–17
Ladder-network type D/A converter. (a) Circuit. (b) Truth table. (c) Output voltage as a function of binary switch settings.

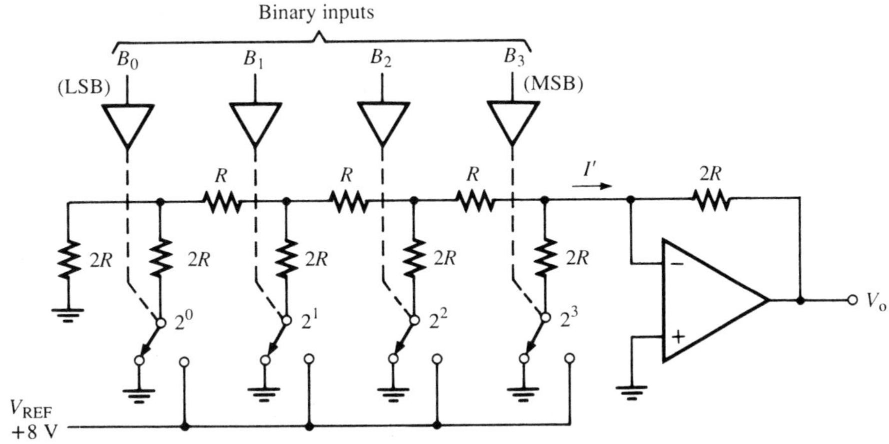

(a)

Switch settings				Decimal equivalent of binary inputs
2^3	2^2	2^1	2^0	
0	0	0	0	0
0	0	0	1	1
0	0	1	0	2
0	0	1	1	3
0	1	0	0	4
0	1	0	1	5
0	1	1	0	6
0	1	1	1	7
1	0	0	0	8
1	0	0	1	9
1	0	1	0	10
1	0	1	1	11
1	1	0	0	12
1	1	0	1	13
1	1	1	0	14
1	1	1	1	15

(b)

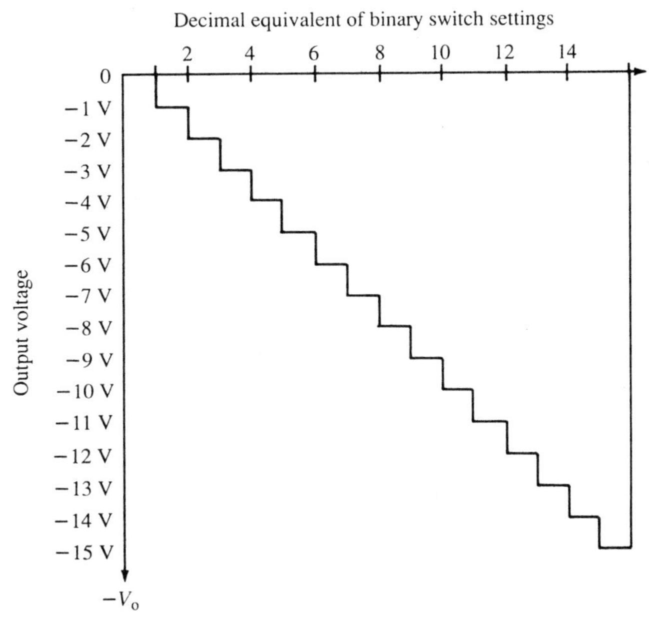

(c)

Integrated-Circuit Digital-to-Analog Converters

The best choice for most DAC applications is the integrated circuit especially designed for this purpose. These are available in 8-, 10-, 12-, 14-, and 16-bit versions. They are available in one of three output configurations:

1. Voltage output only
2. Current output only
3. Both voltage and current outputs

Figure 12–18 shows a block diagram of a DAC-IC8B 8-bit integrated-circuit DAC manufactured by Datel/Intersil. Internally, it consists of eight fast-switching current sources, an *R*-2*R* resistor ladder network, a bias circuit, and a reference control amplifier. An external reference current of 2 mA nominally programs the scaling factor for the DAC. This is usually accomplished by means of an external voltage reference source (such as Zener diode) and a resistor.

As an example, Figure 12–19 shows the DAC-IC8B connected to provide a unipolar output voltage. The Zener diode network connected to pins 14 and 15 supplies the required reference current, whereas the operational amplifier converts the resulting output current to a voltage.

Figure 12–18
Logic diagram of a DAC-IC8B 8-bit integrated-circuit D/A converter.

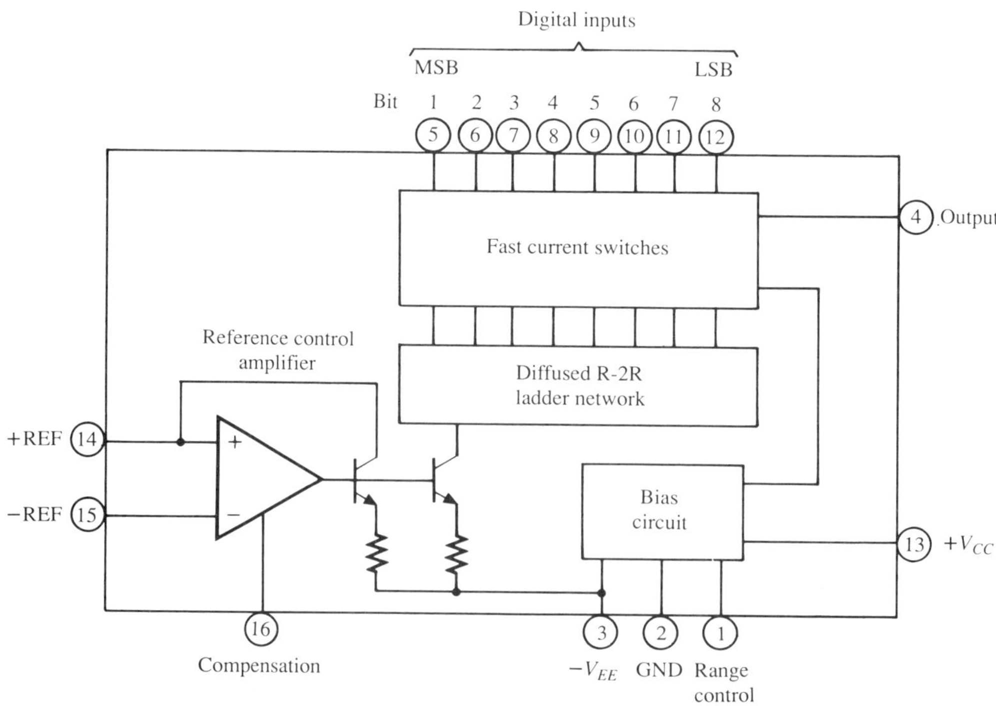

Figure 12–19
Complete D/A converter using a DAC-IC8B.

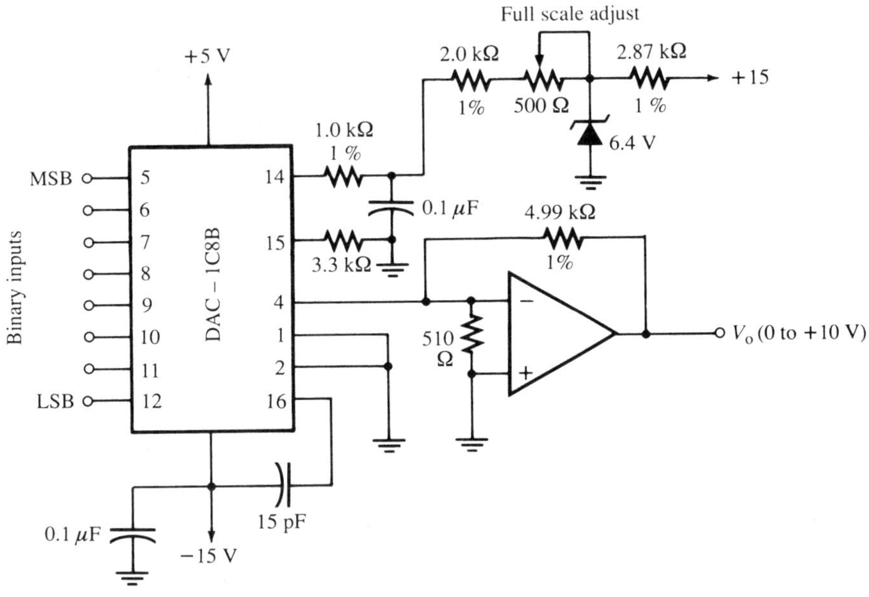

Digital-to-Analog Converter Specifications

Some of the more important specifications that apply to digital-to-analog converters are as follows:

- *Absolute accuracy:* The term *accuracy* is actually a misnomer, since the accuracy parameters are actually errors or deviations and could in effect be called *inaccuracies*. The **absolute accuracy** of a DAC is specified in terms of the difference between the actual analog output and the output that is expected when a given binary input is applied.
- *Relative accuracy:* **Relative accuracy** is specified in terms of a percentage, parts per million, or fractions of one LSB. It is a measure of the deviation from an ideal conversion gain, exclusive of DC offset voltage or full-scale gain calibration. Since the ideal conversion gain transfer is usually a straight line, the relative accuracy of a DAC can then be considered as a measure of nonlinearity.
- *Resolution:* The **resolution** of a DAC is the smallest basic increment of the analog output. The basic resolution of a DAC is given by

$$\text{Resolution} = \frac{V_o(\text{FS})}{2^N} \qquad \textbf{(12–25)}$$

where

$$V_o(\text{FS}) = \text{DAC full-scale output voltage}$$
$$N = \text{number of bits}$$

Manufacturers usually simply specify resolution by the number of binary input bits. For example, a 10-bit DAC has a finer resolution (0.098% of full scale) than an 8-bit DAC (0.39% of full scale).

- *Linearity:* **Linearity** is a measure of the deviation from the ideal straight-line conversion gain drawn between the zero and full-scale calibration points. It usually represents the ultimate limit in converter performance and, as such, could be considered the single most important specification. Other errors, such as gain and DC offset, can be trimmed to zero, but linearity cannot be changed.
- *Monoticity:* The output of a monotonic converter will always increase or remain the same when the decimal equivalent of its binary input increases. Monoticity means that the output voltage will always have a unique output voltage for any given binary input.
- *Settling time:* The **settling time** is the time taken for the DAC to settle to within $\pm\frac{1}{2}$ LSB of the final value for a given input.
- *Offset:* The offset causes the entire conversion gain of a converter to shift by a fixed DC level. This is analogous to the Y-axis intercept of the equation of a straight line. Offset error is usually adjustable, such that a binary 0 is applied to all the input terminals and the DAC is adjusted or nulled to zero the DC output voltage.

12–8 Analog-to-Digital Converters

The **analog-to-digital (A/D) converter,** or **ADC,** is the opposite of a DAC in that it changes an analog input signal to a corresponding weighted binary output. The A/D conversion process is normally more complex than that of D/A conversion; in addition, it requires a finite time to complete a conversion cycle. Although there are a number of methods for making an A/D conversion, only a few of them are commonly used in practical A/D converters. The conversion schemes that will be discussed are as follows:

- Ramp-type (single-slope)
- Staircase ramp
- Dual-slope
- Successive approximation

The Ramp-Type A/D Converter

The **ramp-type** A/D converter concept, sometimes called a *single-slope converter,* is shown by the block diagram of Figure 12–20. It has four basic sections: a ramp generator, a voltage comparator, a clock, and a binary weighted counter.

Figure 12–20
Ramp-type A/D converter.
(a) Block diagram.
(b) Waveforms.

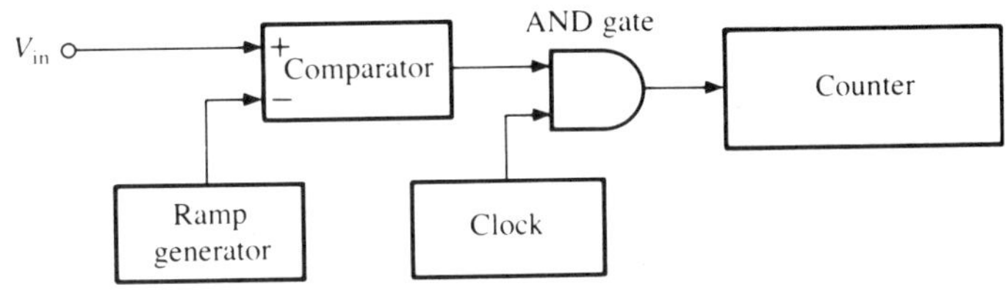

(a)

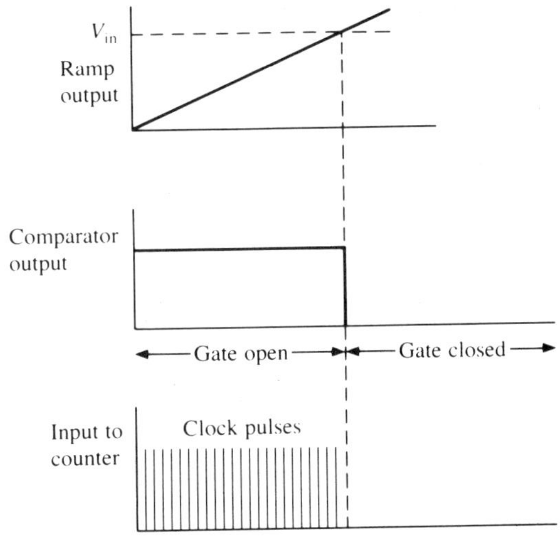

(b)

At some point in the cycle, a signal resets the binary counter to zero and simultaneously starts the ramp generator. As long as the ramp voltage is less than the voltage being measured (V_{in}), the output of the comparator is high, enabling the AND gate, which passes the clock pulses along to the counter. When the output of the ramp generator exceeds the instantaneous value of V_{in}, the output of the comparator goes low, so that the clock pulses are prevented from reaching the counter and the conversion cycle is ended. The total count is then proportional to the value of V_{in}.

Example 12–6

A ramp-type ADC has a ramp generator with a slope of 1 V/ms and a clock frequency of 100 kHz. If the analog input voltage is 2 V, what is the binary count at the end of the conversion cycle?

Solution: With a slope of 1 V/ms, the ramp voltage then takes 2 ms to reach a level of 2 V. During the 2-ms period that the AND gate is enabled, 2 ms × 100 kHz, or 200, clock pulses reach the binary counter. At the end of the conversion cycle, the counter holds a binary count (i.e., **11001000_2**) that is equal to the decimal equivalent of **200**. This might then later be converted to decimal form and displayed as 2.00 V.

Although the ramp-type A/D converter is simple in concept, its accuracy is limited by (1) the linearity and slope of the ramp and (2) the frequency stability of the clock frequency.

Staircase-Ramp A/D Converter

The staircase-ramp A/D converter uses a DAC as part of a feedback loop, as illustrated in Figure 12–21 with the block diagram of a 4-bit system. The principle involved is similar to the ramp-type A/D converter, but the ramp generator is now replaced by a DAC, whose output is a staircase waveform serving the same purpose as the analog ramp in the ramp-type converter.

At the start of a conversion cycle, the binary counter is reset to zero, and the output of the DAC is also zero. At this point the output of the comparator is high

Figure 12–21
A 4-bit staircase ramp A/D converter. (a) Block diagram. (b) The binary counter stops when V_R exceeds V_{in}.

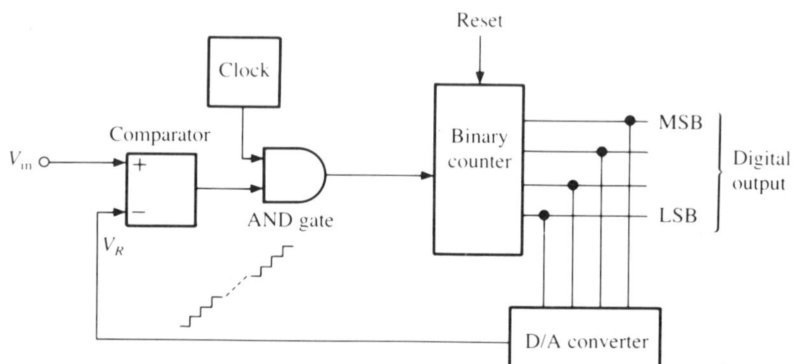

(a)

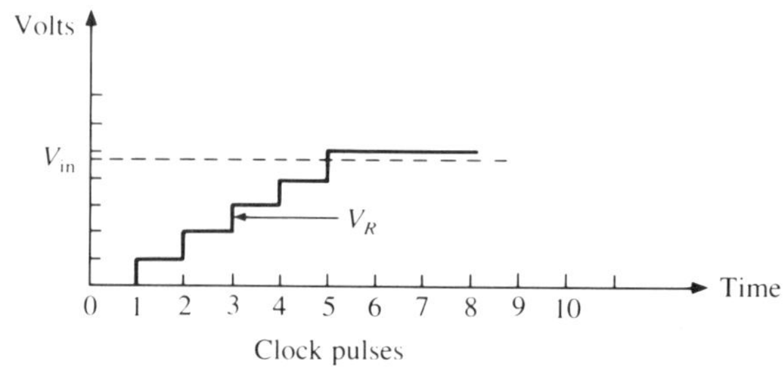

(b)

and the clock pulses go through the gate to the counter. As the binary count increases, the staircase waveform increases in equal steps toward V_{in}. When the staircase voltage exceeds V_{in}, the output of the comparator goes low, stopping the counting process. The binary count may then be converted to decimal form and used to update a digital display or it may be used as is as input data to a computer.

Dual-Slope A/D Converter

Another type of A/D converter, which is often found in hand-held digital multimeters, is the *dual-slope* A/D converter. Figure 12–22a shows a typical dual-slope conversion circuit, which has a ramp generator whose output is switched between the analog input voltage V_{in} and a negative reference voltage, $-V_{REF}$. The analog switch is controlled by the MSB of the counter. When the MSB is a logic 0, the voltage being measured is connected to the ramp generator input. When the MSB is logic 1, the negative reference voltage is connected to the ramp generator.

The conversion cycle can be divided into two distinct steps. First, the counter is reset and the input voltage is connected to the input of the ramp generator. This then causes a positive-going ramp voltage, so—assuming an 8-bit counter—the counter will count upward from 00000000_2 to 01111111_2 (i.e., 0 to 127_{10}). The slope of the ramp during this counting period is given by V_{in}/RC. When the binary count reaches 10000000_2 (128_{10}), $-V_{REF}$ is connected to the input of the ramp generator. This causes the ramp to go negative with a slope equal to $-V_{REF}/RC$, and the counter will continue to count until the ramp voltage again reaches zero. When the ramp voltage passes through zero, the comparator output goes low, which in turn (1) disables the AND gate to the clock signal and (2) stops the counter. The current count may then be processed or converted to a decimal form and displayed.

As shown in Figure 12–22b, a large input signal produces a large slope (line A), whereas a smaller input signal produces a smaller slope (line C). Since the *sample time* (t_1) for the MSB of the counter to reach a logic 1 is constant, the peak value of the ramp then is proportional to the value of V_{in}.

The slope of the negative-going ramp is a constant, meaning that large values of the peak ramp voltage take proportionally longer than smaller values to return to zero. Thus, the *measurement time* (t_2) is a linear function of the value of V_{in}, whereas the number of clock pulses counted during t_2 is proportional to the analog input voltage. The MSB of the count is not included in the final value because it is used only to control the selection of V_{in} or $-V_{REF}$. Time period t_1 always produces a full count. Time period t_2 must not exceed t_1, or the counter's MSB will again change to logic 0 on the way down, preventing the ramp from reaching zero. This then establishes an upper range limit for V_{in} that is equal to the absolute value of $-V_{REF}$. The products of the slopes and times must be equal during t_1 and t_2 because the changes in ramp value are equal. Therefore,

$$\frac{V_{in}}{RC}t_1 = \frac{V_{REF}}{RC}t_2 \qquad (12\text{–}26)$$

Figure 12–22
Dual-slope A/D converter. (a) Block diagram. (b) Charging and discharging slopes.

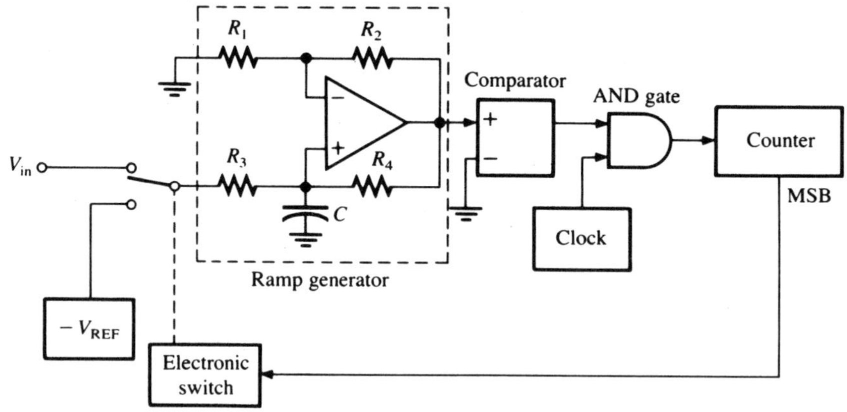

(a)

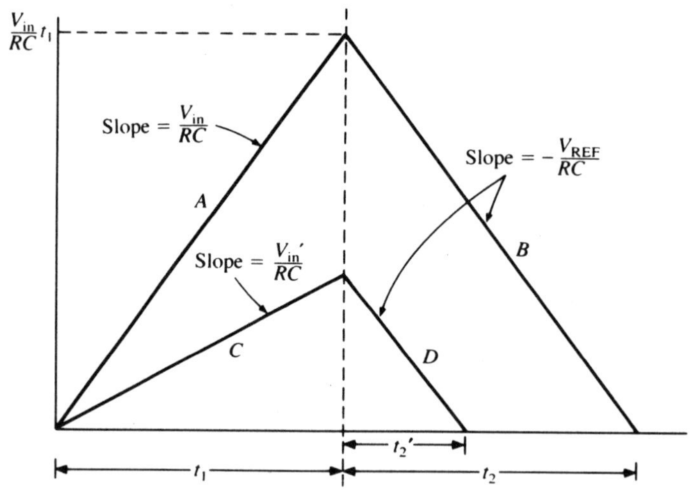

(b)

Since time and binary count are proportional, times t_1 and t_2 may be replaced with their corresponding binary counts. Therefore, Equation 12–22 may be rewritten as

$$\frac{N_2}{N_1} = \frac{V_{in}}{V_{REF}} \qquad (12\text{–}27)$$

Component tolerances, long-term drift in component values, and clock frequency do not significantly affect the accuracy of the dual-slope converter. However, the clock frequency and component values must not change significantly from t_1 to t_2. Consequently, only good short-term stability is required.

Example 12–7

A digital voltmeter using a dual-slope A/D converter has a reference voltage of −10.00 V and the ability to display a full-scale reading of 1000. Find the reading of the voltmeter at the end of time t_2 if the value of analog input voltage is 2.5 V.

Solution: Since the full-scale reading is proportional to N_1 and the actual reading is proportional to N_2, then from Equation 12–27,

$$N_2 = (1000)\frac{2.5\text{ V}}{10\text{ V}}$$
$$= \mathbf{250}$$

Dual-slope converters provide high accuracy, economy, and good immunity to noise. However, dual-slope converters tend to be slower than most other A/D converters. Nevertheless, their advantages make them well suited for use in the laboratory and in portable digital voltmeters, where extreme speed is not required.

Successive Approximation A/D Converter

Two major drawbacks found in most A/D converters are that (1) they are relatively slow and (2) the conversion time is a function of the magnitude of the input voltage. The successive approximation A/D converter overcomes both of these problems to a considerable degree. As a result, the successive approximation–

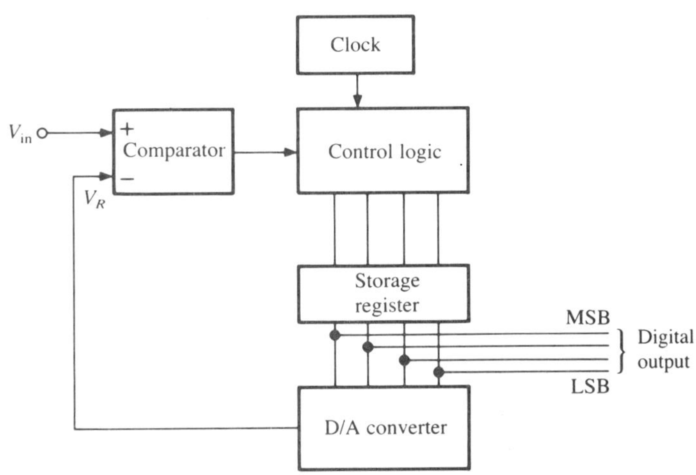

Figure 12–23
Block diagram of a 4-bit successive approximation A/D converter.

type A/D converter has the ability to make conversions in microseconds, compared with the several milliseconds required by the ramp, dual-slope, and V/F converters.

Although the block diagram for the successive approximation A/D converter shown in Figure 12–23 looks very much like that of a staircase ramp A/D converter, it differs in the control logic section. Instead of allowing the staircase to climb from zero through all sequential states up to V_{in}, it first tries a step equivalent to the most significant bit of the converter. If this step does not exceed the value of V_{in}, it is then retained in the storage register. Otherwise, the storage register is set to zero, and the next largest bit is then tried, and so on, until all bits have been tried in order of decreasing significance. A significant feature of this system is that it takes only N steps to produce an N-bit binary output, regardless of the magnitude of the input signal.

The conversion time of the successive approximation converter is a constant and is given by

$$T = \frac{N}{f_{clk}} \tag{12–28}$$

where

N = number of bits in the converter

f_{clk} = clock frequency

Thus a 10-bit converter with a 1-MHz clock makes a conversion every 10/1 MHz, or 10 μs.

Summary

This chapter looked at those operational amplifier and linear integrated-circuit networks that proportionally convert signals from one kind of parameter to another. Specifically, these include voltage-to-current, current-to-voltage, voltage-to-frequency, frequency-to-voltage, analog-voltage-to-digital, and digital-to-analog-voltage converters using either a weighted summing amplifier or an R-$2R$ resistive ladder network. Analog-to-digital converters can be based on one of the following conversion methods: ramp-type (single-slope), staircase ramp, dual-slope, and successive approximation.

As a general rule, the simpler circuit functions such as the current-to-voltage and voltage-to-current converters (floating and grounded loads), as well as current amplifiers, are best made using operational amplifiers. The more complex functions, such as the voltage-to-frequency, frequency-to-voltage, analog-to-digital, and digital-to-analog converters are best achieved using one of the specialized integrated-circuit devices designed for these purposes.

Analog signals that are transmitted over considerable distances by hard-wired cable circuits are first converted from a voltage into a proportional current signal at the transmitting end and then converted back to the corresponding voltage at the receiving end, which eliminates voltage drops in a long circuit path.

Like the voltage-to-current conversion, voltage-to-frequency conversion is also often used where the voltage to be measured is physically located some distance from the actual source, as it is easier to transmit the pulse train with more reliability than a low-level DC signal. In addition, it is well suited for use in tone generators, frequency-shift keying (FSK), FM modulators, clock generators, signal generators, function generators, and phase-locked loops.

Those analog signals that must be read and processed further by a computer are converted to equivalent digital signals by an analog-to-digital converter. On the other hand, computers control external devices with analog inputs via a digital-to-analog converter.

Specific examples of integrated circuit devices were shown using the 566 and 4151 for a V/F converter and either the LM331 or the 4151 for F/V converters.

Glossary

absolute accuracy The difference between the actual analog output of a digital-to-analog converter and the output that is expected when a given binary input is applied.

analog-to-digital converter (ADC) A network in which the magnitude of an analog signal is converted to an equivalent binary value.

conversion gain The proportionality constant that relates the output parameter of a linear converter is related to its input parameter; sometimes referred to as the *transfer function*.

current-to-voltage converter A network in which the input signal in the form of a current is changed to a proportional output voltage.

digital-to-analog (D/A) converter (DAC) A network in which digital data in binary form is changed into an equivalent analog output voltage.

frequency-to-voltage (F/V) converter A network in which the output voltage is proportional to its input frequency.

linearity A measure of the deviation from the ideal straight-line conversion gain, drawn between the zero and full-scale calibration points.

monoticity The property whereby the output voltage of a converter will always have a unique output voltage for any given binary input.

relative accuracy A measure of the deviation from an ideal conversion gain, exclusive of DC offset voltage or full-scale gain calibration; specified in terms of a percentage, parts per million, or fractions of one LSB.

resolution The smallest basic increment of the analog output; specified by the number of binary input bits.

settling time The time taken for the output of a digital-to-analog converter to settle to within $\pm\frac{1}{2}$ of the LSB of the final value for a given input.

voltage-to-current converter A network in which the input signal in the form of a voltage is changed to a proportional output current; also called a *transconductance amplifier*.

voltage-to-frequency (V/F) converter A network in which the output frequency is proportional to the level of the input control voltage; also called a *voltage-controlled oscillator*.

Review Questions

1. Define *conversion gain*.
2. Give the units of the conversion gain for each of the following converter types: voltage-to-current, current-to-voltage, voltage-to-frequency, and frequency-to-voltage.
3. Explain when each of the following type converters may be used: voltage-to-current, current-to-voltage, voltage-to-frequency, and frequency-to-voltage.
4. Draw the schematic diagram of a current-to-voltage converter using an operational amplifier, and give the equation to use in calculating the output voltage.
5. Draw the schematic diagram of a voltage-to-current converter using an operational amplifier, and give the equation to use in calculating the output current.
6. Draw the schematic diagram of a V/F converter using either a 4151 or 566 device; also show how the circuit component values are to be determined.
7. Draw the schematic diagram of an F/V converter using either an LM331 or 4151 device; also show how the circuit component values are to be determined.
8. What is a DAC?
9. Draw the schematic diagram of a 4-bit DAC using a weighted summing amplifier network.
10. Draw the schematic diagram of a 4-bit DAC using an R-$2R$ ladder network.
11. Explain the advantage of using an R-$2R$ network over a weighted resistor network in a DAC.
12. Define the *resolution* of a DAC.
13. Explain the differences among the following analog-to-digital converters: ramp (single slope), dual slope, staircase ramp, and successive approximation.
14. Explain the difference between the sample and measurement times in a dual-slope DAC.

Exercises

1. For the current-to-voltage converter circuit of Figure 12–1, determine the value of R_F needed to produce a maximum output voltage of -2.5 V if the maximum input current is 4 mA. What is the conversion gain?
2. The current-to-voltage converter circuit of Figure 12–3a uses a photoresistor to measure variations in light intensity when it is connected to a 1.5-V source and $R_F = 10\ \text{k}\Omega$. Determine the range of the output voltage if the photocell's resistance varies between 1260 Ω and 9635 Ω.
3. What is the range of the photoresistor current in the converter of Exercise 2?
4. What is the conversion gain for the current-to-voltage converter of Exercise 2?
5. Determine the current flowing through a 500-Ω load for the voltage-to-current circuit of Figure 12–4 for an input voltage of 3.5 V if $R_1 = 10\ \text{k}\Omega$.
6. What is the required value for R_1 in the voltage-to-current circuit of Figure 12–4 to give a load current of 10 mA when the input voltage is 1.5 V? What is the conversion gain?
7. The voltage-to-current converter circuit of Figure 12–4 is to serve as the basis for a high–impedance voltmeter. If a 1-mA meter having an internal resistance of 220 Ω is used for R_L, determine the value of R_1 needed for the meter to give a full-scale meter reading if the maximum input voltage is 10 V.
8. Repeat Exercise 7 for a maximum input voltage of 150 V.
9. Determine the load current for the grounded-load, voltage-to-current circuit of Figure 12–6 for an input voltage of 1.5 V if $R = 300\ \Omega$ and $R_L = 1.5\ \text{k}\Omega$.
10. What is the conversion gain for the voltage-to-current converter of Exercise 9?

11. The current amplifier of Figure 12–7 uses a 1-mA ammeter to measure the current through a solar cell. Determine the current through the solar cell if the ammeter reads 0.72 mA when $R_1 = 10\ \text{k}\Omega$ and $R_2 = 100\ \text{k}\Omega$.

12. Determine suitable component values for the current amplifier circuit of Figure 12–7 so that a 100 mA meter reads half-scale for an input current of 25 μA.

13. A V/F converter has an output frequency of 25.8 kHz when no input signal is applied. Determine its conversion gain if the output frequency is 25.53 kHz when the input voltage is −0.27 V.

14. A 566 device is wired as a V/F converter using the basic circuit of Figure 12–9c. Determine the free-running output frequency at either pins 3 or 4 if $R_1 = 5.6\ \text{k}\Omega$, $R_2 = 2.2\ \text{k}\Omega$, $R_3 = 10\ \text{k}\Omega$, $C = 820$ pF, and $V_{CC} = 9$ V.

15. Determine the output voltage of the binary-weighted digital-to-analog converter circuit of Figure 12–16 when the inputs are set at the following logic levels:
    (a) $2^0 = 0,\ 2^1 = 1,\ 2^2 = 1,\ 2^3 = 1$
    (b) $2^0 = 1,\ 2^1 = 0,\ 2^2 = 1,\ 2^3 = 0$
    (c) $2^0 = 0,\ 2^1 = 1,\ 2^2 = 0,\ 2^3 = 1$
    (d) $2^0 = 0,\ 2^1 = 0,\ 2^2 = 1,\ 2^3 = 1$

16. What is the percent resolution of the DAC of Exercise 15 if the full-scale output voltage is −15 V?

17. How many bits would a DAC need in order to achieve a resolution of at least each value?
    (a) 0.15% of full scale
    (b) 0.07% of full scale
    (c) 0.02% of full scale

18. What is the conversion time of a 12-bit successive approximation ADC if the clock frequency is 75 kHz?

13 The Phase-Locked Loop

Instructional Objectives

Upon completing this chapter, you will be able to

- Draw the block diagram of the basic phase-locked loop.
- Explain the general principle of the phase-locked loop.
- Briefly explain the function of the phase detector, loop filter, and VCO.
- Explain the difference between free-running, capture, and phase-locked states in terms of their frequency ranges.
- Explain the operation of the frequency synthesizer and how it can be used to increase the resolution of a frequency counter.
- Design phase-locked loop circuits built around the following integrated-circuit phase-locked loop devices: 560B, 561B, 562, 565, 567, and 4046.

Introduction

In the early 1930s, the superheterodyne receiver was king, but a simpler method was sought to reduce the number of tuned stages. The new receiver that resulted was called a *synchrodyne,* and it first consisted of a local oscillator, mixer, and an audio amplifier. However, the frequency drift of the local oscillator hampered reception. To combat this drift, the frequency of the local oscillator was then compared with the receiver's input by a phase detector so that a correction voltage would be generated. This electronic feedback circuit began the evolution of the phase-locked loop.

The **phase-locked loop,** or **PLL,** is now a very sophisticated electronic network that is often used for such diverse purposes as motor speed control, Touch-Tone® (DTMF) decoders, modulation, demodulation, filtering, frequency scaling, and frequency control of electronic communication equipment. Although entire books have been written on the phase-locked loop principle, this chapter will serve only as a brief introduction to the basic concept while presenting a number of the commonly used PLL integrated-circuit devices.

Figure 13–1
Block diagram of the basic phase-locked loop.

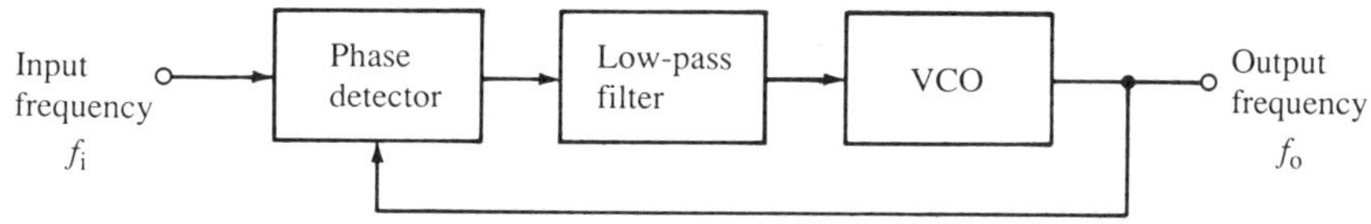

13–1 The Basic Principle

As shown in the block diagram of Figure 13–1, the basic phase-locked loop is basically an electronic feedback loop system consisting of the following components:

1. Phase detector
2. Low-pass filter
3. Voltage-controlled oscillator (VCO)

These three building blocks are said to be in the loop's *forward path;* the single connection between the output of the VCO and one input of the phase detector is the *feedback path*. The characteristics of these individual building blocks together determine the overall response characteristics of the loop.

The Phase Detector

The **phase detector,** also called a *mixer* or *phase comparator,* generates an average (DC) output voltage that is proportional to the phase difference ($\Delta\theta$) between the input signal of the phase-locked loop and that of the output of the VCO. This output voltage (V_o) is referred to as the **error voltage** and is given by

$$V_o = K_\theta \Delta\theta \qquad \textbf{(13–1)}$$

where K_θ is the phase detector conversion gain in volts/radian (V/rad).*

The Voltage-Controlled Oscillator

As discussed in Chapter 12, the voltage-controlled oscillator (VCO) is a voltage-to-frequency converter that changes a DC control voltage input into a pulsed waveform whose output frequency is proportional to the magnitude of the input

*1 radian = 360°/2π, or 57.3°.

control voltage. Its conversion gain (K_{VCO}) relates output frequency to input voltage in the form of the equation for a straight line:

$$f_o = f_{\min} + K_{VCO}V_{\text{in}} \qquad \textbf{(13–2)}$$

where

$$f_{\min} = \text{output frequency when } V_{\text{in}} = 0$$
$$K_{VCO} = \text{VCO conversion gain, Hz/V}$$

The VCO output frequency (f_o) is fed to one input of the phase detector, where it is compared with the frequency of the input signal (f_i) at the other input of the phase detector.

The Loop Filter

The inclusion of an active or passive low-pass filter network in the phase-locked loop has two major functions. First, it removes any noise and high-frequency components from the output voltage of the phase detector, resulting in an average (DC) control voltage that drives the VCO. Secondly, it is the primary building block that determines the dynamic performance of the loop covering the following factors:

- Capture and lock ranges
- Bandwidth
- Transient response

The complete phase-locked loop system exhibits the characteristics of a second-order system (analogous to the transient behavior of a swinging pendulum or vibrating string). As a function of the radian frequency ($\omega = 2\pi f$), the decibel response of a second-order system has the form:

$$A_{\text{dB}} = -20 \log[\omega^4 + 2\omega^2(2\alpha^2 - 1) + 1]^{1/2} \qquad \textbf{(13–3)}$$

where α is a dimensionless *damping factor,* which is also referred to as the *damping ratio.* By plotting Equation 13–3 on semilogarithmic axes, as shown in Figure 13–2, the single parameter that governs the overall shape of the response curve is α. If $\alpha = 0$, then we have a sinusoidal oscillator.

The VCO output frequency (f_o) is fed back to one input of the phase detector, where it is compared with the frequency of the input signal (f_i) at the other input of the phase detector. The output of the phase detector is an error voltage, which is an average (DC) voltage that is proportional to the difference in frequency ($f_i - f_o$) and phase ($\Delta\theta$) between the input and VCO signals. The error voltage is then filtered to remove traces of high-frequency noise and in turn serves as the input control voltage for the VCO.

Figure 13–2
Frequency response of a second-order system for various values of damping ratio.

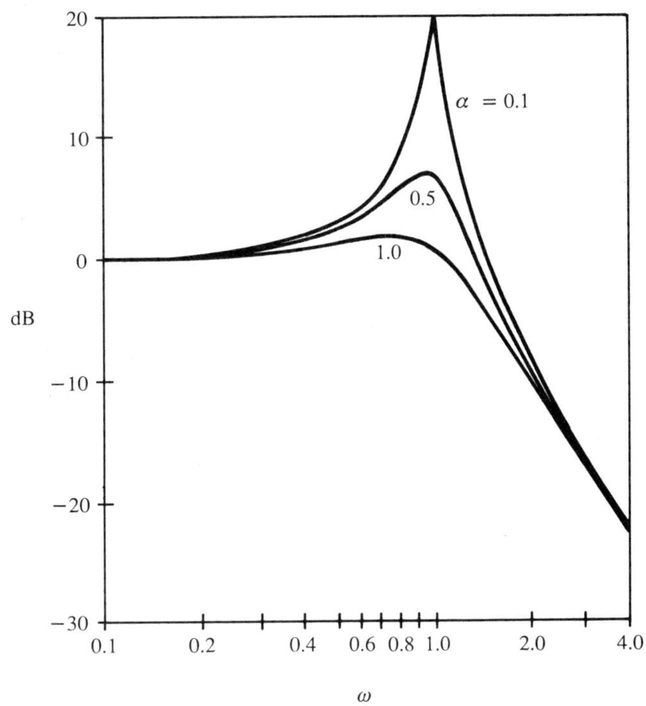

Free-Running, Capture, and Phase-Locked

In operation, the phase-locked loop is in any one of the following three distinct states:

- Free-running
- Capture
- Phase-lock

If we assume that initially the phase-locked loop has no input signal, the VCO then runs at its **free-running frequency,** which is set by either an external RC or LC network. Here the loop is said to be *unlocked*. When an input signal is applied to the phase detector and the VCO frequency starts to change in a direction that reduces the frequency difference between the VCO and the input, the loop is said to have *acquired capture* and is now in the *capture state*. The input and output frequencies are not equal, but in time they will be.

In order for a phase-locked loop that is initially unlocked to respond to an input signal and then be in the capture state, the frequency of the input signal must be within a narrow frequency range called the **capture range,** which is set by the response parameters of the loop circuit. If the input frequency is outside this capture range, the loop will never acquire capture and the output frequency of the

VCO will equal its free-running frequency. On the other hand, if the input frequency is within the loop's capture range, then the VCO frequency starts to change in a direction that reduces the frequency difference between the VCO and the input.

The capture range is often specified as a percentage of the VCO's free-running frequency, so that the free-running frequency is within this range. For example, the capture range of a given phase-locked loop is ±2% of a free-running frequency of 1250 Hz. Consequently, the capture range is ±(0.02)(1250 Hz), or ±25 Hz, about the free-running frequency. If the input frequency is between 1225 Hz and 1275 Hz, then the loop will acquire capture, and the VCO frequency will then shift to equal that of the input frequency. Otherwise, the output frequency of the VCO will remain at 1250 Hz for those input frequencies less than 1225 Hz and above 1275 Hz.

When the loop acquires capture, the input and output frequencies are not equal. However, the process is repeated many times going around the loop. The input and VCO frequencies are compared by the phase detector. Its DC output voltage is proportional to this frequency difference. After filtering, this frequency difference drives the VCO whose output frequency tries to get closer to the input frequency. Eventually the loop will reach the situation where the VCO's frequency and the loop's input frequency are exactly equal, except for a possible phase shift. When the input and output signals have exactly the same frequency, except for a given phase shift, the loop is then said to be *locked,* or *phase-locked.* The phase shift, if any, occurs because the VCO is asynchronous with the input signal. If there is no input signal, the VCO runs along at its own pace (i.e., its free-running frequency). When the loop receives an input signal, it is then asynchronous with the VCO signal.

Because of the loop's transient behavior, it takes a finite amount of time for the loop, once it acquires capture, to become locked; this is referred to as the *settling time.* In a well-designed phase-locked loop system, this may be on the order of several milliseconds; hence the transition from being an unlocked loop to acquiring capture and becoming locked may appear as a virtually instantaneous process.

Once locked, the input frequency can change within certain limits, and the VCO's output frequency will follow this change. When the input frequency changes, even slightly, the phase detector then senses this frequency difference and the shifting process starts all over again until the VCO output and input frequencies are exactly the same (except for a possible phase difference). Once locked, the input frequency range over which the VCO can track is called the **lock,** or *hold-in,* **range.** As with the loop's capture range, it is often specified as a percentage of the VCO's free-running frequency and is always wider than the capture range. Once the loop has acquired capture and phase-lock, if the input frequency then shifts to a frequency outside the lock range, then the loop immediately loses phase-lock and the output frequency equals the VCO's free-running frequency. For the loop to again be phase-locked, the new input signal then has to be within the loop's capture range to start the cycle all over again.

Figure 13–3
Block diagram of the basic phase-locked loop frequency synthesizer or frequency multiplier.

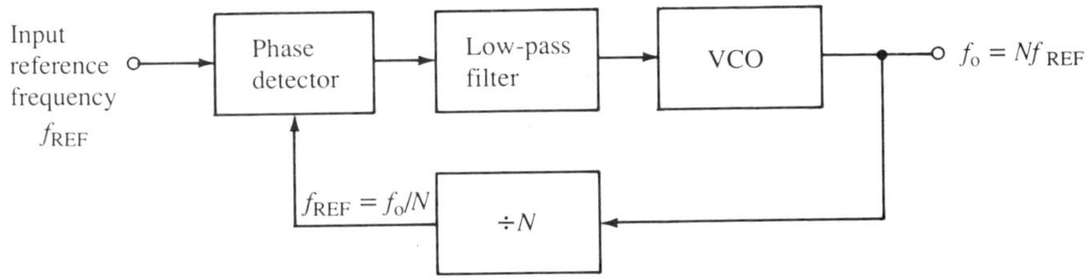

13–2 The Frequency Synthesizer

A **frequency synthesizer** is a frequency source whose output is an integer multiple of an input reference frequency. In its simplest form, it is nothing more than a frequency multiplier and is now an important building block in many communications systems. As shown in Figure 13–3, the basic frequency synthesizer is formed by breaking the feedback path between the VCO and the phase detector of the basic PLL of Figure 13–1 and inserting a divide-by-N counter. In addition, the input signal reference is obtained from a very stable crystal-controlled oscillator (Figure 13–4) in the 1- to 10-MHz range. As in the basic phase-locked loop, the phase detector, loop filter, and VCO make up the loop's forward path, whereas the divide-by-N counter now constitutes the feedback path. If the VCO's output frequency is N times larger than the input reference frequency,

$$f_o = Nf_{REF} \tag{13–4}$$

This output signal is then fed to the divide-by-N counter, which produces one output pulse for every N input pulses. The counter's modulus (N) may be either fixed or programmable via thumbwheel switches in BCD (binary-coded decimal)

Figure 13–4
TTL-based crystal-controlled frequency reference from 1 MHz to 10 MHz.

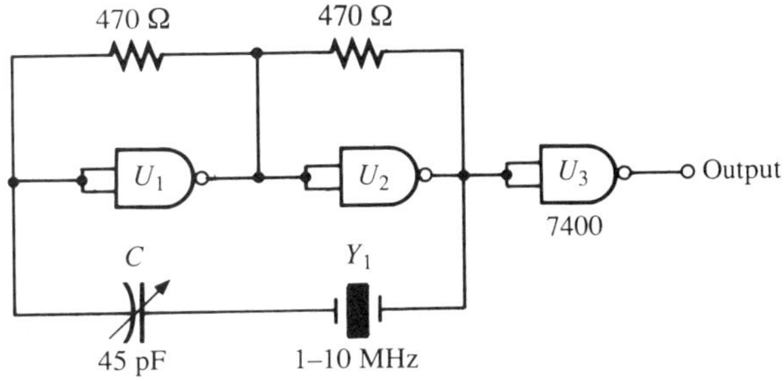

format. The output frequency of the counter is then f_o/N. Using Equation 13–1, this then becomes

$$f_c = \frac{f_o}{N} = f_{REF} \qquad (13\text{–}5)$$

When the loop is phase-locked, the frequencies of the two inputs of the phase detector are equal, except for a possible phase difference.

In communications systems, practical synthesizers generally have their output frequency in the 30- to 300-MHz range (VHF and UHF). However, synthesizers are capable of working with low-frequency signals in the audio frequency range (20 Hz to 20 kHz). Very often, a frequency synthesizer will be used to increase the resolution of a frequency counter by a factor of 10 or 100 without increasing the sampling or display update time. Most counters use a sampling or display update time of 1 s. In order to obtain a resolution of one more decimal place, the frequency counter must then be set to have a sampling or display update time of 10 s, which will take ten times longer to display the input frequency. In this 10-s interval, the input signal might change frequency slightly and give erroneous readings.

Instead of using a very stable input frequency reference, the frequency that would otherwise be measured by the frequency counter is now fed to the input of the synthesizer. The frequency counter is connected to the output of the VCO and reads either 10 or 100 times larger that the input signal (Figure 13–5). For example, a frequency counter measures a given signal and indicates a frequency of 974 Hz. Because of the typical ±1 count error inherent to all digital display, the actual input frequency could be between 973 and 975 Hz. To achieve a greater resolution, a ×10 synthesizer is placed between the frequency source and the frequency

Figure 13–5
Block diagram showing the use of a phase-locked loop to increase the resolution of a frequency counter.

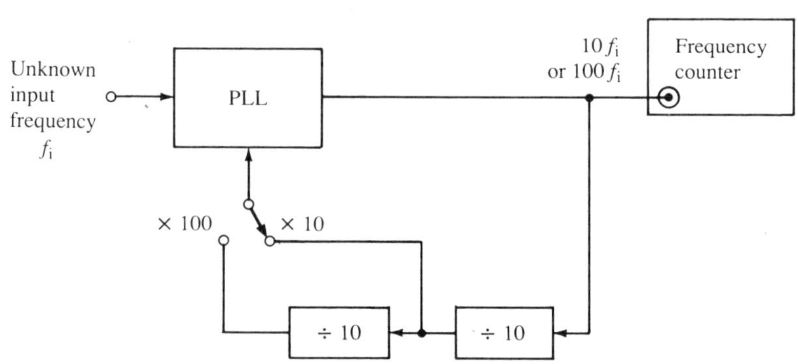

counter. Suppose the counter reads 9746 Hz. The actual input frequency is then 9746 Hz/10, or 974.6 Hz (actually the frequency will be between 974.5 Hz and 974.7 Hz). As mentioned earlier, we could set the frequency counter to have a sampling or display time of 10 s in order to produce the extra decimal place, but it would take ten times longer to display the input frequency. Using the synthesizer method, the extra time lag is eliminated. All that is necessary is mentally to divide the displayed frequency by 10 (or 100, if need be) to obtain the correct input frequency.

13–3 Integrated-Circuit PLL Devices

The following subsections describe several of the popular integrated-circuit phase-locked loop devices that contain a phase detector, VCO, and several specialized functions within a single package. Several external resistors and capacitors are then usually required to set the VCO free-running frequency and the loop filter.

560B Phase-Locked Loop

The 560B is the most fundamental of what is sometimes referred to as the "560 series" of phase-locked loop devices first introduced by Signetics, which include the 560B, 561B, 562, 564, 565, 566, and 567. The devices in the 560 series are commonly referred to as *analog* phase-locked loops and differ from digital phase-locked loops primarily in the type of phase detector that is used. In all cases, the VCO output signal is a square wave.

The 560B, whose internal block and pin diagrams are shown in Figure 13–6, contains a phase detector, amplifier, and VCO in a 16-pin package. When locked onto an input signal, two useful outputs are provided. First, an output voltage that is proportional to the frequency of the incoming signal is available at pin 9 as the demodulated FM output. The second output is the square-wave output signal of the VCO. The value of the external capacitor, connected between pins 2 and 3, sets the VCO free-running frequency according to

$$C_o(\text{pF}) = \frac{300}{f_o} \qquad (13\text{–}6)$$

The 560B is used primarily as an FM demodulator, as shown by the basic circuit of Figure 13–7. The VCO is adjusted by C_o to the center frequency of the FM signal. For most applications, the pair of loop filter capacitors (C_1 and C_2) can be found from

$$C_1 = C_2 = \frac{13.3}{f_{3\ \text{dB}}} \qquad (13\text{–}7)$$

where $f_{3\ \text{dB}}$ is the desired 3-dB bandwidth of the demodulated information.

Figure 13–6
560B integrated-circuit phase-locked loop. (a) Pin diagram; (b) Internal block diagram.

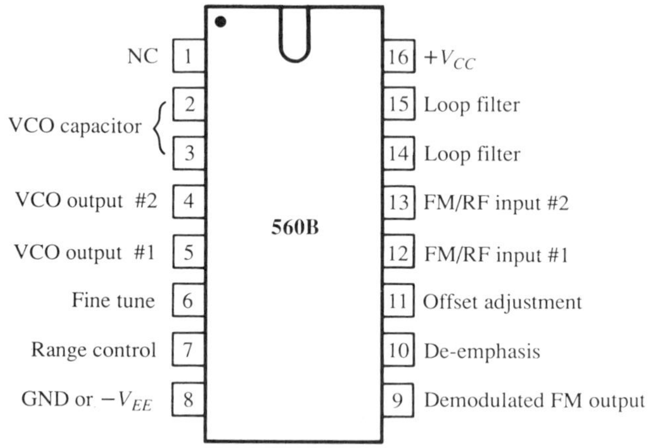

(a)

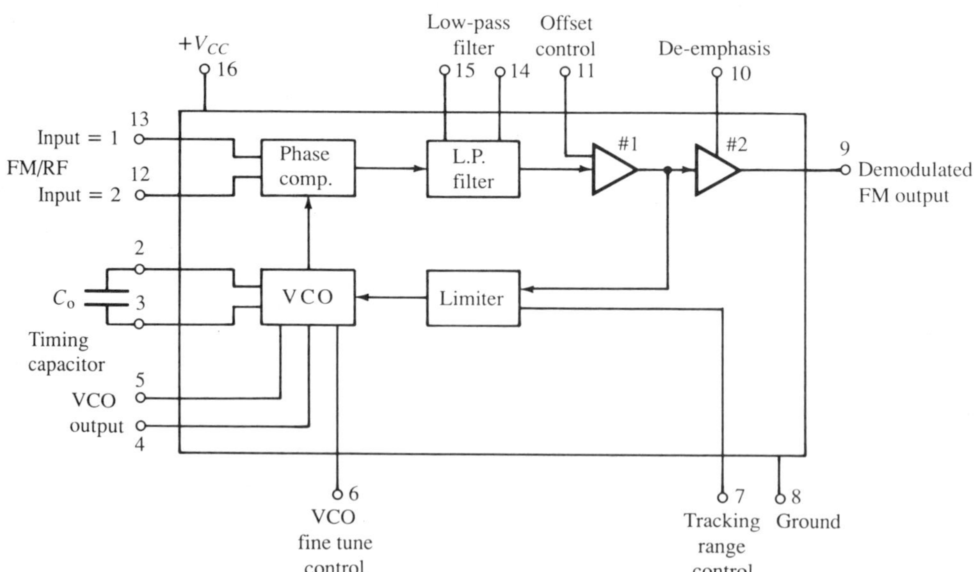

(b)

Figure 13–7
Phase-locked loop FM demodulator using a 560B.

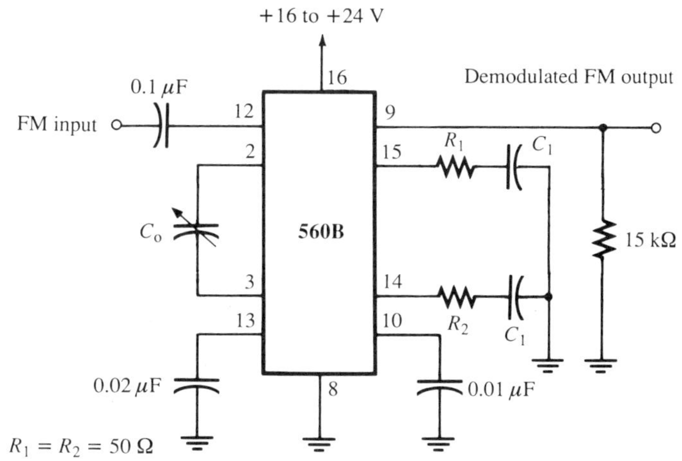

561B Phase-Locked Loop

The 561B, shown in Figure 13–8, is identical to the 560B except that it includes an additional phase detector, which allows the device to be used as a synchronous AM detector as well as an FM demodulator.

The demodulated FM output at pin 9 is an output voltage that is a function of the frequency deviation of the input signal. For a $\pm 1\%$ deviation, the output is approximately 0.3 V peak-to-peak (0.11 V RMS). As an example, a standard 10.7-MHz IF (intermediate frequency) circuit has a deviation of approximately ± 75 kHz. The percent deviation is then

$$
\begin{aligned}
\%\ \text{deviation} &= \frac{\pm 75\ \text{kHz}}{10.7\ \text{MHz}} \times 100\% \\
&= \pm 0.7\%
\end{aligned}
$$

Figure 13–8
561B integrated-circuit phase-locked loop pin diagram.

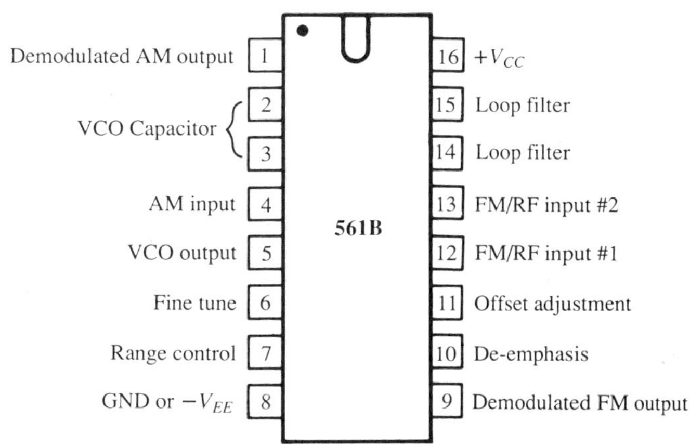

Consequently, the output voltage swing is

$$V_o = \frac{0.3 \text{ V p-p}}{\pm 1\% \text{ deviation}} \times (\pm 0.7\%)$$
$$= 0.21 \text{ V p-p} \quad (0.075 \text{ V RMS})$$

As an illustration of the 561B's demodulation capability, Figure 13–9 shows the circuit for a simple AM receiver, which unlike other AM receivers, requires no capacitor-inductor tank circuit for tuning. Instead, the 365-pF tuning capacitor between pins 2 and 3 is adjusted to make the VCO oscillate at the carrier frequency of the station to be received.

To operate, the receiver circuit requires an outside antenna and a good ground. In addition, sufficient signal must be present at the input of the phase-locked loop. Otherwise, a "swishing" sound may result, which is due to the frequency difference between the incoming carrier and the VCO when the loop lock is unstable. Another drawback to this simple circuit is the "hand-capacitance" effect as a result of the nongrounded capacitive tuning of the VCO. One remedy for this problem is to use a vernier dial and an insulated shaft on the tuning capacitor. Furthermore, the circuit's operation may be enhanced by using an untuned broadband amplifier ahead of the receiver to increase sensitivity while limiting input signal levels to less than 0.5 V RMS.

562 Phase-Locked Loop

The 562 PLL (Figure 13–10) is similar to the 560B, except that the internal connection between the VCO output and the phase detector is broken. Thus, two external connections are provided, which allow for insertion of TTL counters to be placed in the feedback path for frequency synthesis, as shown in Figure 13–11.

Figure 13–9
Simple phase-locked loop AM receiver.

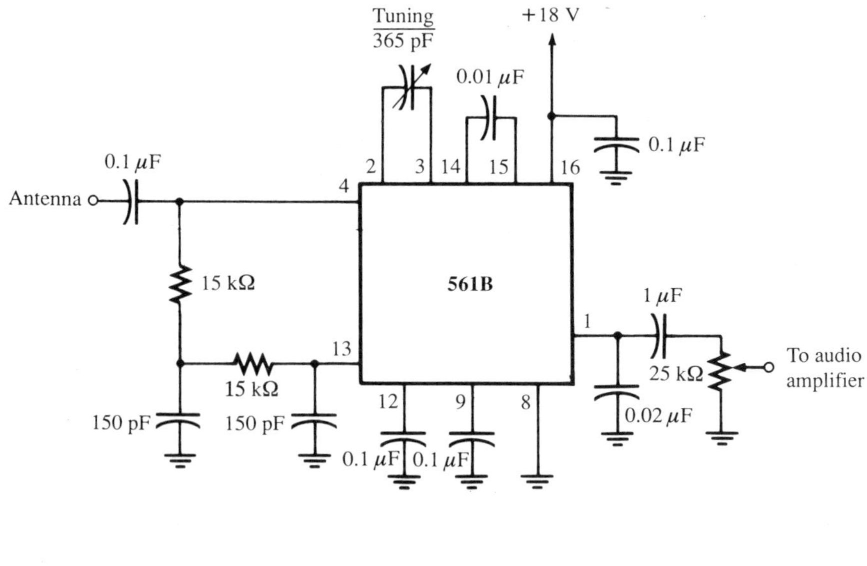

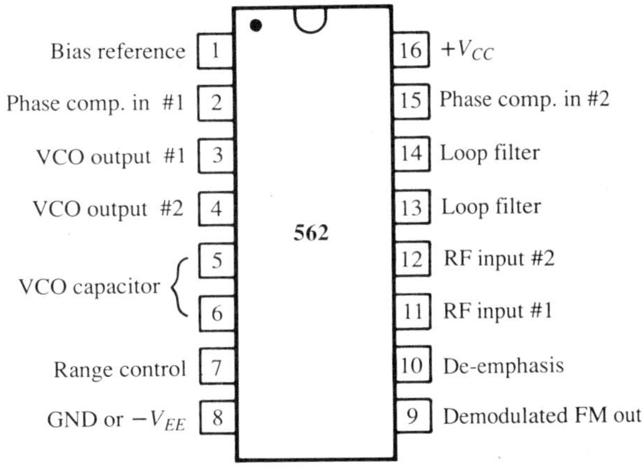

Figure 13–10
562 integrated-circuit phase-locked loop pin diagram.

565 Phase-Locked Loop

The 565 PLL, shown in Figure 13–12, is perhaps the most popular device of the 560 series. It is a general-purpose device similar to the 562. Whereas the 560B, 561B, and 562 devices are limited to frequencies below 30 MHz, the 565 is limited to frequencies below 500 kHz.

A generalized circuit for the 565 is shown in Figure 13–13. The VCO free-running frequency is found from

$$f_o = \frac{1.2}{4R_1C_1} \tag{13–8}$$

Figure 13–11
Phase-locked loop frequency multiplier using TTL divide-by-*N* counter.

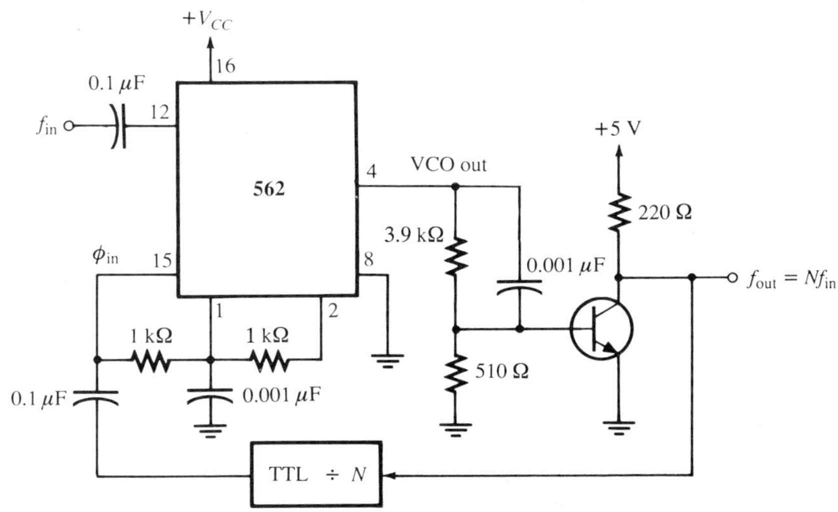

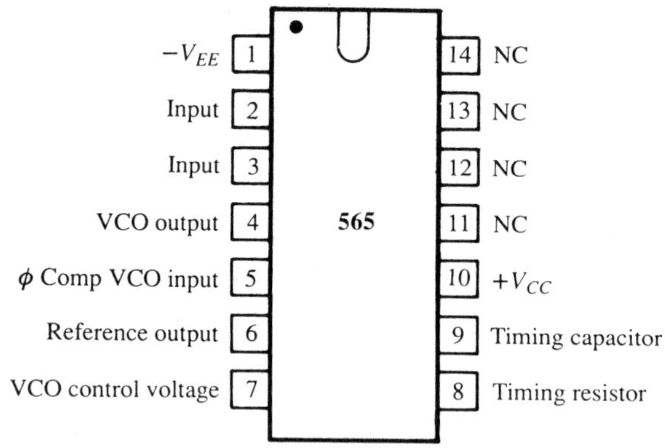

Figure 13–12
565 integrated-circuit phase-locked loop pin diagram.

whereas the lock and capture ranges are determined, respectively, from

$$f_L = \pm \frac{8 f_o}{V_{CC}} \tag{13–9}$$

and

$$f_C = \pm \frac{1}{2\pi} \sqrt{\frac{2\pi f_L}{3600 C_2}} \tag{13–10}$$

Capacitor C_1 can be any value, but R_1 should be between 2 kΩ and 20 kΩ. A small capacitor, typically 0.001 μF, is placed between pins 7 and 8 to eliminate spontaneous oscillations. A simple first-order loop filter is formed by C_2 and an internal resistance of approximately 3.6 kΩ.

One popular application using the 565 is that of a frequency-shift keying (FSK) demodulator. FSK refers to the data transmission by means of a carrier

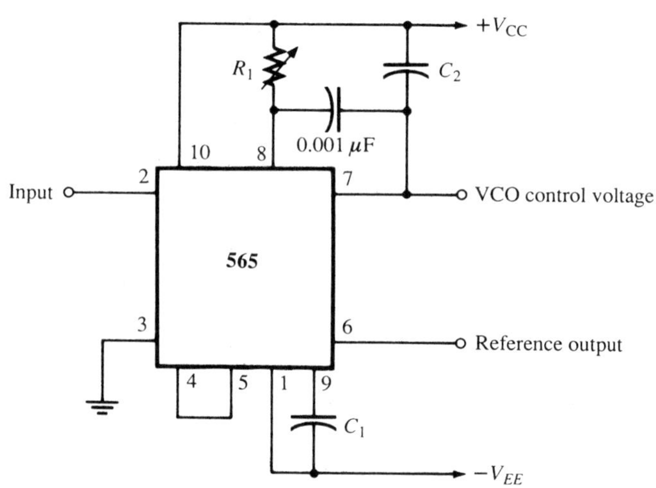

Figure 13–13
565 phase-locked-loop general circuit.

Table 13–1
Common mark-space pairs

Mark	Space	Shift
1070 Hz	1270 Hz	200 Hz
2025 Hz	2225 Hz	200 Hz
2125 Hz	2975 Hz	850 Hz

frequency that is shifted between two preset frequencies. FSK is widely used with teletypewriter (TTY) systems. Over the years, several standards have been used to set the mark and space frequencies, which correspond to the logic 1 and logic 0 states of the binary data signal. Several of these frequency pairs are listed in Table 13–1. The frequency difference between the mark and space frequencies is called the **frequency shift.** For the 1070–1270-Hz pair, the frequency shift is 200 Hz. When transmitting TTY information using a modulator-demodulator, or **modem** system, this frequency pair represents what is called the *originate signal,* whereas the 2025–2225-Hz pair represents the frequencies of the *answer signal.* By law, radio teletype, or *RTTY* (pronounced "rit-tee") must have a frequency shift of less than 900 Hz to meet certain bandwidth requirements. Ham radio operators many years ago adopted the 2125–2975-Hz (850-Hz shift) standard for RTTY.

A simple FSK demodulator using the 565 for a 1070–1270-Hz mark-space pair is shown in Figure 13–14. As the FSK signal appears at the input, the 565 PLL locks onto the input frequency and tracks it between the mark and space frequencies with a corresponding shift in DC voltage at the output. The free-running frequency of the VCO is adjusted with the 5-kΩ potentiometer in order to give a slightly positive output voltage with an input frequency of 1070 Hz (the mark frequency).

Figure 13–14
FSK demodulator.

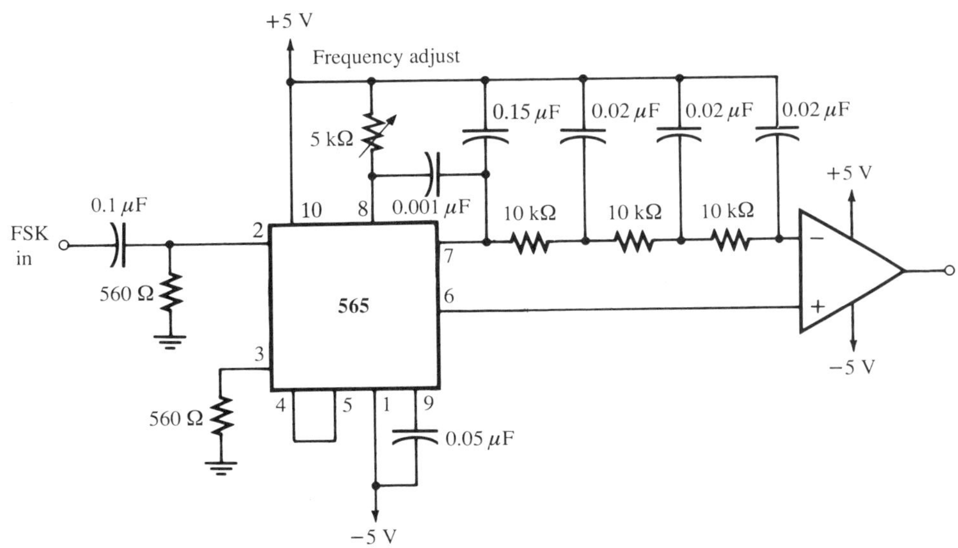

Another application is the SCA (Subsidiary Communications Authorization) decoder circuit of Figure 13–15. Some commercial FM radio stations are authorized by the FCC to broadcast uninterrupted background music for commercial use in stores, factories, offices, and so on. To accomplish this, SCA programming is transmitted on a 67-kHz subcarrier in order not to interfere with the FM station's regular programming on its assigned main carrier channel, which can be either in mono or stereo. In addition, the level of this subcarrier is only 10% of the amplitude of the combined signal.

The input is connected to the FM tuner's detector stage (before the de-emphasis network); the output of the decoder can be connected to any audio amplifier, since the SCA audio response is limited to a maximum of 7 kHz. Once connected, the 5-kΩ potentiometer between pin 8 and the positive supply voltage is tuned to 67 kHz.

567 Tone Decoder

The 567 device (Figure 13–16) is a phase-locked loop system designed specifically to respond to a given tone of constant frequency within its bandwidth. Although similar to the 561B, the 567 also has a power output stage capable of sourcing 100 mA. However, its frequency range is similar to the 565, which is limited to 500 kHz.

Figure 13–17 shows the basic connections for the 567. The free-running VCO frequency, often called the *center frequency,* is set by

$$f_o = \frac{1.1}{R_1 C_1} \qquad \textbf{(13–11)}$$

where R_1 should be between 2 kΩ and 20 kΩ. The value for C_2 is best found from the graph of bandwidth versus input signal amplitude (Figure 13–18) usually given on the 567 data sheet. The value for C_3 is not critical, but it is suggested by the manufacturer to be at least twice C_2.

However, the value for C_2 can be determined based on the level of the input signal (V_i). For input signal levels less than 200 mV RMS, the bandwidth, as a percentage of the center frequency of the 567, is found from

$$\%\text{BW} = 1070 \sqrt{\frac{V_i}{f_o C_2}} \qquad \textbf{(13–12)}$$

However, for input levels greater than 200 mV RMS, the bandwidth is then typically 14% of the center frequency. When the loop is locked, the output at pin 8 is at ground. When the loop is unlocked, pin 8 is at the supply voltage, which can range from +4.75 V to +9 V.

Figure 13–15
67-kHz FSK decoder.

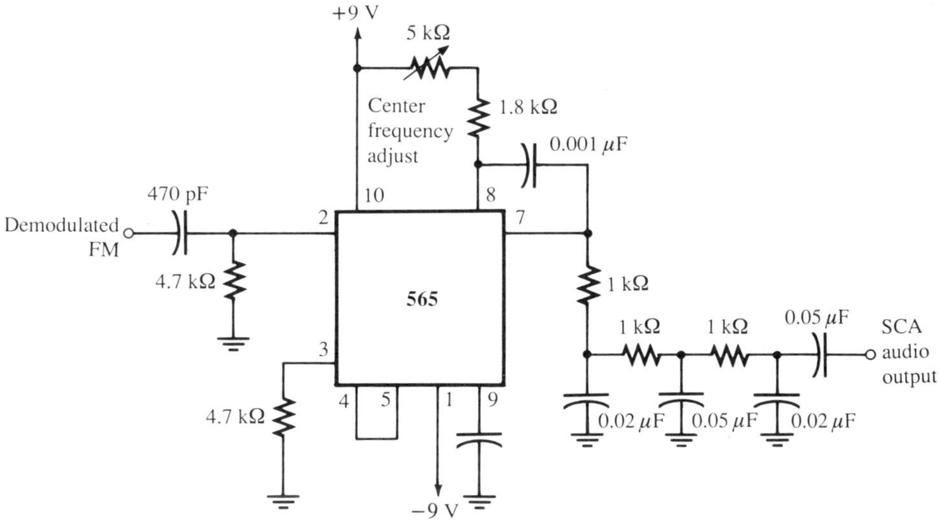

Figure 13–16
567 integrated-circuit phase-locked loop tone decoder pin diagram.

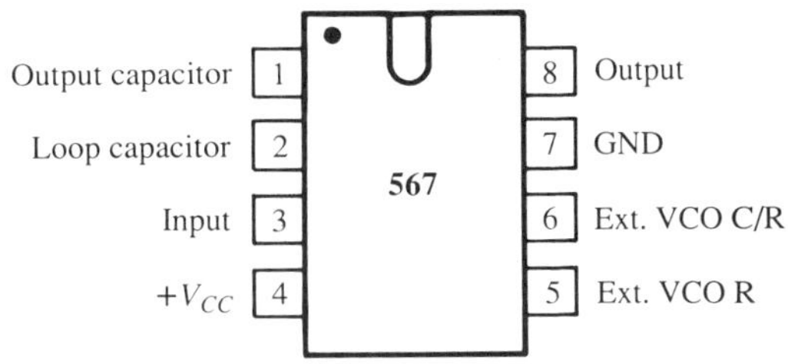

Figure 13–17
Basic tone decoder circuit.

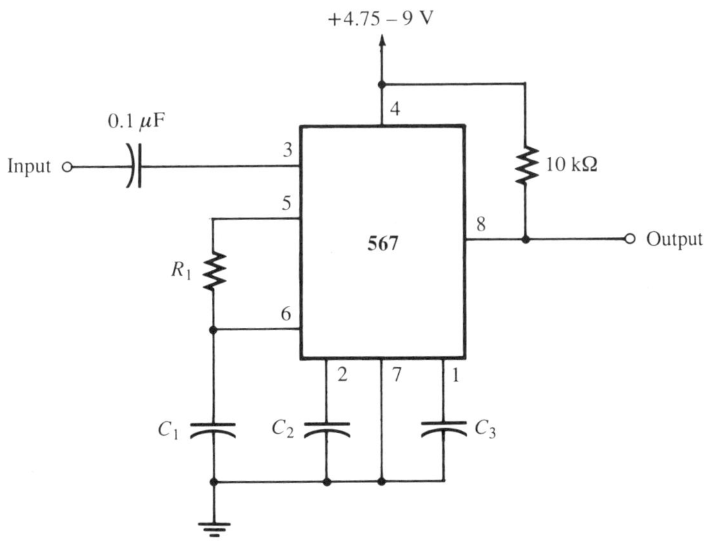

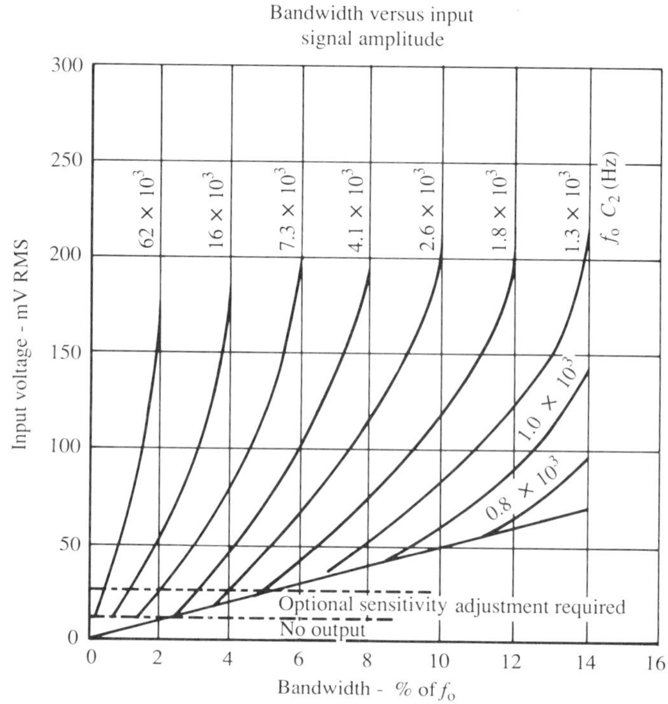

Figure 13–18
567 bandwidth versus signal amplitude curve for the selection of capacitor C_2.

Because of its internal design, the 567 becomes somewhat sensitive to those input frequencies that are *odd harmonics* of the center frequency. As a consequence, the 567 may lock onto frequencies that are $f_o/3$, $f_o/5$, $f_o/7$, etc. Furthermore, the loop may also lock onto frequencies near $(2N + 1)f_o$, where N is any integer. If such frequencies are anticipated, then they should be filtered or attenuated before reaching the 567 input.

Example 13–1

Using the basic 567 tone-decoder circuit of Figure 13–17, determine the component values for R_1, C_1, C_2, and C_3 to meet the following specifications:

Input frequency: 700 Hz
Input signal level: 100 mV RMS
Bandwidth: 12% of center frequency

Solution: To determine R_1 from Equation 13–11, a standard value for C_1 is chosen. Letting C_1 = **0.1 μF,**

$$R_1 = \frac{1.1}{(700\text{ Hz})(0.1\ \mu\text{F})}$$
$$= \mathbf{15.7\ k\Omega} \qquad \text{(use a 15-k}\Omega\text{ standard value)}$$

Capacitor C_2 is determined either by using Figure 13–18 or by rearranging Equation 13–12:

$$C_2 = \frac{(100\text{ mV})}{(700\text{ Hz})}\left(\frac{1070}{12\%}\right)^2$$
$$= \mathbf{1.14\ \mu F} \qquad \text{(use a 1-}\mu\text{F standard value)}$$

Capacitor C_3 should be at least twice C_2, so that C_3 is chosen to be equal to a standard value of **3.3 μF.** The completed circuit is shown in Figure 13–19.

An application for the 567 is the decoding of Touch-Tone signals, such as that used by most push-button telephones. Touch-Tone information is coded in tone-pairs using two of seven possible tone frequencies for the numbers 0 through 9 and the symbols # (pound) and * (star). The seven audio frequencies used for these 12 keys are summarized in matrix form in Table 13–2. The seven frequencies are divided into a *low-tone group* with frequencies of 697, 770, 852, and 941 Hz and a *high-tone group* having frequencies of 1209, 1336, and 1477 Hz. Each of the 12 numbers and symbols are then represented by a unique pair of frequencies. One frequency is represented from each tone group. For example, the number 9 is represented by an 852-Hz tone from the low-tone group and a 1477-Hz tone from the high-tone group.

To decode the tone frequencies properly for the number 9 from the 12 possible choices, two 567 decoders are needed. As shown in Figure 13–20, one 567 decoder is set to a center frequency 852 Hz, and the other is set to 1477 Hz. When the Touch-Tone signal corresponding to the number 9 is present at the circuit's input, both decoders will be locked and the output of both decoders will be at logic 0 (ground). The output of the NOR gate will then be logic 1 ($+V_{CC}$ = 5 V for TTL

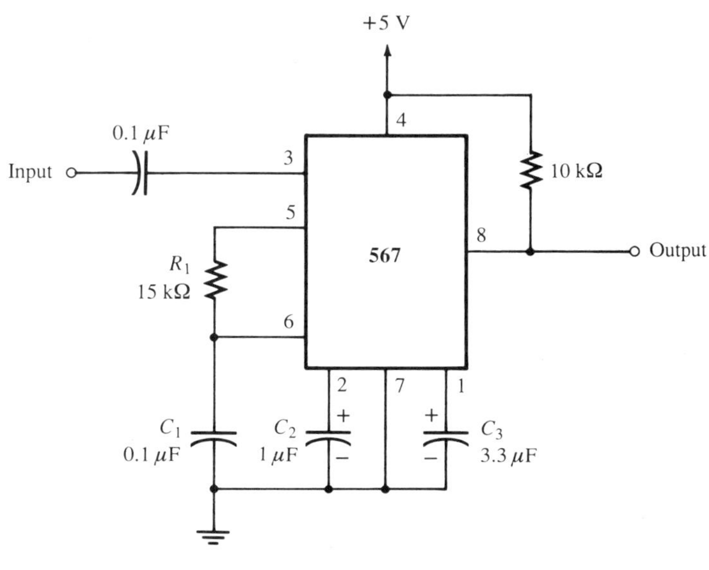

Figure 13–19
Completed 567-tone decoder circuit for Example 13–1.

Table 13–2
Touch-Tone frequencies

	High-Tone Group		
Low-Tone Group	**1209 Hz**	**1336 Hz**	**1477 Hz**
697 Hz	1	2	3
770 Hz	4	5	6
852 Hz	7	8	9
941 Hz	*	0	#

devices). If only one of the required two tone frequencies is present, only one of the decoder outputs will be at logic 0, so that the output of the NOR gate is also logic 0. Expanding on the basic circuit, Figure 13–21 shows the circuit for a complete Touch-Tone decoder that is capable of decoding all 12 possible tone-pairs.

4046 CMOS Phase-Locked Loop

The 4046 is a CMOS phase-locked loop device made by RCA and comes in a 16-pin DIP configuration (Figure 13–22). One of the major differences between the 4046 and the 560 series is that the 4046's phase-detector system is digital rather than analog. Furthermore, the 4046 contains two different types of phase detectors, both of which require that the input signals have 50% duty cycles.

The VCO frequency is at a minimum (f_{min}) when its input control voltage is zero and increases linearly to a maximum (f_{max}) when the control voltage equals the supply voltage ($+V_{DD}$). Typically, the maximum VCO frequency can be 700 kHz when $V_{DD} = 5$ V and 1.9 MHz when $V_{DD} = 15$ V. The frequency range, $f_{max} - f_{min}$, of the VCO is set by the external components R_1 and C_1, whereas the

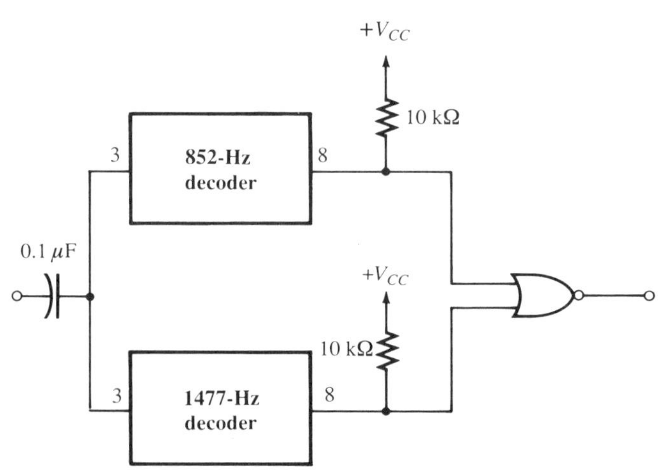

Figure 13–20
Using two 567-tone decoders to properly decode the number 9. The output of the NOR gate is high when both 852-Hz and 1477-Hz tones are simultaneously present at the input.

Figure 13–21
Circuit for a complete 12-digit Touch-Tone decoder using seven 567-tone decoders.

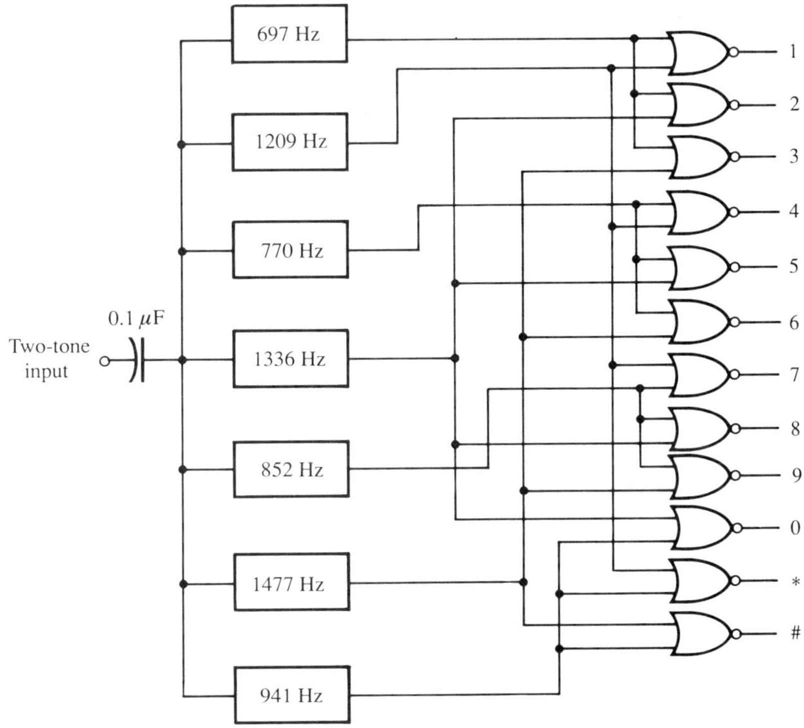

Figure 13–22
4046 CMOS integrated-circuit phase-locked loop pin diagram.

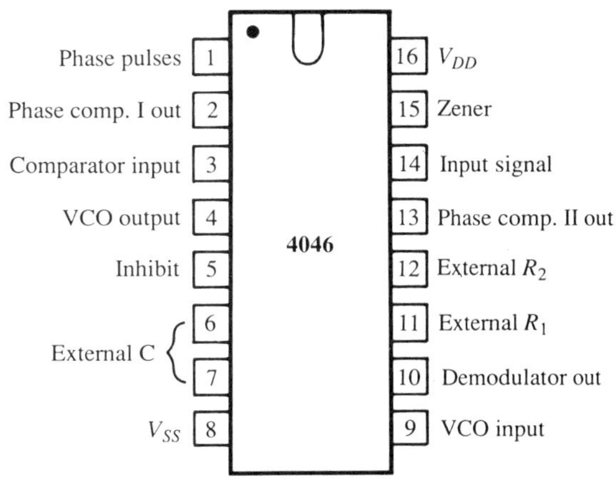

Figure 13–23
4046 loss of lock circuit.

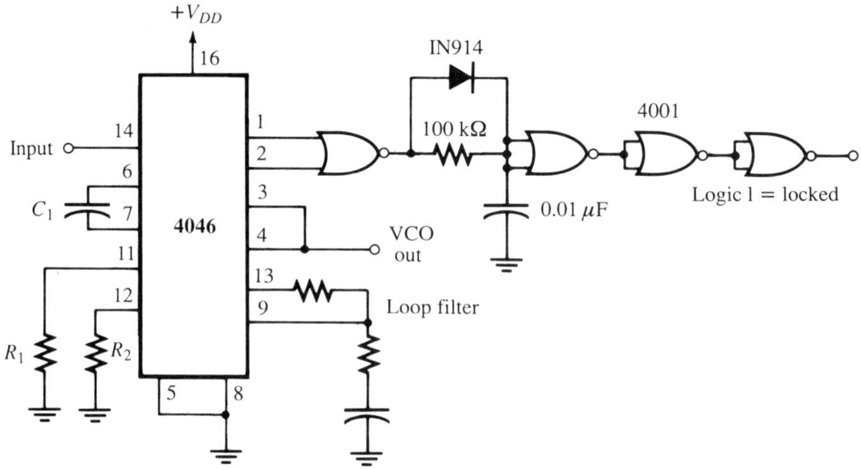

minimum VCO frequency is controlled by R_2 and C_1. Resistors R_1 and R_2 should be between 10 kΩ and 1 MΩ, whereas C_1 should be greater than

$$100 \text{ pF} \quad \text{for } 5\text{ V} < V_{DD} < 10\text{ V}$$
$$50 \text{ pF} \quad \text{for } V_{DD} > 10\text{ V to } 15\text{ V}$$

The VCO can be turned off at any time, reducing power consumption when it is not in use, by connecting the inhibit input (pin 5) to $+V_{DD}$. Otherwise, pin 5 is normally connected to ground.

Figure 13–24
100x CMOS phase-locked loop frequency multiplier for increasing the resolution of frequency counters.

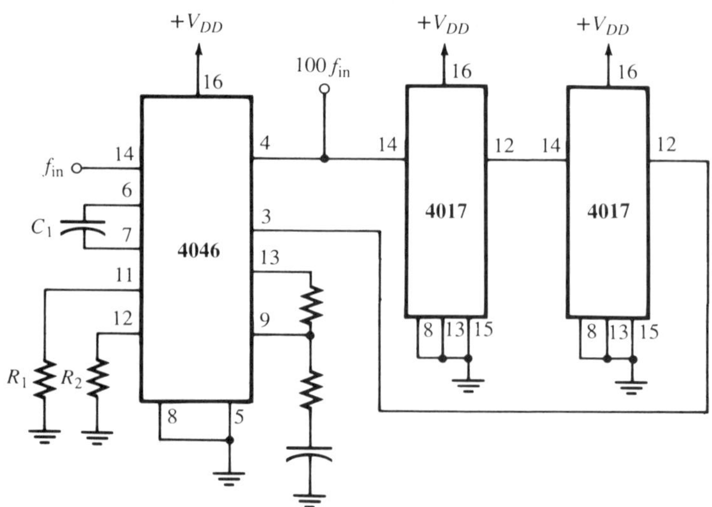

The output of phase detector II (pin 1) can be NORed with the output of phase detector I (pin 2) to form a lock detector, as shown in Figure 13–23. The output of the last 4001 CMOS NOR gate will be a logic 1 when the 4046 PLL is locked.

As with many phase-locked loop integrated-circuit devices, the 4046 can be used as a frequency synthesizer, or multiplier, by inserting a CMOS counter in the feedback path between the VCO (pin 4) and the phase detector (pin 3). As an example, Figure 13–24 shows the addition of two 4017 CMOS decade counters to multiply the input signal by 100. When used with an ordinary frequency counter, the resolution is then increased from ±1 Hz to ±0.01 Hz while still having a timebase sampling or update time of 1 s. Thus an input frequency of 59.72 Hz would be displayed as 5972 Hz. Otherwise, all you would observe is either 59 or 60 Hz. If the counter's modulus is made equal to 60, the same technique can be used to display the input frequency in terms of *cycles per minute*. This is extremely useful when measuring the frequency of physiological events such as respiration or heart rate.

Summary

This chapter discussed the operation of the basic phase-locked loop in terms of a phase detector, low-pass loop filter, and voltage-controlled oscillator arranged in a feedback loop. In operation, the phase-locked loop can be in any one of three distinct states: free-running, capture, and phase-lock. When phase-locked, the input and output frequencies are equal except for a possible phase difference. If the loop is unlocked, the output frequency equals the free-running frequency of the voltage-controlled oscillator. By inserting a divide-by-N counter in the loop's feedback path, a frequency synthesizer, or multiplier, is formed whereby the output frequency is an integer multiple of the input frequency.

The remainder of this chapter illustrated a number of integrated-circuit phase-locked loop devices from the 560-series manufactured by Signetics, in addition to several applications. These included the 560B (FM demodulator), 561B (AM receiver), 562 (basic frequency synthesizer), 565 (FSK demodulator and SCA decoder), and the 567 (Touch-Tone decoder). In addition, the popular 4046 CMOS phase-locked loop device was discussed. An application of its use was a prescaler for a digital frequency counter.

Glossary

capture range The frequency range of a phase-locked loop over which the loop's output frequency begins to change from its free-running frequency to that of the loop's input.

error voltage The DC output voltage of the phase detector, which is proportional to the phase difference between the input signal of the phase-locked loop and that of the output of the VCO.

free-running frequency The output frequency of a phase-locked loop with no input signal.

frequency shift The frequency difference between the mark and space frequencies.

frequency synthesizer A frequency source whose output is an integer multiple of an input reference frequency. It is a phase-locked loop with a divide-by-N counter in its feedback path.

lock range Also called the *hold-in range;* an input frequency range over which the VCO can track the input signal once the loop is locked.

modem The acronym for modulator-demodulator, a device that transforms digital logic levels into corresponding audio tones, and vice versa, for the transmission of data over telephone networks.

phase detector Also called a *mixer* or *phase comparator;* a circuit that generates an average (DC) output voltage that is proportional to the phase difference between the input signal of the phase-locked loop and that of the output of the VCO.

phase-locked loop (PLL) A feedback control circuit consisting of a phase detector, low-pass filter, and a voltage-controlled oscillator, which compares the frequency of the voltage-controlled oscillator with that of an input signal reference.

Review Questions

1. What is a *phase-locked loop?*
2. Draw the block diagram of a phase-locked loop, labeling the loop's forward and feedback paths.
3. Explain the function of the phase detector and give the units of its conversion gain.
4. Explain the function of the voltage-controlled oscillator and give the units of its conversion gain.
5. Explain the function of the loop filter.
6. Explain the differences between the following states of a phase-locked loop: free-running, capture, and phase-locked.
7. How does the loop's capture range compare with its lock range?
8. What is a frequency synthesizer and how does it differ from the basic phase-locked loop?
9. Explain how a frequency-synthesizer may be used to improve the resolution of a digital frequency counter.

Exercises

1. The conversion gain of a phase detector is 0.872 V/rad. Determine the DC output voltage of the phase detector for each phase shift of its two input signals.
   (a) 30°
   (b) 105°
   (c) 311°
2. If the conversion gain of a phase detector is 1.06 V/rad, determine the phase shift (in degrees) of its two input signals for each DC output voltage.
   (a) 6.31 V
   (b) 0.59 V
   (c) 4.22 V
3. A VCO has a conversion gain of 1278 Hz/V. When the DC input voltage to the VCO is zero, the output frequency is set at 35.062 kHz. Determine the VCO output frequency for each DC input voltage.
   (a) 1.57 V
   (b) −3.46 V
4. The phase detector of Exercise 1 is connected to the VCO of Exercise 3. Determine the VCO output frequency if the phase shift of its two inputs signals is 45°.
5. The free-running frequency of a phase-locked loop is 17.5 kHz. The lock and capture ranges are ±20% and ±3% of the free-running frequency, respectively. Determine each value.
   (a) The frequency range over which the loop can acquire capture if initially unlocked.
   (b) If locked, the frequency range over which the loop can track input signals and remain locked.

6. A given phase-locked loop frequency synthesizer has two divide-by-N counters. One counter has a fixed modulus of 10 and is placed ahead of the synthesizer. The other counter is programmable and is placed in the feedback path of the phase-locked loop. Determine the output frequency of the synthesizer if the input frequency is 500 kHz and the modulus of the programmable counter is set at each value.
   (a) 40
   (b) 23
   (c) 95

7. A 565 is wired as a basic phased-locked loop using the circuit of Figure 13–13 with a positive supply voltage of 15 V while pin 1 is grounded. Determine the values for R_1 and C_2 if the free-running frequency is to be 941 Hz, with a capture range of ± 100 Hz. Assume that $C_1 = 0.033\ \mu F$.

8. For the 565 PLL of Exercise 7, what is the lock range?

9. A 567 is to be used as a tone decoder using the circuit of Figure 13–17. Determine suitable values for R_1, C_2, and C_3 for an input signal having an amplitude of 50 mV RMS at 697 Hz. The tone decoder is to have a bandwidth that is 10% of the center frequency. Assume that $C_1 = 0.47\ \mu F$.

Appendix A Standard Resistor and Capacitor Values

Resistors

1. The following ±5% standard decade values are available. Those marked with * are the ones most readily available from electronic suppliers.

1.0*	1.3	1.8*	2.4	3.3*	4.3	5.6*	7.5
1.1	1.5*	2.0	2.7*	3.6	4.7*	6.2	8.2*
1.2*	1.6	2.2*	3.0	3.9*	5.1	6.8*	9.1

To obtain standard resistance values, multiply preferred number from decade table by powers of 10. Standard values are available from 10 Ω to 22 MΩ.

2. The following ±1% values are available, but at a higher cost.

10.0	12.1	14.7	17.8	21.5	26.1	31.6	38.3	46.4	56.2	68.1	82.5
10.2	12.4	15.0	18.2	22.1	26.7	32.4	39.2	47.5	57.6	69.8	84.5
10.5	12.7	15.4	18.7	22.6	27.4	33.2	40.2	48.7	59.0	71.5	86.6
10.7	13.0	15.8	19.1	23.2	28.0	34.0	41.2	49.9	60.4	73.2	88.7
11.0	13.3	16.2	19.6	23.7	28.7	34.8	42.2	51.1	61.9	75.0	90.0
11.3	13.7	16.5	20.0	24.3	29.4	35.7	43.2	52.3	63.4	76.8	93.1
11.5	14.0	16.9	20.5	24.9	30.1	36.5	44.2	53.6	64.9	78.7	95.3
11.8	14.3	17.4	21.0	25.5	30.9	37.4	45.3	54.9	66.5	80.6	97.6

These standard values are available from 10 Ω to 22.1 MΩ.

Capacitors

In general, capacitor values follow the standard decade values for ±10% resistors. The values listed below are in microfarads, and those marked with * are the ones most readily available from electronic suppliers.

.001*	.002	.0033*	.0056	.01*	.02	.033*	.056	.1*	.2	.33*	.56
.0012	.0022*	.0039	.0068*	.012	.022*	.039	.068*	.12	.22*	.39	.68*
.0015	.0025	.0047*	.0075	.015	.025	.047*	.075	.15	.25	.47*	.75
.0018	.0027	.005	.0082	.018	.027	.05	.082	.18	.27	.5	.82

Appendix B Data Sheets

The complete data sheets for the following integrated-circuit devices are included in the instructor's manual.

- LM1900 Quad Norton Amplifier (National Semiconductor)
- CA3080 Operational Transconductance Amplifier (RCA)
- XR-555 Timer (Exar)
- XR-2240 Programmable Timer/Counter (Exar)
- XR-567 Phase-Locked Loop Tone Decoder (Exar)
- LM109 Five-Volt Regulator (Signetics)
- LM117 Three-Terminal Adjustable Regulator (National Semiconductor)
- 4151/4152 Voltage-to-Frequency Converter (Raytheon)

The data sheet for the 741 operational amplifier is presented in Figure 3–1.

National Semiconductor

Operational Amplifiers/Buffers

LM1900/LM2900/LM3900, LM3301, LM3401 quad amplifiers

general description

The LM1900 series consists of four independent, dual input, internally compensated amplifiers which were designed specifically to operate off of a single power supply voltage and to provide a large output voltage swing. These amplifiers make use of a current mirror to achieve the non-inverting input function. Application areas include: ac amplifiers, RC active filters, low frequency triangle, squarewave and pulse waveform generation circuits, tachometers and low speed, high voltage digital logic gates.

features

- Wide single supply voltage range or dual supplies — $4\ V_{DC}$ to $36\ V_{DC}$; $\pm 2\ V_{DC}$ to $\pm 18\ V_{DC}$
- Supply current drain independent of supply voltage
- Low input biasing current — 30 nA
- High open-loop gain — 70 dB
- Wide bandwidth — 2.5 MHz (Unity Gain)
- Large output voltage swing — $(V^+ - 1)$ Vp-p
- Internally frequency compensated for unity gain
- Output short-circuit protection

schematic and connection diagrams

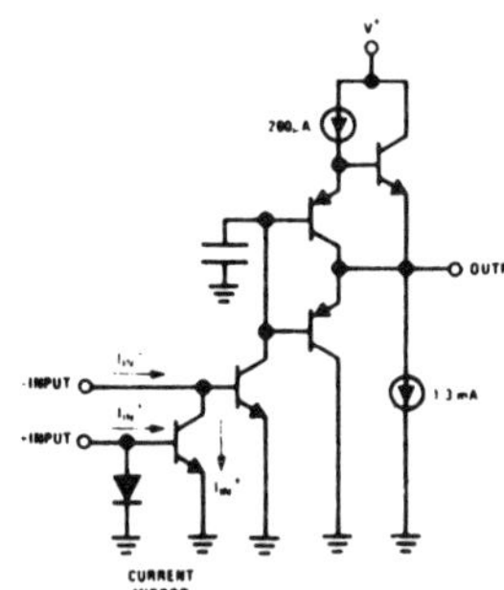

Order Number LM1900D
or LM2900D
See NS Package D14E

Order Number LM1900J
or LM2900J
See NS Package J14A

Order Number LM2900N,
LM3900N, LM3301N
or LM3401N
See NS Package N14A

Dual-In-Line and Flat Package

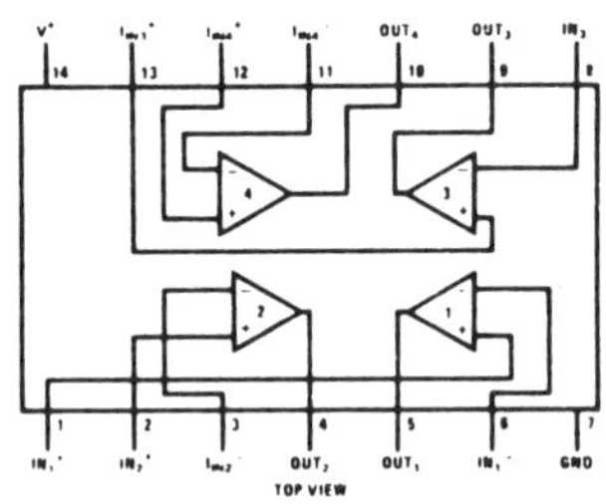

typical applications ($V^+ = 15\ V_{DC}$)

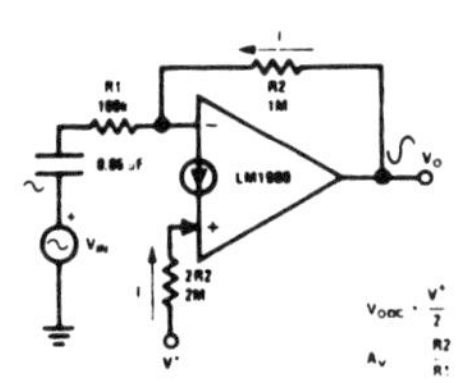

Inverting Amplifier

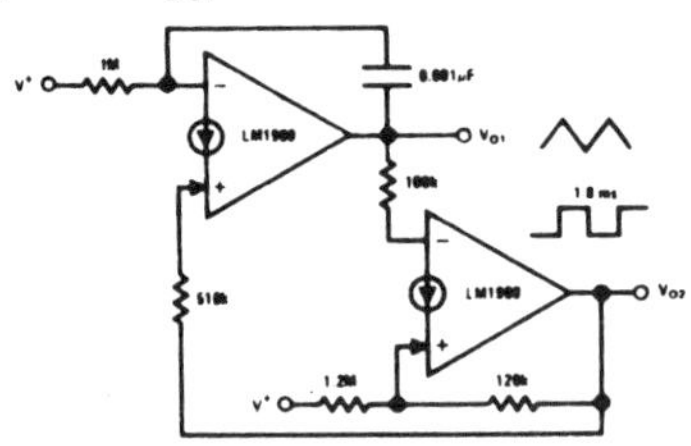

Triangle/Square Generator

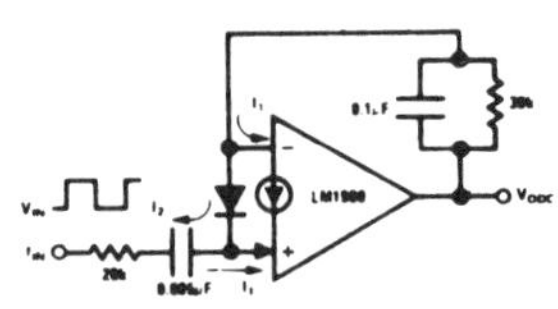

Frequency-Doubling Tachometer

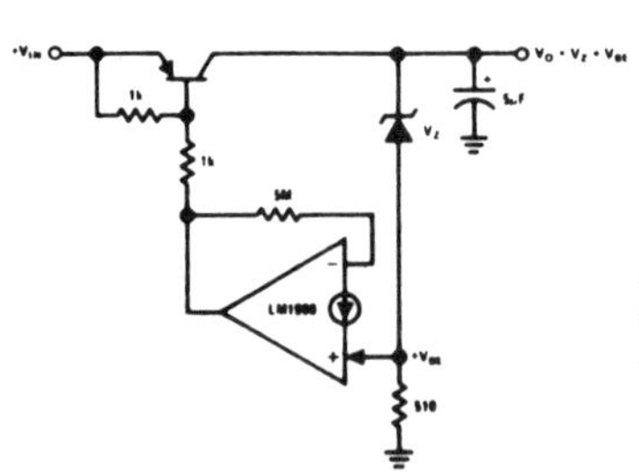

Low $V_{IN} - V_{OUT}$ Voltage Regulator

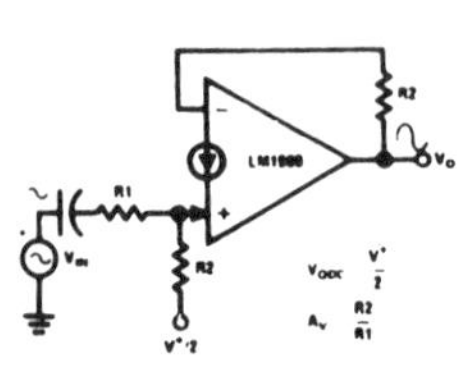

Non-Inverting Amplifier

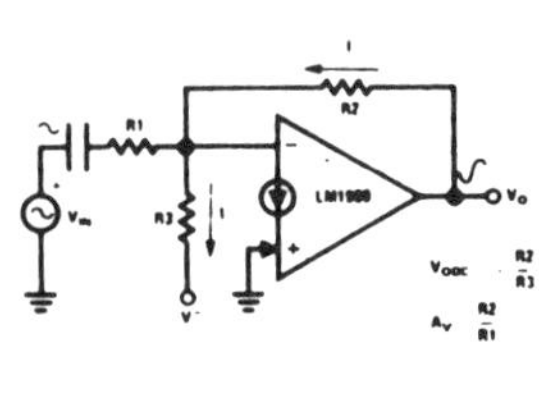

Negative Supply Biasing

RCA
Solid State
Division

Linear Integrated Circuits

Monolithic Silicon

CA3080A, CA3080AE, CA3080AS
CA3080, CA3080E, CA3080S

CA3080S
CA3080AS
8-LEAD TO-5
with Dual-In-Line
Formed Leads
"DIL-CAN"
H-1787

CA3080
CA3080A
8-LEAD TO-5
H-1528

H-1817

CA3080E
CA3080AE
8-LEAD Dual-In-Line Plastic Package
"MINI-DIP"

Operational Transconductance Amplifiers (OTA's)

Gatable-Gain Blocks

Features:

- **Slew rate (unity gain, compensated): 50 V/μs**
- **Adjustable power consumption: 10μW to 30 mW**
- **Flexible supply voltage range: ± 2 V to ± 15 V**
- **Fully adjustable gain: 0 to $g_m R_L$ limit**
- **Tight gm spread: CA3080 (2:1), CA3080A (1.6:1)**
- **Extended g_m linearity: 3 decades**

The RCA-CA3080 and CA3080A types are Gatable-Gain Blocks which utilize the unique operational-transconductance-amplifier (OTA) concept described in Application Note ICAN-6668, "Applications of the CA3080 and CA3080A High-Performance Operational Transconductance Amplifiers".

The CA3080 and CA3080A types have differential input and a single-ended, push-pull, class A output. In addition, these types have an amplifier bias input which may be used either for gating or for linear gain control. These types also have a high output impedance and their transconductance (g_m) is directly proportional to the amplifier bias current (I_{ABC}).

The CA3080 and CA3080A types are notable for their excellent slew rate (50 V/μs), which makes them especially useful for multiplex and fast unity-gain voltage followers. These types are especially applicable for multiplex applications because power is consumed only when the devices are in the "ON" channel state.

The CA3080A is rated for operation over the full military-temperature range (−55 to +125° C) and its characteristics are specifically controlled for applications such as sample-hold, gain-control, multiplex, etc. Operational transconductance amplifiers are also useful in programmable power-switch applications, e.g., as described in Application Note ICAN-6048, "Some Applications of a Programmable Power Switch/Amplifier" (CA3094, CA3094A, CA3094B).

These types are supplied in the 8-lead TO-5-style package (CA3080, CA3080A), and in the 8-lead TO-5-style package with dual-in-line formed leads ("DIL-CAN", CA3080S, CA3080AS). The CA3080 is also supplied in the 8-lead dual-in-line plastic ("MINI-DIP") package (CA3080E, CA3080AE), and in chip form (CA3080H).

Applications:

- **Sample and hold**
- **Multiplex**
- **Voltage follower**
- **Multiplier**
- **Comparator**

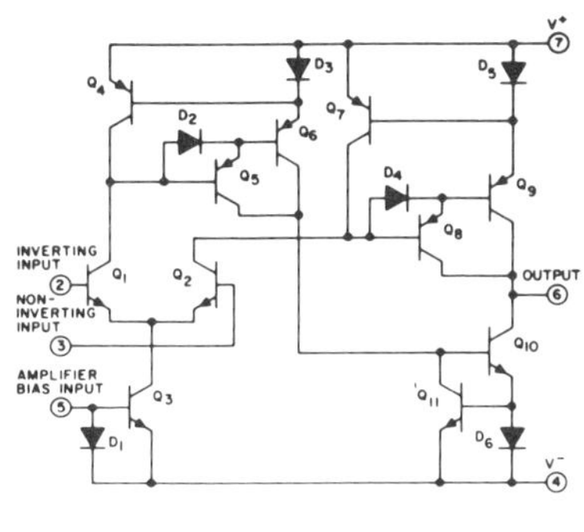

Fig. 1 – Schematic diagram for CA3080 and CA3080A.

EXAR

XR-555

Timing Circuit

GENERAL DESCRIPTION

The XR-555 monolithic timing circuit is a highly stable controller capable of producing accurate timing pulses. It is a direct, pin-for-pin replacement for the SE/NE 555 timer. The circuit contains independent control terminals for triggering or resetting if desired, as shown in the functional block diagram of Figure 1.

In the monostable mode of operation, the time delay is controlled by one external resistor and one capacitor. For astable operation as an oscillator, the free-running frequency and the duty cycle are accurately controlled with two external resistors and one capacitor (as shown in Figure 2).

The XR-555 may be triggered or reset on falling waveforms. Its output can source or sink up to 200 mA or drive TTL circuits.

FEATURES

Direct Replacement for SE/NE 555
Timing from Microseconds Thru Hours
Operates in Both Monostable and Astable Modes
High Current Drive Capability (200 mA)
TTL and DTL Compatible Outputs
Adjustable Duty Cycle
Temperature Stability of 0.005%/°C

ABSOLUTE MAXIMUM RATINGS

Power Supply	18 volts
Power Dissipation (package limitation)	
Plastic Package	300 mW
Derate above +25°C	2.5 mW/°C
Power Dissipation (package limitation)	
To 99	600 mW
Storage Temperature Range	−65°C to +150°C

APPLICATIONS

Precision Timing
Pulse Generation
Sequential Timing
Pulse Shaping
Clock Generation
Missing Pulse Detection
Pulse-Width Modulation
Frequency Division
Pulse-Position Modulation
Appliance Timing

AVAILABLE TYPES

Part Number	Package	Operating Temperature
XR-555MT	TO-99	−55°C to +125°C
XR-555CP	Plastic	0°C to +75°C
XR-555CT	TO-99	-0°C to +75°C

EQUIVALENT SCHEMATIC DIAGRAM

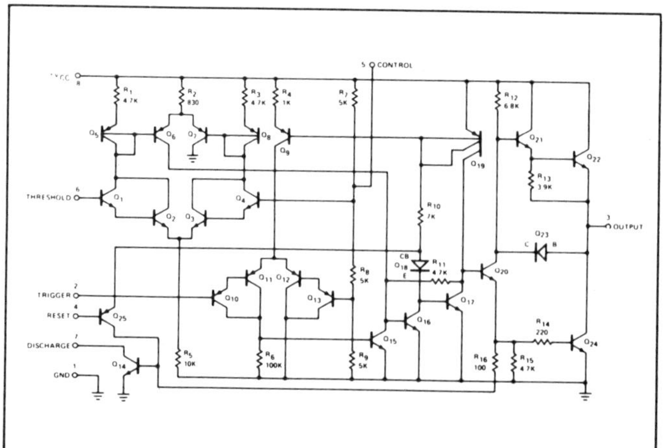

FUNCTIONAL BLOCK DIAGRAM

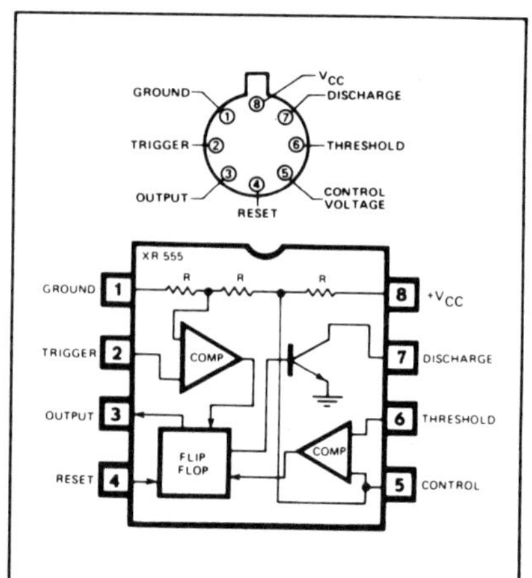

8-78 REV 4

Courtesy EXAR Corporation.

EXAR

XR-2240

Programmable Timer/Counter

GENERAL DESCRIPTION

The XR-2240 Programmable Timer/Counter is a monolithic controller capable of producing ultra-long time delays without sacrificing accuracy. In most applications, it provides a direct replacement for mechanical or electromechanical timing devices and generates programmable time delays from micro-seconds up to five days. Two timing circuits can be cascaded to generate time delays up to three years.

As shown in Figure 1, the circuit is comprised of an internal time-base oscillator, a programmable 8-bit counter and a control flip-flop. The time delay is set by an external R-C network and can be programmed to any value from 1 RC to 255 RC.

In astable operation, the circuit can generate 256 separate frequencies or pulse-patterns from a single RC setting and can be syncronized with external clock signals. Both the control inputs and the outputs are compatible with TTL and DTL logic levels.

FEATURES

Timing from micro-seconds to days
Programmable delays: 1RC to 255 RC
Wide supply range; 4V to 15V
TTL and DTL compatible outputs
High accuracy: 0.5%
External Sync and Modulation Capability
Excellent Supply Rejection: 0.2%/V

APPLICATIONS

Precision Timing	Frequency Synthesis
Long Delay Generation	Pulse Counting/Summing
Sequential Timing	A/D Conversion
Binary Pattern Generation	Digital Sample and Hold

ABSOLUTE MAXIMUM RATINGS

Supply Voltage	18V
Power Dissipation	
Ceramic Package	750 mW
Derate above +25°C	6 mw/°C
Plastic Package	625 mW
Derate above +25°C	5 mW/°C
Operating Temperature	
XR-2240M	−55°C to +125°C
XR-2240C	0°C to +70°C
Storage Temperature	−65°C to +150°C

FUNCTIONAL BLOCK DIAGRAM

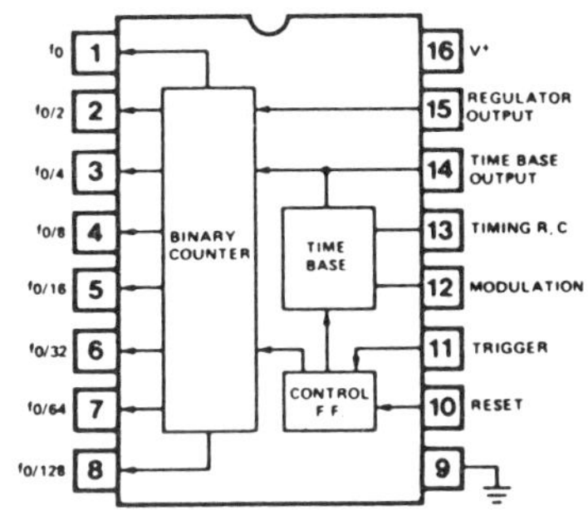

ORDERING INFORMATION

Part Number	Package	Operating Temperature
XR-2240M	Ceramic	−55°C to +125°C
XR-2240N	Ceramic	0°C to +70°C
XR-2240CN	Ceramic	0°C to +70°C
XR-2240P	Plastic	0°C to +70°C
XR-2240CP	Plastic	0°C to +70°C

SYSTEM DESCRIPTION

The XR-2240 is a combination timer/counter capable of generating accurate timing intervals ranging from microseconds through several days. The time base works as an astable multivibrator with a period equal to RC. The eight bit counter can divide the time base output by any integer value from 1 to 255. The wide supply voltage range of 4.5 to 15 V, TTL and DTL logic compatibility, and 0.5% accuracy allow wide applicability. The counter may operate independently of the time base. Counter outputs are open collector and may be wire-OR connected.

The circuit is triggered or reset with positive going pulses. By connecting the reset pin (Pin 10) to one of the counter outputs, the time base will halt at timeout. If none of the outputs are connected to the reset, the circuit will continue to operate in the astable mode. Activating the trigger terminal (Pin 11) while the timebase is stopped will set all counter outputs to the low state and start the timebase.

EXAR

XR-567

Monolithic Tone Decoder

GENERAL DESCRIPTION

The XR-567 is a monolithic phase-locked loop system designed for general purpose tone and frequency decoding. The circuit operates over a wide frequency band of 0.01 Hz to 500 kHz and contains a logic compatible coutput which can sink up to 100 milliamps of load current. The bandwidth, center frequency, and output delay are independently determined by the selection of four external components.

Figure 1 contains a functional block diagram of the complete monolithic system. The circuit consists of a phase detector, low-pass filter, and current-controlled oscillator which comprise the basic phase-locked loop; plus an additional low-pass filter and quadrature detector that enable the system to distinguish between the presence or absence of an input signal at the center frequency.

FEATURES

Bandwidth adjustable from 0 to 14%.

Logic compatible output with 100mA current sinking capability.

Highly stable center frequency.

Center frequency adjustable from 0.01 Hz to 500 kHz.

Inherent immunity to false signals.

High rejection of out-of-band signals and noise.

Frequency range adjustable over 20:1 range by external resistor.

APPLICATIONS

Touch-Tone® Decoding
Sequential Tone Decoding
Communications Paging
Ultrasonic Remote-Control
Telemetry Decoding

ABSOLUTE MAXIMUM RATINGS

Power Supply	10 volts
Power Dissipation (package limitation)	
Ceramic package	385 mW
Plastic Package	300 mW
Derate above +25°C	2.5 mW/°C
Temperature	
Operating	
XR-567M	-55°C to +125°C
XR-567CN/567CP	0°C to +70°C
Storage	-65°C to +150°C

ORDER INFORMATION

Part Number	Package	Operating Temperature
XR-567M	Ceramic	-55°C to +125°C
XR-567CN	Ceramic	0°C to +75°C
XR-567CP	Plastic	0°C to +75°C

EQUIVALENT SCHEMATIC DIAGRAM

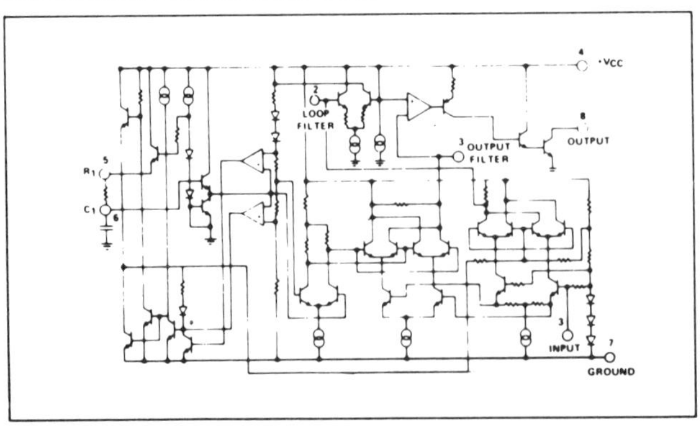

FUNCTIONAL BLOCK DIAGRAM

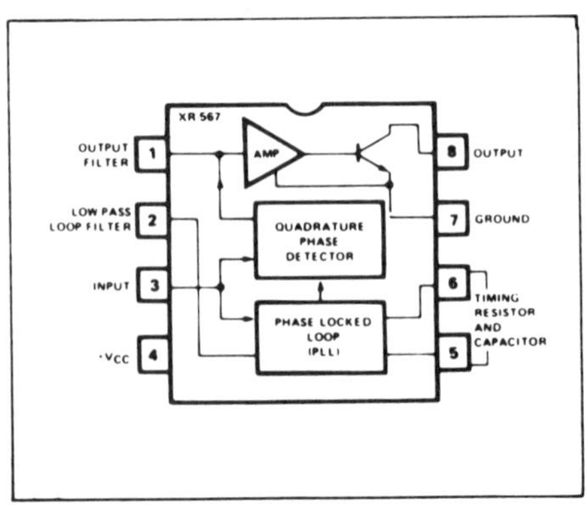

Courtesy EXAR Corporation.

FIVE-VOLT REGULATORS LM109 LM309

LINEAR INTEGRATED CIRCUITS

DESCRIPTION

The LM109 and LM309 are complete 5 volt regulators fabricated on a single silicon chip. These regulators are designed for local "on card" regulation to eliminate many of the noise and ground loop problems associated with single-point regulation. They employ internal current limiting, thermal shutdown, and safe-area compensation which makes the circuitry essentially blow-out proof. If adequate heat sinking is provided, the devices can deliver output currents in excess of 200mA from the TO-5 package, and 1A from the TO-3 package. In addition to their use as fixed 5 volt regulators, these devices may be used with external components to obtain adjustable output levels. They may also be used as the power pass element in precision regulators.

FEATURES

- OUTPUT CURRENTS IN EXCESS OF 1 amp
- INTERNAL THERMAL OVERLOAD PROTECTION
- INTERNAL CURRENT LIMITING
- NO EXTERNAL COMPONENTS REQUIRED

ABSOLUTE MAXIMUM RATINGS

Input Voltage	35V
Power Dissipation	Internally Limited
Operating Junction Temperature Range	
LM109	-55°C to 150°C
LM309	0°C to 125°C
Storage Temperature Range	-65°C to 150°C
Lead Temperature (Soldering, 10 sec)	300°C

PIN CONFIGURATIONS

H PACKAGE
(Bottom View)

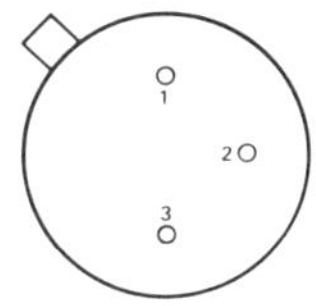

1. Input
2. Output
3. Ground

ORDER PART NOS. LM109H/LM309H

K PACKAGE
(Bottom View)

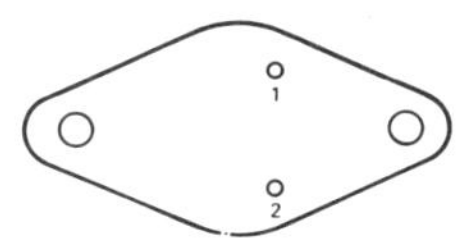

1. Input
2. Output

Case is connected to ground.

ORDER PART NOS. LM109K/LM309K

EQUIVALENT CIRCUIT

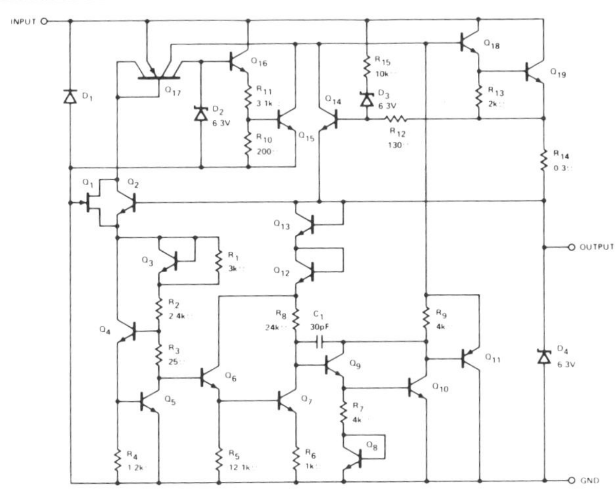

Courtesy Signetics Corporation.

National Semiconductor

JULY 1978

LM117/LM217/LM317 3-terminal adjustable regulator

general description

The LM117/LM217/LM317 are adjustable 3-terminal positive voltage regulators capable of supplying in excess of 1.5A over a 1.2V to 37V output range. They are exceptionally easy to use and require only two external resistors to set the output voltage. Further, both line and load regulation are better than standard fixed regulators. Also, the LM117 is packaged in standard transistor packages which are easily mounted and handled.

In addition to higher performance than fixed regulators, the LM117 series offers full overload protection available only in IC's. Included on the chip are current limit, thermal overload protection and safe area protection. All overload protection circuitry remains fully functional even if the adjustment terminal is disconnected.

features

- Adjustable output down to 1.2V
- Guaranteed 1.5A output current
- Line regulation typically 0.01%/V
- Load regulation typically 0.1%
- Current limit constant with temperature
- **100% electrical burn-in**
- Eliminates the need to stock many voltages
- Standard 3-lead transistor package
- 80 dB ripple rejection

Normally, no capacitors are needed unless the device is situated far from the input filter capacitors in which case an input bypass is needed. An optional output capacitor can be added to improve transient response. The adjustment terminal can be bypassed to achieve very high ripple rejections ratios which are difficult to achieve with standard 3-terminal regulators.

Besides replacing fixed regulators, the LM117 is useful in a wide variety of other applications. Since the regulator is "floating" and sees only the input-to-output differential voltage, supplies of several hundred volts can be regulated as long as the maximum input to output differential is not exceeded.

Also, it makes an especially simple adjustable switching regulator, a programmable output regulator, or by connecting a fixed resistor between the adjustment and output, the LM117 can be used as a precision current regulator. Supplies with electronic shutdown can be achieved by clamping the adjustment terminal to ground which programs the output to 1.2V where most loads draw little current.

The LM117K, LM217K and LM317K are packaged in standard TO-3 transistor packages while the LM117H, LM217H and LM317H are packaged in a solid Kovar base TO-5 transistor package. The LM117 is rated for operation from −55°C to +150°C, the LM217 from −25°C to +150°C and the LM317 from 0°C to +125°C. The LM317T and LM317MP, rated for operation over a 0°C to +125°C range, are available in a TO-220 plastic package and a TO-202 package, respectively.

For applications requiring greater output current in excess of 3A and 5A, see LM150 series and LM138 series data sheets, respectively. For the negative complement, see LM137 series data sheet.

LM117 Series Packages and Power Capability

DEVICE	PACKAGE	RATED POWER DISSIPATION	DESIGN LOAD CURRENT
LM117	TO-3	20W	1.5A
LM217 LM317	TO-5	2W	0.5A
LM317T	TO-220	15W	1.5A
LM317M	TO-202	7.5W	0.5A

typical applications

1.2V–25V Adjustable Regulator

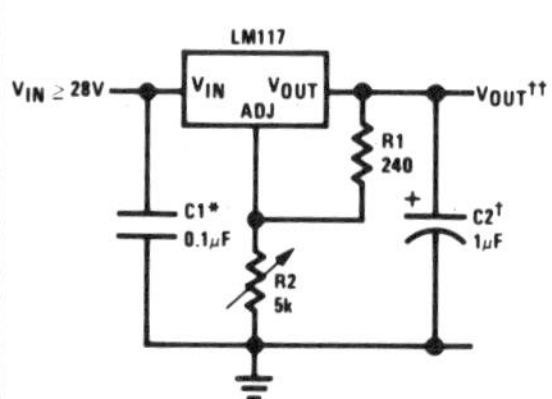

†Optional—improves transient response

*Needed if device is far from filter capacitors

$$^{\dagger\dagger}V_{OUT} = 1.25V\left(1 + \frac{R2}{R1}\right)$$

Digitally Selected Outputs

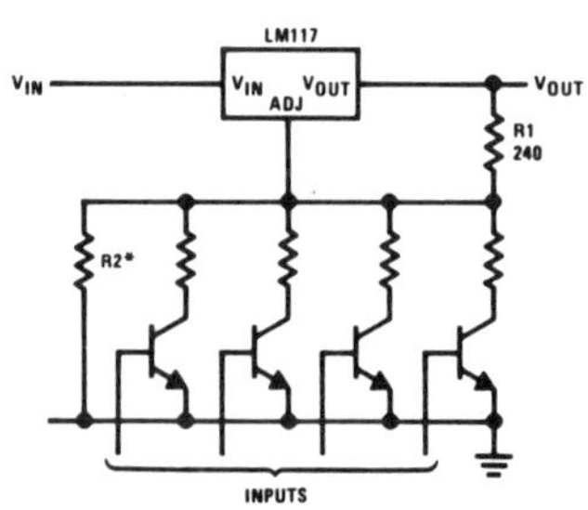

*Sets maximum V_{OUT}

5V Logic Regulator with Electronic Shutdown*

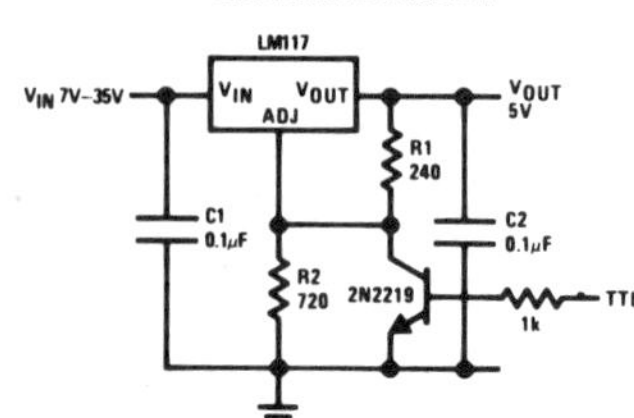

*Min output ≈ 1.2V

PRODUCT SPECIFICATIONS — **LINEAR INTEGRATED CIRCUITS**

RAYTHEON

Voltage-to-Frequency Converters

4151, 4152

Features

- Single Supply Operation
- Pulse Output Compatible with All Logic Forms (DTL/TTL/CMOS)
- Programmable Scale Factor (K)
- High Noise Rejection
- Inherent Monotonicity
- Easily Transmittable Output
- Simple Full Scale Trim
- Single-Ended Input, Referenced to Ground
- V-F or F-V Conversion
- Voltage or Current Input
- Wide Dynamic Range

Description

The 4151 and 4152 are monolithic circuits containing all of the active components needed to build a complete voltage-to-frequency converter. Circuits that convert a DC voltage to a pulse train (VFC) can be built by adding a few resistors and capacitors to the internal comparator, one-shot, voltage reference, and switched current source. Frequency-to-voltage converters (FVC') and many other signal conditioning circuits are also easily created using these converters.

Raytheon was the first company to introduce a monolithic VFC. The low cost 4151 was followed by the 4152, a pin compatible replacement offering guaranteed temperature and accuracy specifications. Both converters are available in a standard 8-pin ceramic or plastic DIP and in an 8-pin metal can.

Applications

- Precision Voltage-to-Frequency Converters
- Pulse-Width Modulators
- Programmable Pulse Generators
- Frequency-to-Voltage Converters
- Integrating Analog-to-Digital Converter
- Long-Term Analog Integrator
- Signal Conversion —
  Current-to-Frequency
  Temperature-to-Frequency
  Pressure-to-Frequency
  Capacitance-to-Frequency
  Frequency-to-Current
- Signal Isolation
  VFC — Opto-Isolation — FVC
  ADC with Opto-Isolation
- Signal Encoding
  FSK Modulation/Demodulation
  Pulse-Width Modulation
- Frequency Scaling
- DC Motor Speed Control

Mask Pattern

4151

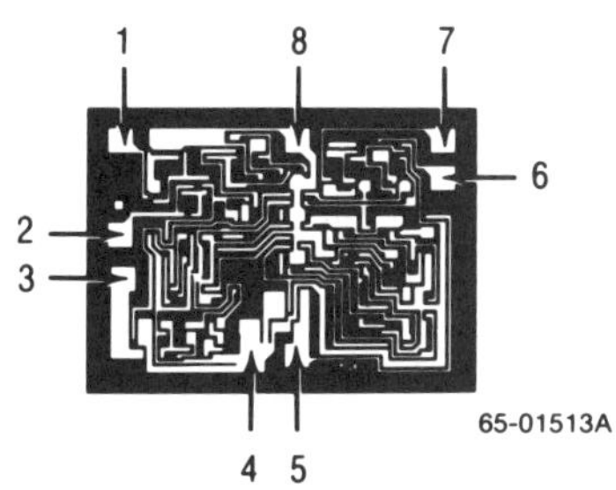

65-01513A

Die Size: 45 x 61 mils
Min. Pad Dimension: 4 x 4 mils

4152

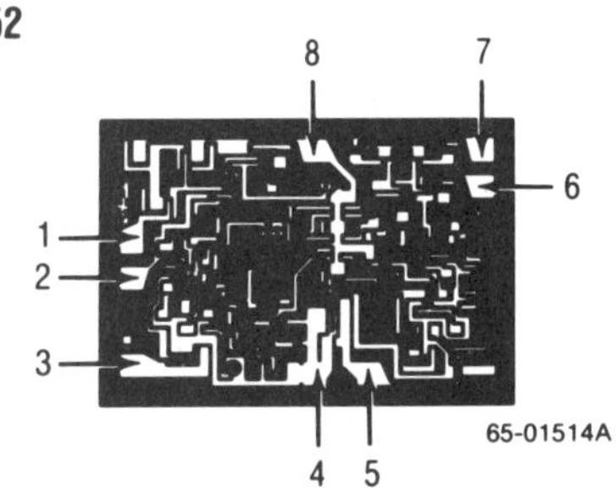

65-01514A

Die Size: 47 x 68 mils
Min. Pad Dimension: 4 x 4 mils

Selected References

Active Filters. Heathkit/Zenith Educational Systems, 1979.

Berlin, H. M. *The 555 Timer Applications Sourcebook with Experiments*. Indianapolis, Ind.: Howard W. Sams & Co., Inc., 1978.

———. *Design of Active Filters, with Experiments*. Indianapolis, Ind.: Howard W. Sams & Co., Inc., 1978.

———. *The Design of Op-Amp Circuits, with Experiments*. Indianapolis, Ind.: Howard W. Sams & Co., Inc., 1978.

———. *Design of Phase-Locked Loop Circuits, with Experiments*. Indianapolis, Ind.: Howard W. Sams & Co., Inc., 1978.

———. (revised by) *CMOS Cookbook* 2d ed. Indianapolis, Ind.: Howard W. Sams & Co., Inc., 1988.

Boyce, J. C. *Operational Amplifiers and Linear Integrated Circuits*. 2d ed. Boston: PWS-Kent Publishing Co., 1988.

Carr, J. J. *IC Timer Handbook*. Blue Ridge Summit, Penn.: Tab Books, Inc., 1981.

Chen, C. *Active Filter Design*. Indianapolis, Ind.: Hayden Books, 1982.

Coughlin, R. F., and F. F. Driscoll. *Operational Amplifiers and Linear Integrated Circuits* 3d ed. Englewood Cliffs, N.J.: Prentice Hall, 1987.

Dungan, F. R. *Linear Integrated Circuits for Technicians*. Breton Publishers, 1984.

Faulkenberry, L. C. *An Introduction to Operational Amplifiers with Linear IC Applications* 2d ed. New York: John Wiley, 1982.

Gayakwad, R. A. *Op-Amps and Linear Integrated Circuits* 2d ed. Englewood Cliffs, N.J.: Prentice Hall, 1988.

Honeycutt, R. A. *Op Amps and Linear Integrated Circuits*. Albany, N.Y.: Delmar Publishers, Inc. 1988.

Jung, W. G. *Audio IC Op-Amp Applications* 3d ed. Indianapolis, Ind.: Howard W. Sams & Co., Inc., 1987.

———. *IC Timer Cookbook* 2d ed. Indianapolis, Ind.: Howard W. Sams & Co., Inc., 1985.

———. *IC Op-Amp Cookbook* 3d ed. Indianapolis, Ind.: Howard W. Sams & Co., Inc., 1986.

Lancaster, D. *Active-Filter Cookbook*. Indianapolis, Ind.: Howard W. Sams & Co., Inc., 1975.

Operational Amplifiers. Heathkit/Zenith Educational Systems, 1979.

Phase-Locked Loops. Heathkit/Zenith Educational Systems, 1979.

Stanley, W. *Operational Amplifiers with Linear Integrated Circuits* 2d ed. Columbus, Ohio: Merrill Publishing Co., 1989.

Young, T. *Linear Integrated Circuits*. New York: John Wiley, 1981.

Answers to Odd-Numbered Exercises

Chapter 2

1. 3000
3. 39.99
5. 9.9995
7. (a) 6.6 (b) 58.4 (c) 2 (d) 1.33
9. (a) 32.4 Ω (b) 1.44 MΩ (c) 45.9 kΩ (d) 99 kΩ
13. (a) 49.5 kΩ (b) 2.7 MΩ (c) 84 kΩ (d) 150 kΩ
17. 30 kΩ
19. −2.25 V
21. −1.58 V

Chapter 3

1. 1000
3. 100 kHz
5. 0.175 μs
7. 12.93 V/μs
9. 796 kHz
11. 0.0001125
13. 10.9 mV RMS
15. R_i(CL) = 5×10^{11} Ω, R_o(CL) = 0.0003 Ω
17. −1.21 mV
19. 14.99 kΩ (use a 15-kΩ standard value)
21. 92.04 dB
23. 150 μV

Chapter 4

1. 454.5 kΩ (use a 470-kΩ standard value)
3. 5.6
5. R_S = 159 Ω, R_F = R_B = 1.59 kΩ
7. 100 kΩ
9. R = 16.9 kΩ, R_F = 169.4 kΩ, R_B = 15.4 kΩ

Chapter 5

1. (a) 100 kΩ (b) 19.4 Hz (input), 17 Hz (output), 3.2 Hz (bias network)
3. (a) 220 kΩ (b) 14.1 (c) 2.2 Hz (input), 15.9 Hz (output), 20.5 Hz (bias network), 20.2 Hz (gain resistors)
5. (a) 15 kΩ (b) 5.5 (c) 2.3 Hz (input), 3.2 Hz (output), 0.97 Hz (bias network)
7. (a) 10 kΩ (b) 6.8 (c) 3.4 Hz (inputs), 3.4 Hz (output), 1.2 Hz (bias network)
9. (a) 15 kΩ (b) 10 (c) 3.2 Hz (inputs), 3.2 Hz (output), 2.9 Hz (bias network)
11. (a) 7.5 V (b) 10 kΩ (c) 15 (d) 3.4 Hz (input), 3.2 Hz (output)
13. (a) 8.4 V (b) 47 kΩ (c) 11.9 (d) 3.4 Hz (input), 1.6 Hz (output)
15. C_1 = 0.0398 μF, C_2 = 3.98 μF
17. $C_1 = C_2 = C_3$ = 8 μF, C_4 = 80 μF, $R_1 = R_2 = R_3$ = 10 kΩ, R_F = 100 kΩ, R_B = 200 kΩ

Chapter 6

1. 234 kΩ
3. 40, assuming g_m = 20,000 μS from graph

Chapter 7

1. (a) 4.7 kΩ (b) 23.1 (c) 5 Hz (input), 6.4 Hz (output), 49.8 Hz (gain resistors)

3. $R_1 = 1\ k\Omega$, $R_2 = 99\ k\Omega$, $R_5 = 20\ k\Omega$, $C_1 = 53\ \mu F$ for 3 Hz
5. $R_1 = 33.3\ k\Omega$, $R_2 = 100\ k\Omega$, $R_3 = 1\ M\Omega$, $C_1 = 0.0018\ \mu F$, $C_2 = 178$ pF
7. 1.2 W

Chapter 8

5. $V_{UT} = 1.28$ V, $V_{LT} = -1.28$ V
7. $R_A = R_B$, such as 10 kΩ
9. $V_{UT} = -1.05$ V, $V_{LT} = -1.86$ V
11. $V_{UT} = 3.28$ V, $V_{LT} = 1.57$ V
13. −4.5 V to +13 V
15. $V_{UT} = 12.30$ V, $V_{LT} = 10.53$ V

Chapter 9

1. $R_1 = 345\ k\Omega$, $R_2 = 6.67\ k\Omega$, $R_3 = 10\ k\Omega$
3. $R_1 = 460\ k\Omega$, $R_2 = 2.9\ k\Omega$, $R_3 = 9.1\ M\Omega$ for 1 mA through R_2 and R_3, $I_C(\max) = 216$ mA, $V_{CE}(\max) = 8.5$ V, $P_D = 864$ mW
5. 2.8 Ω
7. 13.72 V
9. 50% (at 18 V), 69% (at 13 V)
11. 15.21 V
13. 0.9 W
15. $R_1 = 704\ \Omega$, $R_2 = 10\ k\Omega$
17. 2 mA
19. 160 mW

Chapter 10

1. 2462 Hz
3. 605 Hz
5. 7238 Hz
7. $R_1 = 11.6\ k\Omega$, $R_2 = 10\ k\Omega$, $R_B = 26{,}737\ \Omega$
9. $R_2 = 2667\ \Omega$, $R_3 = 2667\ \Omega$
11. 1638 Hz, 55.1%
13. $R_A = 2998\ \Omega$, $R_B = 2071\ \Omega$

Chapter 11

1. (a) −26.6 dB (b) 215 mV peak-to-peak
3. 384 Hz
5. −11.7 dB
7. (a) 734 Hz (b) 185 Hz (c) 3.97
9. (a) −13.02 dB (b) −14.43 dB
11. $f_o = 1457$ Hz, $Q = 2.62$
13. 402 Hz
15. $R = 13.27\ k\Omega$
17. All frequency-determining resistors for all stages are 12.25 kΩ.
    Second stage: $R_B = 0.382\ R_A$
    Third stage: $R_B = 1.382\ R_A$
19. (a) 494 Hz (b) 1.6 (c) 3.49 (d) 142 Hz
21. $R_1 = 36{,}190\ \Omega$, $R_2 = 1508\ \Omega$, $R_3 = 144{,}760\ \Omega$
23. $R = 16{,}762\ \Omega$, $R_A = 2\ k\Omega$, and $R_B = 238\ k\Omega$

Chapter 12

1. 625 Ω, −1.6 mA/V
3. 155.7 μA (at 9635 Ω) to 1.19 mA (at 1260 Ω)
5. 0.35 mA
7. 10 kΩ
9. 5 mA
11. 65.5 ΩA
13. 1 kHz/V
15. (a) −14 V (b) −5 V, (c) −10 V (d) −12 V
17. (a) 10 bits (b) 11 bits (c) 13 bits

Chapter 13

1. (a) 0.456 V (b) 1.597 V (c) 4.731 V
3. (a) 37.068 kHz (b) 30.640 kHz
5. (a) 16.97 to 18.03 kHz (b) 14 to 21 kHz
7. $R_1 = 9661\ \Omega$, $C_2 = 2.22\ \mu F$
9. $R_1 = 3.36\ k\Omega$, $C_2 = 0.82\ \mu F$, $C_3 > 1.64\ \mu F$

Index